Tryptophan in Nutrition and Health

Tryptophan in Nutrition and Health

Editor

Burkhard Poeggeler

MDPI • Basel • Beijing • Wuhan • Barcelona • Belgrade • Manchester • Tokyo • Cluj • Tianjin

Editor
Burkhard Poeggeler
Department of Physiology
Georg August University of
Göttingen
Sassenburg
Germany

Editorial Office
MDPI
St. Alban-Anlage 66
4052 Basel, Switzerland

This is a reprint of articles from the Special Issue published online in the open access journal *International Journal of Molecular Sciences* (ISSN 1422-0067) (available at: www.mdpi.com/journal/ijms/special_issues/Tryptophan_Health).

For citation purposes, cite each article independently as indicated on the article page online and as indicated below:

LastName, A.A.; LastName, B.B.; LastName, C.C. Article Title. *Journal Name* **Year**, *Volume Number*, Page Range.

ISBN 978-3-0365-7077-8 (Hbk)
ISBN 978-3-0365-7076-1 (PDF)

© 2023 by the authors. Articles in this book are Open Access and distributed under the Creative Commons Attribution (CC BY) license, which allows users to download, copy and build upon published articles, as long as the author and publisher are properly credited, which ensures maximum dissemination and a wider impact of our publications.

The book as a whole is distributed by MDPI under the terms and conditions of the Creative Commons license CC BY-NC-ND.

Contents

About the Editor vii

Burkhard Poeggeler, Sandeep Kumar Singh and Miguel A. Pappolla
Tryptophan in Nutrition and Health
Reprinted from: *Int. J. Mol. Sci.* **2022**, *23*, 5455, doi:10.3390/ijms23105455 1

Antonella Calderaro, Alessandro Maugeri, Salvatore Magazù, Giuseppina Laganà, Michele Navarra and Davide Barreca
Molecular Basis of Interactions between the Antibiotic Nitrofurantoin and Human Serum Albumin: A Mechanism for the Rapid Drug Blood Transportation
Reprinted from: *Int. J. Mol. Sci.* **2021**, *22*, 8740, doi:10.3390/ijms22168740 5

Ning Liu, Shiqiang Sun, Pengjie Wang, Yanan Sun, Qingjuan Hu and Xiaoyu Wang
The Mechanism of Secretion and Metabolism of Gut-Derived 5-Hydroxytryptamine
Reprinted from: *Int. J. Mol. Sci.* **2021**, *22*, 7931, doi:10.3390/ijms22157931 19

Alisa Schnellbaecher, Anton Lindig, Maxime Le Mignon, Tim Hofmann, Brit Pardon and Stephanie Bellmaine et al.
Degradation Products of Tryptophan in Cell Culture Media: Contribution to Color and Toxicity
Reprinted from: *Int. J. Mol. Sci.* **2021**, *22*, 6221, doi:10.3390/ijms22126221 35

Ivan V. Gmoshinski, Vladimir A. Shipelin, Nikita V. Trusov, Sergey A. Apryatin, Kristina V. Mzhelskaya and Antonina A. Shumakova et al.
Effects of Tyrosine and Tryptophan Supplements on the Vital Indicators in Mice Differently Prone to Diet-Induced Obesity
Reprinted from: *Int. J. Mol. Sci.* **2021**, *22*, 5956, doi:10.3390/ijms22115956 49

Ibrahim Yusufu, Kehong Ding, Kathryn Smith, Umesh D. Wankhade, Bikash Sahay and G. Taylor Patterson et al.
A Tryptophan-Deficient Diet Induces Gut Microbiota Dysbiosis and Increases Systemic Inflammation in Aged Mice
Reprinted from: *Int. J. Mol. Sci.* **2021**, *22*, 5005, doi:10.3390/ijms22095005 67

Jay Ronel V. Conejos, Jalil Ghassemi Nejad, Jung-Eun Kim, Jun-Ok Moon, Jae-Sung Lee and Hong-Gu Lee
Supplementing with L-Tryptophan Increases Medium Protein and Alters Expression of Genes and Proteins Involved in Milk Protein Synthesis and Energy Metabolism in Bovine Mammary Cells
Reprinted from: *Int. J. Mol. Sci.* **2021**, *22*, 2751, doi:10.3390/ijms22052751 79

Ruta Zulpaite, Povilas Miknevicius, Bettina Leber, Kestutis Strupas, Philipp Stiegler and Peter Schemmer
Tryptophan Metabolism via Kynurenine Pathway: Role in Solid Organ Transplantation
Reprinted from: *Int. J. Mol. Sci.* **2021**, *22*, 1921, doi:10.3390/ijms22041921 93

Anna Birková, Marcela Valko-Rokytovská, Beáta Hubková, Marianna Zábavníková and Mária Mareková
Strong Dependence between Tryptophan-Related Fluorescence of Urine and Malignant Melanoma
Reprinted from: *Int. J. Mol. Sci.* **2021**, *22*, 1884, doi:10.3390/ijms22041884 125

George Anderson, Annalucia Carbone and Gianluigi Mazzoccoli
Tryptophan Metabolites and Aryl Hydrocarbon Receptor in Severe Acute Respiratory Syndrome, Coronavirus-2 (SARS-CoV-2) Pathophysiology
Reprinted from: *Int. J. Mol. Sci.* **2021**, *22*, 1597, doi:10.3390/ijms22041597 . . . . **137**

Monika Szelest, Katarzyna Walczak and Tomasz Plech
A New Insight into the Potential Role of Tryptophan-Derived AhR Ligands in Skin Physiological and Pathological Processes
Reprinted from: *Int. J. Mol. Sci.* **2021**, *22*, 1104, doi:10.3390/ijms22031104 . . . . **153**

Michele Dei Cas, Ileana Vigentini, Sara Vitalini, Antonella Laganaro, Marcello Iriti and Rita Paroni et al.
Tryptophan Derivatives by *Saccharomyces cerevisiae* EC1118: Evaluation, Optimization, and Production in a Soybean-Based Medium
Reprinted from: *Int. J. Mol. Sci.* **2021**, *22*, 472, doi:10.3390/ijms22010472 . . . . **185**

Massimo E. Maffei
5-Hydroxytryptophan (5-HTP): Natural Occurrence, Analysis, Biosynthesis, Biotechnology, Physiology and Toxicology
Reprinted from: *Int. J. Mol. Sci.* **2020**, *22*, 181, doi:10.3390/ijms22010181 . . . . **205**

Paulina Wigner, Ewelina Synowiec, Paweł Jóźwiak, Piotr Czarny, Michał Bijak and Katarzyna Białek et al.
The Effect of Chronic Mild Stress and Escitalopram on the Expression and Methylation Levels of Genes Involved in the Oxidative and Nitrosative Stresses as Well as Tryptophan Catabolites Pathway in the Blood and Brain Structures
Reprinted from: *Int. J. Mol. Sci.* **2020**, *22*, 10, doi:10.3390/ijms22010010 . . . . **231**

Sailen Barik
The Uniqueness of Tryptophan in Biology: Properties, Metabolism, Interactions and Localization in Proteins
Reprinted from: *Int. J. Mol. Sci.* **2020**, *21*, 8776, doi:10.3390/ijms21228776 . . . . **251**

Chien-Ning Hsu and You-Lin Tain
Developmental Programming and Reprogramming of Hypertension and Kidney Disease: Impact of Tryptophan Metabolism
Reprinted from: *Int. J. Mol. Sci.* **2020**, *21*, 8705, doi:10.3390/ijms21228705 . . . . **273**

Katarzyna Walczak, Ewa Langner, Anna Makuch-Kocka, Monika Szelest, Karolina Szalast and Sebastian Marciniak et al.
Effect of Tryptophan-Derived AhR Ligands, Kynurenine, Kynurenic Acid and FICZ, on Proliferation, Cell Cycle Regulation and Cell Death of Melanoma Cells—In Vitro Studies
Reprinted from: *Int. J. Mol. Sci.* **2020**, *21*, 7946, doi:10.3390/ijms21217946 . . . . **295**

About the Editor

Burkhard Poeggeler

Dr. Burkhard Poeggeler is affiliated with the Georg-August-University Göttingen for more than 30 years and works on the biochemistry and pharmacology of regeneration by antioxidant protection and trophic prosurvival signaling. This research focuses on bioenergetic agents, such as nitric oxide, and on redox regulation, such as neurovascular coupling. Neurovascular coupling drains the brain of waste and facilitates the removal of the toxic amyloid peptides by the glymphatic system and the gastrointestinal tract. Dr. Miguel A. Pappolla discovered the neuroprotective activity of melatonin and structurally related tryptophan metabolites with similar functions by amyloid clearance. Based on the exploration of these protective agents, he works on exploring new possibilities to extend human health and life span. His over 100 in PubMed listed publications demonstrate this wide range of research activities covering important bioactive compounds that shape aging and development, such as melatonin, arginine, and kynurenic acid. Dr. Burkhard Poeggeler has successfully orchestrated the development of potent nutraceuticals in the fields of prevention and therapy that enable new approaches to a unique biomatrix precision supplementation that assures healthy aging.

Editorial

Tryptophan in Nutrition and Health

Burkhard Poeggeler [1,*], Sandeep Kumar Singh [2] and Miguel A. Pappolla [3]

[1] Johann-Friedrich-Blumenbach-Institute for Zoology and Anthropology, Faculty of Biology and Psychology, Georg-August-University of Göttingen, Am Türmchen 3, 33332 Gütersloh, Germany
[2] Indian Scientific Education and Technology Foundation, Lucknow 226002, India; sandeeps.bhu@gmail.com
[3] Department of Neurology, University of Texas Medical Branch, 301 University Boulevard, Galveston, TX 77555, USA; pappolla@aol.com
* Correspondence: bpoegge@gwdg.de; Tel.: +49-175-6537935

Tryptophan is a rate-limiting essential amino acid and a unique building block of peptides and proteins. This largest amino acid serves as the precursor for important endogenous indoleamines such as serotonin, N-acetylserotonin and melatonin, which act as neurotransmitters, neuromodulators and neurohormones. An enhanced synthesis of these signaling molecules can improve health, quality-of-life and well-being. The main metabolic pathway of tryptophan is the oxidation to bioactive kynurenines and niacin. Kynurenic acid is the most potent endogenous anti-exitotoxic agent. Other highly relevant pathways of tryptophan are the reversible transamination to indole-3-pyruvate with the formation of the related indolic acids as well as the synthesis of indole compounds and their derivatives by side chain cleavage.

The indolic acids act as potent protective antioxidant agents, whereas the indoles such as indole, indoxyl and indoxyl sulfate are reactive compounds that are primarily studied because of their potential toxicity. Research on the physiology and pathophysiology of tryptophan metabolism has revealed a key role for the amino acid and its metabolites as endogenous molecular master regulators of physiology and plasticity in development and aging. The ratio of tryptophan to kynurenine is a key parameter reflecting endogenous adaptation to stress determining inflammation and degeneration. Tryptophan metabolites such as melatonin and structurally related agents such as indole-3-propionic acid act as potent catalytic antioxidant and bioenergetic agents that facilitate regeneration and protection against stress and aging.

Several indole compounds act as uremic toxins since these agents can induce radical formation that is associated with enhanced oxidative stress and damage. The exploration of the effects of these protective and toxic tryptophan-derived agents has revealed important molecular mechanisms and mediators of adaptation and aging. Research on tryptophan in nutrition and health can facilitate the development of new approaches to extending human health and lifespan. Amino acids are the building blocks of life that enable repair as well as recycling and regeneration in the body and the brain. Research on nutrients, including amino acids such as tryptophan and its metabolites as well as peptides and proteins, or extracts containing this molecular metabolism's modifiers can improve health. Research into the indololome is a new emerging and rapidly growing field of utmost relevance to science and society.

The Special Issue on "Tryptophan in Nutrition and Health" reports on the broad field of tryptophan research and has examined the key tryptophan pathways and their molecular targets that mediate the effects of the amino acid and its metabolites on nutrition and health (Figure 1).

The latest developments with the rapid progress in tryptophan research are the focus of this collection of articles, and the studies herein demonstrate the relevance of tryptophan and its metabolites that form the indobolome on nutrition and health. The discovery of

Citation: Poeggeler, B.; Singh, S.K.; Pappolla, M.A. Tryptophan in Nutrition and Health. *Int. J. Mol. Sci.* **2022**, *23*, 5455. https://doi.org/10.3390/ijms23105455

Received: 3 May 2022
Accepted: 10 May 2022
Published: 13 May 2022

Publisher's Note: MDPI stays neutral with regard to jurisdictional claims in published maps and institutional affiliations.

Copyright: © 2022 by the authors. Licensee MDPI, Basel, Switzerland. This article is an open access article distributed under the terms and conditions of the Creative Commons Attribution (CC BY) license (https://creativecommons.org/licenses/by/4.0/).

a broad range of bioactive compounds derived from tryptophan can enable a better understanding of the unique role of this amino acid in physiology and development. New methods have become available and allow us to establish a biomimetic precision pharmacology to prevent disease and protect health. The complexity of the current activities in exploring the many different effects of tryptophan and its metabolites demonstrates the necessity for new approaches in targeting the physiology and pharmacology of indoleamines, indoles and kynurenines. Bioavailability by food consumption, protein degradation or colonic formation in symbiotic organisms seem to be key issues as tryptophan can induce its own depletion, especially under conditions of stress and disease.

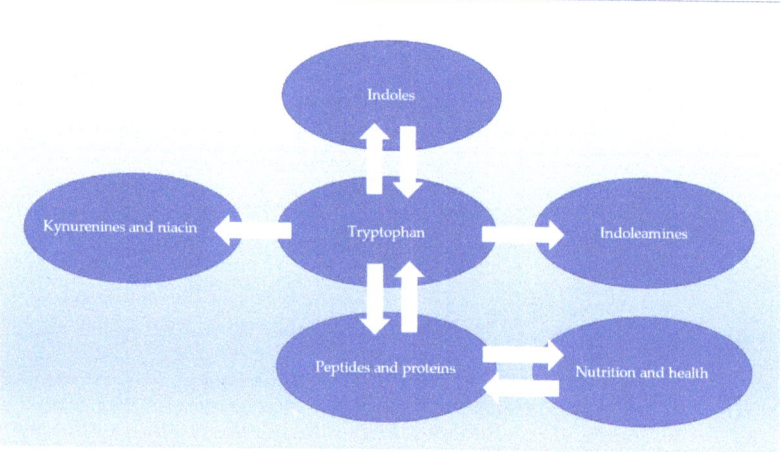

Figure 1. Tryptophan and its main metabolic pathways in nutrition and health.

This Special Issue contains seven reviews and nine original research articles that conclusively demonstrate developmental programming and reprogramming [1], uniqueness of tryptophan [2], the role of 5-hydroxytryptophan [3], tryptophan AhR-ligands in the skin [4], the impacts of tryptophan metabolites on coronavirus pathophysiology [5], tryptophan metabolism in organ transplantation [6], gut-derived 5-hydroxytryptamin [7], effects of AhR-ligands on melanoma cells [8], effects of stress and escitalopram on genes of the tryptophan catabolite pathways [9], formation of tryptophan derivatives by *Saccharomyces cerevisiae* [10], the relationship between tryptophan-related fluorescence in urine and malignant melanoma [11], the effects of tryptophan supplementation on milk protein synthesis and energy metabolism [12], the induction of tryptophan deficiency and dysbiosis with associated increased systemic inflammation in aged mice [13], the effects of tyrosine and tryptophan supplementation on diet-induced obesity [14], degradation products of tryptophan [15], and the molecular interactions of nitrofurantoin and albumin [16].

The research not only demonstrates that only a sufficient supply of tryptophan can improve, sustain and maintain health but also indicates that increased oxidative tryptophan degradation can lead to the formation of toxic compounds that have detrimental effects. Tryptophan is a double-edged sword, and interventions that modify its metabolism have to be carefully designed to address the specific needs of the target population. The selective improvement of tryptophan metabolism constitutes a great chance to meet the urgent challenge of chronic diseases associated with premature aging, inflammation and degeneration. The decisive endogenous molecular mechanisms and mediators that affect and determine the effects of tryptophan metabolism are covered by this Special Issue and allow for the development of effective strategies to implement prevention, protection and therapy.

Author Contributions: B.P. did the writing with the preparation of the original draft. S.K.S. did the writing with reviewing, editing and extending the original draft. M.A.P. equally contributed to the writing with reviewing, editing and extending the original draft. All authors have read and agreed to the published version of the manuscript.

Funding: This research received no external funding.

Institutional Review Board Statement: Not applicable.

Informed Consent Statement: Not applicable.

Data Availability Statement: Not applicable.

Acknowledgments: Burkhard Poeggeler the coauthors for reviewing and extending this editorial on "Tryptophan in Nutrition and Health". As a Guest Editor, Burkhard Poeggeler also appreciates the work of all of the authors and their contributions. The support of the reviewers was decisive in evaluating the manuscripts that were selected for publication.

Conflicts of Interest: The authors declare no conflict of interest.

References

1. Hsu, C.-N.; Tain, Y.-L. Developmental Programming and Reprogramming of Hypertension and Kidney Disease: Impact of Tryptophan Metabolism. *Int. J. Mol. Sci.* **2020**, *21*, 8705. [CrossRef] [PubMed]
2. Barik, S. The Uniqueness of Tryptophan in Biology: Properties, Metabolism, Interactions and Localization in Proteins. *Int. J. Mol. Sci.* **2020**, *21*, 8776. [CrossRef] [PubMed]
3. Maffei, M.E. 5-Hydroxytryptophan (5-HTP): Natural Occurrence, Analysis, Biosynthesis, Biotechnology, Physiology and Toxicology. *Int. J. Mol. Sci.* **2021**, *21*, 181. [CrossRef] [PubMed]
4. Szelest, M.; Walczak, K.; Plech, T. A New Insight into the Potential Role of Tryptophan-Derived AhR Ligands in Skin Physiological and Pathological Processes. *Int. J. Mol. Sci.* **2021**, *22*, 1104. [CrossRef] [PubMed]
5. Anderson, G.; Carbone, A.; Mazzoccoli, G. Tryptophan Metabolites and Aryl Hydrocarbon Receptor in Severe Acute Respiratory Syndrome, Coronavirus-2 (SARS-CoV-2) Pathophysiology. *Int. J. Mol. Sci.* **2021**, *22*, 1597. [CrossRef] [PubMed]
6. Zulpaite, R.; Miknevicius, P.; Leber, B.; Strupas, K.; Stiegler, P.; Schemmer, P. Tryptophan Metabolism via Kynurenine Pathway: Role in Solid Organ Transplantation. *Int. J. Mol. Sci.* **2021**, *22*, 1921. [CrossRef] [PubMed]
7. Liu, N.; Sun, S.; Wang, P.; Sun, Y.; Hu, Q.; Wang, X. The Mechanism of Secretion and Metabolism of Gut-Derived 5-Hydroxytryptamine. *Int. J. Mol. Sci.* **2021**, *22*, 7931. [CrossRef] [PubMed]
8. Walczak, K.; Langner, E.; Makuch-Kocka, A.; Szelest, M.; Szalast, K.; Marciniak, S.; Plech, T. Effect of Tryptophan-Derived AhR Ligands, Kynurenine, Kynurenic Acid and FICZ, on Proliferation, Cell Cycle Regulation and Cell Death of Melanoma Cells—In Vitro Studies. *Int. J. Mol. Sci.* **2020**, *21*, 7946. [CrossRef] [PubMed]
9. Wigner, P.; Synowiec, E.; Jóźwiak, P.; Czarny, P.; Bijak, M.; Białek, K.; Szemraj, J.; Gruca, P.; Papp, M.; Śliwiński, T. The Effect of Chronic Mild Stress and Escitalopram on the Expression and Methylation Levels of Genes Involved in the Oxidative and Nitrosative Stresses as Well as Tryptophan Catabolites Pathway in the Blood and Brain Structures. *Int. J. Mol. Sci.* **2021**, *22*, 10. [CrossRef] [PubMed]
10. Cas, M.D.; Vigentini, I.; Vitalini, S.; Laganaro, A.; Iriti, M.; Paroni, R.; Foschino, R. Tryptophan Derivatives by Saccharomyces cerevisiae EC1118: Evaluation, Optimization, and Production in a Soybean-Based Medium. *Int. J. Mol. Sci.* **2021**, *22*, 472. [CrossRef]
11. Birková, A.; Valko-Rokytovská, M.; Hubková, B.; Zábavníková, M.; Mareková, M. Strong Dependence between Tryptophan-Related Fluorescence of Urine and Malignant Melanoma. *Int. J. Mol. Sci.* **2021**, *22*, 1884. [CrossRef] [PubMed]
12. Conejos, J.R.V.; Nejad, J.G.; Kim, J.-E.; Moon, J.-O.; Lee, J.-S.; Lee, H.-G. Supplementing with L-Tryptophan Increases Medium Protein and Alters Expression of Genes and Proteins Involved in Milk Protein Synthesis and Energy Metabolism in Bovine Mammary Cells. *Int. J. Mol. Sci.* **2021**, *22*, 2751. [CrossRef] [PubMed]
13. Yusufu, I.; Ding, K.; Smith, K.; Wankhade, U.D.; Sahay, B.; Patterson, G.T.; Pacholczyk, R.; Adusumilli, S.; Hamrick, M.W.; Hill, W.D.; et al. A Tryptophan-Deficient Diet Induces Gut Microbiota Dysbiosis and Increases Systemic Inflammation in Aged Mice. *Int. J. Mol. Sci.* **2021**, *22*, 5005. [CrossRef] [PubMed]
14. Gmoshinski, I.V.; Shipelin, V.A.; Trusov, N.V.; Apryatin, S.A.; Mzhelskaya, K.V.; Shumakova, A.A.; Timonin, A.N.; Riger, N.A.; Nikityuk, D.B. Effects of Tyrosine and Tryptophan Supplements on the Vital Indicators in Mice Differently Prone to Diet-Induced Obesity. *Int. J. Mol. Sci.* **2021**, *22*, 5956. [CrossRef] [PubMed]
15. Schnellbaecher, A.; Lindig, A.; Mignon, M.L.; Hofmann, T.; Pardon, B.; Bellmaine, S.; Zimmer, A. Degradation Products of Tryptophan in Cell Culture Media: Contribution to Color and Toxicity. *Int. J. Mol. Sci.* **2021**, *22*, 6221. [CrossRef] [PubMed]
16. Calderaro, A.; Maugeri, A.; Magazù, S.; Laganà, G.; Navarra, M.; Barreca, D. Molecular Basis of Interactions between the Antibiotic Nitrofurantoin and Human Serum Albumin: A Mechanism for the Rapid Drug Blood Transportation. *Int. J. Mol. Sci.* **2021**, *22*, 8740. [CrossRef] [PubMed]

Article

Molecular Basis of Interactions between the Antibiotic Nitrofurantoin and Human Serum Albumin: A Mechanism for the Rapid Drug Blood Transportation

Antonella Calderaro [1], Alessandro Maugeri [1], Salvatore Magazù [2], Giuseppina Laganà [1], Michele Navarra [1,*] and Davide Barreca [1]

[1] Department of Chemical, Biological, Pharmaceutical and Environmental Sciences, University of Messina, 98166 Messina, Italy; anto.calderaro@gmail.it (A.C.); amaugeri@unime.it (A.M.); giuseppina.lagana@unime.it (G.L.); davide.barreca@unime.it (D.B.)

[2] Department of Mathematical and Informatics Sciences, Physical Sciences and Earth Sciences, University of Messina, 98166 Messina, Italy; salvatore.magazu@unime.it

* Correspondence: mnavarra@unime.it

Abstract: Nitrofurantoin is an antimicrobial agent obtained through the addition of a nitro group and a side chain containing hydantoin to a furan ring. The interactions of the antibiotic with human serum albumin (HSA) have been investigated by fluorescence, UV-VIS, Fourier transform infrared spectroscopy (FTIR) spectroscopy, and protein-ligand docking studies. The fluorescence studies indicate that the binding site of the additive involves modifications of the environment around Trp214 at the level of subdomain IIA. Fluorescence and UV-VIS spectroscopy, displacement studies, and FTIR experiments show the association mode of nitrofurantoin to HSA, suggesting that the primary binding site of the antibiotic is located in Sudlow's site I. Molecular modeling suggests that nitrofurantoin is involved in the formation of hydrogen bonds with Trp214, Arg218, and Ser454, and is located in the hydrophobic cavity of subdomain IIA. Moreover, the curve-fitting results of the infrared Amide I' band indicate that the binding of nitrofurantoin induces little change in the protein secondary structure. Overall, these data clarify the blood transportation process of nitrofurantoin and its rapid transfer to the kidney for its elimination, hence leading to a better understanding of its biological effects and being able to design other molecules, based on nitrofurantoin, with a higher biological potential.

Keywords: nitrofurantoin; antibiotics; human serum albumin; molecular interactions; FTIR; fluorescence

1. Introduction

The use of antibiotics is growing every year to such an extent that there is a global effort for the development of novel therapeutics, from both a natural and synthetic origin, to combat bacterial, fungal, and viral resistance, as well as for rediscovery of the so-called "old drugs" [1]. Nitrofurantoin is a synthetic antibacterial agent widely used in the treatment of urinary tract infections [2–4]. It has a bacteriostatic, and, at higher concentrations, a bactericidal action on a wide range of Gram-positive and Gram-negative organisms; in particular, >90% of clinical strains of *E. coli* and *Citrobacter* spp. are sensitive [2,3,5]. After oral administration, ~40–50% of the drug is absorbed, but this value increases when nitrofurantoin is taken with food. Inside the organism, the drug is in part rapidly metabolized by the liver, and in part excreted unmodified in the urine, where it reaches a high concentration (50–250 mg L^{-1}) [2,3,5,6]. Moreover, the trypanocidal activity of some synthetic analogs has been recently described [7]. The nitrofurantoin mechanism of action is complex and not well-understood, but appears to be linked to its rapid reduction by nitrofuran reductase to multiple reactive intermediates that indiscriminately attack cellular macromolecules (such us proteins and deoxyribonucleic acid) and influence metabolic energetic pathways inside the bacterial cells. On the other hand, nitrofurantoin also shows an antibacterial activity when the nitro-reductase activity is inhibited, suggesting that it may act, albeit in part, without this enzymatic reduction [8–10]. Some microorganisms, such as several members of the Enterobacterales order (e.g., members of the Proteae) are intrinsically resistant to nitrofurantoin, thus limiting the number of therapeutic alternatives, especially for outpatients. Nevertheless, this resistance has become a parameter utilized for their phenotypic identification/differentiation in the routine practice of microbiological laboratories [11].

Human serum albumin (HSA) is the most abundant plasma protein and is the main human carrier of several endogenous and exogenous compounds circulating in blood [12]. Therefore, the investigation of molecules based on albumin binding is of essential importance, taking into account that the bioavailability of many biologically active compounds is correlated with this affinity. The binding oftentimes induces modifications of weak interactions, which, in turn, can change the macromolecular structure and, sometimes, increase its stability [13–20]. HSA is a globular protein with three homologous domains (domains I-III) and a molecular weight of 66.5 kDa [21]. Each domain is divided into two subdomains (A and B), characterized by the presence of a α-helices structure with multiple ligand-binding sites localized in each of these subdomains [22–25]. Overall, the protein is composed of 585 amino acids, and contains 17 disulfide bridges and one free cysteine (Cys34). In this paper, the interactions between nitrofurantoin and HSA, carried out by means of a multi-spectroscopic method, are shown, providing important information such as the association and changes of the protein secondary structure. The potential site of interaction and the residues involved have been analyzed by docking studies, corroborating the data obtained experimentally.

2. Results

2.1. UV-VIS Absorption Spectra of the Nitrofurantoin−HSA Complex

UV-VIS spectroscopy supplies the first evidence of a nitrofurantoin-HSA interaction. In the range of 220–480 nm, the nitrofurantoin spectrum is characterized by three main absorption bands at 230 (shoulder), 270, and 380 nm (Figure 1A). The band at 270 nm is due to the absorption of the conjugated C=N−N unit, while those at 230 and 380 nm are due to π-π* transitions involving the nitro-substituted furan ring. The titration of nitrofurantoin (60 µM) with a solution of HSA up to 60 µM showed remarkable changes within the range 260–460 nm. Indeed, we observed a clear hypochromic shift of the maximum of absorption for the band at 380 nm, with the formation of two isosbestic points at 359 and 397 nm (Figure 1B).

This intensity decrease at 380 nm may be due to the progressive inclusion of nitrofurantoin to the HSA binding site, with the possible formation of a direct interaction between the nitro-substituted furan ring and amino acid residues able to form hydrogen bonding and apolar interactions. The titration of nitrofurantoin with HSA concentrations higher than 60 µM did not show any significant change in the absorption band, suggesting that all of the nitrofurantoin molecules are already involved in the interaction with HSA (data not shown). These results let us to suppose that the interaction between the antibiotic and the protein can be described with 1:1 stoichiometry. Less evident and of difficult interpretation are the changes in absorption at 230 and 270 nm, due to the net increase of absorbance at 280, attributable to HSA that overlaps with the two bands.

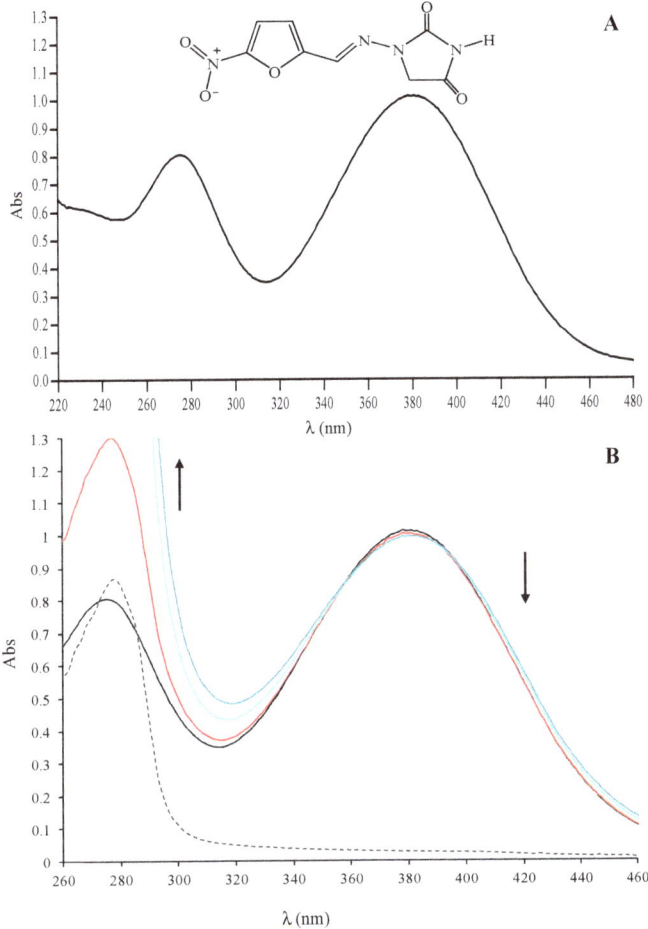

Figure 1. Interaction of nitrofurantoin with HSA monitored by UV-VIS spectroscopy. (**A**) UV-VIS absorption spectrum of nitrofurantoin 60 µM; (**B**) UV-VIS absorption spectra of nitrofurantoin (60 µM) in the absence (dark line) or in the presence of HSA 30 (red line), 45 (cyan line), or 60 (blue line) µM. In the graph, the spectrum of HSA 30 µM alone is also reported (dashed grey line).

2.2. Fluorescence Characterization of Nitrofurantoin-HSA Complex

The intrinsic fluorescence of HSA and the nitrofurantoin-HSA complex is depicted in Figure 2A, as well as that of HSA and warfarin, employed as the reference compound (Figure 2B). HSA shows a well-defined fluorescence emission with a maximum at 350 nm, due to the tryptophan residue (Trp214) present in the cavity of subdomain IIA (Sudlow's site I). In fact, although HSA has more tyrosine than tryptophan residues, it belongs to the protein of class B, whose fluorescence emission derived from tryptophan by Forster's resonance transfer process. The antimicrobial agent was almost non-fluorescent under the present experimental conditions. Its addition to the HSA solution gives a net decrease in the fluorescence intensity, accompanied by a shift of the wavelength emission maximum (a blue shift) in the albumin spectrum, as shown in Figure 2A. We also performed the same experiment with warfarin (a well-known compound able to bind to HSA), highlighting a similar fluorescence decease and blue shift (Figure 2B).

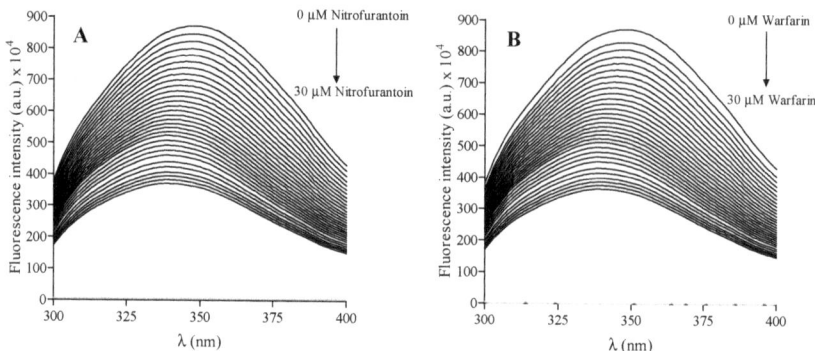

Figure 2. Fluorescence emission spectra of HSA in the absence or presence of increasing concentrations of nitrofurantoin (from 0 to 30 μM) (**A**) and warfarin (from 0 to 30 μM) (**B**) at 310 K.

Therefore, the area surrounding the tryptophan residue is highly hydrophobic. Following this, the well-known Stern-Volmer equation was employed to define the mechanism of fluorescence quenching of nitrofurantoin:

$$F_0/F = 1 + K_{SV}[Q] = 1 + K_q \tau_0 [Q]$$

where F_0 and F are the fluorescence intensities of HSA in the absence and in the presence of nitrofurantoin, K_q is the quenching rate constant, K_{SV} is the Stern−Volmer dynamic quenching constant, τ_0 is the average lifetime of the fluorophore in the absence of quenchers, and [Q] is the concentration of the quencher. The Figure 3A depicts the curves of F_0/F versus [Q] at different temperatures. The K_{SV} values obtained at 293, 298, 304, 310, and 315 K are 2.48 ($\pm$0.32) $\times$ 10^4, 2.63 ($\pm$0.22) $\times$ 10^4, 2.7 ($\pm$0.18) $\times$ 10^4, 2.86 ($\pm$0.30) $\times$ 10^4, and 2.98 ($\pm$0.27) $\times$ 10^4 M^{-1}, respectively. Nitrofurantoin is soluble in a buffer solution and this may accelerate its diffusion rate and hence its collision with fluorophore.

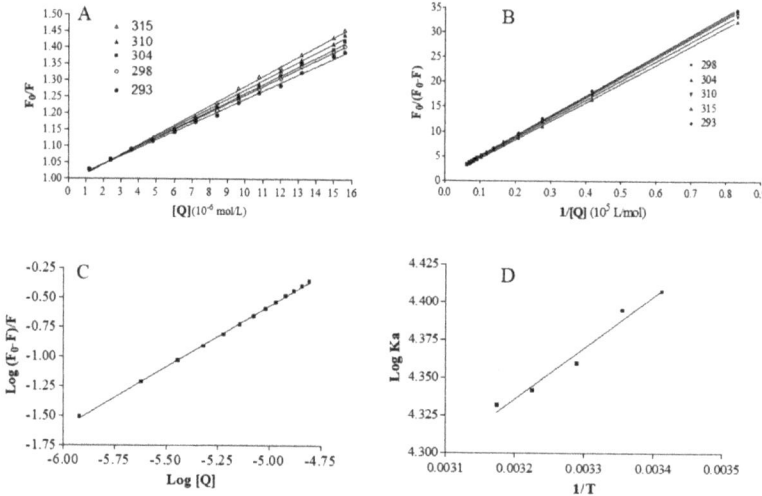

Figure 3. Fluorescence analysis. (**A**) The Stern−Volmer plots for the HSA-nitrofurantoin system. (**B**) Modified Stern-Volmer plots for the HSA−nitrofurantoin system. (**C**) Plots of log(F_0-F)/F as a function of log [Q] for the binding of nitrofurantoin with HSA at the temperature of 310 K. (**D**) Van't Hoff plot for the binding of nitrofurantoin to HSA.

The fluorescence data were further examined using the modified Stern−Volmer equation:

$$\frac{F_0}{F_0 - F} = \frac{1}{f_a K_a [Q]} + \frac{1}{f_a}$$

where K_a is the modified Stern−Volmer association constant for the accessible fluorophores, and f_a is the fraction of accessible fluorescence. The linear regression analysis of $F_0/(F_0\text{-}F)$ versus $1/[Q]$ is shown in Figure 3B. The obtained K_a values are 2.55×10^4, 2.48×10^4, 2.28×10^4, 2.19×10^4, and 2.15×10^4 L mol^{-1} at 293, 298, 304, 310, and 315 K, respectively. This shows that the binding constant is moderate and the effect of temperature is not significant. The analysis of the plots showed that, at the tested concentrations, there is a good linear relationship, indicating that the quenching mechanism is driven by the formation of a complex.

2.3. Analysis of Binding Equilibrium

The binding ability of a compound to HSA is very useful for evaluating its biological and, eventual, therapeutic potential, because it can also influence its stability and toxicity, as well as the rapidity of the renal excretion. The independent bind of small molecules to a set of equivalent sites on a macromolecule can be described by analyzing the equilibrium between free and bound molecules using the following equation:

$$\text{Log}(F_0 - F)/F = \text{Log}K + n\text{log}[Q]$$

where K is the observed binding constant to a site and n is the number of binding sites per HSA. In Figure 3C, the linear plot of log $(F_0\text{-}F)/F$ as a function of log[Q] at 310 K is shown. The values of n are approximately equal to 1, which demonstrates that there is a single class-binding site for nitrofurantoin in the proximity of the tryptophan residue.

The binding strength of a compound to HSA is one of the main elements in its availability to diffuse in the organism and, through the circulation system, to reach its target organ or to be eliminated by the organism [26]. The binding of a ligands to HSA and proteins, in general, is often reversible with moderate affinities (binding constants in the range of $1\text{--}15 \times 10^4$ L/mol) [27]. The binding constant (K) obtained at the physiological temperature value (310 K) is $4.10 \pm 0.02 \times 10^4$ L/mol, highlighting that the binding between nitrofurantoin and HSA is moderate in strength, and the formation of the complex is reversible. In this way, the antibiotic can be stored and carried around the body by HSA. The moderate value of K indicates that the drug does not remain so much time linked to HSA because other molecules with higher affinity compete with nitrofurantoin for the binding, and this decreases its availability in the circulating system and probably is responsible also of its rapid elimination.

2.4. Thermodynamics and Acting Forces

The binding of molecules to macromolecules involves mainly non-covalent interactions (such as hydrogen bonds, van der Waals forces, hydrophobic, and electrostatic interactions). These types of forces can be analyzed by determining the thermodynamic parameters of the binding reaction [13,28–35]. Therefore, the thermodynamic parameters dependent on temperature were calculated from the van't Hoff plot to characterize the forces acting between nitrofurantoin and HSA. When the temperature change is not very large, the enthalpy change (ΔH) of a system can be regarded as a constant. Under these conditions, both the enthalpy (ΔH) and entropy (ΔS) changes can be evaluated from the van't Hoff equation:

$$\text{Ln}K_a = -\frac{\Delta H}{RT} + \frac{\Delta S}{R}$$

where R is the gas constant. The enthalpy change (ΔH) is calculated from the slope of the van't Hoff plot (Figure 3D), while the entropy change (ΔS) is calculated from the intercept. The free energy change (ΔG) is then estimated from the following equation:

$$\Delta G = \Delta H - T\Delta S$$

The van't Hoff plot for the evaluation of the thermodynamic parameters due to the nitrofurantoin-HSA interaction is depicted in Figure 3D. The ΔH, ΔS, and ΔG values are depicted in Table 1.

Table 1. Analysis of the thermodynamic parameters.

ΔH.	ΔS	ΔG				
(kJ mol^{-1})	(J mol^{-1}K^{-1})	(kJ mol^{-1})				
		293	298	304	310	315
−3.93 ± 0.26	27.05 ± 0.85	−11.32 ± 1.25	−11.45 ± 1.11	−11.62 ± 1.13	−11.78 ± 1.41	−11.91 ± 1.72

The positive ΔS value is evidence for the formation of hydrophobic interactions, while the negative value of ΔH suggests that the binding process is predominately enthalpy driven and is probably due to hydrogen binding interactions. The negative ΔG values, accompanied by positive entropy change (ΔS), are indicative of a spontaneous process during the binding of nitrofurantoin to HSA. Moreover, the free energy change in the binding is a direct consequence of the strength of the interaction (such as protein molecules). Therefore, both hydrophobic interactions and hydrogen bonds play a major role in the binding of an antibiotic agent to HSA according to above mentioned data.

2.5. Fluorescence Displacement Binding Experiments

Warfarin, ibuprofen, and digitoxin, namely markers of HSA binding sites I, II, and III, respectively, were employed to carry out displacement binding experiments and to gather further information on the nitrofurantoin binding site. The fluorescence emission spectra of nitrofurantoin-HSA complex spectra were recorded both in the absence and presence of increasing concentrations of warfarin, ibuprofen, and digitoxin. As displayed in Figure 4, The fluorescence of HSA-nitrofurantoin rapidly decreased after warfarin addition, whereas it remained almost unvaried regardless of the presence of both ibuprofen and digitoxin. As warfarin competes with nitrofurantoin for the same HSA binding site, this supports the fact that the latter probably binds to the hydrophobic pocket situated in subdomain IIA (Sudlow's site I). These results are in line with those above reported, and suggest that the residue Trp214 should be close or inside the binding site of nitrofurantoin.

2.6. Conformation Investigation by Fourier Transform Infrared Spectroscopy (FTIR) Spectroscopy

In order to investigate the modification of the HSA secondary structure after binding to nitrofurantoin, as well as its mechanisms of interaction to the protein, we exploited FTIR spectroscopy [36–41]. The HSA secondary structure is characterized by 67% α-helix, 10% turn, and 23% extended chains. As shown in Figure 5, HSA exhibited a strong Amide I' band centered at around 1652 cm^{-1}, corresponding mainly to the α-helix. In addition, the interaction between nitrofurantoin and HSA brought about a change, as well as a slight shift in band intensity, at both 1652 cm^{-1} and 1680 cm^{-1} of the FTIR spectra, thus suggesting a rapid variation in the association and dissociation of nitrofurantoin. This was further reinforced by the observed increase of absorbance values of both amide II' and amide II bands.

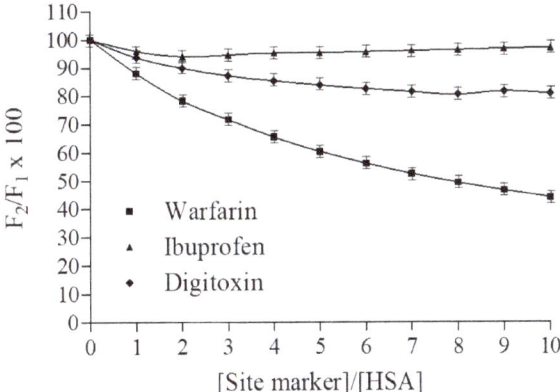

Figure 4. Effect of site-specific ligands on the fluorescence of the HSA-nitrofurantoin complex. F_1 and F_2 are the fluorescence of HSA−nitrofurantoin in the absence and presence of the site markers, respectively. Data represent mean ± SD (N = 3).

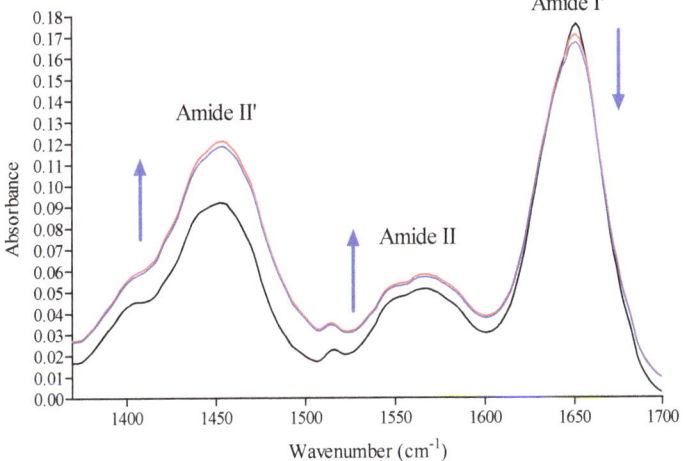

Figure 5. FTIR absorption spectra of HSA at 310 K in absence (black line) or presence of nitrofurantoin (red line) and warfarin (blue line). The arrows indicate the changes in the main amide bands after interaction with nitrofurantoin or warfarin.

The absorption of these two bands arises from the amide bonds that link the amino acids. The absorption is primarily due to bending vibrations of the N−H bond and, because they are involved in hydrogen bonding occurring among the different elements of a secondary structure (as well as in H−D exchange), they are sensitive to the changes in secondary structure and to the interaction with the surrounding environment of the protein.

2.7. Molecular Modeling Study

Our computational modeling study was performed on a crystal structure of HSA taken from the Protein Data Bank (entry PDB code 1GNI) in order to identify the possible binding site of nitrofurantoin. Based on the displace experimental results and to further define the binding site, the molecule docking simulation box was set on Sudlow's site I. The best energy ranked result (−5.43 kcal/mol) showed that nitrofurantoin may be situated within subdomain IIA, formed by six helices, which is consistent with the supposition on the basis of our experimental results. The nitro group and the oxygen of the furan ring established a hydrogen bond with Trp214 and Arg218, respectively, and the C=N−N unit of hydantoin established a further one with Ser454 (Figure 6). The nitrofurantoin ring is located within the binding pocket with the nitro group and the furan ring, which protrudes from it. The hydrogen bond formed with Arg218 and Ser454, as well as the one with Trp214, indicates that the interaction between nitrofurantoin and HSA is not exclusively hydrophobic in nature, but involves ionic and polar interactions. Although nitrofurantoin and warfarin bind to the Sudlow's site I within in the subdomain IIA, they do not share the same binding region. The warfarin binding region is located inside the hydrophobic pocket of the IIA subdomain and, in its vicinity, there are four positively charged residues (Lys199, Arg222, His242, and Arg257) and three apolar residues (Tyr150, Leu238, and Leu260). Warfarin also forms three hydrogen bonds with Tyr150, Arg222, and His242 [42–44].

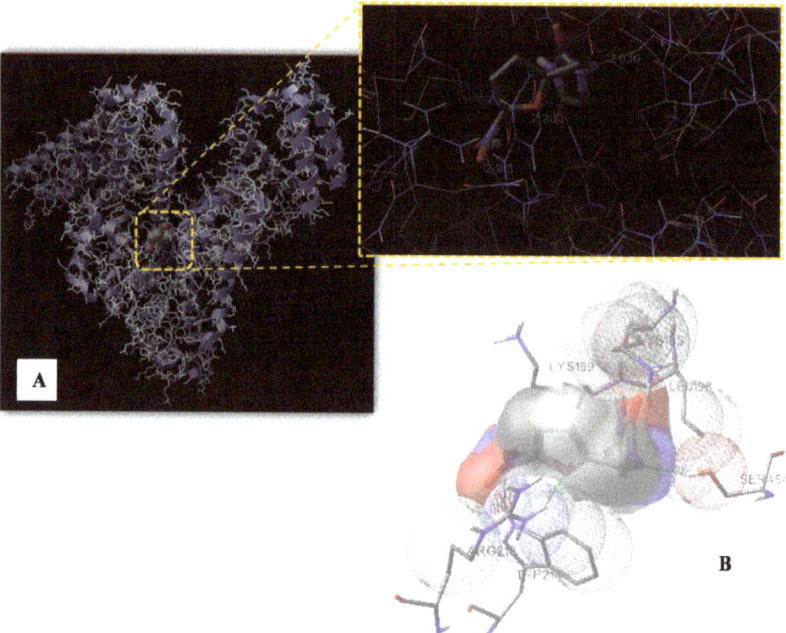

Figure 6. Binding site of nitrofurantoin to HSA. (**A**) The HSA structure is rendered with ribbons and lines, while nitrofurantoin is rendered as space fill. The inset shows a magnification of the binding site with nitrofurantoin represented using a stick model. The hydrogen bonds between the ligand and the protein are shown in green and the distance is expressed in Å. (**B**) Representation of the amino acid residues in the binding site with their van der Waals radii.

3. Discussion

The identification and characterization of novel antibiotics for clinical utilization is one of the greatest challenges of current basic experimentation. Indeed, discoveries in this field have always represented a remarkable achievement in the history of medicine. Moreover, the development and diffusion of bacterial antibiotic-resistance, due to the modification of different proteins and hence cellular processes (i.e., enzymatic degradation, molecular target alteration, decrement of drug uptake, overexpression of specific efflux pump proteins, etc.), is a major public health issue. Therefore, researchers are continuously trying to develop new agents of both a natural and synthetic origin that are able to overcome this resistance and fight the major issue bacterial infections still represent [45–47]. This may be achieved by enhancing the basic knowledge of biological processes and, sometimes, by repurposing "old" drugs in order to overcome the most difficult periods. Nitrofurantoin is an antibiotic widely used in medicine, but the analysis of the literature allowed us to notice a lack of information regarding its blood transportation and its interaction/binding with either serum/plasma proteins. The interaction and eventually binding with "carrier proteins" (such as HSA) is fundamental for the evaluation of the biological potentials of drugs, in general, and antibiotics, in particular, as it is actually one of the steps during antibiotic drug development [48]. In our study, the obtained spectroscopic results clearly highlighted the potentiality of HSA to bind nitrofurantoin and to drive it around the circulating system. The binding involved the formation of weak interactions, in particular between the nitro-substituted furan ring and amino acid residues able to form hydrogen bonds and apolar interactions at the level of the hydrophobic pocket located in subdomain IIA (Sudlow's site I), as shown by UV-VIS and fluorescence spectroscopy. Moreover, the data obtained for K_{SV} in the Stern-Volmer dynamic quenching equation and the evaluation of the other thermodynamic and kinetic parameters indicate that there is the formation of a binary complex in a single site in the neighboring of a tryptophan residue. Both hydrophobic interactions and hydrogen bonds are acknowledged to have a crucial role in the binding of nitrofurantoin to HSA, as in the case of other antibiotics [49]. The binding clearly affects the secondary structure of HSA, given the decrease in the percentage of the α-helix structure, as well as the change in absorbance values of both amides I and II after FTIR analysis. The vibration modes of these two bands are affected by both the conformation and environment of the amide group. This increase is given by the H–D exchange, which involves hydrogen within the core, thus justifying the formation of transitory protein conformational states during the HSA-nitrofurantoin complex formation. The increment of the amide II band may be a characteristic of hydrogen bonding formation during this process. Indeed, this affects mostly N-H bending vibration, which contributes exclusively to amide II vibration, which is more sensitive to H-D exchange. Starting from these observations and following molecular modelling data, we have shown that weak interactions play an important role in stabilizing the complex HSA−nitrofurantoin, and shed light on the results obtained by UV-VIS and fluorescence spectroscopy. Moreover, the hydrogen bond at the level of the C=N−N unit of nitrofurantoin with Ser454 contributes to its orientation and spatial arrangement inside the Sudlow's site I and provides experimental evidence to explain the fluorescence quenching of HSA emission in the presence of the antibiotic and the decrease in the absorbance of the band at 380 nm. Both changes are probably attributable to the formation of the hydrogen bond of the nitro-substituted furan ring with Trp214 and Arg218, influencing not only the quenching of HSA fluorescence, but also the π-π* transitions observed by UV-VIS spectroscopy. As reported in the results, both nitrofurantoin and warfarin bind to the same binding site and induce a change in the secondary structure content of the protein, although both the region of interaction and the mode of binding are different. This is due to the flexibility of this site and the possibility to bind different molecules (such as large heterocyclic and negatively charged compounds) in a specific zone, characterized by the presence of selected amino acid clusters.

4. Materials and Methods

4.1. Materials

Both HSA and nitrofurantoin were purchased from MERCH (Darmstadt, Germany, Europe). We employed no fat-free HSA so as to resemble a much closer situation closer to what occurs in vivo. Double distilled water was employed to prepare all of the solutions.

4.2. UV-VIS Spectra

UV-VIS spectroscopy was employed to study the interaction between HSA and nitrofurantoin, adding increasing amounts of the former (up to 90 µM) to 60 µM of the latter in a 20 mM phosphate buffer with pH 7.4. The absorption spectra were recorded from 220 to

480 nm through a spectrophotometer with a quartz cuvette. The spectra of each buffer solution were subtracted from the sample ones.

4.3. FTIR Spectra

Lyophilized HSA was dissolved in D_2O at 298 K for 1 day and was lyophilized again. This procedure was carried out twice for each sample. The FTIR spectra (1350 cm^{-1} and 1750 cm^{-1}) were measured through a Bruker Vertex 80 spectrometer under a vacuum, after dissolving the samples in 20 mM phosphate buffer with pH 7.4 in D_2O. The pD value of the HSA solution was measured and corrected according to pD = pH + 0.4 for deuterium isotope effects. A pair of CaF_2 windows divided by a 25 μm Teflon spacers were employed to place the protein solution (30 mg/mL), with or without nitrofurantoin or warfarin. Sixty-four interferograms with a spectral resolution of 4 cm^{-1} were collected for each sample. The IR spectra of the buffer with or without nitrofurantoin or warfarin were subtracted from the spectra of the corresponding sample.

4.4. Data Analysis of FTIR Spectra

The HSA spectra with or without nitrofurantoin or warfarin were smoothed by employing the Loess algorithm, and the deconvolved spectra were fitted with Gaussian band profiles. From these spectra, the initial values for the peak heights and widths were assessed. For the final fits, the positions, heights, and widths of each band were varied simultaneously. The curve fitting process was calculated through Seasolve PeakFit v4.12 software.

4.5. Fluorescence Spectra

The fluorescence spectra were recorded using a FluoroMax-4 spectrofluorometer by Horiba Jobin-Yvon, equipped with a pulsed xenon lamp and an F-3006 Autotitration Injector with two Hamilton Syringes (mods. Gastight 1725 and 1001 TLLX, with a 250 μL and 1.0 mL capacity, respectively). The resolutions of the wavelength selectors and titrant additions were 0.3 nm and 0.25 μL, respectively. The instrument was also equipped with a Peltier Sample Cooler (mod. F-3004) controlled by a Peltier Thermoelectric Temperature Controller model LFI-3751 (5 A–40 W). The whole system was controlled by the FluorEssence 2.1 software by Horiba Jobin-Yvon. The titrations were performed directly in a Hellma type 101-OS precision cell (Light Path 10 mm), where a magnetic stirrer and the anti-diffusion burette tip were placed in a position that would not interfere with the light beam. The automatic data acquisition (fluorescence intensity vs. λ(nm) for each titrant addition) was performed using the same FluorEssence 2.1 software. The excitation and emission bandwidths were both 5 nm. The protein samples were excited at 280 and 295 nm in order to characterize the possible different behavior of tryptophan and/or tyrosine residues. It was observed that both spectra were similar. The rest of the experiment was acquired by excitation at 280 nm and the emission spectra were recorded in the range of 300–400 nm. The potential interaction between nitrofurantoin and HSA was performed by fluorimetric titration. A 2.0 mL solution containing 1.5×10^{-5} mol/L HSA in 20 mM sodium phosphate buffer (pH 7.4) was titrated by a successive additions of nitrofurantoin stock solution (1.0×10^{-3} mol/L) to give a final concentration ranging from 0 to 3.0×10^{-5} mol/L. The fluorescence spectra were recorded at 293, 298, 304, 310, and 315 K in the wavelength range of 300−400 nm with an excitation wavelength at 280 nm.

The displacement studies were carried out utilizing different site markers (warfarin, ibuprofen, and digitoxin for sites I, II, and III, respectively). A solution of HSA (1.5×10^{-5} mol/L) in a 20 mM sodium phosphate buffer (pH 7.4) containing nitrofurantoin at the same final concentration (1.5×10^{-5} mol/L) was titrated by successive additions of site markers solution (1.0×10^{-3} mol/L) to obtain overall site markers concentrations ranging from 0 to 9×10^{-5} mol/L. Fluorescence spectra were recorded at 310 K in the range of 300-400 nm with an excitation wavelength at 280 nm. The fluorescence of the ternary mixture as a percentage of the initial fluorescence was determined according to the method of Sudlow et al. [50]:

$$\frac{F_2}{F_1} \times 100$$

where F_1 and F_2 are the fluorescence of HSA−nitrofurantoin in the absence and in the presence of the site markers, respectively.

4.6. Molecular Modeling Study

Docking calculations were carried out using AutoDock 4.2, and were graphically rendered by PyMOL 2.5 [51,52]. The MMFF94 force field was used for energy minimization of the ligand molecule (TC) using DockingServer. Gasteiger partial charges were added to the ligand atoms. Nonpolar hydrogen atoms were merged and rotatable bonds were defined. Docking calculations were carried out on an HSA protein model (PDB code 1GNI) and on nitrofurantoin, whose coordinates were taken from the ZINC database (entry code 7997568). Essential hydrogen atoms, Kollman united atom type charges, and solvation parameters were added with the aid of AutoDock tools. Affinity (grid) maps of 50 × 32 × 40 Å grid points and 0.375 Å spacing were generated using the Autogrid program. The AutoDock parameter set and distance-dependent dielectric functions were used in the calculation of the van der Waals and the electrostatic terms, respectively. Docking simulations were performed using the Lamarckian genetic algorithm (LGA). Initial positions, orientations, and torsions of the ligand molecules were set randomly. All of the rotatable torsions were released during docking. Each run of the docking experiment was set to terminate after a maximum of 250,000 energy evaluations. The population size was set to 150. The conformer with the lowest binding free energy was analyzed and used for further determinations.

5. Conclusions

The present experimental work describes, for the first time, the binding of nitrofurantoin to HSA. Our results may describe one of the mechanisms underlying the blood transportation of this compound through our body up to the kidney for its elimination. The binding reaction is spontaneous and involves hydrogen bond and hydrophobic interaction, as highlighted by the changes in the thermodynamic parameters. The interactions were confirmed through alterations in the UV−VIS absorption spectra of nitrofurantoin and albumin and the quenching of the intrinsic fluorescence of the protein, probably due to the hydrogen bond and apolar interactions with the nitro-substituted furan ring. In addition, the nitrofurantoin binding site is located in the hydrophobic pocket of subdomain IIA according to the site competitive study. Altogether, our results help to achieve a better understanding of the pharmacokinetic of nitrofurantoin, in order to design analogous molecules with a greater biological potential.

Author Contributions: Conceptualization, D.B.; G.L., D.B., and A.C. performed the experiments, elaborated the data, and assisted in their interpretation; G.L., D.B., A.C., M.N., A.M., and S.M. contributed to writing, reviewing, and editing the draft version of the manuscript; supervision, D.B.; project administration, D.B.; funding acquisition, D.B. All authors have read and agreed to the published version of the manuscript.

Funding: D.B. was supported by FFABR 2017 grants (ANVUR/MIUR, Italian Ministry of University and Research).

Institutional Review Board Statement: Not applicable.

Informed Consent Statement: Not applicable.

Data Availability Statement: Not applicable.

Conflicts of Interest: The authors declare no conflict of interest.

References

1. Zhu, N.J.; McLeod, M.; McNulty, C.A.M.; Lecky, D.M.; Holmes, A.H.; Ahmad, R. Trends in Antibiotic Prescribing in Out-of-Hours Primary Care in England from January 2016 to June 2020 to Understand Behaviours during the First Wave of COVID-19. *Antibiotics* **2021**, *10*, 32. [CrossRef]
2. Konwar, M.; Gogtay, N.J.; Ravi, R.; Thatte, U.M.; Bose, D. Evaluation of efficacy and safety of fosfomycin versus nitrofurantoin for the treatment of uncomplicated lower urinary tract infection (UTI) in women—A systematic review and meta-analysis. *J. Chemother.* **2021**, *3*, 1–10. [CrossRef]
3. Ten Doesschate, T.; Hendriks, K.; van Werkhoven, C.H.; van der Hout, E.C.; Platteel, T.N.; Groenewegen, I.A.M.; Muller, A.E.; Hoepelman, A.I.M.; Bonten, M.J.M.; Geerlings, S.E. Nitrofurantoin 100 mg versus 50 mg prophylaxis for urinary tract infections, a cohort study. *Clin. Microbiol. Infect. Off. Publ. Eur. Soc. Clin. Microbiol. Infect. Dis.* **2021**, *5*, 100. [CrossRef]
4. Gajdacs, M.; Abrok, M.; Lazar, A.; Burian, K. Comparative Epidemiology and Resistance Trends of Common Urinary Pathogens in a Tertiary-Care Hospital: A 10-Year Surveillance Study. *Medicina* **2019**, *55*, 356. [CrossRef] [PubMed]
5. Garau, J. Other antimicrobials of interest in the era of extended-spectrum beta-lactamases: Fosfomycin, nitrofurantoin and tigecycline. *Clin. Microbiol. Infect. Off. Publ. Eur. Soc. Clin. Microbiol. Infect. Dis.* **2008**, *14* (Suppl. 1), 198–202. [CrossRef]
6. Hooper, D.C. Urinary tract agents: Nitrofurantoin and methenamine. In *Mandell, Douglas, and Bennett's Principles and Practice of Infectious Diseases*, 6th ed.; Churchill Livingstone: Philadelphia, PA, USA, 2005; pp. 423–428.
7. Munsimbwe, L.; Seetsi, A.; Namangala, B.; N'Da, D.D.; Inoue, N.; Suganuma, K. In Vitro and In Vivo Trypanocidal Efficacy of Synthesized Nitrofurantoin Analogs. *Molecules* **2021**, *26*, 3372. [CrossRef] [PubMed]
8. Manin, A.N.; Drozd, K.V.; Voronin, A.P.; Churakov, A.V.; Perlovich, G.L. A Combined Experimental and Theoretical Study of Nitrofuran Antibiotics: Crystal Structures, DFT Computations, Sublimation and Solution Thermodynamics. *Molecules* **2021**, *26*, 3444. [CrossRef] [PubMed]
9. Day, M.A.; Jarrom, D.; Christofferson, A.J.; Graziano, A.E.; Anderson, J.L.R.; Searle, P.F.; Hyde, E.I.; White, S.A. The structures of E. coli NfsA bound to the antibiotic nitrofurantoin; to 1,4-benzoquinone and to FMN. *Biochem. J.* **2021**, *478*, 2601–2617. [CrossRef]
10. Edowik, Y.; Caspari, T.; Williams, H.M. The Amino Acid Changes T55A, A273P and R277C in the Beta-Lactamase CTX-M-14 Render *E. coli* Resistant to the Antibiotic Nitrofurantoin, a First-Line Treatment of Urinary Tract Infections. *Microorganisms* **2020**, *8*, 1983. [CrossRef]
11. Gajdacs, M.; Urban, E. Comparative Epidemiology and Resistance Trends of Proteae in Urinary Tract Infections of Inpatients and Outpatients: A 10-Year Retrospective Study. *Antibiotics* **2019**, *8*, 91. [CrossRef]
12. Kragh-Hansen, U.; Chuang, V.T.; Otagiri, M. Practical aspects of the ligand-binding and enzymatic properties of human serum albumin. *Biol. Pharm. Bull.* **2002**, *25*, 695–704. [CrossRef]
13. Barreca, D.; Lagana, G.; Toscano, G.; Calandra, P.; Kiselev, M.A.; Lombardo, D.; Bellocco, E. The interaction and binding of flavonoids to human serum albumin modify its conformation, stability and resistance against aggregation and oxidative injuries. *Biochim. Biophys. Acta Gen. Subj.* **2017**, *1861*, 3531–3539. [CrossRef]
14. Barreca, D.; Lagana, G.; Ficarra, S.; Tellone, E.; Leuzzi, U.; Galtieri, A.; Bellocco, E. Ornithine carbamoyltransferase unfolding states in the presence of urea and guanidine hydrochloride. *Appl. Biochem. Biotechnol.* **2014**, *172*, 854–866. [CrossRef]
15. Barreca, D.; Lagana, G.; Bruno, G.; Magazu, S.; Bellocco, E. Diosmin binding to human serum albumin and its preventive action against degradation due to oxidative injuries. *Biochimie* **2013**, *95*, 2042–2049. [CrossRef]
16. Di Salvo, C.; Barreca, D.; Lagana, G.; di Bella, M.; Tellone, E.; Ficarra, S.; Bellocco, E. Myelin basic protein: Structural characterization of spherulites formation and preventive action of trehalose. *Int. J. Biol. Macromol.* **2013**, *57*, 63–68. [CrossRef]
17. Barreca, D.; Lagana, G.; Ficarra, S.; Gattuso, G.; Magazu, S.; La Torre, R.; Tellone, E.; Bellocco, E. Spectroscopic determination of lysozyme conformational changes in the presence of trehalose and guanidine. *Cell Biochem. Biophys.* **2013**, *66*, 297–307. [CrossRef] [PubMed]
18. Barreca, D.; Lagana, G.; Ficarra, S.; Tellone, E.; Leuzzi, U.; Magazu, S.; Galtieri, A.; Bellocco, E. Anti-aggregation properties of trehalose on heat-induced secondary structure and conformation changes of bovine serum albumin. *Biophys. Chem.* **2010**, *147*, 146–152. [CrossRef]
19. Curry, S. Lessons from the crystallographic analysis of small molecule binding to human serum albumin. *Drug Metab. Pharmacokinet.* **2009**, *24*, 342–357. [CrossRef] [PubMed]
20. Barreca, D.; Laganà, G.; Magazù, S.; Migliardo, F.; Bellocco, E. Glycerol, trehalose and glycerol–trehalose mixture effects on thermal stabilization of OCT. *Chem. Phys.* **2013**, *424*, 100–104. [CrossRef]
21. Mandeville, J.S.; Tajmir-Riahi, H.A. Complexes of dendrimers with bovine serum albumin. *Biomacromolecules* **2010**, *11*, 465–472. [CrossRef]
22. Ascenzi, P.; Fasano, M. Allostery in a monomeric protein: The case of human serum albumin. *Biophys. Chem.* **2010**, *148*, 16–22. [CrossRef]
23. Fasano, M.; Curry, S.; Terreno, E.; Galliano, M.; Fanali, G.; Narciso, P.; Notari, S.; Ascenzi, P. The extraordinary ligand binding properties of human serum albumin. *IUBMB Life* **2005**, *57*, 787–796. [CrossRef]
24. Curry, S. Beyond expansion: Structural studies on the transport roles of human serum albumin. *Vox Sang.* **2002**, *83* (Suppl. 1), 315–319. [CrossRef]
25. Peters, T. *All about Albumin: Biochemistry, Genetics, and Medical Applications*; Academic Press: San Diego, CA, USA, 1996.

26. Colmenarejo, G.; Alvarez-Pedraglio, A.; Lavandera, J.L. Cheminformatic models to predict binding affinities to human serum albumin. *J. Med. Chem.* **2001**, *44*, 4370–4378. [CrossRef] [PubMed]
27. Dufour, C.; Dangles, O. Flavonoid-serum albumin complexation: Determination of binding constants and binding sites by fluorescence spectroscopy. *Biochim. Biophys. Acta* **2005**, *1721*, 164–173. [CrossRef]
28. Timaseff, S.N. Thermodynamics of Protein Interactions. In *Proteins of Biological Fluids*; Peeters, H., Ed.; Pergamon Press: Oxford, UK, 1972; pp. 511–519.
29. Barreca, D.; Lagana, G.; Magazu, S.; Migliardo, F.; Gattuso, G.; Bellocco, E. FTIR, ESI-MS, VT-NMR and SANS study of trehalose thermal stabilization of lysozyme. *Int. J. Biol. Macromol.* **2014**, *63*, 225–232. [CrossRef] [PubMed]
30. Hu, Y.J.; Ou-Yang, Y.; Dai, C.M.; Liu, Y.; Xiao, X.H. Site-selective binding of human serum albumin by palmatine: Spectroscopic approach. *Biomacromolecules* **2010**, *11*, 106–112. [CrossRef]
31. Barreca, D.; Bellocco, E.; Galli, G.; Lagana, G.; Leuzzi, U.; Magazu, S.; Migliardo, F.; Galtieri, A.; Telling, M.T. Stabilization effects of kosmotrope systems on ornithine carbamoyltransferase. *Int. J. Biol. Macromol.* **2009**, *45*, 120–128. [CrossRef]
32. Hu, Y.J.; Liu, Y.; Xiao, X.H. Investigation of the interaction between Berberine and human serum albumin. *Biomacromolecules* **2009**, *10*, 517–521. [CrossRef] [PubMed]
33. Bellocco, E.; Barreca, D.; Lagana, G.; Tellone, E.; Ficarra, S.; Migliardo, F.; Leuzzi, U.; Magazu, S.; Galtieri, A. Influences of temperature and threshold effect of NaCl concentration on Alpias vulpinus OCT. *Int. J. Biol. Macromol.* **2008**, *43*, 474–480. [CrossRef]
34. Belloco, E.; Lagana, G.; Barreca, D.; Ficarra, S.; Tellone, E.; Magazu, S.; Branca, C.; Kotyk, A.; Galtieri, A.; Leuzzi, U. Role of polyols in thermal inactivation of shark ornithine transcarbamoylase. *Physiol. Res.* **2005**, *54*, 395–402. [PubMed]
35. Ross, P.D.; Subramanian, S. Thermodynamics of protein association reactions: Forces contributing to stability. *Biochemistry* **1981**, *20*, 3096–3102. [CrossRef]
36. Yang, H.; Yang, S.; Kong, J.; Dong, A.; Yu, S. Obtaining information about protein secondary structures in aqueous solution using Fourier transform IR spectroscopy. *Nat. Protoc.* **2015**, *10*, 382–396. [CrossRef]
37. Barth, A. Infrared spectroscopy of proteins. *Biochim. Biophys. Acta* **2007**, *1767*, 1073–1101. [CrossRef]
38. Kong, J.; Yu, S. Fourier transform infrared spectroscopic analysis of protein secondary structures. *Acta Biochim. Biophys. Sin.* **2007**, *39*, 549–559. [CrossRef]
39. Fabian, H.M.W. Infrared spectroscopy of proteins. In *Handbook of Vibrational Spectroscopy*; John Wiley & Sons: Chichester, UK, 2002; Volume 5.
40. Mantsch, H.H.; Chapman, D. *Infrared Spectroscopy of Biomolecules*; Wiley-Liss: New York, NY, USA, 1996.
41. Havel, H.A. *Spectroscopic Methods for Determining Protein Structure in Solution*; John Wiley & Sons: Chichester, UK, 1995.
42. Petitpas, I.; Bhattacharya, A.A.; Twine, S.; East, M.; Curry, S. Crystal structure analysis of warfarin binding to human serum albumin: Anatomy of drug site I. *J. Biol. Chem.* **2001**, *276*, 22804–22809. [CrossRef]
43. Ghuman, J.; Zunszain, P.A.; Petitpas, I.; Bhattacharya, A.A.; Otagiri, M.; Curry, S. Structural basis of the drug-binding specificity of human serum albumin. *J. Mol. Biol.* **2005**, *353*, 38–52. [CrossRef] [PubMed]
44. Rimac, H.; Dufour, C.; Debeljak, Z.; Zorc, B.; Bojic, M. Warfarin and Flavonoids Do Not Share the Same Binding Region in Binding to the IIA Subdomain of Human Serum Albumin. *Molecules* **2017**, *22*, 1153. [CrossRef]
45. Atanasov, A.G.; Zotchev, S.B.; Dirsch, V.M.; The International Natural Product Sciences Taskforce; Supuran, C.T. Natural products in drug discovery: Advances and opportunities. *Nat. Rev. Drug Discov.* **2021**, *20*, 200–216. [CrossRef]
46. Gajdacs, M. The Concept of an Ideal Antibiotic: Implications for Drug Design. *Molecules* **2019**, *24*, 892. [CrossRef] [PubMed]
47. Barreca, D.; Trombetta, D.; Smeriglio, A.; Mandalari, G.; Romeo, O.; Felice, M.R.; Gattuso, G.; Nabavi, S.M. Food flavonols: Nutraceuticals with complex health benefits and functionalities. *Trends Food Sci. Technol.* **2021**, *3*, 26. [CrossRef]
48. Schmidt, S.; Rock, K.; Sahre, M.; Burkhardt, O.; Brunner, M.; Lobmeyer, M.T.; Derendorf, H. Effect of protein binding on the pharmacological activity of highly bound antibiotics. *Antimicrob. Agents Chemother.* **2008**, *52*, 3994–4000. [CrossRef]
49. Yasmeen, S.; Rabbani, G. Calorimetric and spectroscopic binding studies of amoxicillin with human serum albumin. *J. Therm. Anal. Calorim.* **2017**, *127*, 1445–1455. [CrossRef]
50. Lehrer, S.S. Solute perturbation of protein fluorescence. The quenching of the tryptophyl fluorescence of model compounds and of lysozyme by iodide ion. *Biochemistry* **1971**, *10*, 3254–3263. [CrossRef] [PubMed]
51. Maugeri, A.; Ferlazzo, N.; de Luca, L.; Gitto, R.; Navarra, M. The link between the AMPK/SIRT1 axis and a flavonoid-rich extract of Citrus bergamia juice: A cell-free, in silico, and in vitro study. *Phytother. Res. PTR* **2019**, *33*, 1805–1814. [CrossRef] [PubMed]
52. Sanner, M.F. Python: A programming language for software integration and development. *J. Mol. Graph. Model.* **1999**, *17*, 57–61. [PubMed]

Review

The Mechanism of Secretion and Metabolism of Gut-Derived 5-Hydroxytryptamine

Ning Liu [1,2,3,†], Shiqiang Sun [4,5,†], Pengjie Wang [2,3], Yanan Sun [2,3], Qingjuan Hu [2,3] and Xiaoyu Wang [1,*]

1. Key Laboratory of Precision Nutrition and Food Quality, College of Food Science and Nutritional Engineering, China Agricultural University, Beijing 100083, China; nuli982390@163.com
2. Department of Nutrition and Health, China Agricultural University, Beijing 100193, China; Wpjl019@cau.edu.cn (P.W.); 15153515695@163.com (Y.S.); wshqlsh@163.com (Q.H.)
3. Beijing Advanced Innovation Center for Food Nutrition and Human Health, China Agricultural University, Beijing 100193, China
4. Department of Gastroenterology and Hepatology, University Medical Center Groningen, University of Groningen, 9713ZG Groningen, The Netherlands; sqsun@hotmail.com
5. Department of Genetics, University Medical Center Groningen, University of Groningen, 9713ZG Groningen, The Netherlands
* Correspondence: xy.wang@cau.edu.cn; Tel.: +86-10-6273-8589
† These authors contributed equally to this work.

Citation: Liu, N.; Sun, S.; Wang, P.; Sun, Y.; Hu, Q.; Wang, X. The Mechanism of Secretion and Metabolism of Gut-Derived 5-Hydroxytryptamine. *Int. J. Mol. Sci.* **2021**, *22*, 7931. https://doi.org/10.3390/ijms22157931

Academic Editors: Hisamitsu Ishihara and Burkhard Poeggeler

Received: 30 April 2021
Accepted: 19 June 2021
Published: 25 July 2021

Publisher's Note: MDPI stays neutral with regard to jurisdictional claims in published maps and institutional affiliations.

Copyright: © 2021 by the authors. Licensee MDPI, Basel, Switzerland. This article is an open access article distributed under the terms and conditions of the Creative Commons Attribution (CC BY) license (https://creativecommons.org/licenses/by/4.0/).

Abstract: Serotonin, also known as 5-hydroxytryptamine (5-HT), is a metabolite of tryptophan and is reported to modulate the development and neurogenesis of the enteric nervous system, gut motility, secretion, inflammation, sensation, and epithelial development. Approximately 95% of 5-HT in the body is synthesized and secreted by enterochromaffin (EC) cells, the most common type of neuroendocrine cells in the gastrointestinal (GI) tract, through sensing signals from the intestinal lumen and the circulatory system. Gut microbiota, nutrients, and hormones are the main factors that play a vital role in regulating 5-HT secretion by EC cells. Apart from being an important neurotransmitter and a paracrine signaling molecule in the gut, gut-derived 5-HT was also shown to exert other biological functions (in autism and depression) far beyond the gut. Moreover, studies conducted on the regulation of 5-HT in the immune system demonstrated that 5-HT exerts anti-inflammatory and proinflammatory effects on the gut by binding to different receptors under intestinal inflammatory conditions. Understanding the regulatory mechanisms through which 5-HT participates in cell metabolism and physiology can provide potential therapeutic strategies for treating intestinal diseases. Herein, we review recent evidence to recapitulate the mechanisms of synthesis, secretion, regulation, and biofunction of 5-HT to improve the nutrition and health of humans.

Keywords: 5-hydroxytryptamine; serotonin; secretion; metabolism

1. Introduction

Serotonin, or 5-Hydroxytryptamine (5-HT), a metabolite of tryptophan (Trp), is an important gastrointestinal (GI) regulatory factor with a wide range of physiological effects on humans and animals [1–4]. Approximately 95% of 5-HT in the body is synthesized and secreted by enterochromaffin (EC) cells in the GI tract. Once 5-HT is released into the lamina propria, it is taken up by the epithelial cells through the serotonin reuptake transporter (SERT). Next, 5-HT diffuses into the bloodstream, where it is taken up by platelets and transported to peripheral target tissues. The physiological effects of 5-HT have been considerably investigated, and 5-HT has been reported to play a crucial role in GI regulation, particularly in intestinal motility and secretion [2]. The role of 5-HT in gut inflammation has also been widely investigated [4–7]. An increased concentration of 5-HT in the mucosa contributes to severe colitis. Serotonin has been shown to exert anti-inflammatory and proinflammatory effects on the gut by binding to different 5-HT receptors in animal models of inflammatory bowel disease (IBD) and colitis [8]. New clues have demonstrated that 5-HT exerts an anti-inflammatory effect on the gut by regulating the expression of the 5-HT4 receptor, with beneficial effects on intestinal epithelial cell barrier functions [4].

EC cells, which are specialized enteroendocrine (EE) cells that reside alongside the epithelium lining the lumen of the digestive tract, can synthesize and secrete 5-HT [9–13]. As a chemosensor, EC cells convert physiological and chemical signals from the lumen into biochemical endocrine signals through microvilli extending into the lumen and enzymes and transporters stored in the apical parts of the enterocytes. Importantly, 5-HT secretion in the gut is influenced by various factors such as nutrients, microbial community, host-derived signaling hormones, and peptides, which in turn directly or indirectly affect immune responses, nutrient metabolism, and intestinal homeostasis [14–16].

Gut-derived 5-HT and its various biological functions are receiving great interest from investigators. Gut-derived 5-HT possesses a range of protective effects, such as modulating gut motility and secretion, gut inflammation, liver regeneration, metabolic homeostasis, and bone remodeling, etc. This review aims to elucidate the functional role of 5-HT in and beyond the gut. We also provide an in-depth review highlighting the understanding of various factors (gut microbiota, nutrition, and hormones) in the regulation of 5-HT secretion. We hope that this review could lay a theoretical foundation for the application of 5-HT in nutrition, clinical medicine, and health.

2. Synthesis and Secretion of Gut-Derived 5-HT

2.1. 5-HT Synthesis

Only 20 of more than 700 amino acids (AAs) in nature are building blocks for proteins in cells and traditionally categorized as nutritionally essential or nonessential for humans and animals on the basis of growth or nitrogen balance [17,18]. Trp is one of nine nutritionally essential AAs [19]. In addition to its role as a substrate for protein synthesis, Trp is an important precursor for many compounds such as 5-HT, melatonin, and kynurenine [20]. Correspondingly, Trp and its metabolites play a key role in nutrition, reproduction, immune system, and anti-stress responses [21–27]. The kynurenine and 5-HT pathways are two main metabolic routes for Trp metabolism in mammals. Approximately 95% of the ingested Trp is degraded into kynurenine, kynurenic acid, xanthurenic acid, quinolinic acid, and picolinic acid through the kynurenine pathway. Additionally, approximately 1–2% of the ingested Trp is degraded into 5-HT and melatonin through the 5-HT pathway [28]. There are two major synthetic routes of 5-HT in the brain stem and peripheral neurons. Moreover, approximately 95% 5-HT in human body is synthesized in the peripheral system, especially in the GI tract [3,29]. Serotonergic neurons of the enteric nervous system and EC cells are two separate sources of gut-derived 5-HT in the GI tract mucosa, of which 90% 5-HT is synthesized in gut-resident EC cells, a subset of EE cells in the GI tract [30,31].

Trp hydroxylase (*TPH*), the specific serotonin-synthesizing gene, exists in two isoforms (*TPH1* and *TPH2*) [32,33]. Both *TPH1* and *TPH2* show Trp hydroxylating activity. *TPH1* is predominantly found in EC cells in the GI tract, whereas *TPH2* is mainly expressed in the central nervous system and serotonergic neurons [34]. TPH, a rate-limiting enzyme for 5-HT production, plays a key role in the conversion of Trp to 5-hydroxytryptophan (5-HTP) [1,35]. 5-HT is rapidly converted to 5-HT by aromatic L-amino acid decarboxylase (L-AADC) in the next enzymatic step [36]. Vesicular monoamine transporter 1 (VMAT1), which participates in 5-HT storage, is expressed by granules/vesicles in EC cells [37]. Newly produced 5-HT compounded with chromogranin A (CGA), an acidic protein expressed in response to 5-HT secretion, is stored in the VMAT1 vesicles of EC cells [38] (Figure 1). 5-HT stored in the dense granules/vesicles near the basal border or apical membrane of EC cells is released into the lamina propria or lumen when EC cells are exposed to intraluminal pressure or chemical and mechanical stimulation [10,15]. The biosynthesis and metabolism of gut-derived 5-HT are illustrated in Figure 1.

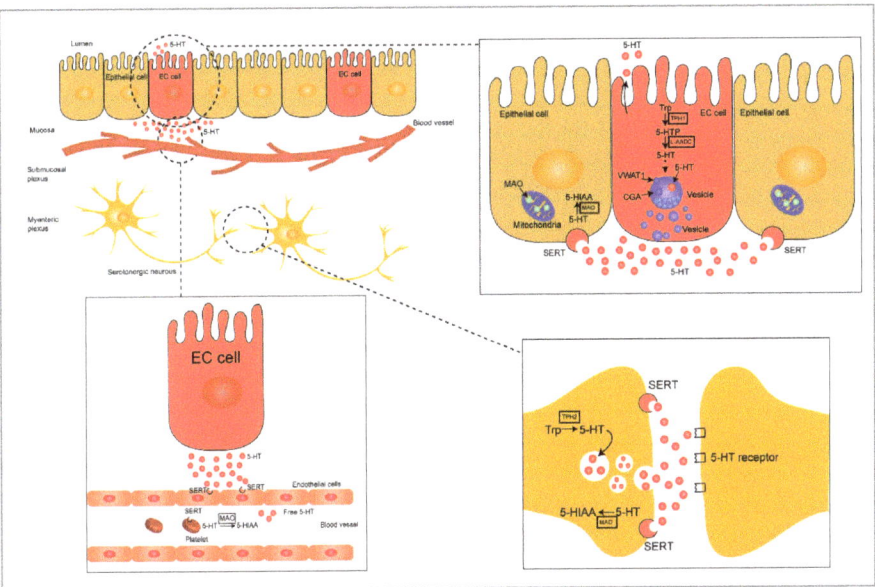

Figure 1. Schematic representation of gut-derived 5-HT biosynthesis and metabolism. Enterochromaffin (EC) cells and serotonergic neurons convert tryptophan into 5-HTP through the rate-limiting enzymes, TPH1 and TPH2, respectively, and the newly formed 5-HTP is rapidly degraded into 5-HT by L-AADC. The synthesized 5-HT and CGA are rapidly packaged into vesicles through VMAT1. EC cells express sensory receptors by acting as chemosensors to continuously release 5-HT in response to stimuli in the luminal environment, including chemical and mechanical stimulation, luminal pressure, and nutritional and intestinal microbial metabolites and hormones. Most of the 5-HT is released into the extracellular space from the bottom of EC cells, and a comparatively smaller amount of 5-HT is released into the lumen through the apical membrane. The surrounding enterocytes take up 5-HT by SERT, and 5-HT is then metabolized to 5-HIAA through MAO in the mitochondria. As a neurotransmitter, the synthetized 5-HT is packaged in synaptic vesicles in serotonergic neurons and released into the synapse cleft. Serotonin exerts a high effect on the postsynaptic membrane through the 5-HT receptors and reabsorbs on the presynaptic membrane through SERT. SERT is also detected in endothelial cells and platelets where 5-HT entering into the lamina propria is taken up and then converted into 5-HIAA or transported to peripheral target tissues. Trp, tryptophan; 5-HT, 5-hydroxytryptamine; TPH1, tryptophan hydroxylase 1; TPH2, tryptophan hydroxylase 2; 5-HTP, 5-hydroxytryptophan; 5-HIAA, 5-hydroxyindoleacetic acid; L-AADC, L-amino acid decarboxylase; CGA, chromogranin A; VMAT1, vesicular monoamine transporter 1; MAO, monoamine oxidase; SERT, serotonin reuptake transporter.

2.2. 5-HT Release and Inactivation

In serotonergic neurons, 5-HT is packaged in synaptic vesicles and then released into synapse cleft [31]. In the gut, 5-HT is mainly released from the granules stored near the basal border of the EC cell, and small amounts of 5-HT are released into lumen through the apical membrane [35]. Once released by the EC cells, there are several possible routes that 5-HT may take. 5-HT released into the lamina propria interacts with nerve terminals, epithelial cells, and immune cells or may also be taken up into the enterocytes by SERT or may enter the general circulation [39] (Figure 1).

5-HT is a positively charged molecule at physiological pH, which results in the failure of transmembrane-mediated transport. SERT relies on Na^+ and Cl^- to reuptake 5-HT released from serotonergic neurons. The driving force of the reuptake process is the transmembrane ion gradient produced by Na^+/K^+-ATPase [40,41]. In the gut, 5-HT is also transported into surrounding enterocytes through SERT and may then be degraded into 5-hydroxyindole acetaldehyde (5-HIAL) by monoamine oxidase (MAO); 5-HIAL is, then, further transformed into 5-hydroxyindoleacetic acid (5-HIAA), which is finally excreted in urine [42–44]. MAO is found in the mitochondria and exists in two forms: MAO-A and MAO-B. MAO-A has a higher affinity for 5-HT [35,45,46].

5-HT released from EC cells also enter in the general circulation and are taken up by platelets via SERT. Approximately 95% of 5-HT in the blood is stored in platelets [8] in granules together with ATP, ADP, and Ca^{2+} [35]. 5-HT absorbed by platelets reaches the liver through portal circulation and is transported to peripheral target tissues through bloodstream to regulate bone density [47], liver regeneration [48,49], obesity and energy metabolism [50], and allergic airway inflammation [51]. One-third of 5-HT is converted into 5-HIAA by MAO and excreted in urine, and the remaining 5-HT is degraded into 5-HTOglucuronide through glucosidase [35].

2.3. 5-HT Receptors

5-HT acts in the lamina propria or lumen in a paracrine manner. In the intestinal epithelial cells or the mucosal afferent nerve of the lamina propria, 5-HT promotes intestinal motility, peristalsis, and secretion through binding to 5-HT-specific receptors (5-HTRs); 5-HTRs are classified into seven families according to structure, function, and effectiveness (5-HTR1–5-HTR7) [52,53]. Of note, the 5-HTR3 receptor is a ligand ion channel, and the other six receptors are G-protein-coupled receptors (GPCRs) [54]. In the gut, compelling evidence has shown that 5-HT regulates GI function by binding to different receptors (5-HTR1, 5-HTR2, 5-HTR3, 5-HTR4, and 5-HTR7). The conventional actions of 5-HT and its receptors in the GI tract are summarized in Table 1.

Table 1. Conventional effects of 5-HT and its receptors in the gastrointestinal tract.

Conventional Effect	Pathway	Mediated Receptors	References
Motility and peristaltic reflex	Activate ascending and descending interneurons	5-HT3 and 5-HT4 receptor	[2,55–57]
Secretion (bicarbonate and electrolyte)	Neural mediated or through paracrine pathway acts on nearby enterocytes	5-HT2, 5-HT3 and 5-HT4 receptor	[58,59]
Pancreatic secretion and gastric emptying	Activate vago-vagal reflex and act in synergy with cholecystokinin (CCK)	5-HT2 and 5-HT3 receptor	[60–62]
Vasodilation	Locally regulate blood vessel diameter through intrinsic reflex circuits	5-HT3 and 5-HT4 receptor	[2,63]
Inflammation	The pro-inflammatory actions by promoting an inflammatory offensive to protect the gut from invasion and the anti-inflammatory actions by inducing neurogenesis	5-HT1A, 2A, 2B, 2C, 5-HT3, 5-HT4, and 5-HT7 receptor	[4–6,64–66]
Neurogenesis and enteric protection	Play an important role though Neuronal 5-HT	5-HT4 receptor	[67–69]
Mucosal growth	Serotonergic neurons project submucosal cholinergic neurons	5-HT2A receptor	[70,71]

3. 5-HT in the Gut

EC cells are considered as "sensor cells" that have the ability to sense the luminal nutrients and non-nutrient chemicals, mechanical stimulations, and signals from the gut microbiota to release 5-HT [12,35,72]. Additionally, EC cells are stimulated to trigger the release of 5-HT in response to high intraluminal pressure changes in pH in the gut lumen [73–75]. Over the past decade, many studies have demonstrated that the stimulation of the intestinal cavity by gut microbiota, nutrients, and hormones could stimulate EC cells to release 5-HT. Therefore, we have described the effect of gut microbiota, nutrients, and host-derived hormones on the secretion of 5-HT in greater details.

3.1. Gut Microbiota and 5-HT Release

Gut microbiota are a complex and dynamic population of microorganisms that inhabit the GI tract of humans and other mammals [76]. Over the past decade, gut microbiota have received considerable attention because of their functional role in regulating host physiology, metabolism, and immunity [77,78]. Emerging evidence has also shown that the gut microbiota play a critical role in regulating host 5-HT secretion in EC cells by interacting with various compounds produced by the host or gut microorganisms [76]. Short-chain fatty acids (SCFAs), as markers of bacterial metabolism [79], enhance colonic *TPH1* mRNA expression by interacting with EC cells [16]. This finding is consistent with previous research that intraluminal administration of SCFAs into the proximal colon significantly augments the release and production of 5-HT by accelerating colonic transit through stimulating 5-HTR3, and thereby, promoting colonic contraction [80]. In contrast, 5-HT production following the stimulation of EC cells by SCFAs via triggering the entry of extracellular Ca^{2+} is unchanged [14], which indicates that the interaction between microorganisms and the host plays an indispensable role in 5-HT secretion. A study conducted by Yano et al. demonstrated that microbial-specific metabolites such as SCFAs, α-tocopherol, tyramine, and p-aminobenzoate promote TPH1 expression and 5-HT release [81]. These results suggest an association between gut microbiota communities and host in regulating the basic biological processes through 5-HT [81]. Many different types of GPCR sensors of microbial metabolites are expressed in colonic EC cells, including olfactory receptor 558 (Olfr558), free fatty acid receptor 2 (FFAR2), olfactory receptor 78 (OLF78) that senses SCFAs, G-protein-coupled receptor 35 (GPR35) that senses small aromatic acids, G-protein-coupled bile acid receptor 1 (GPBAR1) that senses secondary bile acids, and G-protein-coupled receptor 132 (GPR132) that senses lactate and acyl amides. These receptors are activated in the process of 5-HT secretion by various gut microbial metabolites [10,13].

Most 5-HT is produced by EC cells, and a small amount of 5-HT is synthesized by a deconjugation process of glucuronide-conjugated 5-HT by a bacterial enzyme such as β-glucuronidase [12]. The gut microorganisms metabolize various substances through their interaction with the host, thereby affecting the release of 5-HT. The elucidation of the gut microbiome and host genetics in the past 10 years has helped to clarify the relationship between gut microbiota and the physiological and pathological conditions of the host. Consequently, the mechanism of microbial dependence that affects the physiological function of the host is likely to be elucidated, which would be beneficial to find methods for using gut microbial intervention to improve body health. Jonathan et al. reported that *Escherichia coli* Nissle 1917, one of the currently available probiotic bacteria, regulates THP1 through the interaction between host and probiotics by enhancing 5-HT level and its bioavailability in ileal tissues [42]. Because of the complexity of the gut microflora, beneficial bacteria promote body health through the 5-HT system, while pathogenic bacteria may cause intestinal diseases by damaging the intake of 5-HT. Enteropathogenic *E. coli*, a foodborne pathogen, inhibits SERT activity by reducing protein tyrosine phosphatase, and the damaged SERT function is associated with infectious diarrheal diseases [82]. Many studies have confirmed that the gut microbial flora and its particular metabolites influence the biosynthesis of 5-HT. However, it is largely unknown whether the alteration of 5-HT level caused by host–microbiota interaction in turn affects the colonization, growth, or adaptation of enteric microorganisms. Therefore, much work is required to investigate the metabolic pathway and molecular mechanisms of microbial metabolites in regulating 5-HT levels in the gut.

3.2. Nutrients and 5-HT Release

The specialized EE cells are dispersed as single cells scattered throughout the epithelium of the GI tract from the stomach to the rectum and are considered as the largest endocrine system of the human body [83]. EE cells regulate various physiological and homeostatic functions both within and outside the gut by secreting various hormones and peptides [84]. EC cells represent around 50% of all EE cells that sense diverse dietary nutrients and metabolites to produce ~95% of total body 5-HT [14]. Studies conducted on human primary colonic EC cells and BON cells (immortalized cell line models of EC cells) found that 5-HT is released from EC cells in response to stimulation of luminal nutrients [11,14,15,85]. Unlike human primary colonic EC cells, BON cells release 5-HT following the stimulation of luminal D-glucose through sodium-glucose-linked transporter 1 (SGLT1) [86]. Additionally, the sensing of glucose in the lumen is related to the expression of SGLT3 in EC cells, which results in the release of 5-HT [87,88]. Ingested food components are digested by digestive enzymes into a form that can be absorbed into the bloodstream. Glucose is the main form of carbohydrate absorbed by mammals and serves as a luminal substance to trigger several key events in the physiological regulation of the intestinal tract [89].

As a chemosensor in the GI mucosa, EC cells release 5-HT by sensing the presence of glucose, thereby inhibiting gastric emptying and food intake by activating 5-HTR3 on exogenous afferent nerves of rodents and humans [60,90]. Intriguingly, the nutrient sensing capacity of EC cells in 5-HT secretion from the mouse duodenum and colon is region-specific. Carbohydrate absorption is generally achieved over the entire small intestine, and only a small amount of glucose reaches the colon. Correspondingly, EC cells in the colon are more sensitive to glucose than those in the duodenum [14], as glucose transporter 1 (GLUT1) is highly expressed in colonic EC cells, while glucose transporter 2 (GLUT2) is abundantly expressed in duodenal EC cells [11]. The low glucose availability leads to the upregulation of GLUT1, which is a high-affinity and low-capacity glucose transporter. On the other hand, GLUT2, a low-affinity and high-capacity glucose transporter, is upregulated in a high concentration of glucose [91].

Zelkas et al. reported that 5-HT-secreting EC cells show enormous diversity in response to acute and chronic changes in glucose availability [92]. Acute exposure to high concentration of glucose results in 5-HT release from EC cells, which involves the entry of Ca^{2+} and an increment in the number of vesicles for exocytosis. Chronic exposure to fasting-related levels of glucose leads to the enhancement of 5-HT synthesis through transcriptional regulation of TPH1. Consistently, food deprivation enhances gut-derived 5-HT synthesis accompanied by enhancement of lipolysis in adipocytes and liver gluconeogenesis, as well as prevention of glucose uptake in hepatocytes [93]. The enhancement of 5-HT synthesis in response to the elevated level of luminal glucose after feeding promotes gut motility and peristaltic reflex through the activation of ascending and descending interneurons to facilitate digestion. 5-HT also plays a pivotal role in enhancing body fat degradation and liver gluconeogenesis during fasting, which contributes to the maintenance of the blood glucose level.

3.3. Hormonal Control of 5-HT Release

The enteroendocrine system is responsible for secreting a diverse range of gut hormones, which play a highly important role in the physiological regulation of the GI tract [84]. EC cells coexist closely with other EE cells, instead of existing in "one cell type" solitarily along the length of the GI tract. A recent discovery is that EE cells communicate with EC cells locally through paracrine action in the gut [94]. The glucagon-like peptide 1 (GLP-1) receptor is particularly highly expressed in EC cells. The neighboring GLP-1-storing EE cells secrete GLP-1, and GLP-1 then stimulates EC cells to release 5-HT through the activation of GLP-1 receptors [13]. The GLP-1 receptor agonist has been reported to release 5-HT in both small intestine and colon. Of note, the spontaneous secretion of 5-HT was higher in the duodenum when compared with that in the colon. However, a significant enhancement in 5-HT release was detected with the treatment of a GLP-1 receptor agonist both in the duodenum and colon [13]. Moreover, it has been reported that 5-HT enhances nutrient-induced GLP-1 release from ileal segments through a process involving interactions with 5-HT receptors [95]. Using the intestinal secretin tumor cell line (STC-1) for further exploration in vitro, results revealed that 5-HT (30 or 100 µM) significantly enhanced GLP-1 secretion in STC-1 cells when compared with control group. Additionally, the non-specific 5-HT receptor antagonist asenapine inhibited the 5-HT-promoted GLP-1 release, which supports the 5-HT receptor-mediated mechanism [95]. Further studies are needed to investigate the interaction and mechanisms between the secretion of 5-HT and GLP-1. EC cells are also sensitive to endogenous regulatory molecules. Norepinephrine-mediated stimulation of EC cells activates alpha-2A adrenergic (Adrα2A) receptors through catecholamines, which leads to chronic visceral hypersensitivity [10]. Hormone crosstalk exists between gut mucosal EC cells and the neighboring enterocytes within the epithelium, but its complex effects remain unknown.

4. Physiological and Pathophysiological Role of Gut-Derived 5-HT
4.1. 5-HT and Gut Inflammation

Accumulating evidence through clinical and animal studies indicates that 5-HT, as a signaling molecule in the intestine, plays a pivotal role in intestinal inflammation (Figure 2). 5-HT signaling has been investigated in an animal model of intestinal inflammation, including 2,4,6-trinitrobenzene sulfonic acid (TNBS)-induced colitis [96] and ileitis [5,97–99], dextran sodium sulfate (DSS)-induced colitis [4,100], and trichinella spiralis infection-induced intestinal inflammation [101]. Several studies have revealed that 5-HT is a key proinflammatory signaling molecule in gut inflammation because of the enhanced concentration of intestinal 5-HT and downregulation of SERT expression under intestinal inflammation [6,65,102,103]. Because of the knockout of *TPH1*, the concentration of 5-HT in the GI tract was significantly reduced, followed by alleviation in clinical severity and histological damage scores by pharmacological adjustment of mucosal 5-HT in DSS- or dinitrobenzene sulfonic acid (DNBS)-induced colitis [104]. Consistently, several studies have reported that a THP1 or TPH inhibitor alleviates the severity of colitis and plays a protective role in colitis [105,106]. Additionally, SERT transcription is reduced during intestinal inflammation, which contributes to impaired absorption of 5-HT [2,107].

5-HT plays an important role in the proinflammatory or anti-inflammatory process through binding to different receptors [5,64]. The anti-inflammatory role of 5-HT is accompanied by the activation of epithelial 5-HTR1A and 5-HTR4. Compared to control, the severity of experimental colitis in mice was enhanced through intraluminal administration of a 5-HTR1A and 5-HTR4 antagonist [4,64]. Upregulation of 5-HTR4 expression protects the large intestine from DSS- or TNBS-induced colitis by maintaining epithelial integrity, stimulating the proliferation of crypt epithelial cells, and reducing apoptosis [64]. A study reported that treatment with a 5-HTR2A antagonist (Ketanserin) alleviates intestinal inflammation by improving gut integrity, reducing the production of inflammatory cytokines in macrophages, and inhibiting the activation of nuclear factor-κB (NF-κB) in experimental colitis; this result further confirmed the deleterious role of 5-HTR2A on intestinal inflammation [108]. However, the 5-HT receptors involved in the proinflammatory and anti-inflammatory processes reported in the current literature are contradictory. Spohn et al. demonstrated that chemical activation of 5-HTR4 reduced the severity of TNBS- and DSS-induced colitis [64]. In contrast, Rapalli et al. found that the inhibition of 5-HTR4 improves the progression and pathological outcome of TNBS-induced colitis, thus suggesting the detrimental effect of 5-HTR4 on TNBS-induced colitis [5]. Kim et al. also reported that the inhibition of 5-HTR7 signaling reversed acute and chronic colitis induced by DSS or TNBS [109]. In contrast, several research studies have demonstrated that the development of colitis was not affected by 5-HTR7 [4,5]. Thus, further studies are required to determine the role of 5-HT receptors on experimental colitis to reveal the association between 5-HT receptors and the downstream signaling pathways under inflammatory conditions (Figure 2).

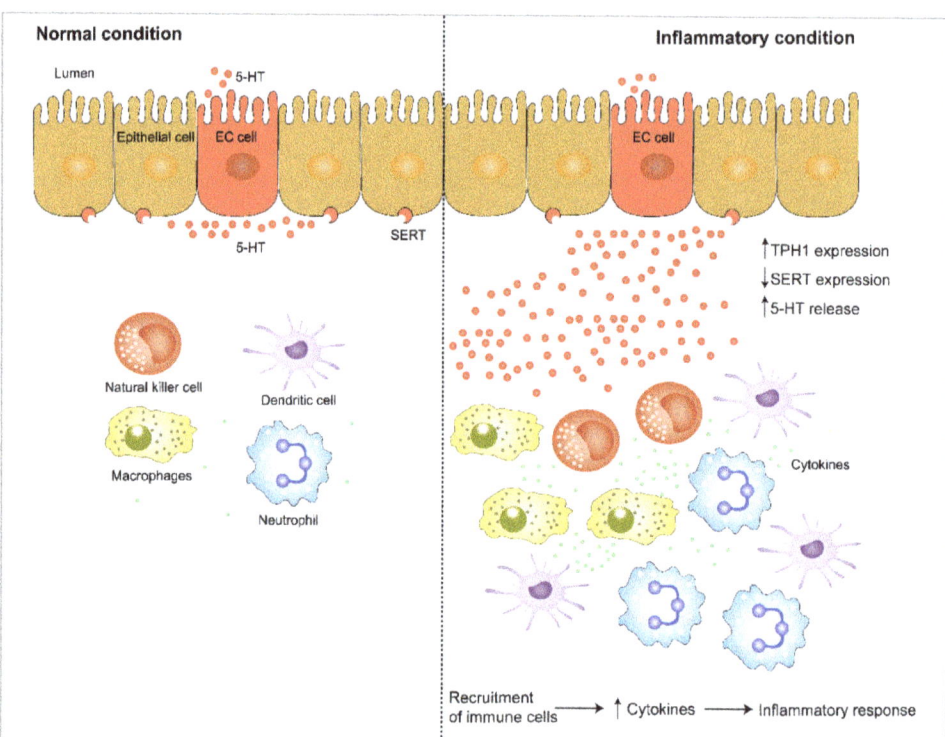

Figure 2. The role of gut-derived 5-HT under inflammatory condition and inflammatory condition. THP1 and SERT expression, as well as 5-HT release can be altered under inflammatory condition. Enhanced 5-HT promotes recruitment of immune cells, such as natural killer cells, dendritic cell, macrophages, and neutrophil during inflammation. Subsequently, enhanced cytokines production is released from immune cells, which can promote inflammatory response. Upwards pointing arrows indicate an enhancement, and downwards pointing arrows indicate a decrease. 5-HT, 5-hydroxytryptamine; TPH1, tryptophan hydroxylase 1; SERT, serotonin reuptake transporter.

The immune response to inflammation involves the extensive proliferation of immune cells and aberrant production of immune mediators and cytokines such as tumor necrosis factor (TNF)-α, interferon (IFN)-γ, interleukin (IL)-1β, IL-6, and IL-8 and their related signaling pathways [110,111]. 5-HT receptors have been identified in human and rodent immune cells [44]. EC cells are in close proximity with immune cells in the gut mucosa, suggesting the existence of interaction between EC cells and immune cells [112]. Immune cells, including dendritic cells, macrophages, neutrophils, lymphocytes, and B lymphocytes, proliferate in the 5-HT-mediated proinflammatory response [4,113], suggesting that 5-HT plays a vital part in the immune response. Recent studies have shown that 5-HT signaling is altered by proinflammatory cytokines such as TNF-α, IL-1β, IL-6, and IFN-γ, as well as the anti-inflammatory cytokine IL-10 by regulating the expression and function of SERT. Intriguingly, several studies have found that IFN-γ, TNF-α, and IL-6 and a low concentration of IL-10 caused a significant decrease in the function and activity of epithelial SERT [4,114–116].

4.2. 5-HT and Liver Regeneration

5-HT acting on the liver is entirely derived from the gut because of the lack of 5-HT synthesis capacity in hepatocytes [117]. 5-HT activated in platelets is released in the liver and mediates liver regeneration after partial hepatectomy and inhibits liver regeneration in the *TPH1* gene knockout mice [48,49]. Liver regeneration was mediated by promoting DNA synthesis and cell proliferation through acting on 5-HTR2 [118] and 5-HTR7 [119,120]. Another study revealed that SERT knockout in platelets has no effect on liver regeneration, thus indicating that the extremely low level of 5-HT in plasma is sufficient for liver regeneration [121].

4.3. 5-HT and Energy Homeostasis

Metabolic homeostasis is regulated by nerves and hormones. Several recent studies have shown that 5-HT, an important endocrine substance and hormone, regulates the metabolic function of many tissues and influences obesity and energy metabolism [1,50,122,123]. The liver, a pivotal organ in an organism's metabolism, plays a central role in regulating plasma glucose metabolism and energy metabolism [124]. 5-HT cannot be produced by hepatocytes; hence, all the peripheral 5-HT in the liver is derived from the gut. A previous study revealed that 5-HT produced during fasting promotes gluconeogenesis by enhancing the activity of two key gluconeogenesis rate-limiting enzymes (glucose 6-phosphatase and fructose 1,6-bisphosphatase) through 5-HTR2B [93]. The cyclic AMP that is the downstream of 5-HTR2B is enhanced at transcriptional level after the elevated activity of two key enzymes; subsequently, cAMP-dependent protein kinase A (PKA) and CREB are activated [125]. Additionally, gut-derived 5-HT in hepatocytes prevents glucose uptake in a GLUT2-dependent manner, thereby further favoring the maintenance of blood glucose levels [93]. Because *TPH1* is expressed in adipocytes, the regulation of 5-HT in adipose tissue is more complicated than that in the liver. TPH1-produced 5-HT in adipocytes regulates the metabolism of adipose tissue through local autocrine signals [50,126,127]. In white adipocytes, 5-HT synthesized in EC cells enhances the phosphorylation and activity of hormone-sensitive lipase (HSL) through binding to the 5-HT2B receptor, therefore elevating circulatory free fatty acids and glycerol [93]. There are two possible pathways to promote lipolysis and inhibit lipogenesis: (1) HSL is activated indirectly by cAMP and cAMP-dependent protein kinase A (PKA); (2) perilipin is phosphorylated by PKA and, consequently, stimulates phosphorylation of HSL [128]. Enhanced glycerol acts as a fuel of gluconeogenesis and is converted into acetyl-CoA by β-oxidation for the synthesis of ketone bodies [1]. Because of the complexity of the serotonergic system in adipose tissue, more studies are required to elucidate the underlying responsible role for 5-HT in the future.

4.4. 5-HT and Bone Remodeling

5-HT and its role in bone metabolism are receiving great interest from researchers. Bone remodeling and renewal is a highly integrated process, which includes bone resorption through osteoclasts and bone formation through osteoblasts. These two processes are dynamically balanced, which contributes to the maintenance of bone [129]. Low-density lipoprotein receptor-related protein-5 (Lrp5) is essential for Wnt signaling to form bones [130–132]. Previous studies have reported that Lrp5 is expressed in osteoblasts and EC cells in the GI tract [133]. However, Lrp5 could act in EC cells in the gut, not in osteoblast, to regulate bone-mass accrual via a Wnt-independent pathway [134]. Lrp5 inhibits the expression of TPH1, thereby reducing 5-HT concentration in the blood. Less 5-HT binds to 5-HTR1B in osteoblasts and 5-HTR1B signaling is reduced in osteoblasts. As a result, the expression and function of cyclic AMP response element binding protein (CREB) is enhanced, which promotes cyclin expression and results in enhanced osteoblasts differentiation and proliferation [134]. In this process, 5-HT derived from the GI tract and transported through the circulation is detrimental to bone formation through inhibiting osteoblast proliferation [134] (Figure 3). Consistently, some studies have supported that gut-derived 5-HT could suppress bone growth in rats [135,136]. Thus, pharmacological inhibition of gut-derived 5-HT synthesis through the inhibitor of THP1 may be a potential bone anabolic treatment for low bone mass [137,138]. Additionally, there are conflicting results in the model where Lrp5 regulates bone mass through duodenal 5-HT. A study conducted by Cui et al. demonstrated that gut-derived 5-HT synthesis is not associated with Lrp5 [131]. Growing evidence has shown that 5-HT plays an important role in bone metabolism. However, because of the different synthesis sites of 5-HT, including brain-derived 5-HT [139], gut-derived 5-HT [134,137,140], and bone-derived 5-HT [141], 5-HT has different roles in bone metabolism (Figure 3).

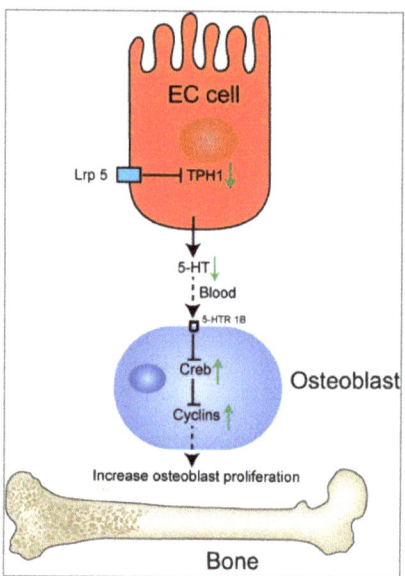

Figure 3. The action of Lrp5 and 5-HT on the regulation of bone formation. Lrp5 inhibits the expression of TPH1 in cells. As a result, reduced 5-HT concentration in circulation reduces 5-HTR1B signaling in osteoblasts. Cyclic AMP response element binding protein (CREB) and cyclin expression is enhanced, which favors osteoblasts proliferation and bone formation. Lrp5, low-density lipoprotein receptor-related protein-5; 5-HT, 5-hydroxytryptamine; CREB, cyclic AMP response element binding protein.

5. Conclusions

5-HT synthesized in EC cells has been recognized for decades as an important signaling molecule in the gut. It is well known that 5-HT derived from neurons and EC cells is involved in the regulation of GI peristalsis, sensation, and secretion. Approximately 95% of 5-HT in the body is synthesized and secreted by EC cells in the GI tract. The findings of several studies have suggested that gut microbiota, nutrients, and hormones could stimulate EC cells to release 5-HT. New clues from recent studies expand our understanding of the functional role of gut-derived 5-HT in and far beyond the gut. As an important neurotransmitter and hormone in the GI tract, research on 5-HT is increasing, but the underlying mechanisms of the relationship between 5-HT and physiological actions in the body remain largely unclear. Therefore, it is essential to highlight the functional role of 5-HT and various factors (gut microbiota, nutrients, and hormones) in the regulation of 5-HT secretion in order to facilitate the application for 5-HT in nutrition, clinical medicine, and health.

Author Contributions: Writing—review and editing, N.L., S.S., P.W., Y.S. and Q.H.; supervision, X.W. All authors have read and agreed to the published version of the manuscript.

Funding: This study was supported by the National Natural Science Foundation of China (No. 32000082, 32001676), and State Key Laboratory of Animal Nutrition (2004DA125184F1909).

Institutional Review Board Statement: Not applicable.

Informed Consent Statement: Not applicable.

Data Availability Statement: Not applicable.

Conflicts of Interest: The authors declare no conflict of interest.

Abbreviations

5-HIAA, 5-hydroxy indole acetic acid; 5-HIAL, 5-hydroxyindole acetaldehyde; 5-HT, 5-hydroxytryptamine; 5-HTP, 5-hydroxytryptophan; 5-HTR, 5-HT receptor; AAs, amino acids; Adrα2A, alpha-2A adrenergic receptor; CGA, chromogranin A; MAO, monoamine oxidase; DNBS, dinitrobenzene sulfonic acid; DSS, dextran sodium sulfate; EC, enterochromaffin; EE, enteroendocrine; FFAR2, free fatty acid receptor; GI, gastrointestinal; GLP-1, glucagon-like peptide 1; GLUT1, glucose transporter 1; GLUT2, glucose transporter 2; GPBAR1, G-protein-coupled bile acid receptor 1; GPCRs, G-protein-coupled receptors; GPR132, G-protein-coupled receptors 132; GPR35, G-protein-coupled receptors 35; HSL, hormone-sensitive lipase; IBD, inflammatory bowel disease; IFN, interferon; IL, interleukin; L-AADC, L-amino acid decarboxylase; Lrp5, low-density lipoprotein receptor-related protein-5; NF-κB, nuclear factor-κB; Olf558, olfactory receptor 558; OLF78, olfactory receptor 78; SCFAs, short-chain fatty acids; SERT, serotonin reuptake transporter; SGLT1, sodium-glucose-linked transporter 1; SGLT3, sodium-glucose-linked transporter 3; TPH 1, tryptophan hydroxylase 1; THP2, tryptophan hydroxylase 2; TNBS, 2,4,6-trinitrobenzene sulfonic acid; TNF, tumor necrosis factor; Trp, tryptophan; VMAT1, vesicular monoamine transporter1.

References

1. Martin, A.M.; Young, R.L.; Leong, L.; Rogers, G.B.; Spencer, N.J.; Jessup, C.F.; Keating, D.J. The diverse metabolic roles of peripheral serotonin. *Endocrinology* **2017**, *158*, 1049–1063. [CrossRef]
2. Mawe, G.M.; Hoffman, J.M. Serotonin signalling in the gut—functions, dysfunctions and therapeutic targets. *Nat. Rev. Gastroenterol. Hepatol.* **2013**, *10*, 473–486. [CrossRef]
3. Spohn, S.N.; Mawe, G.M. Non-conventional features of peripheral serotonin signalling—The gut and beyond. *Nat. Rev. Gastroenterol. Hepatol.* **2017**, *14*, 412–420. [CrossRef]
4. Wang, B.; Sun, S.; Liu, M.; Chen, H.; Liu, N.; Wu, Z.; Wu, G.; Dai, Z. Dietary L-tryptophan regulates colonic serotonin homeostasis in mice with dextran sodium sulfate-induced colitis. *J. Nutr.* **2020**, *150*, 1966–1976. [CrossRef] [PubMed]
5. Rapalli, A.; Bertoni, S.; Arcaro, V.; Saccani, F.; Grandi, A.; Vivo, V.; Cantoni, A.M.; Barocelli, E. Dual role of endogenous serotonin in 2,4,6-trinitrobenzene sulfonic acid-induced colitis. *Front. Pharmacol.* **2016**, *7*, 68. [CrossRef]
6. MacEachern, S.J.; Keenan, C.M.; Papakonstantinou, E.; Sharkey, K.A.; Patel, B.A. Alterations in melatonin and 5-HT signalling in the colonic mucosa of mice with dextran-sodium sulfate-induced colitis. *Br. J. Pharmacol.* **2018**, *175*, 1535–1547. [CrossRef]
7. Coates, M.D.; Tekin, I.; Vrana, K.E.; Mawe, G.M. Review article: The many potential roles of intestinal serotonin (5-hydroxytryptamine, 5-HT) signalling in inflammatory bowel disease. *Aliment. Pharmacol. Ther.* **2017**, *46*, 569–580. [CrossRef]
8. Terry, N.; Margolis, K.G. Serotonergic Mechanisms Regulating the GI Tract: Experimental Evidence and Therapeutic Relevance. *Handb. Exp. Pharmacol.* **2017**, *239*, 319–342. [CrossRef]
9. Wang, F.; Knutson, K.; Alcaino, C.; Linden, D.R.; Gibbons, S.J.; Kashyap, P.; Grover, M.; Oeckler, R.; Gottlieb, P.A.; Li, H.J.; et al. Mechanosensitive ion channel Piezo2 is important for enterochromaffin cell response to mechanical forces. *J. Physiol.* **2017**, *595*, 79–91. [CrossRef]
10. Bellono, N.W.; Bayrer, J.R.; Leitch, D.B.; Castro, J.; Zhang, C.; O'Donnell, T.A.; Brierley, S.M.; Ingraham, H.A.; Julius, D. Enterochromaffin cells are gut chemosensors that couple to sensory neural pathways. *Cell* **2017**, *170*, 185–198.e116. [CrossRef]
11. Martin, A.M.; Lumsden, A.L.; Young, R.L.; Jessup, C.F.; Spencer, N.J.; Keating, D.J. The nutrient-sensing repertoires of mouse enterochromaffin cells differ between duodenum and colon. *Neurogastroenterol. Motil.* **2017**, *29*. [CrossRef] [PubMed]
12. Hata, T.; Asano, Y.; Yoshihara, K.; Kimura-Todani, T.; Miyata, N.; Zhang, X.T.; Takakura, S.; Aiba, Y.; Koga, Y.; Sudo, N. Regulation of gut luminal serotonin by commensal microbiota in mice. *PLoS ONE* **2017**, *12*, e0180745. [CrossRef] [PubMed]
13. Lund, M.L.; Egerod, K.L.; Engelstoft, M.S.; Dmytriyeva, O.; Theodorsson, E.; Patel, B.A.; Schwartz, T.W. Enterochromaffin 5-HT cells—A major target for GLP-1 and gut microbial metabolites. *Mol. Metab.* **2018**, *11*, 70–83. [CrossRef]

14. Martin, A.M.; Lumsden, A.L.; Young, R.L.; Jessup, C.F.; Spencer, N.J.; Keating, D.J. Regional differences in nutrient-induced secretion of gut serotonin. *Physiol. Rep.* **2017**, *5*. [CrossRef] [PubMed]
15. Braun, T.; Voland, P.; Kunz, L.; Prinz, C.; Gratzl, M. Enterochromaffin cells of the human gut: Sensors for spices and odorants. *Gastroenterology* **2007**, *132*, 1890–1901. [CrossRef]
16. Reigstad, C.S.; Salmonson, C.E.; Rainey, J.F., 3rd; Szurszewski, J.H.; Linden, D.R.; Sonnenburg, J.L.; Farrugia, G.; Kashyap, P.C. Gut microbes promote colonic serotonin production through an effect of short-chain fatty acids on enterochromaffin cells. *Faseb. J.* **2015**, *29*, 1395–1403. [CrossRef] [PubMed]
17. Hou, Y.; Wu, G. Nutritionally essential amino acids. *Adv. Nutr.* **2018**, *9*, 849–851. [CrossRef]
18. Lopez, M.J.; Mohiuddin, S.S. Biochemistry, essential amino acids. In *StatPearls*; StatPearls Publishing LLC.: Treasure Island, FL, USA, 2021.
19. Comai, S.; Bertazzo, A.; Brughera, M.; Crotti, S. Tryptophan in health and disease. *Adv. Clin. Chem.* **2020**, *95*, 165–218. [CrossRef]
20. Roth, W.; Zadeh, K.; Vekariya, R.; Ge, Y.; Mohamadzadeh, M. Tryptophan metabolism and gut-brain homeostasis. *Int. J. Mol. Sci.* **2021**, *22*, 2973. [CrossRef] [PubMed]
21. Vivi, D.D.; Bentley, G.E. Seasonal reproduction in vertebrates: Melatonin synthesis, binding, and functionality using tinbergen's four questions. *Molecules* **2018**, *23*, 652. [CrossRef]
22. Wu, G.; Song, D.; Wei, Q.; Xing, J.; Shi, X.; Shi, F. Melatonin mitigates bisphenol A-induced estradiol production and proliferation by porcine ovarian granulosa cells in vitro. *Anim. Reprod. Sci.* **2018**, *192*, 91–98. [CrossRef] [PubMed]
23. Rode, J.; Yang, L.; König, J.; Hutchinson, A.N.; Wall, R.; Venizelos, N.; Brummer, R.J.; Rangel, I.; Vumma, R. Butyrate rescues oxidative stress-induced transport deficits of tryptophan: Potential implication in affective or gut-brain axis disorders. *Neuropsychobiology* **2020**, 1–11. [CrossRef]
24. Gao, J.; Xu, K.; Liu, H.; Liu, G.; Bai, M.; Peng, C.; Li, T.; Yin, Y. Impact of the gut microbiota on intestinal immunity mediated by tryptophan metabolism. *Front. Cell Infect. Microbiol.* **2018**, *8*, 13. [CrossRef]
25. Badawy, A.A. Kynurenine pathway of tryptophan metabolism: Regulatory and functional aspects. *Int. J. Tryptophan Res.* **2017**, *10*, 1178646917691938. [CrossRef]
26. Dehhaghi, M.; Kazemi Shariat Panahi, H.; Guillemin, G.J. Microorganisms, tryptophan metabolism, and kynurenine pathway: A complex interconnected loop influencing human health status. *Int. J. Tryptophan Res.* **2019**, *12*, 1178646919852996. [CrossRef] [PubMed]
27. Agus, A.; Planchais, J.; Sokol, H. Gut microbiota regulation of tryptophan metabolism in health and disease. *Cell Host Microbe* **2018**, *23*, 716–724. [CrossRef] [PubMed]
28. Badawy, A.A. Tryptophan availability for kynurenine pathway metabolism across the life span: Control mechanisms and focus on aging, exercise, diet and nutritional supplements. *Neuropharmacology* **2017**, *112*, 248–263. [CrossRef] [PubMed]
29. Jones, L.A.; Sun, E.W.; Martin, A.M.; Keating, D.J. The ever-changing roles of serotonin. *Int. J. Biochem. Cell Biol.* **2020**, *125*, 105776. [CrossRef] [PubMed]
30. Shajib, M.S.; Baranov, A.; Khan, W.I. Diverse effects of gut-derived serotonin in intestinal inflammation. *ACS Chem. Neurosci.* **2017**, *8*, 920–931. [CrossRef]
31. Yabut, J.M.; Crane, J.D.; Green, A.E.; Keating, D.J.; Khan, W.I.; Steinberg, G.R. Emerging roles for serotonin in regulating metabolism: New implications for an ancient molecule. *Endocr. Rev.* **2019**, *40*, 1092–1107. [CrossRef]
32. Walther, D.J.; Peter, J.U.; Bashammakh, S.; Hörtnagl, H.; Voits, M.; Fink, H.; Bader, M. Synthesis of serotonin by a second tryptophan hydroxylase isoform. *Science* **2003**, *299*, 76. [CrossRef]
33. Park, S.; Kim, Y.; Lee, J.; Lee, J.Y.; Kim, H.; Lee, S.; Oh, C.M. A systems biology approach to investigating the interaction between serotonin synthesis by tryptophan hydroxylase and the metabolic homeostasis. *Int. J. Mol. Sci.* **2021**, *22*, 2452. [CrossRef] [PubMed]
34. Swami, T.; Weber, H.C. Updates on the biology of serotonin and tryptophan hydroxylase. *Curr. Opin. Endocrinol. Diabetes Obes.* **2018**, *25*, 12–21. [CrossRef] [PubMed]
35. Bertrand, P.P.; Bertrand, R.L. Serotonin release and uptake in the gastrointestinal tract. *Auton. Neurosci.* **2010**, *153*, 47–57. [CrossRef]
36. Penuelas, A.; Tashima, K.; Tsuchiya, S.; Matsumoto, K.; Nakamura, T.; Horie, S.; Yano, S. Contractile effect of TRPA1 receptor agonists in the isolated mouse intestine. *Eur. J. Pharmacol.* **2007**, *576*, 143–150. [CrossRef]
37. Rindi, G.; Leiter, A.B.; Kopin, A.S.; Bordi, C.; Solcia, E. The "normal" endocrine cell of the gut: Changing concepts and new evidences. *Ann. N. Y. Acad. Sci.* **2004**, *1014*, 1–12. [CrossRef]
38. Montesinos, M.S.; Machado, J.D.; Camacho, M.; Diaz, J.; Morales, Y.G.; Alvarez de la Rosa, D.; Carmona, E.; Castañeyra, A.; Viveros, O.H.; O'Connor, D.T.; et al. The crucial role of chromogranins in storage and exocytosis revealed using chromaffin cells from chromogranin A null mouse. *J. Neurosci.* **2008**, *28*, 3350–3358. [CrossRef]
39. O'Hara, J.R.; Sharkey, K.A. Proliferative capacity of enterochromaffin cells in guinea-pigs with experimental ileitis. *Cell Tissue Res.* **2007**, *329*, 433–441. [CrossRef]
40. El Aidy, S.; Ramsteijn, A.S.; Dini-Andreote, F.; van Eijk, R.; Houwing, D.J.; Salles, J.F.; Olivier, J.D.A. Serotonin transporter genotype modulates the gut microbiota composition in young rats, an effect augmented by early life stress. *Front. Cell Neurosci.* **2017**, *11*, 222. [CrossRef]
41. Gill, R.K.; Pant, N.; Saksena, S.; Singla, A.; Nazir, T.M.; Vohwinkel, L.; Turner, J.R.; Goldstein, J.; Alrefai, W.A.; Dudeja, P.K. Function, expression, and characterization of the serotonin transporter in the native human intestine. *Am. J. Physiol. Gastrointest. Liver Physiol.* **2008**, *294*, G254–G262. [CrossRef]

42. Nzakizwanayo, J.; Dedi, C.; Standen, G.; Macfarlane, W.M.; Patel, B.A.; Jones, B.V. Escherichia coli Nissle 1917 enhances bioavailability of serotonin in gut tissues through modulation of synthesis and clearance. *Sci. Rep.* **2015**, *5*, 17324. [CrossRef]
43. Saraf, M.K.; Piccolo, B.D.; Bowlin, A.K.; Mercer, K.E.; LeRoith, T.; Chintapalli, S.V.; Shankar, K.; Badger, T.M.; Yeruva, L. Formula diet driven microbiota shifts tryptophan metabolism from serotonin to tryptamine in neonatal porcine colon. *Microbiome* **2017**, *5*, 77. [CrossRef] [PubMed]
44. Shajib, M.S.; Khan, W.I. The role of serotonin and its receptors in activation of immune responses and inflammation. *Acta Physiol.* **2015**, *213*, 561–574. [CrossRef] [PubMed]
45. Keszthelyi, D.; Troost, F.J.; Masclee, A.A. Understanding the role of tryptophan and serotonin metabolism in gastrointestinal function. *Neurogastroenterol. Motil.* **2009**, *21*, 1239–1249. [CrossRef]
46. Edmondson, D.E.; Binda, C. Monoamine oxidases. *Subcell Biochem.* **2018**, *87*, 117–139. [CrossRef] [PubMed]
47. Lavoie, B.; Lian, J.B.; Mawe, G.M. Regulation of bone metabolism by serotonin. *Adv. Exp. Med. Biol.* **2017**, *1033*, 35–46. [CrossRef]
48. Lesurtel, M.; Graf, R.; Aleil, B.; Walther, D.J.; Tian, Y.; Jochum, W.; Gachet, C.; Bader, M.; Clavien, P.A. Platelet-derived serotonin mediates liver regeneration. *Science* **2006**, *312*, 104–107. [CrossRef] [PubMed]
49. Fang, Y.; Liu, C.; Shu, B.; Zhai, M.; Deng, C.; He, C.; Luo, M.; Han, T.; Zheng, W.; Zhang, J.; et al. Axis of serotonin -pERK-YAP in liver regeneration. *Life Sci.* **2018**, *209*, 490–497. [CrossRef] [PubMed]
50. Wyler, S.C.; Lord, C.C.; Lee, S.; Elmquist, J.K.; Liu, C. Serotonergic control of metabolic homeostasis. *Front. Cell Neurosci.* **2017**, *11*, 277. [CrossRef]
51. Flanagan, T.W.; Sebastian, M.N.; Battaglia, D.M.; Foster, T.P.; Cormier, S.A.; Nichols, C.D. 5-HT(2) receptor activation alleviates airway inflammation and structural remodeling in a chronic mouse asthma model. *Life Sci.* **2019**, *236*, 116790. [CrossRef]
52. Hannon, J.; Hoyer, D. Molecular biology of 5-HT receptors. *Behav. Brain. Res.* **2008**, *195*, 198–213. [CrossRef]
53. Pytliak, M.; Vargová, V.; Mechírová, V.; Felšöci, M. Serotonin receptors—From molecular biology to clinical applications. *Physiol. Res.* **2011**, *60*, 15–25. [CrossRef]
54. Hoyer, D.; Hannon, J.P.; Martin, G.R. Molecular, pharmacological and functional diversity of 5-HT receptors. *Pharmacol. Biochem. Behav.* **2002**, *71*, 533–554. [CrossRef]
55. Margolis, K.G.; Li, Z.; Stevanovic, K.; Saurman, V.; Israelyan, N.; Anderson, G.M.; Snyder, I.; Veenstra-VanderWeele, J.; Blakely, R.D.; Gershon, M.D. Serotonin transporter variant drives preventable gastrointestinal abnormalities in development and function. *J. Clin. Investig.* **2016**, *126*, 2221–2235. [CrossRef] [PubMed]
56. Kendig, D.M.; Grider, J.R. Serotonin and colonic motility. *Neurogastroenterol. Motil.* **2015**, *27*, 899–905. [CrossRef]
57. Keating, D.J.; Spencer, N.J. What is the role of endogenous gut serotonin in the control of gastrointestinal motility? *Pharmacol. Res.* **2019**, *140*, 50–55. [CrossRef]
58. Tuo, B.G.; Isenberg, J.I. Effect of 5-hydroxytryptamine on duodenal mucosal bicarbonate secretion in mice. *Gastroenterology* **2003**, *125*, 805–814. [CrossRef]
59. Tuo, B.G.; Sellers, Z.; Paulus, P.; Barrett, K.E.; Isenberg, J.I. 5-HT induces duodenal mucosal bicarbonate secretion via cAMP- and Ca2+-dependent signaling pathways and 5-HT4 receptors in mice. *Am. J. Physiol. Gastrointest. Liver Physiol.* **2004**, *286*, G444–G451. [CrossRef]
60. Raybould, H.E.; Glatzle, J.; Robin, C.; Meyer, J.H.; Phan, T.; Wong, H.; Sternini, C. Expression of 5-HT3 receptors by extrinsic duodenal afferents contribute to intestinal inhibition of gastric emptying. *Am. J. Physiol. Gastrointest. Liver Physiol.* **2003**, *284*, G367–G372. [CrossRef]
61. Browning, K.N. Role of central vagal 5-HT3 receptors in gastrointestinal physiology and pathophysiology. *Front. Neurosci.* **2015**, *9*, 413. [CrossRef]
62. Li, Y.; Hao, Y.; Zhu, J.; Owyang, C. Serotonin released from intestinal enterochromaffin cells mediates luminal non-cholecystokinin-stimulated pancreatic secretion in rats. *Gastroenterology* **2000**, *118*, 1197–1207. [CrossRef]
63. Vanner, S.; Macnaughton, W.K. Submucosal secretomotor and vasodilator reflexes. *Neurogastroenterol. Motil.* **2004**, *16* (Suppl. 1), 39–43. [CrossRef]
64. Spohn, S.N.; Bianco, F.; Scott, R.B.; Keenan, C.M.; Linton, A.A.; O'Neill, C.H.; Bonora, E.; Dicay, M.; Lavoie, B.; Wilcox, R.L.; et al. Protective actions of epithelial 5-Hydroxytryptamine 4 receptors in normal and inflamed colon. *Gastroenterology* **2016**, *151*, 933–944.e933. [CrossRef] [PubMed]
65. Tada, Y.; Ishihara, S.; Kawashima, K.; Fukuba, N.; Sonoyama, H.; Kusunoki, R.; Oka, A.; Mishima, Y.; Oshima, N.; Moriyama, I.; et al. Downregulation of serotonin reuptake transporter gene expression in healing colonic mucosa in presence of remaining low-grade inflammation in ulcerative colitis. *J. Gastroenterol. Hepatol.* **2016**, *31*, 1443–1452. [CrossRef]
66. Iacomino, G.; Rotondi Aufiero, V.; Iannaccone, N.; Melina, R.; Giardullo, N.; De Chiara, G.; Venezia, A.; Taccone, F.S.; Iaquinto, G.; Mazzarella, G. IBD: Role of intestinal compartments in the mucosal immune response. *Immunobiology* **2020**, *225*, 151849. [CrossRef]
67. Fiorica-Howells, E.; Maroteaux, L.; Gershon, M.D. Serotonin and the 5-HT(2B) receptor in the development of enteric neurons. *J. Neurosci.* **2000**, *20*, 294–305. [CrossRef]
68. Liu, M.T.; Kuan, Y.H.; Wang, J.; Hen, R.; Gershon, M.D. 5-HT4 receptor-mediated neuroprotection and neurogenesis in the enteric nervous system of adult mice. *J. Neurosci.* **2009**, *29*, 9683–9699. [CrossRef]

69. Matsuyoshi, H.; Kuniyasu, H.; Okumura, M.; Misawa, H.; Katsui, R.; Zhang, G.X.; Obata, K.; Takaki, M. A 5-HT(4)-receptor activation-induced neural plasticity enhances in vivo reconstructs of enteric nerve circuit insult. *Neurogastroenterol. Motil.* **2010**, *22*, 806–813.e226. [CrossRef]
70. Gross, E.R.; Gershon, M.D.; Margolis, K.G.; Gertsberg, Z.V.; Li, Z.; Cowles, R.A. Neuronal serotonin regulates growth of the intestinal mucosa in mice. *Gastroenterology* **2012**, *143*, 408–417.e402. [CrossRef]
71. Tackett, J.J.; Gandotra, N.; Bamdad, M.C.; Muise, E.D.; Cowles, R.A. Enhanced serotonin signaling stimulates ordered intestinal mucosal growth. *J. Surg. Res.* **2017**, *208*, 198–203. [CrossRef]
72. Nozawa, K.; Kawabata-Shoda, E.; Doihara, H.; Kojima, R.; Okada, H.; Mochizuki, S.; Sano, Y.; Inamura, K.; Matsushime, H.; Koizumi, T.; et al. TRPA1 regulates gastrointestinal motility through serotonin release from enterochromaffin cells. *Proc. Natl. Acad. Sci. USA* **2009**, *106*, 3408–3413. [CrossRef]
73. Fujimiya, M.; Okumiya, K.; Kuwahara, A. Immunoelectron microscopic study of the luminal release of serotonin from rat enterochromaffin cells induced by high intraluminal pressure. *Histochem. Cell Biol.* **1997**, *108*, 105–113. [CrossRef]
74. Bulbring, E.; Lin, R.C. The effect of intraluminal application of 5-hydroxytryptamine and 5-hydroxytryptophan on peristalsis; the local production of 5-HT and its release in relation to intraluminal pressure and propulsive activity. *J. Physiol.* **1958**, *140*, 381–407.
75. Gershon, M.D. Review article: Serotonin receptors and transporters—Roles in normal and abnormal gastrointestinal motility. *Aliment. Pharmacol. Ther.* **2004**, *20* (Suppl. 7), 3–14. [CrossRef]
76. Thursby, E.; Juge, N. Introduction to the human gut microbiota. *Biochem. J.* **2017**, *474*, 1823–1836. [CrossRef]
77. Lavelle, A.; Sokol, H. Gut microbiota-derived metabolites as key actors in inflammatory bowel disease. *Nat. Rev. Gastroenterol. Hepatol.* **2020**, *17*, 223–237. [CrossRef]
78. Mentella, M.C.; Scaldaferri, F.; Pizzoferrato, M.; Gasbarrini, A.; Miggiano, G.A.D. Nutrition, IBD and gut microbiota: A review. *Nutrients* **2020**, *12*, 944. [CrossRef]
79. Morrison, D.J.; Preston, T. Formation of short chain fatty acids by the gut microbiota and their impact on human metabolism. *Gut Microbes* **2016**, *7*, 189–200. [CrossRef] [PubMed]
80. Fukumoto, S.; Tatewaki, M.; Yamada, T.; Fujimiya, M.; Mantyh, C.; Voss, M.; Eubanks, S.; Harris, M.; Pappas, T.N.; Takahashi, T. Short-chain fatty acids stimulate colonic transit via intraluminal 5-HT release in rats. *Am. J. Physiol. Regul. Integr. Comp. Physiol.* **2003**, *284*, R1269–R1276. [CrossRef]
81. Yano, J.M.; Yu, K.; Donaldson, G.P.; Shastri, G.G.; Ann, P.; Ma, L.; Nagler, C.R.; Ismagilov, R.F.; Mazmanian, S.K.; Hsiao, E.Y. Indigenous bacteria from the gut microbiota regulate host serotonin biosynthesis. *Cell* **2015**, *161*, 264–276. [CrossRef] [PubMed]
82. Esmaili, A.; Nazir, S.F.; Borthakur, A.; Yu, D.; Turner, J.R.; Saksena, S.; Singla, A.; Hecht, G.A.; Alrefai, W.A.; Gill, R.K. Enteropathogenic Escherichia coli infection inhibits intestinal serotonin transporter function and expression. *Gastroenterology* **2009**, *137*, 2074–2083. [CrossRef] [PubMed]
83. Yu, Y.; Yang, W.; Li, Y.; Cong, Y. Enteroendocrine cells: Sensing gut microbiota and regulating inflammatory bowel diseases. *Inflamm. Bowel. Dis.* **2020**, *26*, 11–20. [CrossRef] [PubMed]
84. Gribble, F.M.; Reimann, F. Enteroendocrine cells: Chemosensors in the intestinal epithelium. *Annu. Rev. Physiol.* **2016**, *78*, 277–299. [CrossRef] [PubMed]
85. Liñán-Rico, A.; Ochoa-Cortes, F.; Zuleta-Alarcon, A.; Alhaj, M.; Tili, E.; Enneking, J.; Harzman, A.; Grants, I.; Bergese, S.; Christofi, F.L. UTP—Gated signaling pathways of 5-HT release from BON Cells as a model of human enterochromaffin cells. *Front. Pharmacol.* **2017**, *8*, 429. [CrossRef] [PubMed]
86. Kim, M.; Cooke, H.J.; Javed, N.H.; Carey, H.V.; Christofi, F.; Raybould, H.E. D-glucose releases 5-hydroxytryptamine from human BON cells as a model of enterochromaffin cells. *Gastroenterology* **2001**, *121*, 1400–1406. [CrossRef] [PubMed]
87. Freeman, S.L.; Bohan, D.; Darcel, N.; Raybould, H.E. Luminal glucose sensing in the rat intestine has characteristics of a sodium-glucose cotransporter. *Am. J. Physiol. Gastrointest. Liver Physiol.* **2006**, *291*, G439–G445. [CrossRef] [PubMed]
88. Vincent, K.M.; Sharp, J.W.; Raybould, H.E. Intestinal glucose-induced calcium-calmodulin kinase signaling in the gut-brain axis in awake rats. *Neurogastroenterol. Motil.* **2011**, *23*, e282–e293. [CrossRef]
89. Merino, B.; Fernández-Díaz, C.M.; Cózar-Castellano, I.; Perdomo, G. intestinal fructose and glucose metabolism in health and disease. *Nutrients* **2019**, *12*, 94. [CrossRef] [PubMed]
90. Savastano, D.M.; Carelle, M.; Covasa, M. Serotonin-type 3 receptors mediate intestinal Polycose- and glucose-induced suppression of intake. *Am. J. Physiol. Regul. Integr. Comp. Physiol.* **2005**, *288*, R1499–R1508. [CrossRef]
91. Yoshikawa, T.; Inoue, R.; Matsumoto, M.; Yajima, T.; Ushida, K.; Iwanaga, T. Comparative expression of hexose transporters (SGLT1, GLUT1, GLUT2 and GLUT5) throughout the mouse gastrointestinal tract. *Histochem. Cell Biol.* **2011**, *135*, 183–194. [CrossRef]
92. Zelkas, L.; Raghupathi, R.; Lumsden, A.L.; Martin, A.M.; Sun, E.; Spencer, N.J.; Young, R.L.; Keating, D.J. Serotonin-secreting enteroendocrine cells respond via diverse mechanisms to acute and chronic changes in glucose availability. *Nutr. Metab.* **2015**, *12*, 55. [CrossRef]
93. Sumara, G.; Sumara, O.; Kim, J.K.; Karsenty, G. Gut-derived serotonin is a multifunctional determinant to fasting adaptation. *Cell Metab.* **2012**, *16*, 588–600. [CrossRef]
94. Liddle, R.A. Neuropods. *Cell Mol. Gastroenterol. Hepatol.* **2019**, *7*, 739–747. [CrossRef]
95. Ripken, D.; van der Wielen, N.; Wortelboer, H.M.; Meijerink, J.; Witkamp, R.F.; Hendriks, H.F. Nutrient-induced glucagon like peptide-1 release is modulated by serotonin. *J. Nutr. Biochem.* **2016**, *32*, 142–150. [CrossRef]

96. Linden, D.R.; Chen, J.X.; Gershon, M.D.; Sharkey, K.A.; Mawe, G.M. Serotonin availability is increased in mucosa of guinea pigs with TNBS-induced colitis. *Am. J. Physiol. Gastrointest. Liver Physiol.* **2003**, *285*, G207–G216. [CrossRef] [PubMed]
97. O'Hara, J.R.; Lomax, A.E.; Mawe, G.M.; Sharkey, K.A. Ileitis alters neuronal and enteroendocrine signalling in guinea pig distal colon. *Gut* **2007**, *56*, 186–194. [CrossRef] [PubMed]
98. Kwon, Y.H.; Wang, H.; Denou, E.; Ghia, J.E.; Rossi, L.; Fontes, M.E.; Bernier, S.P.; Shajib, M.S.; Banskota, S.; Collins, S.M.; et al. Modulation of Gut Microbiota Composition by Serotonin Signaling Influences Intestinal Immune Response and Susceptibility to Colitis. *Cell Mol. Gastroenterol. Hepatol.* **2019**, *7*, 709–728. [CrossRef]
99. Yang, Y.; Zhu, X.; Qin, Y.; Chen, G.; Zhou, J.; Li, L.; Guan, J.; Ma, L.; Xue, Y.; Li, C. The anti-inflammatory effect of guchangzhixiepill by reducing colonic EC cell hyperplasia and serotonin availability in an ulcerative colitis rat model. *Evid. Based Complement. Alternat Med.* **2017**, *2017*, 8547257. [CrossRef] [PubMed]
100. Chen, M.; Gao, L.; Chen, P.; Feng, D.; Jiang, Y.; Chang, Y.; Jin, J.; Chu, F.F.; Gao, Q. Serotonin-exacerbated DSS-induced colitis is associated with increase in MMP-3 and MMP-9 expression in the mouse colon. *Mediat. Inflamm.* **2016**, *2016*, 5359768. [CrossRef] [PubMed]
101. Wheatcroft, J.; Wakelin, D.; Smith, A.; Mahoney, C.R.; Mawe, G.; Spiller, R. Enterochromaffin cell hyperplasia and decreased serotonin transporter in a mouse model of postinfectious bowel dysfunction. *Neurogastroenterol. Motil.* **2005**, *17*, 863–870. [CrossRef] [PubMed]
102. Shajib, M.S.; Wang, H.; Kim, J.J.; Sunjic, I.; Ghia, J.E.; Denou, E.; Collins, M.; Denburg, J.A.; Khan, W.I. Interleukin 13 and serotonin: Linking the immune and endocrine systems in murine models of intestinal inflammation. *PLoS ONE* **2013**, *8*, e72774. [CrossRef]
103. Zang, K.H.; Rao, Z.; Zhang, G.Q.; Qin, H.Y. Anticolitis activity of Chinese herbal formula yupingfeng powder via regulating colonic enterochromaffin cells and serotonin. *Indian J. Pharmacol.* **2015**, *47*, 632–637. [CrossRef]
104. Margolis, K.G.; Pothoulakis, C. Serotonin has a critical role in the pathogenesis of experimental colitis. *Gastroenterology* **2009**, *137*, 1562–1566. [CrossRef] [PubMed]
105. Margolis, K.G.; Stevanovic, K.; Li, Z.; Yang, Q.M.; Oravecz, T.; Zambrowicz, B.; Jhaver, K.G.; Diacou, A.; Gershon, M.D. Pharmacological reduction of mucosal but not neuronal serotonin opposes inflammation in mouse intestine. *Gut* **2014**, *63*, 928–937. [CrossRef]
106. Kim, J.J.; Wang, H.; Terc, J.D.; Zambrowicz, B.; Yang, Q.M.; Khan, W.I. Blocking peripheral serotonin synthesis by telotristat etiprate (LX1032/LX1606) reduces severity of both chemical- and infection-induced intestinal inflammation. *Am. J. Physiol. Gastrointest. Liver Physiol.* **2015**, *309*, G455–G465. [CrossRef]
107. Bischoff, S.C.; Mailer, R.; Pabst, O.; Weier, G.; Sedlik, W.; Li, Z.; Chen, J.J.; Murphy, D.L.; Gershon, M.D. Role of serotonin in intestinal inflammation: Knockout of serotonin reuptake transporter exacerbates 2,4,6-trinitrobenzene sulfonic acid colitis in mice. *Am. J. Physiol. Gastrointest. Liver Physiol.* **2009**, *296*, G685–G695. [CrossRef]
108. Xiao, J.; Shao, L.; Shen, J.; Jiang, W.; Feng, Y.; Zheng, P.; Liu, F. Effects of ketanserin on experimental colitis in mice and macrophage function. *Int. J. Mol. Med.* **2016**, *37*, 659–668. [CrossRef] [PubMed]
109. Kim, J.J.; Bridle, B.W.; Ghia, J.E.; Wang, H.; Syed, S.N.; Manocha, M.M.; Rengasamy, P.; Shajib, M.S.; Wan, Y.; Hedlund, P.B.; et al. Targeted inhibition of serotonin type 7 (5-HT7) receptor function modulates immune responses and reduces the severity of intestinal inflammation. *J. Immunol.* **2013**, *190*, 4795–4804. [CrossRef]
110. Zhang, Y.Z.; Li, Y.Y. Inflammatory bowel disease: Pathogenesis. *World J. Gastroenterol.* **2014**, *20*, 91–99. [CrossRef]
111. Cheng, H.Y.; Ning, M.X.; Chen, D.K.; Ma, W.T. Interactions between the gut microbiota and the host innate immune response against pathogens. *Front. Immunol.* **2019**, *10*, 607. [CrossRef]
112. Khan, W.I.; Ghia, J.E. Gut hormones: Emerging role in immune activation and inflammation. *Clin. Exp. Immunol.* **2010**, *161*, 19–27. [CrossRef] [PubMed]
113. Herr, N.; Bode, C.; Duerschmied, D. The effects of serotonin in immune cells. *Front. Cardiovasc Med.* **2017**, *4*, 48. [CrossRef] [PubMed]
114. Barbaro, M.R.; Di Sabatino, A.; Cremon, C.; Giuffrida, P.; Fiorentino, M.; Altimari, A.; Bellacosa, L.; Stanghellini, V.; Barbara, G. Interferon-γ is increased in the gut of patients with irritable bowel syndrome and modulates serotonin metabolism. *Am. J. Physiol. Gastrointest. Liver Physiol.* **2016**, *310*, G439–G447. [CrossRef]
115. Latorre, E.; Mendoza, C.; Matheus, N.; Castro, M.; Grasa, L.; Mesonero, J.E.; Alcalde, A.I. IL-10 modulates serotonin transporter activity and molecular expression in intestinal epithelial cells. *Cytokine* **2013**, *61*, 778–784. [CrossRef]
116. Kong, E.; Sucic, S.; Monje, F.J.; Savalli, G.; Diao, W.; Khan, D.; Ronovsky, M.; Cabatic, M.; Koban, F.; Freissmuth, M.; et al. STAT3 controls IL6-dependent regulation of serotonin transporter function and depression-like behavior. *Sci. Rep.* **2015**, *5*, 9009. [CrossRef]
117. Ruddell, R.G.; Mann, D.A.; Ramm, G.A. The function of serotonin within the liver. *J. Hepatol.* **2008**, *48*, 666–675. [CrossRef]
118. Papadimas, G.K.; Tzirogiannis, K.N.; Panoutsopoulos, G.I.; Demonakou, M.D.; Skaltsas, S.D.; Hereti, R.I.; Papadopoulou-Daifoti, Z.; Mykoniatis, M.G. Effect of serotonin receptor 2 blockage on liver regeneration after partial hepatectomy in the rat liver. *Liver Int.* **2006**, *26*, 352–361. [CrossRef]
119. Tzirogiannis, K.N.; Kourentzi, K.T.; Zyga, S.; Papalimneou, V.; Tsironi, M.; Grypioti, A.D.; Protopsaltis, I.; Panidis, D.; Panoutsopoulos, G.I. Effect of 5-HT7 receptor blockade on liver regeneration after 60–70% partial hepatectomy. *BMC Gastroenterol.* **2014**, *14*, 201. [CrossRef]

120. Balasubramanian, S.; Paulose, C.S. Induction of DNA synthesis in primary cultures of rat hepatocytes by serotonin: Possible involvement of serotonin S2 receptor. *Hepatology* **1998**, *27*, 62–66. [CrossRef]
121. Matondo, R.B.; Punt, C.; Homberg, J.; Toussaint, M.J.; Kisjes, R.; Korporaal, S.J.; Akkerman, J.W.; Cuppen, E.; de Bruin, A. Deletion of the serotonin transporter in rats disturbs serotonin homeostasis without impairing liver regeneration. *Am. J. Physiol. Gastrointest. Liver Physiol.* **2009**, *296*, G963–G968. [CrossRef]
122. Berger, M.; Gray, J.A.; Roth, B.L. The expanded biology of serotonin. *Annu. Rev. Med.* **2009**, *60*, 355–366. [CrossRef]
123. El-Merahbi, R.; Löffler, M.; Mayer, A.; Sumara, G. The roles of peripheral serotonin in metabolic homeostasis. *FEBS Lett.* **2015**, *589*, 1728–1734. [CrossRef]
124. Lefort, C.; Cani, P.D. The Liver under the spotlight: Bile acids and oxysterols as pivotal actors controlling metabolism. *Cells* **2021**, *10*, 400. [CrossRef] [PubMed]
125. Lin, B.; Morris, D.W.; Chou, J.Y. The role of HNF1alpha, HNF3gamma, and cyclic AMP in glucose-6-phosphatase gene activation. *Biochemistry* **1997**, *36*, 14096–14106. [CrossRef]
126. Rozenblit-Susan, S.; Chapnik, N.; Froy, O. Serotonin Prevents Differentiation of Brown Adipocytes by Interfering with Their Clock. *Obesity (Silver Spring)* **2019**, *27*, 2018–2024. [CrossRef]
127. Shong, K.E.; Oh, C.M.; Namkung, J.; Park, S.; Kim, H. Serotonin regulates de novo lipogenesis in adipose tissues through serotonin receptor 2A. *Endocrinol. Metab.* **2020**, *35*, 470–479. [CrossRef] [PubMed]
128. Kraemer, F.B.; Shen, W.J. Hormone-sensitive lipase: Control of intracellular tri-(di-)acylglycerol and cholesteryl ester hydrolysis. *J. Lipid. Res.* **2002**, *43*, 1585–1594. [CrossRef]
129. Siddiqui, J.A.; Partridge, N.C. Physiological bone remodeling: Systemic regulation and growth factor involvement. *Physiology* **2016**, *31*, 233–245. [CrossRef]
130. Kobayashi, Y.; Uehara, S.; Udagawa, N.; Takahashi, N. Regulation of bone metabolism by Wnt signals. *J. Biochem.* **2016**, *159*, 387–392. [CrossRef]
131. Cui, Y.; Niziolek, P.J.; MacDonald, B.T.; Zylstra, C.R.; Alenina, N.; Robinson, D.R.; Zhong, Z.; Matthes, S.; Jacobsen, C.M.; Conlon, R.A.; et al. Lrp5 functions in bone to regulate bone mass. *Nat. Med.* **2011**, *17*, 684–691. [CrossRef]
132. Williams, B.O. LRP5: From bedside to bench to bone. *Bone* **2017**, *102*, 26–30. [CrossRef]
133. Yadav, V.K.; Ducy, P. Lrp5 and bone formation: A serotonin-dependent pathway. *Ann. N. Y. Acad. Sci.* **2010**, *1192*, 103–109. [CrossRef] [PubMed]
134. Yadav, V.K.; Ryu, J.H.; Suda, N.; Tanaka, K.F.; Gingrich, J.A.; Schütz, G.; Glorieux, F.H.; Chiang, C.Y.; Zajac, J.D.; Insogna, K.L.; et al. Lrp5 controls bone formation by inhibiting serotonin synthesis in the duodenum. *Cell* **2008**, *135*, 825–837. [CrossRef]
135. Erjavec, I.; Bordukalo-Niksic, T.; Brkljacic, J.; Grcevic, D.; Mokrovic, G.; Kesic, M.; Rogic, D.; Zavadoski, W.; Paralkar, V.M.; Grgurevic, L.; et al. Constitutively elevated blood serotonin is associated with bone loss and Type 2 diabetes in rats. *PLoS ONE* **2016**, *11*, e0150102. [CrossRef]
136. Blazevic, S.; Erjavec, I.; Brizic, M.; Vukicevic, S.; Hranilovic, D. Molecular background and physiological consequences of altered peripheral serotonin homeostasis in adult rats perinatally treated with tranylcypromine. *J. Physiol. Pharmacol.* **2015**, *66*, 529–537. [PubMed]
137. Yadav, V.K.; Balaji, S.; Suresh, P.S.; Liu, X.S.; Lu, X.; Li, Z.; Guo, X.E.; Mann, J.J.; Balapure, A.K.; Gershon, M.D.; et al. Pharmacological inhibition of gut-derived serotonin synthesis is a potential bone anabolic treatment for osteoporosis. *Nat. Med.* **2010**, *16*, 308–312. [CrossRef]
138. Inose, H.; Zhou, B.; Yadav, V.K.; Guo, X.E.; Karsenty, G.; Ducy, P. Efficacy of serotonin inhibition in mouse models of bone loss. *J. Bone Miner. Res.* **2011**, *26*, 2002–2011. [CrossRef]
139. Yadav, V.K.; Oury, F.; Suda, N.; Liu, Z.W.; Gao, X.B.; Confavreux, C.; Klemenhagen, K.C.; Tanaka, K.F.; Gingrich, J.A.; Guo, X.E.; et al. A serotonin-dependent mechanism explains the leptin regulation of bone mass, appetite, and energy expenditure. *Cell* **2009**, *138*, 976–989. [CrossRef]
140. Kode, A.; Mosialou, I.; Silva, B.C.; Rached, M.T.; Zhou, B.; Wang, J.; Townes, T.M.; Hen, R.; DePinho, R.A.; Guo, X.E.; et al. FOXO1 orchestrates the bone-suppressing function of gut-derived serotonin. *J. Clin. Investig.* **2012**, *122*, 3490–3503. [CrossRef]
141. Chabbi-Achengli, Y.; Coudert, A.E.; Callebert, J.; Geoffroy, V.; Côté, F.; Collet, C.; de Vernejoul, M.C. Decreased osteoclastogenesis in serotonin-deficient mice. *Proc. Natl. Acad. Sci. USA* **2012**, *109*, 2567–2572. [CrossRef]

Article

Degradation Products of Tryptophan in Cell Culture Media: Contribution to Color and Toxicity

Alisa Schnellbaecher, Anton Lindig, Maxime Le Mignon, Tim Hofmann, Brit Pardon, Stephanie Bellmaine *[ID] and Aline Zimmer [ID]

Upstream R&D, Merck KGaA, Frankfurter Strasse 250, 64293 Darmstadt, Germany; alisa.schnellbaecher@merckgroup.com (A.S.); anton.lindig@tu-dortmund.de (A.L.); maxime.le-mignon@merckgroup.com (M.L.M.); tim.hofmann@merckgroup.com (T.H.); brit.pardon@merckgroup.com (B.P.); aline.zimmer@merckgroup.com (A.Z.)
* Correspondence: stephanie.bellmaine@merckgroup.com

Abstract: Biomanufacturing processes may be optimized by storing cell culture media at room temperature, but this is currently limited by their instability and change in color upon long-term storage. This study demonstrates that one of the critical contributing factors toward media browning is tryptophan. LC-MS technology was utilized to identify tryptophan degradation products, which are likely formed primarily from oxidation reactions. Several of the identified compounds were shown to contribute significantly to color in solutions but also to exhibit toxicity against CHO cells. A cell-culture-compatible antioxidant, a-ketoglutaric acid, was found to be an efficient cell culture media additive for stabilizing components against degradation, inhibiting the browning of media formulations, and decreasing ammonia production, thus providing a viable method for developing room-temperature stable cell culture media.

Keywords: tryptophan; color; cell culture media; LC-MS; antioxidant; cytotoxicity; biomanufacturing

Citation: Schnellbaecher, A.; Lindig, A.; Le Mignon, M.; Hofmann, T.; Pardon, B.; Bellmaine, S.; Zimmer, A. Degradation Products of Tryptophan in Cell Culture Media: Contribution to Color and Toxicity. *Int. J. Mol. Sci.* **2021**, *22*, 6221. https://doi.org/10.3390/ijms22126221

Academic Editor: Burkhard Poeggeler

Received: 30 April 2021
Accepted: 5 June 2021
Published: 9 June 2021

Publisher's Note: MDPI stays neutral with regard to jurisdictional claims in published maps and institutional affiliations.

Copyright: © 2021 by the authors. Licensee MDPI, Basel, Switzerland. This article is an open access article distributed under the terms and conditions of the Creative Commons Attribution (CC BY) license (https://creativecommons.org/licenses/by/4.0/).

1. Introduction

One strategy for improving biopharmaceutical manufacturing processes is to optimize cell culture media (CCM). The latest drive in the industry is to improve the chemical stability of hydrated media against degrading conditions such as exposure to light and increased temperatures. Detrimental light exposure can be encountered unintentionally in the transport and storage of CCM or during the cell culture itself, as modern cell culture bioprocesses are often carried out in reaction vessels that are exposed to light [1,2]. It is standard practice to maintain CCM at 4 °C prior to use, but the design of hydrated CCM that are stable at room temperature (RT) may decrease the energy consumption by eliminating the need for refrigeration. This optimization of CCM stability involves both developing methods to detect critical changes to CCM that can negatively impact cell culture [3,4] and using this information to discover strategies to mitigate such changes.

When exposed to stressed conditions (e.g., light, heat), sensitive CCM components decompose and new degradation products are generated. This chemical change in the media composition can often be directly observed by a change in the color of the media. This is a greater issue for feeds of a fed-batch process than for basal media, due to the higher concentration of components. This color change is not only a warning signal for a chemically altered CCM but is of added concern for the potential to alter the color of the drug product, a critical quality attribute (cQA) of therapeutic monoclonal antibodies (mAbs). The ability of degraded CCM components to alter the color of the mAb product was best demonstrated by Prentice et al., who showed that the light-induced degradation of cyanocobalamin (vitamin B12) to hydroxocobalamin caused the mAb product to become pink-colored [5]. However, while many reports explore the contributions of particular CCM components and specific degrading conditions to mAb coloration [5–8], researchers have yet to explore the coloration of CCM itself. The color change associated with the degradation of some cell culture media components has been reported, such as for the Maillard reaction of amino acids (AAs) and glucose, which leads to highly oxidized, highly colored organic compounds [9–11]. Other common organic CCM components shown to produce colored degradation products are thiamine [12,13] and tryptophan [14,15]. The instability of CCM components is not only an issue for color production, but also for the potential production of toxic degradation products, which would be detrimental to cell culture. Methods to inhibit color change in CCM when exposed to stressed conditions (e.g., heat, light) may also then mitigate the production of toxic degradation products, which would be an additional benefit for the biomanufacturing industry.

In this study, the major organic compounds (i.e., amino acids and vitamins) in a feed, developed for CHO-based bioproduction, were investigated for their contribution to coloration (also known as browning) under stressed conditions, with a focus on attaining knowledge that may be used to develop RT stable CCM. Tryptophan (Trp) was found to be the major contributor. LC-MS technology was used to identify tryptophan degradation products which were built up in the stressed conditions that correlated with browning, and these products were tested for their ability to contribute to feed coloration and cellular toxicity. An antioxidant known to be tolerated in high concentrations in cell culture was investigated for its ability to reduce browning and/or the generation of detrimental Trp degradation products. This work contributes heavily to the understanding of Trp degradation in cell culture media and other aqueous solutions, including advanced knowledge of Trp degradation products and methods toward inhibiting problematic degradation, especially the generation of toxic or highly colored species.

2. Results

2.1. Tryptophan Is a Major Inducer of Browning in a Feed Used for CHO-Based Bioproduction

Experiments investigating the effects of increased temperature and light exposure in the long-term storage of feed showed no color change at 4 °C storage, but storage at RT in the presence of light or at 37 °C in the absence of light both caused a stark browning of the medium (Figure 1A).

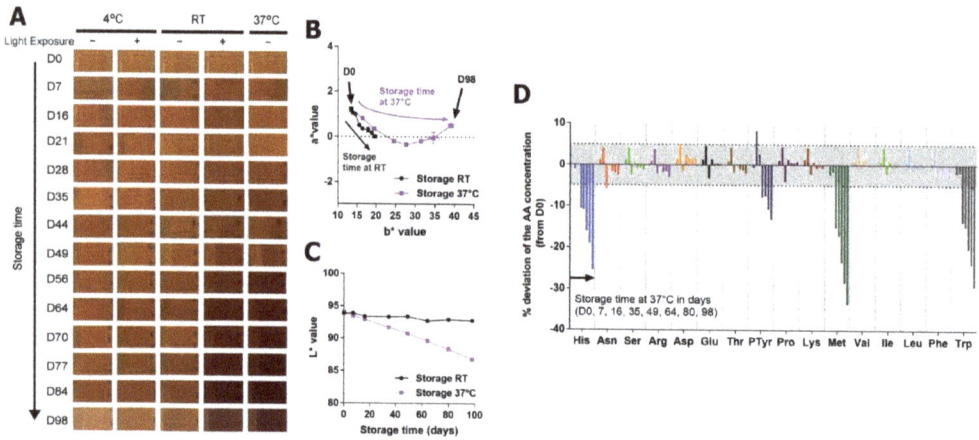

Figure 1. Discoloration and amino acid degradation of feed upon 98 days of storage in stressed conditions: (**A**) images of feed throughout the stability study; changes in (**B**) a* and b* values and (**C**) L* values of the RT and 37 °C stored samples (no light exposure); (**D**) changes in amino acid content of feed upon 37 °C storage (no light exposure). Gray zones indicate the maximum variability due to the analytic method.

A more precise measure of the color change was determined using the CIE L*a*b* color system (See Supplementary Materials Section S1 for details) [16,17]. In the sample stored at 37 °C, the b* value shifted consistently in a positive direction over time (becoming more yellow), while the a* value decreased slightly at first (became more green) and then increased again (more red). The RT sample did not exhibit such a strong color shift in these axes but followed the same trajectory as the 37 °C sample (Figure 1B). The RT sample showed no change in L*, while for the 37 °C sample, the L* value decreased notably, which is representative of the darkening of the solution over time (Figure 1C). The similar a*b* trajectory in the two samples but stronger L* changes in the 37 °C sample may indicate that the increased temperature is causing the same degradation, but at a faster rate. Investigations into the changes in the amino acid levels of these two solutions showed that at RT, only methionine (Met) and Trp were significantly affected (data not shown), while at 37 °C histidine (His) and phosphotyrosine (PTyr) were also degraded (Figure 1D). The PTyr decrease at 37 °C is known to be the result of PTyr being converted to Tyr, as opposed to being degraded to other components [18]. The vitamins were also investigated for their contribution to color, but this was found to be negligible (data not shown).

Due to the considerable amount of literature regarding the production of yellow degradation products from tryptophan [8,19,20], initial experiments focused on the possible contribution of Trp to the color of the degraded feed, by comparing the complete feed to a version lacking Trp. The presence of Trp in the feed caused a clearly visible increase in browning (Figure 2A). The increase in b* with Trp inclusion suggested a greater increase of color in the yellow coordinate (Figure 2B), while the decrease in L* indicated a much stronger darkening of color over time (Figure 2C). This shows that there is a significant color contribution to the solutions as a result of the presence of Trp. The degradation profile of the sensitive AAs His, Met, and PTyr were all comparable in both solutions (data not shown), demonstrating that the degradation of these AAs has no bearing on color change in the solution. The degradation of these AAs even in the absence of Trp indicates that Trp degradation is not inducing the degradation of any of the other AAs, and vice-versa. These experiments confirm that Trp is the major component producing brown products in this feed.

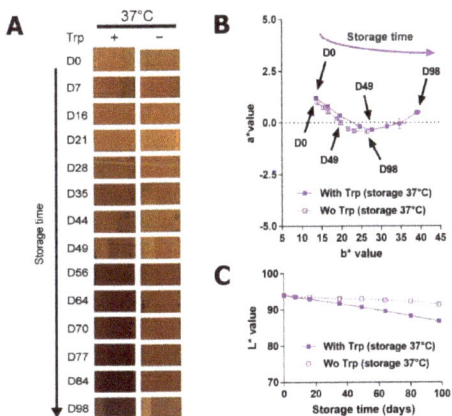

Figure 2. Stability study of the feed with or without (wo) Trp at 37 °C over 98 days. (**A**) Images of the solutions. (**B**) Changes in a*/b* and (**C**) L* values in the solutions.

2.2. Identification of Degradation Products from Trp in Water Stored under Stressed Conditions

Having established that Trp degradation is the lead cause of browning in this feed, the next step in understanding the color change was to determine which degradation products are being generated. Due to the complexity of the solutions, LC-MS was chosen as the optimal method for identifying degradation products of interest. The method was first applied to degraded solutions of Trp in water, considering this might also help identify components in the feed that are coming exclusively from Trp self-reactions. Additional conditions also known to degrade Trp were included in this part of the study, namely exposure to hydrogen peroxide, exposure to UV light, and storage at various temperatures [14]. This experiment was shorter in duration (24 days) but included a 70 °C stressed condition, as representative of an accelerated degradation condition. The measurement of the Trp concentration changes showed that the UV light exposure and 70 °C conditions resulted in the greatest Trp degradation (37% and 42% total loss after 24 days, respectively), followed by hydrogen peroxide incubation, while the storage at 4 °C, RT, and 37 °C induced minimal Trp degradation (Figure 3A). Changes in the parameter ΔE*, which broadly quantifies the overall color change in the L*a*b* system (see Supplementary Materials Section S1) [17] showed that the 70 °C condition changed color most significantly, followed by UV, while the other solutions did not change at all (Figure 3B). For the temperature and UV conditions, there is clearly a correlation between the extent of Trp degradation and the increase in color. Interestingly, the degradation caused by hydrogen peroxide did not lead to a correlating increase in color.

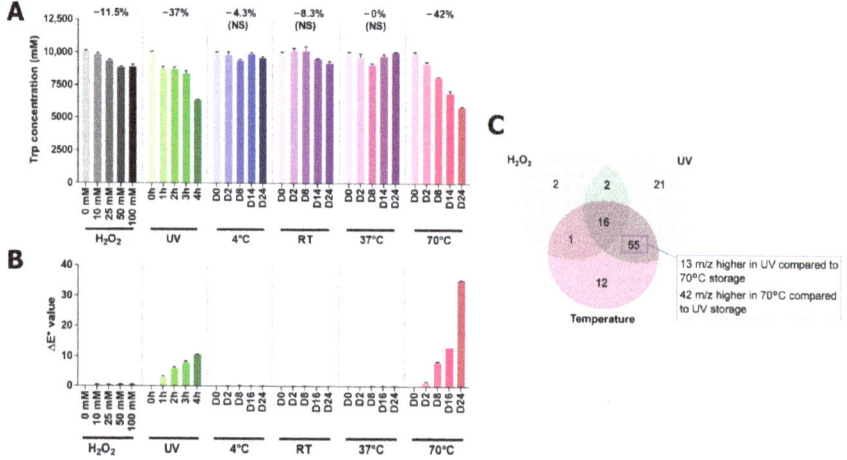

Figure 3. Degradation of tryptophan in water (10 mM, pH 7) under stressed conditions. The conditions tested were stored with varying concentrations of hydrogen peroxide, exposure to UV light, and storage at varying temperatures for 24 days. (**A**) Changes in Trp concentration (NS, non significant). (**B**) Color change displayed as ΔE*. (**C**) Venn diagram of the features identified in LC-MS above an abundance threshold of 30,000.

With the knowledge of which stress conditions result in an increase in color, LC-MS was employed to determine the major degradation products in each condition, and ultimately which products were contributing to browning. LC-MS features that corresponded to unique compounds were characterized with unique identifying information (retention time (rt) and m/z for a given parent ion) and focus was given to analyzing the most prominent features, defined as those generated above a chosen abundance threshold (>30,000) (see Supplementary Materials Section S2 for detailed data processing). Fifty-five features were found to be increased in both the UV and the 70 °C conditions, i.e., the conditions in which browning occurred. This indicates that these features are those most likely to represent small molecules that are contributing to browning. In addition, 42 of these features presented an abundance significantly higher in the 70 °C compared to the UV condition, corresponding to the relative magnitude of browning in these conditions, and as such the small molecules corresponding to these features may be those most significantly contributing to browning. Structure suggestions for these features were made based on literature knowledge or by the analysis of the MSMS fragmentation pattern. Where possible, standards were purchased or synthesized to attain a confirmation for the proposed structure (see Supplementary Materials Figure S1 for an example of matching MSMS profiles for standard and experimental data). The structure proposals for each feature were assigned a confidence level (a tier) based on best practices for small molecule structure identification using LC-MS [21,22]. A compound was assigned tier 1 when the MSMS data and rt matched with a pure standard and tier 1* if the confirmation was via a standard that could not be purified, but for which the structure assignment based on the synthetic route was supported by the literature (see Supplementary Materials Section S2 for detailed tier descriptions). Of the 109 features detected as significantly modulated in these aqueous solutions of Trp, 30 features were assigned to tier 1 or 1*, corresponding to a total of 28 confirmed compound structures, as some features were assigned as diastereomers (absolute configuration unknown—features with earlier rt are designated as 'a' and later rt as 'b') (Figure 4).

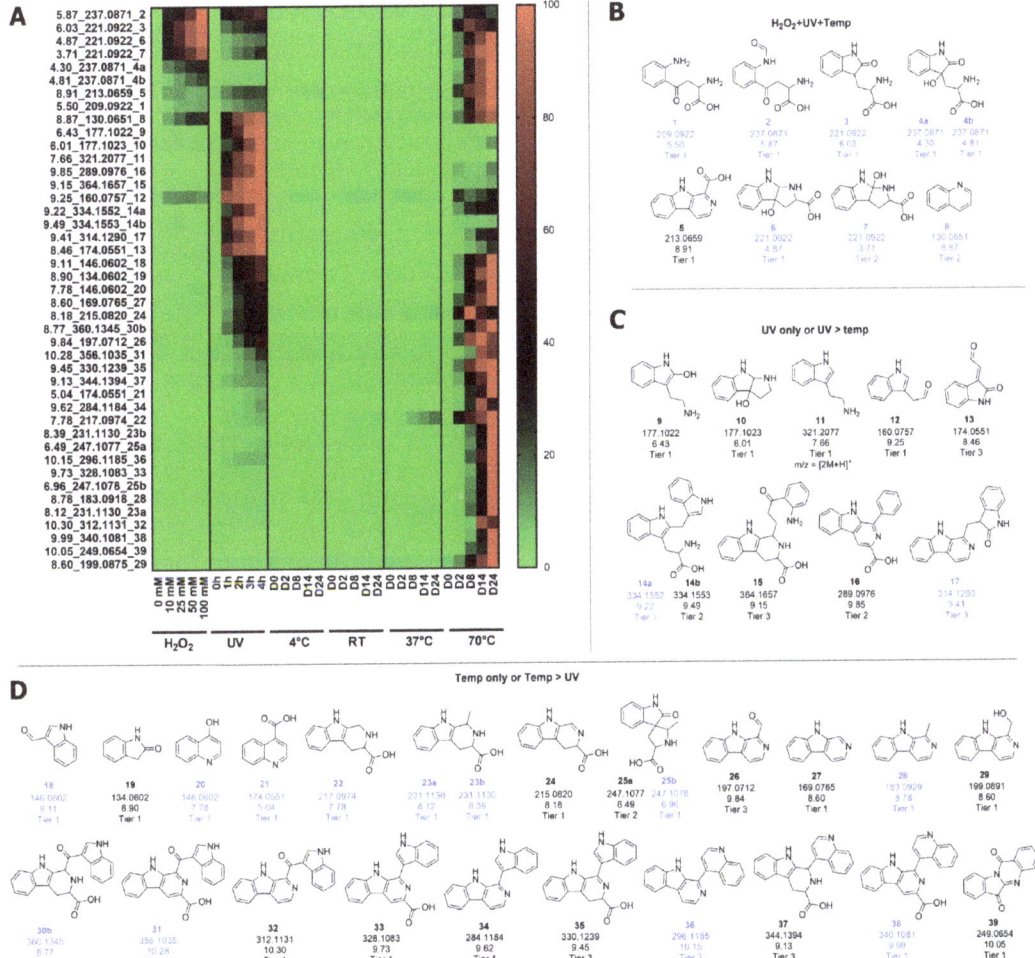

Figure 4. Tier 1–3 features in the aqueous Trp samples which were significantly modulated in at least one of the stress conditions. (**A**) Normalized abundance profiles of each feature (rt_m/z_structure number), presented in a heatmap. Green represents the lowest abundance and red the highest abundance for each feature independently. Structures for features that were increased (**B**) in all stress conditions, (**C**) in UV only or UV > 70 °C conditions, and (**D**) in 70 °C only or 70 °C > UV conditions. All m/z represent [M+H]$^+$ unless specified otherwise (shown with structures). Features that are also present in the 37 °C feed experiment (Table S1) have the compound number and feature data displayed in blue.

Five of these compounds were found in all degradation conditions and correspond to major known degradation products of Trp: kynurenine (KYN, **1**), *N*-formylkynurenine (NFK, **2**), oxindolylalanine (Oia, **3**), dioxindolylalanine (diOia, **4a** and **4b**), and pyrroloindole-3-carboxylic acid (PIC, **6**) [23]. Within the tier 1 features that increased in abundance in the UV and/or 70 °C condition and may therefore be contributing to browning, three main structure classes were identified—b-carbolines, quinolines, and indoles. It is well established that tetrahydro-b-carboline structures can be easily formed from Trp via a condensation reaction with aldehydes and ketones known as the Pictet-Spengler reaction [24], and that these structures can readily undergo subsequent oxidation, optionally with concomitant decarboxylation [25]. This pathway easily explains the presence of the many b-carboline products, and in some cases the direct aldehyde precursor was also detected, as is the case for aldehyde **18**, which is potentially a direct precursor for carbolines **33** and **34**. Quinolines have been shown to be formed from Trp via the initial formation of KYN (**1**) [26], which provides a sensible explanation for the appearance of compounds **8**, **20**, and **21**. While the indole moiety in the products is carried over from Trp, most indole structures exhibit additional oxidation (e.g., **9**, **19**, **25b**). High levels of oxidation are prevalent in most of the degradation products, especially in the products formed at higher temperatures, such as the fully unsaturated b-carbolines (e.g., **26–29**, **31–34**, **36**, **38**). The oxidation of tetrahydro-b-carbolines by atmospheric dissolved oxygen has been reported [25,27]. The presence of spiro[pyrrolidineoxindole] **25** also indicates that the storage conditions are highly oxidizing. This spiro compound may be formed either via a Mannich reaction (another type of condensation reaction) between acetaldehyde and the oxidation product Oia (**3**) or from the oxidative intramolecular rearrangement of the tetrahydro-b-carboline **23** [28]. Tryptanthrin (**39**) is a natural product produced by yeast and plant species. Despite being a natural product, many methods have been developed to synthesize this structure chemically, usually involving the oxidation of an indole-containing precursor such as indole or isatin, which indicates that this compound is also probably formed in these samples via oxidative reactions [29]. The only products exhibiting signs of reduction formed predominantly in the UV condition and are likely caused by radical/photochemical reactions, as has been shown in the literature for compounds **11** and **14** [30,31]. Structures were also proposed for 13 other features, falling into the tier 2 or 3 categories (Figure 4). These contain the same functional groups as seen in the tier 1 compounds and therefore the generation of these compounds in aqueous solutions under these degradation conditions is highly likely. The remaining features in this experiment fell into the tier 4 and 5 categories, and their corresponding abundance data are located in Supplementary Materials, Figure S2.

2.3. Identification of Degradation Products Originating from Trp in Feed Stored at Elevated Temperature

While the industry is predominantly interested in developing RT stable feed, for the purposes of detecting degradation products, 37 °C was chosen as representative of an accelerated degradation condition. An LC-MS analysis of the degraded solutions previously analyzed for color (Figure 2A) identified 85 features that were highly abundant only in the feed containing Trp (Table S1). Of the 47 features that were categorized into tiers 1–3 (55% annotation of features), 22 features were also found in the aqueous Trp samples (corresponding features have the compound number and feature data in blue in Figure 4 and are indicated with an asterisk next to the compound number in Table S1). Additional stability studies of feeds selectively depleted in other organic CCM components were used to establish the origins of some of the features (CCM component origin shown in Table S1). Some features were identified as having arisen solely from Trp from the water-only study (e.g., **40**, **41**) (Figure 5). A total of 25 features were clearly derived from AAs, and 13 of these were classified as tier 1–3. Seven features were derived from non-AA organic CCM components (namely choline, thiamine or pyridoxine) and five of these were categorized into tiers 1–3 (**42** from pyridoxine and **43a/b** and **44a/b** from choline). As with the water solutions of Trp, the structures of tier 1 features in feed contain many of the same functional groups (e.g., ketone, imine, alkene) and structural skeletons (e.g., b-carboline, spirane), indicating that the same kinds of degradative chemical processes (e.g., condensation, oxidation) occur in feed and water samples. Sixty-seven features (39 in tiers 1–3) increased in abundance over time and thus correlated with the increase in browning (Table S1). The compounds correlating with browning show many common features—a high level of oxidation and almost always the presence of an aromatic ring. It is well understood that increasing aromaticity and conjugation in an organic compound leads to a greater ability to absorb light [16,17], so it is possible that these compounds are contributing directly to the color increase. However, an examination of the direct contribution of these compounds to color in solution is required to make such a claim. It is important to note that features that were prominent in the water study but not in the feed study are not necessarily absent from the feed. The same features may be generated in feed below the cutoff threshold chosen for data analysis or may be difficult to detect in the complex matrix due to signal suppression.

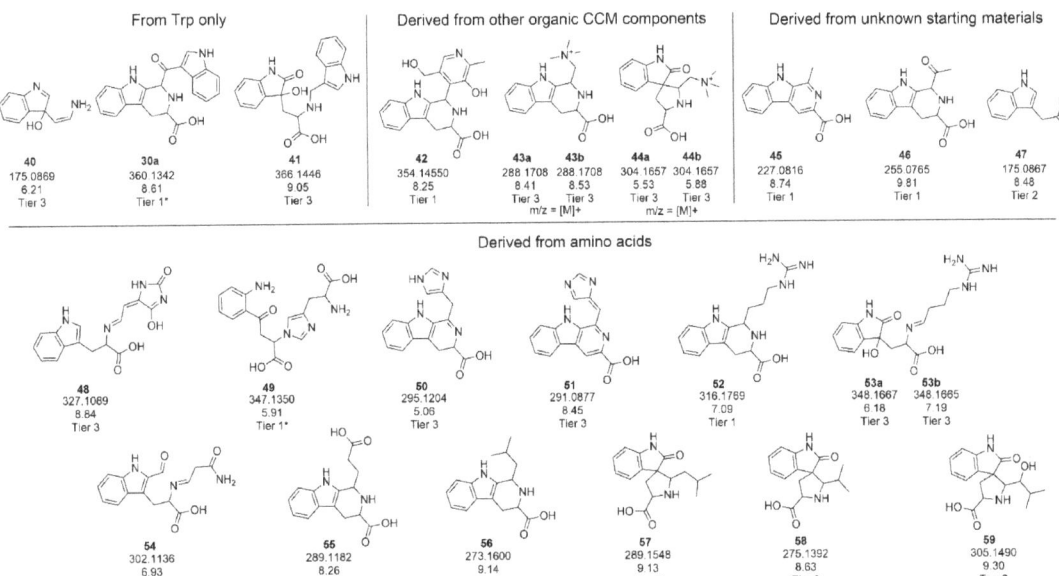

Figure 5. Structures for novel Trp degradants detected in feed studies (those features detected in both water and feed studies are shown in Figure 4 with their feature data marked in blue). Structures are provided for tier 1–3 features that were significantly modulated in the Trp-containing feed compared to the Trp-depleted feed. All m/z represent the parent peaks detected as [M + H]+ ions, unless indicated otherwise.

2.4. Color Contribution of Tier 1 Compounds to Water and Feed Solutions

Standards available in significant enough quantities were used for spiking experiments to see whether they contributed to the coloration of either pure water or feed. ΔE* > 1 was deemed as a significant change (Figure 6A). Some of these compounds have already been identified as possessing notable color. Tryptanthrin (**39**) and pityriacitrin (**32**) are both natural products characterized as yellow solids [29,32], and pityriacitrin was shown to have broad absorption in the UV spectrum [32].

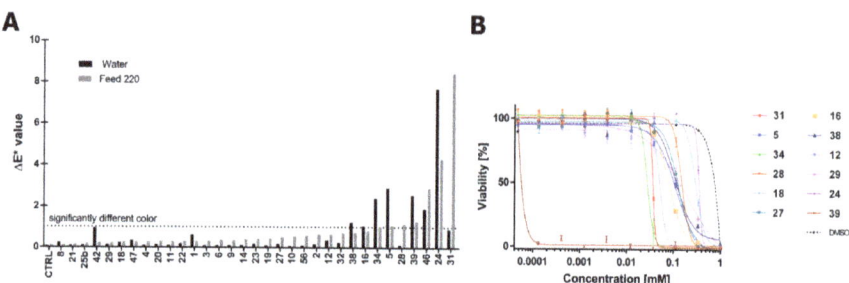

Figure 6. (**A**) Color change as measured by ΔE* in both water and feed upon addition of degradation products. Solutions of degradation products in DMSO were spiked into aqueous solutions to yield a theoretical concentration of 500 μM (1% DMSO). (**B**) Toxicity data for the Trp degradation products that produced a significant toxicity in CHOK1 GS cells as compared to the DMSO control. Error bars, S.D.

Compounds were dissolved in DMSO at 50 mM and then spiked into the solutions to achieve a final compound concentration of 500μM. Some of the tested compounds formed a precipitate immediately upon spiking of the DMSO solution. As the precipitates in solution can interfere with the absorption measurement, each sample was centrifuged prior to taking absorption measurements, meaning that the final concentrations were not all at 500 μM. This interference from precipitation was shown not to occur in the complete feed samples, as the L*a*b* measurements in the 37 °C samples at 98 days of storage were shown to be identical before and after the filtration of the medium through a 0.22 μm filter (data not shown). A significant color change was produced by compounds **16** and **38** in water, by compounds **28** and **31** in feed, and by compounds **5**, **24**, **34**, **39**, and **46** in both conditions (Figure 6A). In stability experiments, eight of these nine compounds (all except **46**) were present in at least one condition in the degraded water samples, and four of these nine compounds (**28**, **31**, **38**, and **46**) were present in the degraded feed. Compounds **28**, **31**, and **46** all had significant abundances in the degraded feed samples and displayed a significant production of color in the feed, which would suggest that these compounds may be critical to enhance color in feed stored at enhanced temperatures.

2.5. Toxicity of Trp-Derived Degradation Products in CHO Cells

Four of the established Trp degradants are natural products and have been tested previously for various types of cellular toxicity. Pityriacitrin (**32**) and pityriacitrin B (**31**) have demonstrated toxicity against some cancer cell lines [33], and while eudistomin U (**34**) displayed some toxicity in human cancer cell lines, it was most potent as an antibacterial agent [34]. Tryptanthrin (**39**) is known to exhibit low micromolar toxicity against a wide range of both cancerous and non-cancerous human cell lines [29,35,36]. As we have demonstrated that these compounds are generated from degraded Trp, it was of interest to investigate the toxicity of these and the other established Trp degradants in CHO cells, the most prevalent cell type used in the biomanufacturing industry. The same panel of standards previously tested for color effects were tested for toxicity in a CHOK1 GS cell line. Twelve compounds were found to elicit notable toxicity in this cell line (GI_{50} < 1 mM) (Figure 6B) and were therefore tested for toxicity in three other CHO cell clones. The compounds showed roughly the same toxicity trends across the four cell lines (see Supplementary Materials, Table S2). Tryptanthrin (**39**) displayed extremely high toxicity across all cell lines ($GI_{50} \leq 1.1$ μM), while eudistomin U (**34**) and pityriacitrin B (**31**) were fairly toxic (10 < GI_{50} < 22 μM across all cell lines) and compounds **18** and **24** displayed the lowest toxicity of the twelve (GI_{50} > 226 μM).

2.6. Inhibition of Browning and Degradation Product Formation Using Alpha-Ketoglutaric Acid

With the observation that the structures correlating with increased browning showed distinct characteristics of oxidation (incorporation of oxygen atoms, increase in aromaticity), the application of a known antioxidant as a method to reduce the browning and concomitant production of these degradants was tested. a-Ketoglutaric acid is a common CCM component with antioxidative properties, which has been shown to have no detrimental effect on the CHO cell culture up to 50 mM and to protect against the temperature-induced degradation of some organic CCM components [37]. The ketoacid was tested for its ability to minimize browning and the production of Trp degradation products in feed stored at RT or 37 °C for 98 days. The ammonia levels in the solutions were also measured, as this could arise from several of the identified Trp degradation products in which nitrogen loss is observed (e.g., **12**, **18**, **19**). Increased ammonia in CCM is undesirable, as it can adversely affect cell growth and product quality [38–40]. The addition of aKG induced a 39% decrease in browning for feed stored at 37 °C, and a 77% decrease at RT (Figure 7A). This demonstrates that the addition of aKG is highly effective in decreasing browning, as the color change at RT was lower in the +aKG condition than in the no Trp control.

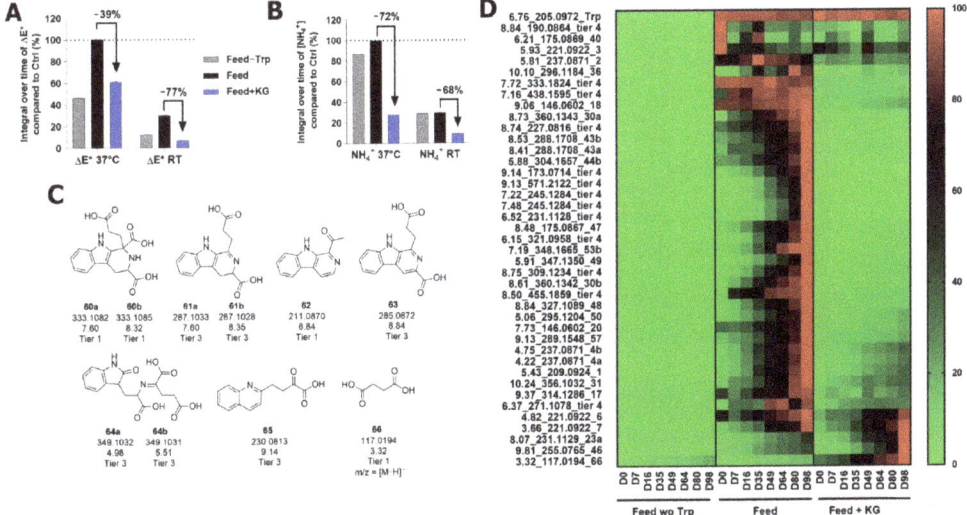

Figure 7. Changes in the feed, feed-Trp, and feed+aKG solutions during long-term storage. (**A**) Color change measured as ΔE*. (**B**) Changes in ammonia concentration. (**C**) Structure identifications for tier 1–3 features which are newly formed upon aKG addition to feed. (**D**) Heatmap showing abundance changes of features of interest (denoted with rt_m/z_structure number for tiers 1–3, or else denoted as tier 4) at RT.

In both conditions, the decrease in ammonia (72% at 37 °C, 68% at RT) far exceeded the decrease brought about when Trp was excluded (13% at 37 °C and 0% at RT) (Figure 7B). This suggests that the ammonia production in the feed is not coming exclusively from Trp, and that aKG is also inhibiting the ammonia-generating degradation of other organic CCM compounds, most probably other AAs, as AA deamination at higher temperatures is a well-established cause of ammonia production [41].

In the 37 °C condition, 62 features in total (see Supplementary Materials Figures S3 and S4) decreased in abundance upon aKG supplementation, five features exhibited similar abundance profiles, and the formation of nine features was inhibited completely. However, 12 features had increased abundances with the addition of aKG, and another 14 features were detected only in the feed+aKG condition. Some of the newly formed features were identified as compounds formed from the combination of Trp and aKG, e.g., tetrahydro-b-carboline **60** and its oxidative decarboxylation product, **61** (Figure 7C). For aKG addition at RT storage, none of these new Trp+aKG features were detected, and 36 features were markedly reduced or eliminated completely (Figure 7D). Only one remained at a comparable abundance with the control (tier 2, corresponding to pyrroloindole **7**), and only two features had increased abundance (tier 1, corresponding to structures **23** and **46**), which were both also increased at 37 °C. Many of the features that had lower abundances in the +aKG condition were those shown to derive from other CCM components (e.g., **43**, **44**, **49**, **53**, **57**). Many features corresponding to structures with high oxidation levels were decreased down to levels comparable to the no Trp control (e.g., **1**, **4**, **18**, **20**, **30**, **36**). The decreased abundance of these Trp products combined with the reduction in browning and ammonia production to below the no Trp control level suggests that aKG may be protecting multiple CCM components from degradation via an antioxidant function. Not much is known about the function of aKG as an antioxidant, although it is known that the reaction with the oxidant hydrogen peroxide causes the oxidative decarboxylation of the a-ketoacid, forming succinic acid (**66**) [42]. The feature corresponding to this compound was detected in the feed in negative mode (Figure 7D), and the abundance was increased in the feed+aKG condition, supporting the hypothesis that aKG is functioning as an antioxidant. It is curious to note that the Trp degradation was not decreased by aKG supplementation, but a complete understanding of the chemical changes induced by aKG addition is beyond the scope of this study.

3. Discussion

This study demonstrated that tryptophan was the major contributor to the browning of a selected feed. The identification of several Trp degradation products demonstrated that the condensation reactions of CCM components and oxidation reactions were the major chemical processes taking place at elevated temperatures. These combinations of reactions often yielded products with fully oxidized heterocycles and a high level of conjugation, which are typical properties in colored organic compounds, and correlates with the color increase in the solutions. As these compounds are significantly different in structure from Trp and as protein translation involves many quality control systems to avoid the incorporation of incorrect AAs [43], it was deemed unlikely that the incorporation of these degradation products in the drug product would be a concern. However, the discovery that some of these compounds can contribute directly to CHO toxicity and feed browning indicates that these degradation processes can be detrimental in biomanufacturing, and therefore methods need to be established to limit such degradation in CCM that are intended to be stored at elevated temperatures. The submicromolar toxicity of tryptanthrin (**39**) is especially noteworthy. Its chemical structure is significantly different from that of all of the other compounds tested (it is the only compound with a indolo[2,1-*b*]quinazoline core), and is therefore likely the fundamental reason behind its uniquely high toxicity. It is not clear by which mechanism tryptanthrin achieves such high toxicity in CHO cells, but in other cellular toxicity studies it commonly bound to ATP-binding sites of crucial enzymes in pathways involved in apoptosis regulation, cellular proliferation, and cell survival [25]. Tryptanthrin could therefore be eliciting toxicity via a similar means in CHO cells. For studies such as this one, which investigate the effects of organic compounds on cell culture, it is relevant to note that very small amounts of transition metal contaminants can be the cause of the observed changes (e.g., color and toxicity) [44–46], and as such follow-up studies may be required to confirm that any detrimental effects are indeed due to the organic component, though this is currently beyond the scope of this work. The addition of a-ketoglutaric acid to the cell culture feed was shown to decrease the detrimental chemical changes in the feed stored at higher temperatures, including browning, ammonia production, and the generation of Trp-derived degradation products. The extent of ammonia reduction suggests that aKG is inhibiting the deamination of other AAs, and the decrease in major Trp oxidation products, along with the production of succinic acid, suggest that this may result from aKG functioning as an antioxidant. The production of succinic acid as a byproduct is possibly even advantageous, as succinic acid is not only colorless in solution and not toxic to CHO cells, but has even shown potential benefits for mAb production in CHO cell cultures [47,48]. This work demonstrates the first known application of an antioxidant to stabilize CCM stored at higher temperatures, and thus represents perhaps the beginning of the next major CCM optimization strategy for the bioprocessing industry.

4. Materials and Methods

4.1. Reagents

Raw materials and cell culture media/feed in the stability studies were all purchased from Merck KGaA, Darmstadt, Germany, if not stated otherwise. Chemical standards for LC-MS feature identification were purchased from a commercial supplier and synthesized externally under contract or in-house. Details of the origins of the standards are provided in the Supplementary Materials in Table S3. The feed formulation used in this study was a customized dry powder derived from Cellvento® Feed220 (Merck, Darmstadt, Germany), depleted in amino acids, and was reconstituted according to the manufacturer's guidelines, with amino acids added back as desired.

4.2. Stability Studies

For the Trp solutions, 10 mM Trp solutions were prepared in water with a pH adjustment to pH 7.0 ± 0.3 using NaOH(aq.) and sterile filtered with Steriflip® filter units (Merck, Darmstadt, Germany). For H_2O_2 exposure testing, the solutions were spiked with hydrogen peroxide (30%, 8.82 M) to 0, 10, 25, 50, or 100 mM, and stored for 4h at RT (~19–23 °C), light-protected. For the UV exposure effects, the solutions were stored for 4 h under a UV-lamp set to an intensity of 540 V (UV-A, -B and -C radiation). Samples tested for temperature effects were stored light-protected for 24 days at 4 °C, room temperature (RT, i.e., ~19–23 °C), 37 °C, or 70 °C. Feed solutions were sterile-filtered as above and stored for 98 days at 4 °C, RT (~19–23 °C) or 37 °C. The samples were either stored light-protected or exposed to laboratory fluorescent lighting with partial exposure to glass-filtered daylight (light-exposed). For testing the a-ketoglutaric acid effects in feed, dry powder feed was reconstituted with a-ketoglutaric acid, pH adjusted to 7.0 ± 0.3 with NaOH (aq.), filtered as above, and stored light-protected at RT (~19–23 °C) or 37 °C for 98 days. Ammonia was quantified using Cedex Bio HT (Roche, Mannheim, Germany). Aliquots were frozen at −20 °C for subsequent amino acid analysis and LC-MS.

4.3. CIELAB Color Analysis

Absorbance measurements were taken in the range of 380 nm to 780 nm (20 nm steps) of 100 µL of the samples in four technical replicates (Envision, Perkin Elmer, Waltham, MA, USA). From the absorbance values, the change in color was calculated using L*a*b* and ΔE* values according to DIN 5033-3 and DIN 6174 norms. The CIE 1931 standard observer (2°) and the D65 illuminant were selected.

4.4. Amino Acid Analysis

Amino acid analysis was performed using a pre-column derivatization using the AccQ Tag® Ultra Reagent kit. Derivatization, chromatography, and data analysis were performed according to the supplier recommendations (Waters, Milford, MA, USA).

4.5. LC-MS Feature Determination and Structure Elucidation

Time point series from stability study experiments were analyzed using UHPLC (Vanquish, Thermo Fisher, Waltham, MA, USA) coupled with an ESI-Q-ToF mass spectrometer (Impact II, Bruker, Bremen, Germany). Briefly, the samples were diluted 10 times in water prior to LC-MS/MS analysis. Five microliters of sample were loaded in 99.9% buffer A (20 mM ammonium formate/0.1% FA) onto a XSelect HSS T3 column (2.1 × 150 mm, 3.5 µm, Waters, Milford, MA, USA) thermostated at 40 °C with a flowrate of 300 µL/min. An optimized 12 min linear gradient was applied using 100% methanol (buffer B) as follows (minute/% B): 0/0.1, 2/0.1, 4/20, 6/30, 8/80, 8.5/100, 9.5/100, 9.6/0.1, 12/0.1.

LC-MS/MS analyses were performed in triplicate using the Impact II mass spectrometer equipped with an ESI source (Bruker, Bremen, Germany). MS acquisition was performed in positive and negative modes with capillary voltages set at 4500 V and 3500 V, respectively, and the end plate offset set at 500 V. Nebulizer and dry gas (250 °C) were set at 1.4 bar and 9.0 L/min, respectively. MS spectra were acquired over the m/z range 20–1000 with a scan rate of 12 Hz followed by data-dependent auto-MS/MS acquisitions using a fixed MS-MSMS cycle time of 0.5 s and a summation time adjusted to the precursor intensity. The data analysis was performed using Data Analysis 4.0 (Bruker, Bremen, Germany) and Progenesis QI (Non-linear Dynamics—Waters, Newcastle, UK). Briefly, raw data were processed as centroid data using the automatic peak detection algorithm of Progenesis QI. No normalization was applied to the dataset due to the nature of the study. Unique ions (retention time (rt) and *m/z* pairs) were grouped, and their abundancies were summed to generate unique features. Only features satisfying the following criteria were considered: abundance >30,000; fold change >3; corrected ANOVA *q*-value < 0.05. Hierarchical clustering was used to identify groups (e.g., identify all Trp related features by using a feed depleted in Trp and selecting the features that are not found in that condition). For the representation in the heatmap, the data were normalized between the minimum and the maximum abundance for each replicate independently. Feature annotations were performed within Progenesis QI using precursor mass, fragment mass, and retention time (if available) with tolerances set at 5 ppm, 20 ppm, and 0.3 min, respectively. Comprehensive LC-MS data is provided in Supplementary Materials 2.

4.6. Color Measurements of Standards

Chemical standards were prepared as 50 mM stocks in DMSO and then spiked into the aqueous solutions (either water or feed) to achieve a theoretical compound concentration of 500 µM (1% DMSO). The solutions were centrifuged, and the absorbance measurements of these solutions was measured and converted to ΔE^* values as described above. The control condition was pure DMSO (1% *v/v*).

4.7. Cytotoxicity Assay

To assay the potential toxicity induced in the CHO cell culture by the Trp degradation products found in this study, a cytotoxicity assay CellTiter-Glo® Luminescent Cell Viability Assay (Cat #G7572, Promega, Madison, WI, USA) was used, following the manufacturer's instructions. Briefly, the cells were seeded with 25,000 viable cells per cavity of a 96-well opaque white cell culture plate (Thermo Fisher, Waltham, MA, USA) in 80 µL of a pre-warmed medium. Tryptophan degradation compounds were added in triplicates to the cells, ranging from 1 mM to 46 nM. A DMSO control was added in the same concentration range as the tryptophan degradation compounds were diluted. The dose response curves were fitted with a 4PL model by GraphPad Prism v.7 (GraphPad Software, San Diego, CA, USA) after luminescence readout with an EnVision 2104 Multilabel Reader (Perkin Elmer, Waltham, MA, USA).

5. Conclusions

Tryptophan is capable of degrading in cell culture media under elevated temperatures to afford colored and toxic degradation products. This is a concern for the development of room-temperature stable CCM, in which medium color change and a buildup of toxic compounds are undesirable. The storage of tryptophan-containing CCM at higher temperatures is best undertaken only if precautions are made to inhibit degradation into colored or toxic compounds. This study demonstrated that the addition of an antioxidant, namely alpha-ketoglutaric acid, is a viable means to achieve this goal of stabilizing CCM against problematic tryptophan degradation processes.

Supplementary Materials: The following are available online at https://www.mdpi.com/article/10.3390/ijms22126221/s1.

Author Contributions: Conceptualization, A.Z.; methodology, A.S., A.L., M.L.M., T.H., and S.B.; formal analysis, A.S., A.L., M.L.M., and T.H.; investigation, A.S., A.L., M.L.M., T.H., and B.P.; data curation, A.Z. and M.L.M.; writing—original draft preparation, S.B.; writing—review and editing, A.Z.; visualization, S.B., A.Z., and T.H.; supervision, A.Z.; project administration, A.Z. All authors have read and agreed to the published version of the manuscript.

Funding: This research received no external funding.

Institutional Review Board Statement: Not applicable.

Informed Consent Statement: Not applicable.

Data Availability Statement: Not applicable.

Acknowledgments: This study was funded by Merck KGaA, Darmstadt, Germany. All Authors are employees of Merck KGaA, except Anton Lindig, who worked on this study while employed at Merck KGaA but is now a student at TU Dortmund, Germany.

Conflicts of Interest: The authors declare no conflict of interest.

References

1. Neutsch, L.; Kroll, P.; Brunner, M.; Pansy, A.; Kovar, M.; Herwig, C.; Klein, T. Media photo-degradation in pharmaceutical biotechnology—Impact of ambient light on media quality, cell physiology, and IgG production in CHO cultures. *J. Chem. Technol. Biotechnol.* **2018**, *93*, 2141–2151. [CrossRef] [PubMed]
2. Mallaney, M.; Wang, S.-H.; Sreedhara, A. Effect of ambient light on monoclonal antibody product quality during small-scale mammalian cell culture process in clear glass bioreactors. *Biotechnol. Prog.* **2014**, *30*, 562–570. [CrossRef]
3. Mayrhofer, P.; Reinhart, D.; Castan, A.; Kunert, R. Monitoring of heat- and light exposure of cell culture media by RAMAN spectroscopy: Towards an analytical tool for cell culture media quality control. *Biochem. Eng. J.* **2020**, 107845. [CrossRef]
4. Floris, P.; McGillicuddy, N.; Albrecht, S.; Morrissey, B.; Kaisermayer, C.; Lindeberg, A.; Bones, J. Untargeted LC-MS/MS Profiling of Cell Culture Media Formulations for Evaluation of High Temperature Short Time Treatment Effects. *Anal. Chem.* **2017**, *89*, 9953–9960. [CrossRef] [PubMed]
5. Prentice, K.M.; Gillespie, R.; Lewis, N.; Fujimori, K.; McCoy, R.; Bach, J.; Connell-Crowley, L.; Eakin, C.M. Hydroxocobalamin association during cell culture results in pink therapeutic proteins. *MAbs* **2013**, *5*, 974–981. [CrossRef]
6. Vijayasankaran, N.; Varma, S.; Yang, Y.; Mun, M.; Arevalo, S.; Gawlitzek, M.; Swartz, T.; Lim, A.; Li, F.; Zhang, B.; et al. Effect of cell culture medium components on color of formulated monoclonal antibody drug substance. *Biotechnol. Prog.* **2013**, *29*, 1270–1277. [CrossRef] [PubMed]
7. Xu, J.; Jin, M.; Song, H.; Huang, C.; Xu, X.; Tian, J.; Qian, N.-X.; Steger, K.; Lewen, N.S.; Tao, L.; et al. Brown drug substance color investigation in cell culture manufacturing using chemically defined media: A case study. *Process Biochem.* **2014**, *49*, 130–139. [CrossRef]
8. Li, Y.; Polozova, A.; Gruia, F.; Feng, J. Characterization of the Degradation Products of a Color-Changed Monoclonal Antibody: Tryptophan-Derived Chromophores. *Anal. Chem.* **2014**, *86*, 6850–6857. [CrossRef]
9. Hofmann, T. Structure, Color, and Formation of Low- and High-Molecular Weight Products Formed by Food-Related Maillard-Type Reactions. In *Chemistry and Physiology of Selected Food Colorants*; ACS Symposium Series; American Chemical Society: Washington, DC, USA, 2001; Volume 775, pp. 134–151.
10. Wang, H.-Y.; Qian, H.; Yao, W.-R. Melanoidins produced by the Maillard reaction: Structure and biological activity. *Food Chem.* **2011**, *128*, 573–584. [CrossRef]
11. Echavarría, A.; Pagán, J.; Ibarz, A. Kinetics of color development in glucose/Amino Acid model systems at different temperatures. *Sci. Agropecu.* **2016**, *7*, 15–21. [CrossRef]
12. Schnellbaecher, A.; Binder, D.; Bellmaine, S.; Zimmer, A. Vitamins in cell culture media: Stability and stabilization strategies. *Biotechnol. Bioeng.* **2019**, *116*, 1537–1555. [CrossRef]
13. Voelker, A.L.; Miller, J.; Running, C.A.; Taylor, L.S.; Mauer, L.J. Chemical stability and reaction kinetics of two thiamine salts (thiamine mononitrate and thiamine chloride hydrochloride) in solution. *Food Res. Int.* **2018**, *112*, 443–456. [CrossRef]
14. Bellmaine, S.; Schnellbaecher, A.; Zimmer, A. Reactivity and degradation products of tryptophan in solution and proteins. *Free Radic. Biol. Med.* **2020**, *160*, 696–718. [CrossRef]
15. Unger, N.; Ferraro, A.; Holzgrabe, U. Investigation of tryptophan-related yellowing in parenteral amino acid solution: Development of a stability-indicating method and assessment of degradation products in pharmaceutical formulations. *J. Pharm. Biomed. Anal.* **2020**, *177*, 112839. [CrossRef] [PubMed]
16. Zollinger, H. *Color Chemistry. Synthesis, Properties and Applications of Organic Dyes and Pigments*, 3rd ed.; Wiley VCH: Zürich, Switzerland, 2001.
17. Nassau, K. *The Physics and Chemistry of Color: The Fifteen Causes of Color*, 2nd ed.; Wiley: Hoboken, NJ, USA, 2001.
18. Zimmer, A.; Mueller, R.; Wehsling, M.; Schnellbaecher, A.; von Hagen, J. Improvement and simplification of fed-batch bioprocesses with a highly soluble phosphotyrosine sodium salt. *J. Biotechnol.* **2014**, *186*, 110–118. [CrossRef]
19. Maskos, Z.; Rush, J.D.; Koppenol, W.H. The hydroxylation of tryptophan. *Arch. Biochem. Biophys.* **1992**, *296*, 514–520. [CrossRef]
20. Wickern, B.V.; Müller, B.; Simat, T.; Steinhart, H. Determination of γ-radiation induced proudcts in aqueous solutions of tryptophan and synthesis of 4-, 6- and 7-hydroxytryptophan. *J. Chromatogr. A* **1997**, *786*, 57–65. [CrossRef]

21. Sumner, L.W.; Amberg, A.; Barrett, D.; Beale, M.H.; Beger, R.; Daykin, C.A.; Fan, T.W.M.; Fiehn, O.; Goodacre, R.; Griffin, J.L.; et al. Proposed minimum reporting standards for chemical analysis. *Metabolomics* **2007**, *3*, 211–221. [CrossRef] [PubMed]
22. Blaženović, I.; Kind, T.; Sa, M.R.; Ji, J.; Vaniya, A.; Wancewicz, B.; Roberts, B.S.; Torbašinović, H.; Lee, T.; Mehta, S.S.; et al. Structure Annotation of All Mass Spectra in Untargeted Metabolomics. *Anal. Chem.* **2019**, *91*, 2155–2162. [CrossRef]
23. Gracanin, M.; Hawkins, C.L.; Pattison, D.I.; Davies, M.J. Singlet-oxygen-mediated amino acid and protein oxidation: Formation of tryptophan peroxides and decomposition products. *Free Radic. Biol. Med.* **2009**, *47*, 92–102. [CrossRef]
24. Stöckigt, J.; Antonchick, A.P.; Wu, F.; Waldmann, H. The Pictet–Spengler Reaction in Nature and in Organic Chemistry. *Angew. Chem. Int. Ed.* **2011**, *50*, 8538–8564. [CrossRef]
25. Nemet, I.; Varga-Defterdarović, L. Methylglyoxal-derived β-carbolines formed from tryptophan and its derivates in the Maillard reaction. *Amino Acids* **2007**, *32*, 291–293. [CrossRef]
26. Tsentalovich, Y.P.; Snytnikova, O.A.; Forbes, M.D.E.; Chernyak, E.I.; Morozov, S.V. Photochemical and thermal reactivity of kynurenine. *Exp. Eye Res.* **2006**, *83*, 1439–1445. [CrossRef]
27. Chu, N.T.; Clydesdale, F.M. Reactions between amino acids and organic acids: Reaction of tryptophan and pyruvic acid. *J. Food Sci.* **1976**, *41*, 891–894. [CrossRef]
28. White, J.D.; Li, Y.; Ihle, D.C. Tandem Intramolecular Photocycloaddition−Retro-Mannich Fragmentation as a Route to Spiro[pyrrolidine-3,3′-oxindoles]. Total Synthesis of (±)-Coerulescine, (±)-Horsfiline, (±)-Elacomine, and (±)-6-Deoxyelacomine. *J. Org. Chem.* **2010**, *75*, 3569–3577. [CrossRef]
29. Kaur, R.; Manjal, S.K.; Rawal, R.K.; Kumar, K. Recent synthetic and medicinal perspectives of tryptanthrin. *Biorg. Med. Chem.* **2017**, *25*, 4533–4552. [CrossRef]
30. Creed, D. The photophysics and photochemistry of the near-UV absorbing amino acids—I. Tryptophan and its simple derivatives. *Photochem. Photobiol.* **1984**, *39*, 537–562. [CrossRef]
31. Domingues, M.R.M.; Domingues, P.; Reis, A.; Fonseca, C.; Amado, F.M.L.; Ferrer-Correia, A.J.V. Identification of oxidation products and free radicals of tryptophan by mass spectrometry. *J. Am. Soc. Mass. Spectrom.* **2003**, *14*, 406–416. [CrossRef]
32. Mayser, P.; Schäfer, U.; Krämer, H.-J.; Irlinger, B.; Steglich, W. Pityriacitrin—An ultraviolet-absorbing indole alkaloid from the yeast Malassezia furfur. *Arch. Dermatol. Res.* **2002**, *294*, 131–134. [CrossRef]
33. Zhang, P.; Sun, X.; Xu, B.; Bijian, K.; Wan, S.; Li, G.; Alaoui-Jamali, M.; Jiang, T. Total synthesis and bioactivity of the marine alkaloid pityriacitrin and some of its derivatives. *Eur. J. Med. Chem.* **2011**, *46*, 6089–6097. [CrossRef]
34. Roggero, C.M.; Giulietti, J.M.; Mulcahy, S.P. Efficient synthesis of eudistomin U and evaluation of its cytotoxicity. *Bioorg. Med. Chem. Lett.* **2014**, *24*, 3549–3551. [CrossRef]
35. Liang, J.L.; Park, S.-E.; Kwon, Y.; Jahng, Y. Synthesis of benzo-annulated tryptanthrins and their biological properties. *Bioorg. Med. Chem.* **2012**, *20*, 4962–4967. [CrossRef]
36. Kirpotina, L.N.; Schepetkin, I.A.; Hammaker, D.; Kuhs, A.; Khlebnikov, A.I.; Quinn, M.T. Therapeutic Effects of Tryptanthrin and Tryptanthrin-6-Oxime in Models of Rheumatoid Arthritis. *Front. Pharmacol.* **2020**, *11*, 1145. [CrossRef]
37. Kuschelewski, J.; Schnellbaecher, A.; Pering, S.; Wehsling, M.; Zimmer, A. Antioxidant effect of thiazolidine molecules in cell culture media improves stability and performance. *Biotechnol. Prog.* **2017**, *33*, 759–770. [CrossRef] [PubMed]
38. Schneider, M.; Marison, I.W.; von Stockar, U. The importance of ammonia in mammalian cell culture. *J. Biotechnol.* **1996**, *46*, 161–185. [CrossRef]
39. Kim, D.Y.; Chaudhry, M.A.; Kennard, M.L.; Jardon, M.A.; Braasch, K.; Dionne, B.; Butler, M.; Piret, J.M. Fed-batch CHO cell t-PA production and feed glutamine replacement to reduce ammonia production. *Biotechnol. Prog.* **2013**, *29*, 165–175. [CrossRef]
40. Yang, M.; Butler, M. Effect of Ammonia on the Glycosylation of Human Recombinant Erythropoietin in Culture. *Biotechnol. Prog.* **2000**, *16*, 751–759. [CrossRef]
41. Sohn, M.; Ho, C.-T. Ammonia Generation during Thermal Degradation of Amino Acids. *J. Agric. Food Chem.* **1995**, *43*, 3001–3003. [CrossRef]
42. Rice, B.; Yerabolu, J.R.; Krishnamurthy, R.; Springsteen, G. The Abiotic Oxidation of Organic Acids to Malonate. *Synlett* **2017**, *28*, 98–102. [CrossRef]
43. Ling, J.; Reynolds, N.; Ibba, M. Aminoacyl-tRNA Synthesis and Translational Quality Control. *Annu. Rev. Microbiol.* **2009**, *63*, 61–78. [CrossRef]
44. Graham, R.J.; Bhatia, H.; Yoon, S. Consequences of trace metal variability and supplementation on Chinese hamster ovary (CHO) cell culture performance: A review of key mechanisms and considerations. *Biotechnol. Bioeng.* **2019**, *116*, 3446–3456. [CrossRef]
45. Mohammad, A.; Agarabi, C.; Rogstad, S.; DiCioccio, E.; Brorson, K.; Ashraf, M.; Faustino, P.J.; Madhavarao, C.N. An ICP-MS platform for metal content assessment of cell culture media and evaluation of spikes in metal concentration on the quality of an IgG3:κ monoclonal antibody during production. *J. Pharm. Biomed. Anal.* **2019**, *162*, 91–100. [CrossRef] [PubMed]
46. Grinnell, C.; Bareford, L.; Matthews, T.E.; Brantley, T.; Moore, B.; Kolwyck, D. Elemental metal variance in cell culture raw materials for process risk profiling. *Biotechnol. Prog.* **2020**, *36*, e3004. [CrossRef] [PubMed]
47. Argentova, V.; Aliev, T.; Dolgikh, D.; Kirpichnikov, M. Effects of Succinic Acid Supplementation on Stable Cell Line Growth, Aggregation, and IgG and IgA Production. *Curr. Pharm. Biotechnol.* **2020**, *21*, 990–996. [CrossRef] [PubMed]
48. Zhang, X.; Jiang, R.; Lin, H.; Xu, S. Feeding tricarboxylic acid cycle intermediates improves lactate consumption and antibody production in Chinese hamster ovary cell cultures. *Biotechnol. Prog.* **2020**, *36*, e2975. [CrossRef]

Article

Effects of Tyrosine and Tryptophan Supplements on the Vital Indicators in Mice Differently Prone to Diet-Induced Obesity

Ivan V. Gmoshinski [1], Vladimir A. Shipelin [1,2,*], Nikita V. Trusov [1], Sergey A. Apryatin [1], Kristina V. Mzhelskaya [1], Antonina A. Shumakova [1], Andrey N. Timonin [1], Nikolay A. Riger [1] and Dmitry B. Nikityuk [1,3]

[1] Federal Research Centre of Nutrition and Biotechnology, 109240 Moscow, Russia; gmosh@ion.ru (I.V.G.); nikkitosu@yandex.ru (N.V.T.); apryatin@mail.ru (S.A.A.); kristik13@yandex.ru (K.V.M.); antonina_sh@list.ru (A.A.S.); andrey8407@mail.ru (A.N.T.); n_rieger@icloud.com (N.A.R.); dimitrynik@mail.ru (D.B.N.)
[2] Academic Department of Innovational Materials and Technologies Chemistry, Plekhanov Russian University of Economics, 115093 Moscow, Russia
[3] Department of Operative Surgery and Topographic Anatomy, I.M. Sechenov First Moscow State Medical University (Sechenov University), 119991 Moscow, Russia
* Correspondence: v.shipelin@yandex.ru; Tel.: +7-495-698-5371

Abstract: We studied the effects of the addition of large neutral amino acids, such as tyrosine (Tyr) and tryptophan (Trp), in mice DBA/2J and tetrahybrid mice DBCB receiving a high-fat, high-carbohydrate diet (HFCD) for 65 days. The locomotor activity, anxiety, muscle tone, mass of internal organs, liver morphology, adipokines, cytokines, and biochemical indices of animals were assessed. The Tyr supplementation potentiated increased anxiety in EPM and contributed to a muscle tone increase, a decrease in the AST/ALT ratio, and an increase in protein anabolism in both mice strains. Tyr contributed to a decrease in liver fatty degeneration and ALT reduction only in DBCB that were sensitive to the development of obesity. The addition of Trp caused an increase in muscle tone and potentiated an increase in anxiety with age in animals of both genotypes. Trp had toxic effects on the livers of mice, which was manifested in increased fatty degeneration in DBCB, edema, and the appearance of micronuclei in DBA/2J. The main identified effects of Tyr on mice are considered in the light of its modulating effect on the dopamine neurotransmitter metabolism, while for the Trp supplement, effects were presumably associated with the synthesis of its toxic metabolites by representatives of the intestinal microflora.

Keywords: obesity; mice; tyrosine; tryptophan; cytokines; behavior; cytokines; inflammation; liver morphology

1. Introduction

Genomic and post-genomic factors that determine the level of physical and metabolic activity play an important role in the development of obesity [1,2]. The development of methods for effective dietary correction of obesity in humans requires an understanding of the biological causes of decreased physical activity which lead to an imbalance between energy consumption and expenditure [3].

In this area, the study of the neuroendocrine regulation of metabolic processes [4] and the possible influence on them by pharmacological and dietary factors [5,6] is of particular interest.

The neurotransmitters dopamine [7], serotonin [8], and their minor metabolites (trace amines, etc.) [9] play a critical role in the regulation of energy metabolism, locomotor activity, and feeding behavior. Large neutral amino acids—tyrosine (Tyr) and tryptophan (Trp)—are precursors of these active metabolites. There is a definite link between changes in the signaling of dopamine in the striatum and motor activity [10]. In turn, the effect

of serotonin on metabolic processes is due to the activation of the signaling pathway in the hypothalamus neurons that are part of the proopiomelanocortin system (POMC). Serotonergic neurons receive information on the composition of the diet through the amount of Trp entering the brain, which depends on the competition in transport across the blood-brain barrier between Trp, Tyr, and other large neutral amino acids (LNAA) (phenylalanine, leucine, isoleucine, methionine) [11].

The catabolism of tissue proteins containing a relatively small amount of Trp decreases under conditions of excessive carbohydrate and fatty nutrition. This leads in turn to an increase in the specific content of Trp in the whole pool of free amino acids of blood plasma and consequently an increase in its transport to the brain, where Trp is transformed into serotonin, which has anorexigenic and sedative effects [12].

In this regard, the question arises about the possibility of directed modulation of metabolic processes, physical activity, and eating behavior by changing the ratio between Trp and other LNAA (especially tyrosine) coming from food. This work aimed to study the effect of Tyr and Trp in the diet on the neuroendocrine regulation of metabolic processes in laboratory mice (*Mus domesticus*), using the assessment of behavioral reactions and integral, morphological, biochemical, and immunological parameters. Two strains of mice were used, of which the first, DBA/2J, is characterized by resistance to the development of diet-induced obesity. The second strain—a complex hybrid of the second generation obtained earlier in our laboratory—easily developed signs of diet-induced obesity and fatty liver degeneration when consuming a diet with excess fat and fructose, as was shown earlier in the study [13].

2. Results

2.1. Amino Acids Characterization

The possible presence of trace impurities in amino acids, especially in tryptophan preparation [14], that could influence their nutritive value and toxicity was checked by HPLC. As shown by the analysis, samples of tyrosine and tryptophan used in animal feeding did not contain any impurities detected within the method sensitivity (less than 0.1% by mass), which complies with the manufacturer's specifications.

2.2. Mice Integral Indices

Assessment of integral indicators, such as specific energy consumption, body weight, internal organs, and tissues, shows the degree of diet-induced obesity in mice and allows assessment of the possibility of its alleviation by consuming the studied supplements. During almost the entire period of the experiment, DBA/2J mice of all groups stably increased their body weight (bw), had a normal appearance, and were mobile; morbidity and mortality were not observed. Among DBCB mice in group 2 fed with HFCD, 4 animals died during the experiment, with signs of necrotic myocardial changes; in groups 3 and 4, which received HFCD with Tyr and Trp, 1 animal died in each one. As follows from the data on Figure 1, DBA/2J mice of all groups receiving HFCD were characterized by higher energy consumption compared with the control group. In contrast, in tetrahybrid mice fed with HFCD, the total energy consumption did not increase compared to the control group since the mass of feed consumed by these animals was reduced in comparison with the control value, but the addition of tyrosine caused a significant increase in energy consumption in animals of group 3.

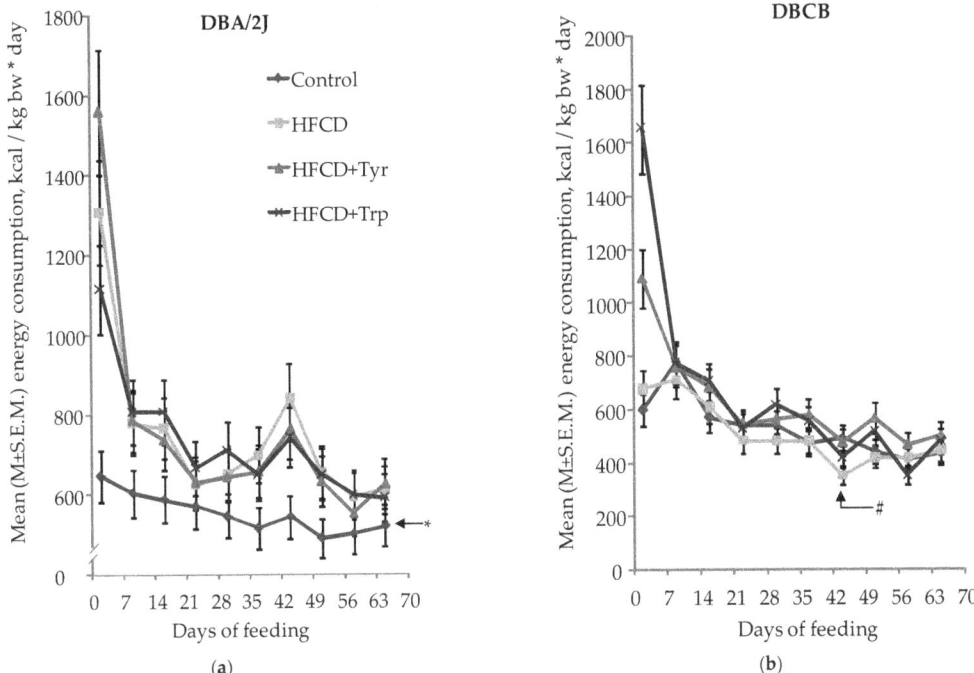

Figure 1. Specific energy consumption (M ± S.E.M.) of mice during the experiment: (**a**) DBA/2J mouse line; (**b**) DBCB mice. * The difference is significant with all groups receiving HFCD; # The difference is significant with group receiving HFCD with Tyr group, $p < 0.05$; Student's t-test for pair-related, group-averaged indicators.

At the end of the experiment, a significantly larger bw was observed in DBCB mice receiving HFCD compared with control, ANOVA, $p < 0.05$ by the diet (D) factor (Figure 2). The relative weight indexes of the spleen, heart, kidney, and gonads in the tetrahybrid mice were significantly lower, and retroperitoneal white fat was higher in comparison with mice DBA/2J, ANOVA, $p < 0.05$ by the genotype (G) factor. The addition of amino acids did not have a significant effect on bw, spleen, and gonad weight indexes, but in DBCB mice, Tyr consumption increased the relative weight of the heart, kidneys, and gonad to the level of the control group, and Trp caused a significant decrease in the weight of the kidneys and brain. Both amino acids in these mice had a potentiating effect on the accumulation of white fat and a decrease in the ratio of brown fat to white fat (see Figure 2h). In linear DBA/2J mice, there was no effect of amino acid additions on the weight indexes of organs and adipose tissue.

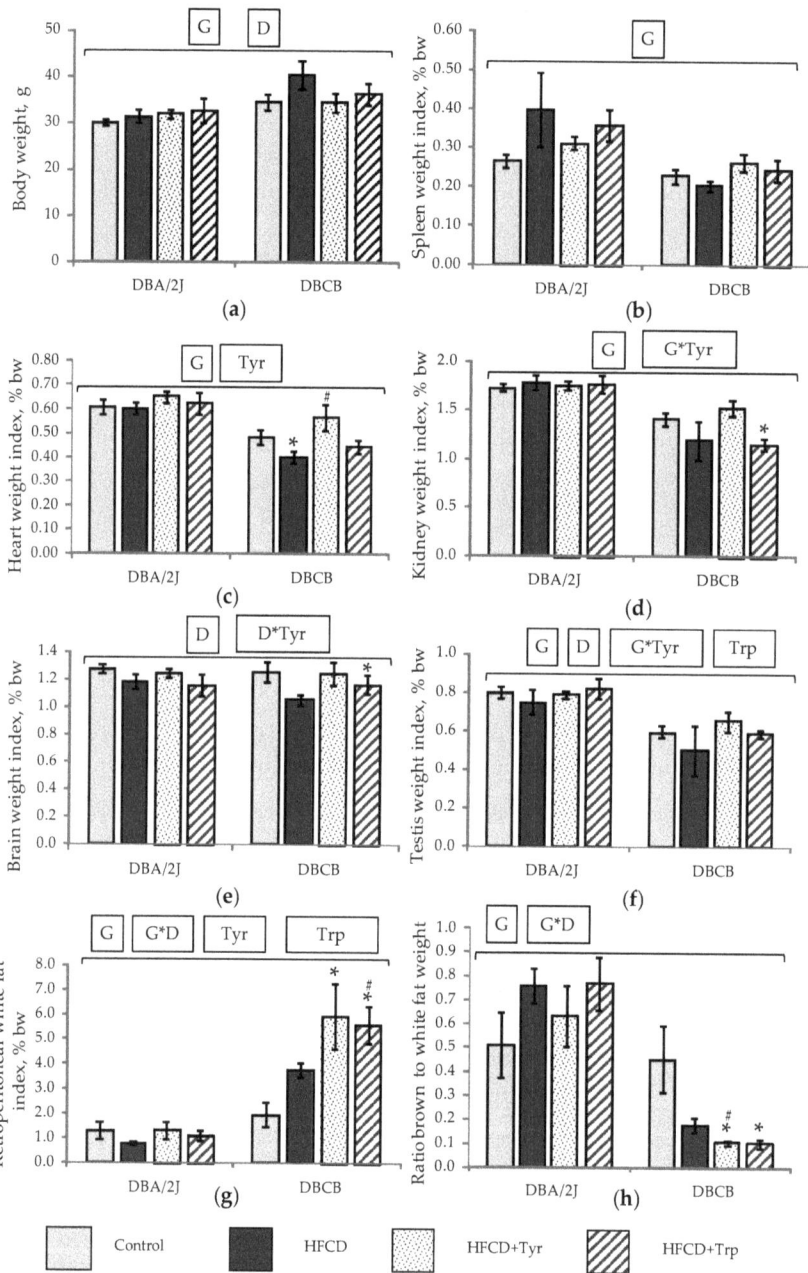

Figure 2. Integral indicators (M ± S.E.M.) of mice at the end of the experiment on the 66th day: (**a**) Bw; relative weight index of organs and tissues: (**b**) Spleen, (**c**) Heart, (**d**) Kidneys, (**e**) Brain, (**f**) Gonads, (**g**) Retroperitoneal white fat; (**h**) The ratio of the weight of interscapular brown fat to the retroperitoneal white fat. * The difference with the control group is significant; # The difference with the HFCD-only fed group is significant, $p < 0.05$, Mann–Whitney U-test. Horizontal bracket—distribution is non-uniform ($p < 0.05$, ANOVA) by the factors genotype (G), diet (D), tyrosine (Tyr), tryptophan, (Trp) and their combinations for the covered range of samples. The number of animals in groups—see Table 1.

Table 1. Biochemical parameters of the blood plasma of mice fed experimental diets, M ± S.E.M *.

Genotype	DBA/2J				DBCB				Factor ***
Group No	1	2	3	4	5	6	7	8	
Diet	Control	HFCD	HFCD + Tyr	HRCD + Trp	Control	HFCD	HFCD + Tyr	HRCD + Trp	
Number of Animals	8	7	8	8	7	4	7	7	
Glucose, mmol/L	12.8 ± 0.7 [2-8]*	16.2 ± 0.9 [1]	17.8 ± 0.8 [1]	19.8 ± 1.9 [1]	20.6 ± 1.8 [1]	22.1 ± 2.5 [1]	20.2 ± 1.8 [1]	20.2 ± 0.7 [1]	G
Protein, g/L	52.0 ± 2.5 [3,4]	52.9 ± 2.0 [3,4]	61.1 ± 1.2 [1,2]	59.8 ± 2.4 [1,2]	58.0 ± 3.6	56.1 ± 4.3	61.5 ± 1.2	60.1 ± 1.0	Tyr
Albumin, g/L	25.7 ± 1.2 [3,4]	25.0 ± 1.9 [3]	30.1 ± 0.7 [1,2]	29.6 ± 1.4 [1]	31.0 ± 1.2	29.7 ± 2.0	32.1 ± 0.6	32.7 ± 0.5	Tyr, Trp
Urea, mmol/L	5.15 ± 0.45 [3,4]	5.18 ± 0.34 [3,4]	7.27 ± 0.43 [1,2]	6.54 ± 0.32 [1,2]	6.63 ± 0.42	5.57 ± 0.32 [7,8]	7.03 ± 0.46 [6]	6.59 ± 0.22 [6]	G****, Tyr, Trp
Calcium, mmol/L	1.86 ± 0.09 [3,4]	1.86 ± 0.10 [3,4]	2.35 ± 0.12 [1,2]	2.34 ± 0.14 [1,2]	2.15 ± 0.06	1.95 ± 0.19 [8]	2.30 ± 0.08	2.26 ± 0.06 [6]	Tyr, Trp
Phosphorus, mmol/L	2.51 ± 0.19 [4]	2.95 ± 0.19	3.06 ± 0.26	3.72 ± 0.32 [1]	3.30 ± 0.20	3.03 ± 0.36	3.14 ± 0.23	3.37 ± 0.19	Trp
AlAT, U/mL	22.0 ± 3.8 [3]	27.5 ± 11.1	50.9 ± 13.4 [1]	35.3 ± 8.6	19.6 ± 4.9 [6]	62.8 ± 20.5 [5]	26.9 ± 6.2	47.7 ± 16.2	D, G × Tyr
AsAT, U/mL	220 ± 43	201 ± 39	205 ± 29	258 ± 66	180 ± 35	454 ± 253	191 ± 39	220 ± 38	G × D
AsAT/AlAT	11.2 ± 2.0 [3]	10.6 ± 2.5 [3]	5.0 ± 0.8 [1,2]	9.9 ± 2.6	12.5 ± 5.1	10.1 ± 4.6	5.1 ± 0.3	12.3 ± 4.7	Tyr
KPK kU/mL	5.7 ± 1.7	5.9 ± 1.6	5.9 ± 0.9	7.1 ± 1.9	5.1 ± 0.7	10.7 ± 3.7 [7]	3.6 ± 1.0 [6]	5.6 ± 1.4	Tyr
Lipase U/mL	129 ± 13	110 ± 4 [4]	120 ± 3	138 ± 9 [2]	119 ± 15	111 ± 7 [8]	148 ± 20	140 ± 8 [6]	Trp
GPO mmol/min/mg protein **	0.71 ± 0.02 [2,3,4]	0.39 ± 0.03 [1]	0.40 ± 0.03 [1]	0.46 ± 0.01 [1]	0.68 ± 0.03 [6,7]	0.44 ± 0.03 [5,8]	0.49 ± 0.06 [5]	0.60 ± 0.02 [6]	G, D, Trp

* Superscripts are group numbers, the difference with which is pairwise significant, $p < 0.05$, non-parametric Mann–Whitney test. ** In erythrocytes, the number of animals in groups of 6. *** Factors significantly influencing the distribution of indicator presented between groups of animals, $p < 0.05$, ANOVA, such as genotype of the animals (G), diet type (D), and amino acid added (Tyr, Trp). **** Only in experiment with Trp. See all abbreviations in Materials and Methods section.

2.3. Muscle Tone, Search, and Anxiety-Like Behavioral Activity

Additional consumption of the amino acids tyrosine and tryptophan can lead to an increase in their transport to the brain, which may result in changes in the synthesis and metabolism of the neurotransmitter amines dopamine and serotonin as well as their derivatives. This is reflected by changes in the level of neuromotor activity, eating behavior, and anxiety. The definition of muscle-compressive force (Figure 3) showed that, in DBA/2J mice, even at the first test (the 3rd day of feeding with rations), the addition of both amino acids leads to an increase in muscle tone compared to the control and HFCD-fed group. Further, this effect is preserved only in the form of an insignificant trend. In DBCD tetrahybrid mice, when testing for the first time, the muscle-compressive force did not differ between the groups, while at the second test, it decreased significantly with the use of HFCD. Both amino acid supplements return this indicator to the control level.

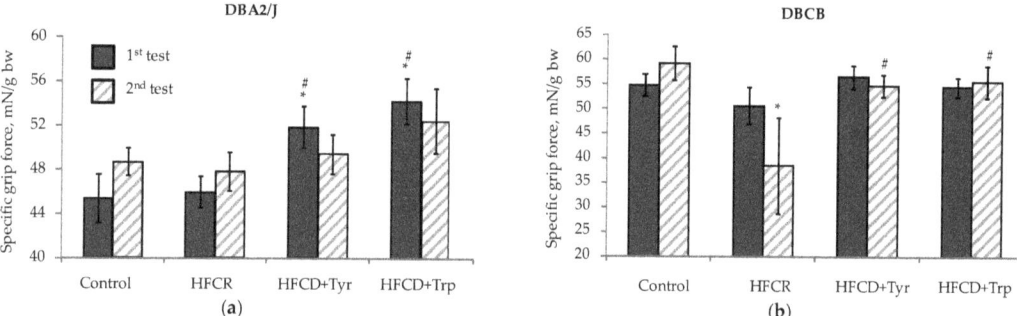

Figure 3. Muscle-contraction strength (grip force, M ± S.E.M.): (**a**) DBA/2J mice; (**b**) Tetrahybrid DBCB mice. * The difference with the control group is significant; # The difference with the HFCD-only group is significant, Mann–Whitney U-test.

Study of locomotor/search activity in elevated plus maze test (EPM) (Figure 4) showed no effect of HFCD and amino acid additives on these indicators in DBA/2J mice (Figure 4a,c,e) except for the general decrease in the distance traveled in open arms (OA) (Figure 4a) for all animals that received HFCD. In DBCB mice, locomotor activity in the OA decreased significantly with repeated testing (Figure 4b), with the most pronounced change in the control group. The addition of Trp in these mice led to a significant decrease in the average movement speed (Figure 4d) and maximum speed in the OA (Figure 4f) during the first testing. When comparing two animal lines, attention is generally paid to the lower mobility of tetrahybrid mice compared to DBA/2J, which is confirmed by the data of factor analysis ($p < 0.05$ by the genotype factor for speed in the OA at 2nd test).

The anxiety-like behavior in mice, detected in EPM (Figure 5), increased in the 2nd testing according to parameters of the center entrance latency, time spent in OA, and the ratio of closed arms (CA) to OA times. These findings coincide with the previously obtained data from C57Black/6J mice, which represent a parental line for DBCB tetrahybrid [15]. Anxiety increase most clearly manifested in a decrease in the latency time to the first exit to the maze center, the time spent in the OA in DBA/2J mice, and an increase in the CA/OA ratio in DBCB tetrahybrid ($p < 0.05$, ANOVA, by the test factor). At the same time, the addition of Tyr to the diet caused a significant decrease in the OA time in mice of both strains during pairwise comparison (according to the Wilcoxon criterion). The addition of Trp led to a significant increase of anxiety in terms of CA/OA ratio only in tetrahybrid mice but not in DBA/2J mice in the pairwise comparison between the 1st and the 2nd test.

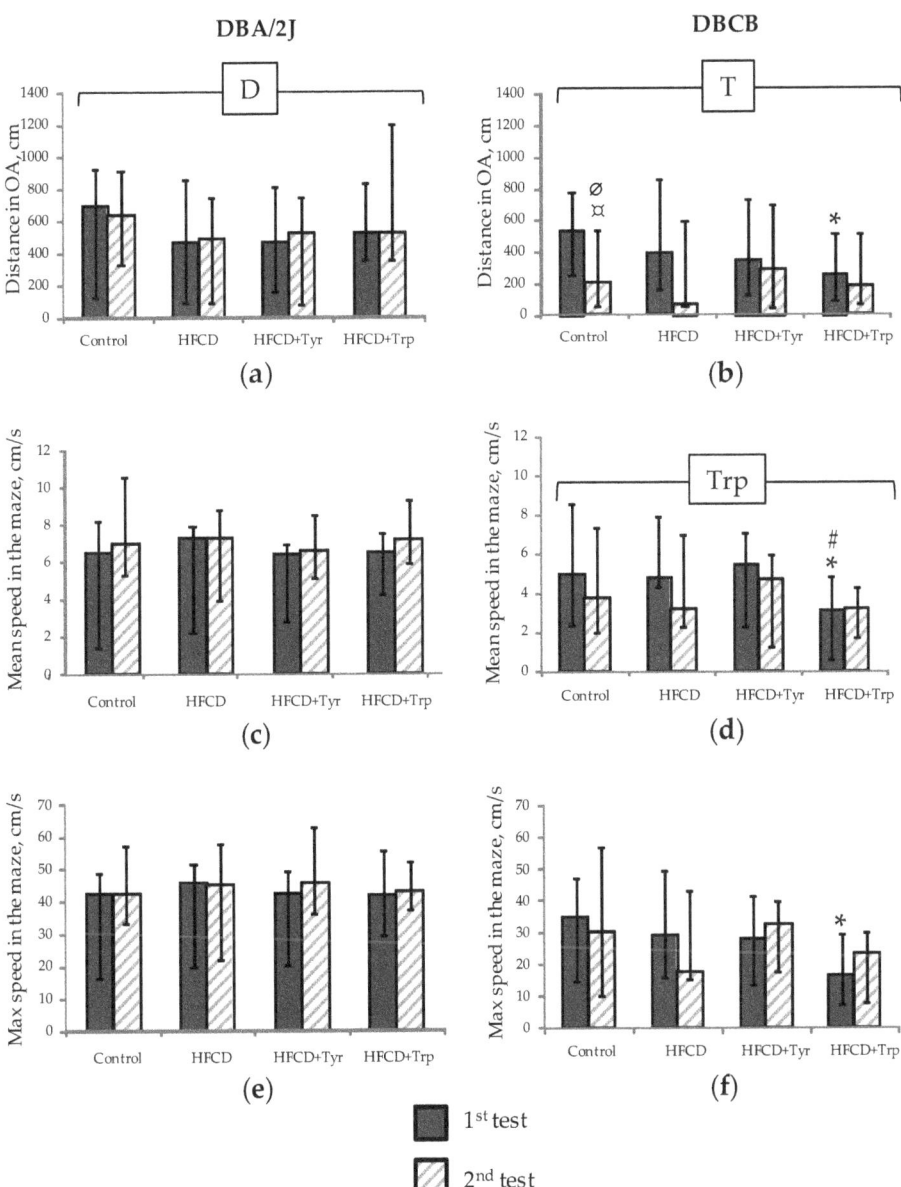

Figure 4. Mobility/locomotor activity indicators (median; change intervals) in the EPM test: (**a**,**b**) Distance traveled in the OA, cm; (**c**,**d**) Average speed in the maze, cm/s; (**e**,**f**) Maximum speed in the OA, cm/s. DBA/2J mice (**a**,**c**,**e**); Tetrahybrid DBCB mice (**b**,**d**,**f**). * The difference with the control group is significant, # the difference with the HFCD-only group is significant, ⌀ the difference with DBA/2J mice is significant, $p < 0.05$, Mann–Whitney U-test; ¤ difference between 1st and 2nd test is significant, $p < 0.05$, Wilcoxon test for pairwise related values. Horizontal bracket—distribution is non-uniform ($p < 0.05$, ANOVA) by factors diet (D), test (T), and tryptophan (Trp) for the covered sample range. The number of animals in groups—see Table 1.

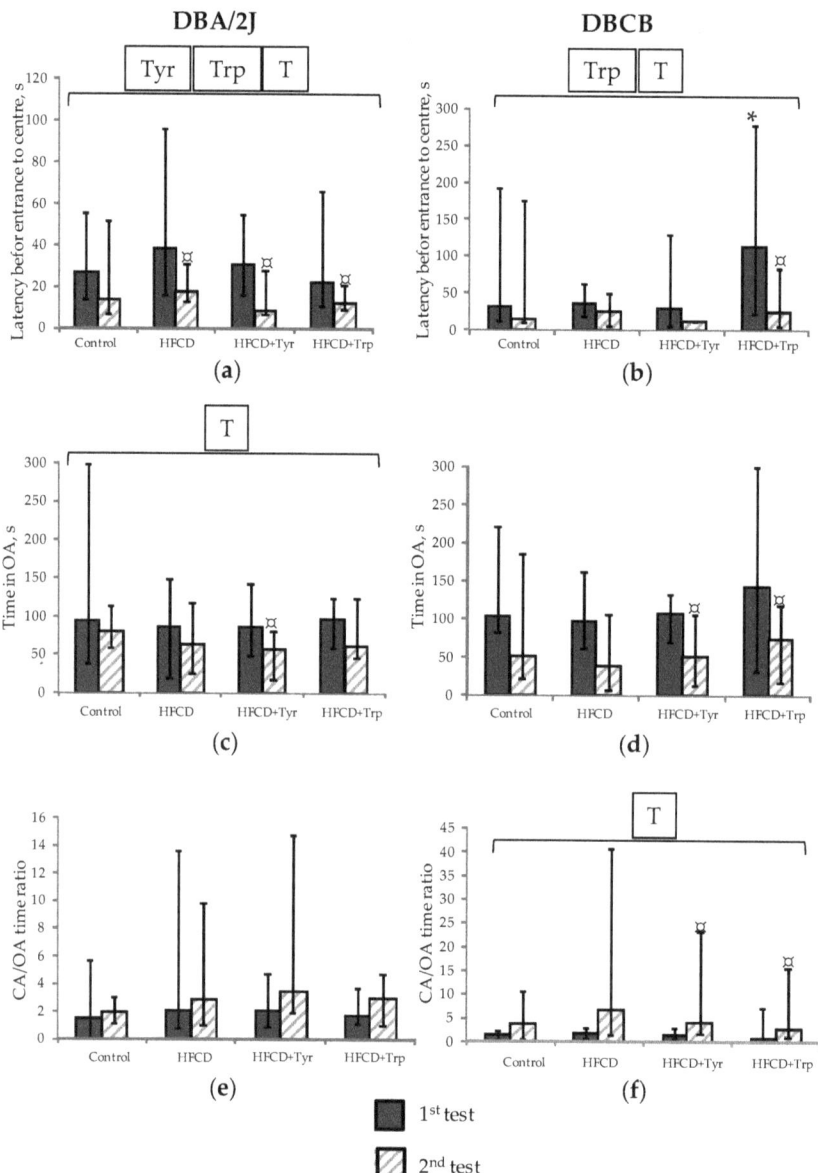

Figure 5. Indicators of anxiety (median; change intervals) of mice in the EPM test: (**a**,**b**) Latency before the first exit to the center of the maze, s; (**c**,**d**) Time in OA, s; (**e**,**f**) The ratio of CA/OA times (dimensionless). DBA/2J mice (**a**,**c**,**e**); Mice DBCB tetrahybrid (**b**,**d**,**f**). * The difference with the control group is significant; # The difference with the group fed HFCD only is significant, $p < 0.05$, Mann–Whitney U-test; ¤ The difference between the 1st and the 2nd tests is significant, $p < 0.05$, Wilcoxon test for pairwise related values. Horizontal bracket—distribution is non-uniform ($p < 0.05$, ANOVA) by the factors test (T), tyrosine (Tyr), and tryptophan (Trp) for the covered range of samples. The number of animals in groups—see Table 1.

2.4. Liver Histology

One of the consequences of the consumption of HFCD in rodents may be the development of fatty degeneration of the liver. Microscopic examination of the liver (Figure 6) revealed a generally orthodox structure of the tissue in DBA2/J mice fed both the control diet and HFCD. The addition of Tyr increased fat accumulation in the perinuclear compartment of hepatocytes in these animals. In contrast, Trp consumption corresponded to the almost complete disappearance of fatty inclusions in the liver and the development of tissue edema with the closure of most of the lumen of the bile ducts. Moreover, the addition of Trp has contributed to the appearance of a large number of degrading micronucleus cells. In the tetrahybrids, signs of increased fat accumulation in the liver were observed even in the control group. At the same time, HFCD in tetrahybrids led to massive fatty degeneration of hepatocytes with the formation of rounded fatty vacuoles lacking internal structure. The addition of Tyr to HFCD seemed to weaken these changes, and the addition of Trp, on the contrary, aggravated them: along with the vacuoles, many cells appeared with deposition of fat in the perinuclear district.

2.5. Blood Biochemical Indices

The development of diet-induced obesity is often accompanied with signs of dyslipidemia, changes in nitrogen exchange, and protein catabolism. Analysis of the biochemical blood plasma parameters showed that HFCD consumption led to an increase in glucose level (Table 1) in DBA/2J mice. Tetrahybrid mice were characterized by an initially elevated glycemia already in the control group ($p < 0.001$, ANOVA, by the genotype factor). The addition of amino acids did not cause further changes in this indicator. According to the factor analysis, the addition of Tyr and, to a lesser extent, Trp to the diets led to a significant increase in indexes of the protein metabolism, such as levels of total protein, albumin, and urea ($p < 0.05$, ANOVA). In DBA/2J mice, these effects were also manifested in paired comparisons of groups. The consumption of Tyr and Trp was accompanied by an increase in the concentration of total calcium and in DBA/2J mice, also phosphorus (in the case of Trp). Under the action of Tyr, an increase in alanine aminotransferase (AlAT) activity was noted in DBA/2J mice, whereas in tetrahybrids, it was, conversely, decreased in comparison with the group that consumed HFCD without additives. Any effects of Trp on this indicator were not observed. Aspartate aminotransferase (AsAT) activity was increased under the action of HFCD only in DBCB mice (ANOVA, $p < 0.05$ by the genotype*diet factor in the tyrosine experiment). As shown by the calculation of the AsAT/AlAT activity ratio (De Ritis ratio), this indicator was significantly reduced in mice of both strains fed with Tyr. Trp led to a significant decrease in the activity of creatine phosphokinase (CPK) in DBCB mice consuming HFCD and to a significant increase in the lipolytic activity in animals of both strains. In erythrocytes of mice treated with HFCD, there was a significant decrease in glutathione peroxidase (GPO) activity, and the addition of Trp led to the normalization of this indicator in the tetrahybrid mice.

2.6. Blood Cytokines and Adipokines Levels

The development of diet-induced obesity is a complex process involving local (in adipose tissue) and systemic inflammation, as evidenced by circulating levels of adipokines and cytokines. The consumption of HFCD led to a significant increase in mice insulin levels (Figure 7a); $p < 0.05$, ANOVA by diet (D) factor. At the same time, only in DBA/2J mice, this was combined with a significant increase in glucose levels (see Table 1), wherein DBCB mice were initially hyperglycemic, as was mentioned above (see Section 2.4). This could be a sign that glycemic control in tetrahybrid mice has been compromised due to their genetic background. No effect of the studied amino acids on the level of insulin and glycemia was found in mice of both genotypes.

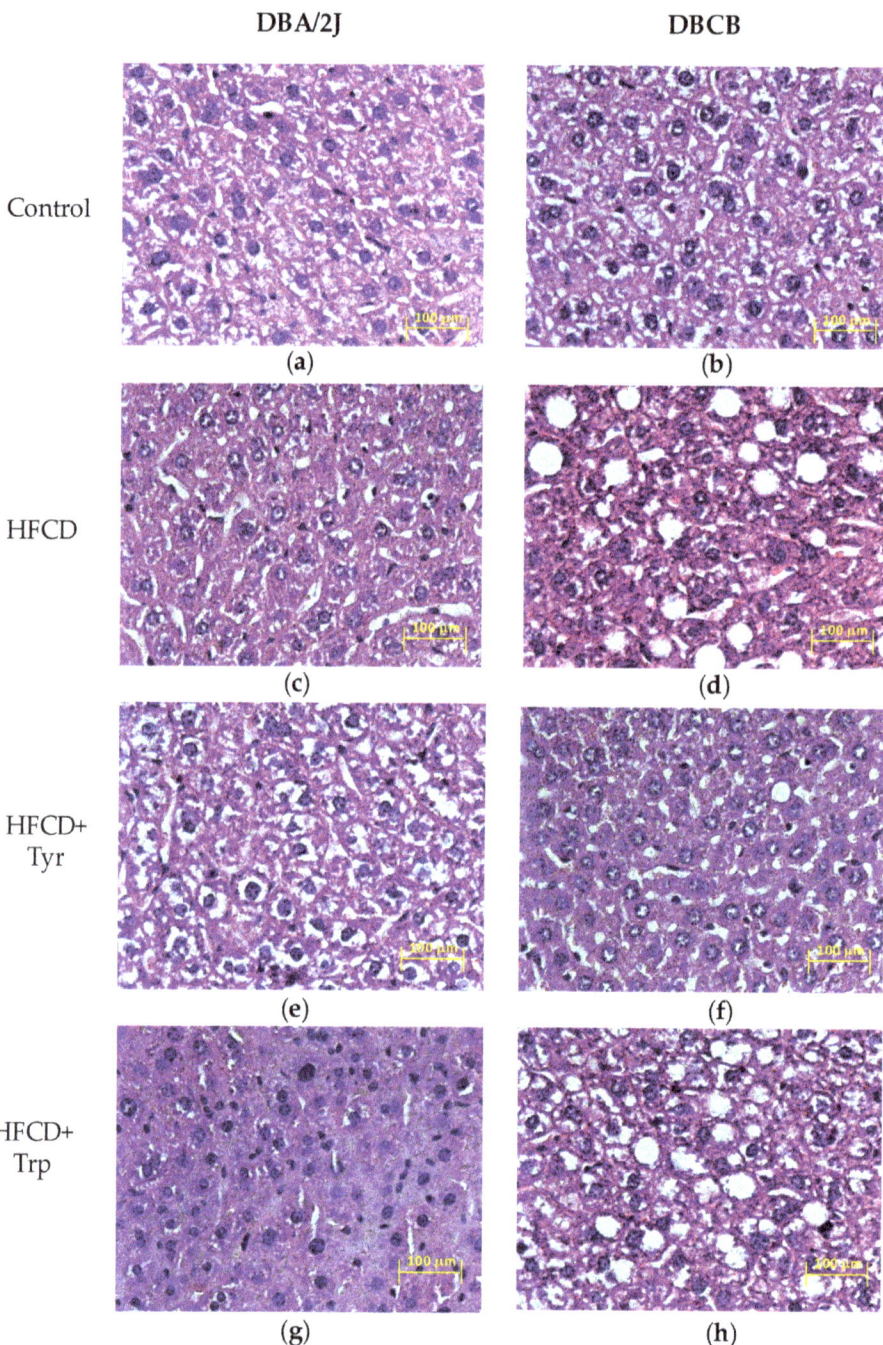

Figure 6. Representative light-optical micrographs of liver sections of mice fed: (**a**,**b**) Control diet; (**c**,**d**) HFCD; (**e**,**f**) HFCD with Tyr; (**g**,**h**) HFCD with Trp. DBA/2J mice (**a**,**c**,**e**,**g**); DBCB tetrahybrid mice (**b**,**d**,**f**,**h**). Stained with hematoxylin-eosin, magnification ×400.

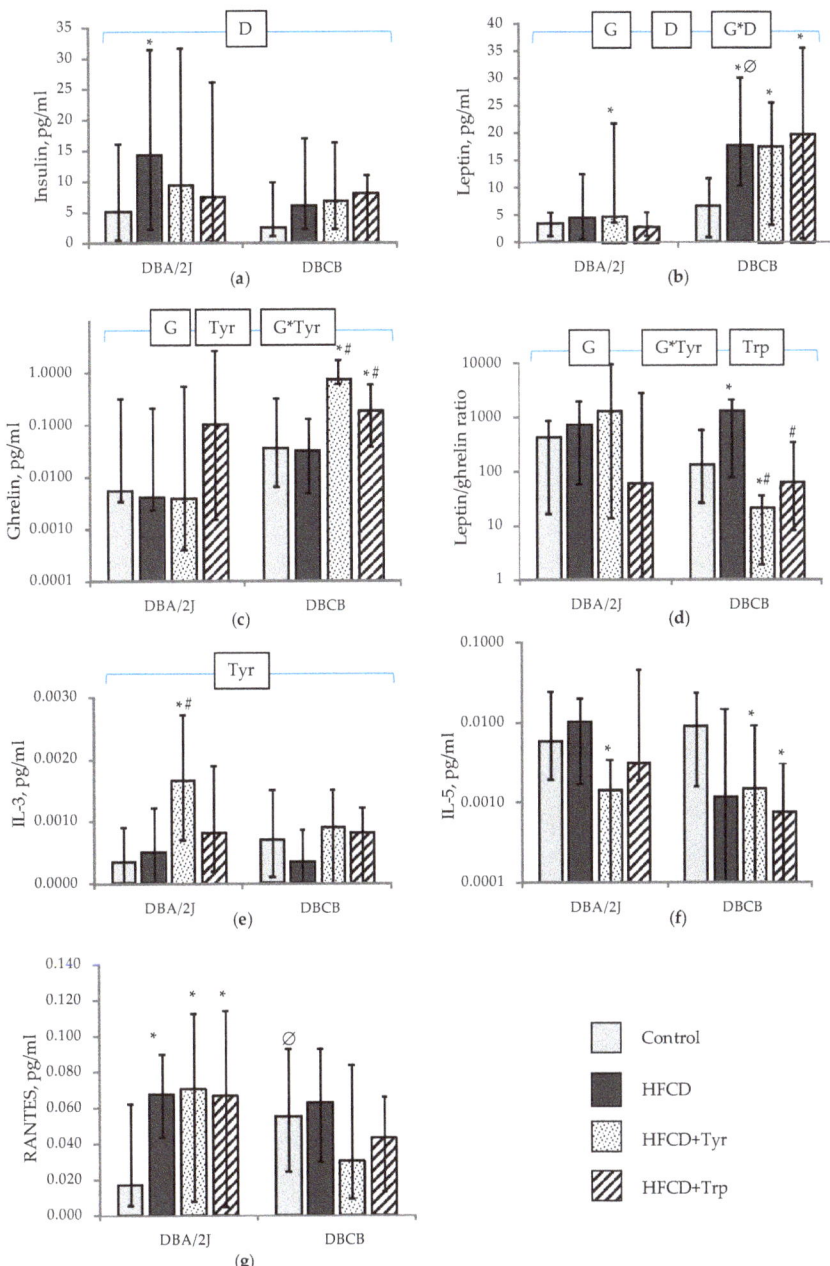

Figure 7. Levels of adipokines and cytokines (median; change intervals) in the blood plasma of mice when removed from the experiment on the 66th day: (**a**) Insulin; (**b**) Leptin; (**c**) Ghrelin; (**d**) Leptin/ghrelin ratio; (**e**) IL-3; (**f**) IL-5; (**g**) RANTES. * The difference with the control group is significant; # The difference with HFCD-only fed group is significant; ∅ The difference with DBA/2J mice is significant, $p < 0.05$, Mann–Whitney U-test. Horizontal bracket—distribution is non-uniform ($p < 0.05$, ANOVA) by the factors genotype (G), diet (D), tyrosine (Tyr), and tryptophan (Trp) and their combinations for the covered range of samples. The number of animals in groups—see Table 1.

The data on (Figure 7b) showed that all mice fed with HFCD showed a significant increase in the levels of circulating leptin ($p < 0.05$, ANOVA, by the diet factor), and the level of the latter was significantly higher in DBCB compared to DBA/2J mice ($p < 0.05$, ANOVA, by the genotype factor). The level of ghrelin (Figure 7c) increased in DBCB mice treated with Tyr and Trp supplements, and, consequently, the leptin/ghrelin ratio (Figure 7d) was the lowest in these animals. No significant effect of amino acids on the concentrations of IL-10, IL-12P (70), IL-17A, and MCP-1 was detected on the background of HFCD consumption (data not shown). At the same time, Tyr caused a significant increase in the IL-3 content (Figure 7e) and a decrease in IL-5 in DBA/2J mice (Figure 7f); in DBCB mice, both amino acids potentiated the effect of reducing the IL-5 level caused by HFCD. The level of RANTES (Figure 7g) was uniformly increased in all groups of DBA/2J mice receiving HFCD; in tetrahybrid mice, RANTES was significantly increased against DBA/2J when comparing control groups and did not significantly respond to the dietary interventions used.

3. Discussion

As a result of the research, differences were revealed in the effects of Tyr and Trp supplements in mice of two genotypes differing in the response to excess fat and carbohydrate intake (HFCD). DBCB tetrahybrid mice reacted to HFCD consumption differently than DBA/2J mice, which are one of their parent lines, due to the greater allelic diversity of the genome. In particular, HFCD possessing the increased energy density caused a decrease in palatability of food in tetrahybrid mice that led to the alignment of their specific energy consumption with the control group. At the same time, these mice had reduced mobility in the EPM test compared to DBA/2J and responded to the HFCD consumption by increasing fat deposition in the depot of white adipose tissue and hepatocytes (up to the development of balloon dystrophy). This also corresponded to the development of hyperleptinemia in these animals. A decrease in the relative proportion of brown fat in tetrahybrid mice indirectly indicates a possible inhibition of thermogenesis processes in them during the use of HFCD, and, therefore, a decrease in energy consumption during the consumption of a hypercaloric diet and, as a result, an increase in the relative weight of white fat. In contrast, DBA/2J mice that received HFCD were characterized by increased energy consumption but did not respond to a significant increase in bw and white fat, which may be a consequence of their increased energy spending. This is indicated by the greater relative proportion of thermogenic brown fat in total fat depots and increased activity in the EPM test. For DBA/2J mice, in contrast to DBCB, there was also a pronounced response of insulin and RANTES to the use of HFCD, which probably indicates a compensatory activation of fat catabolism processes in adipose tissue [16].

The study of the influence of tyrosine and tryptophan supplements on the vital indicators in mice was carried out taking into account the possible mechanisms mediated by biogenic amines—dopamine, catecholamines (in the case of tyrosine), and serotonin (in the case of tryptophan) both in the brain of animals and in peripheral organs. When assessing the central effects of amino acids, we used indicators of anxiety and search activity in the EPM test as well as muscle tone, measured in the grip force test. The latter indicator also may be indirectly related to the intensity of energy metabolism in muscle tissues, as is known from the literature [17]. Possible peripheral effects, allowing consideration of the influence of amino acids on the development of diet-induced obesity, were investigated using the assessment of integral, biochemical, and morphological parameters on the development of systemic inflammation accompanying the development of obesity using the analysis of adipokines and cytokines.

Taking into account the data received, the addition of Tyr did not have a significant effect in DBA/2J mice on feeding behavior (feed consumption) and accumulation of white and brown fat but contributed to an increase in muscle tone and potentiated anxiety enhancement in EPM. The reduction of the AsAT/AlAT ratio (De-Ritis [18]) under the influence of Tyr can be considered as a sign of a decrease in the intensity of protein catabolism when the transamination is reduced of aspartate with the formation of oxaloacetic acid (OAA) [19]. This is consistent with the increase in total protein levels in mice of both genotypes and albumin in DBA/2J mice receiving Tyr. An increase in urea levels in the same animals can be explained by increased deamination of an excess of aspartate in the urea cycle due to aspartate excess forming as a result of inhibition of its transamination into OAA.

The effect of Tyr on the liver was different in DBA/2J mice and tetrahybrids. If, in the first strain, the supplement led to an increase in AlAT activity and increased fat accumulation, in the second one, these changes had the opposite direction. Only in DBA/2J was a marked increase noticed in the level of IL-3 in combination with a decrease in IL-5. The discrepancy of these immune parameters, according to the literature, can occur at the level of a change in the ratio of specific receptors for IL-3 and IL-5 on eosinophils, which can be one of the signs of eosinophilic liver inflammation [20,21].

In DBCB mice, Tyr supplementation significantly prevented the development of fatty degeneration of hepatocytes caused by HFCD and contributed to the normalization of AlAT levels. This corresponded to the minimum value of the leptin/ghrelin ratio, which may indicate a weakening of the processes of lipogenesis [22]. Besides, Tyr exerted an influence on the state of the heart muscle in these animals, contributing to the normalization of organ mass and a decrease in CPK activity in the circulation.

The Trp supplementation did not affect the energy consumption of animals of both lines but increased muscle-contraction force and activated the processes of mineral metabolism, causing an increase in the levels of calcium and phosphorus in the blood plasma. The effect of Trp on the liver was ambiguous. In DBA/2J mice, it manifested, on the one hand, in the reduction of fat deposits and, on the other hand, in the development of edema and the appearance of micronucleus cells. In DBCB mice, Trp caused increased fat accumulation in the liver.

Thus, the features of the effect of Tyr supplementation on mice of the two lines used had both similarities and differences. In DBA/2J and DBCB mice, this supplement caused an increase in muscle tone, an increase in protein anabolic processes, and potentiated the effect of anxiety increase with the age of animals. Besides, in DBA/2J mice resistant to the development of diet-induced obesity, Tyr increased the toxic effect and fat accumulation in the liver, whereas, in spontaneously hyperglycemic and obesity-prone DBCB mice, these effects were opposite. These data are consistent with our previous results of a comparative assessment of integral, biochemical, and physiological parameters in DAT-KO knockout rats with genetic dopamine reuptake disorder and wild-type animals [23] and indicate, apparently, the ability of dietary Tyr to modulate dopamine exchange in the central nervous system.

Peculiarities of the action of the Trp supplement common to mice of both genotypes consisted in a muscle tone increase and potentiation of anxiety increase with the age of animals. The latter is in contradiction with a sedative effect postulated for Trp as a precursor of serotonin [11]. Trp influenced the mobility in the EPM test only in DBCB but not in DBA/2J mice. The differences between the two lines were manifested in the greater effect of Trp on the protein metabolism in DBA/2J mice and the increased accumulation of fat in the liver against the background of an increase in lipolytic activity in DBCB mice. Only in the latter, Trp showed signs of an antioxidant effect in terms of GPO activity and a partial cardioprotective effect (decrease in mortality, decrease in CPK activity, normalization of heart mass). The reasons for the Trp effects on behavioral reactions, metabolic processes, and toxic liver damage, which are complex and unequal for different mouse strains, should be sought apparently in the interaction of this amino acid with the intestinal microbiome of the animals. As it is known, indole and indolyl-3-propionic acid are among the main microbial metabolites of Trp [24]. The first one, entering the liver, is transformed there by the action of microsomal monooxygenases and sulfates into indoxyl sulfate, which has a toxic and prooxidant action [25]. The second metabolite, on the contrary, is considered as a trap of free radicals and can have a different, organ-protective effect under oxidative stress [26]. The ratio in the activity of different microbiota populations, alternatively synthesizing these metabolites, can vary significantly in mice of different strains [27], which partially explains the identified interlinear differences.

4. Materials and Methods

4.1. Animals and Experimental Design

The amino acids Tyr and Trp were purchased from Wirud Co (Bad Homburg, Germany), with a purity of 99.6% according to the manufacturer. The degree of amino acid purity was additionally checked by HPLC according to [28].

Male mice of inbred line DBA/2J, which were obtained from the nursery Stolbovaya (Stolbovaya, Moscow region, Russia), and a complex hybrid of the 2nd generation (referred to here as tetrahybrid) DBCB bred independently by crossing 4 different inbred lines of mice (DBA/2J, BALB/c, CBA/lac, and C57Black/6J) were used. The age of the animals at the beginning of the experiment was 8–9 weeks; the initial body weight (bw) was 24 ± 2 g. The method for breeding hybrid mice was presented earlier [15]. The work with animals was carried out following the rules of good laboratory practice [29] and in accordance with the Order of the Ministry of Health of the Russian Federation No. 199n of 1 April 2016, "On the approval of rules of good laboratory practice". The design of the experiment was approved by the Ethics Committee of the Federal Research Centre of Nutrition and Biotechnology (protocol No. 4 of 20 April 2017).

DBA/2J mice and DBCB tetrahybrid mice were divided into 4 groups of 8 individuals not differing in the mean initial bw within each genotype ($p > 0.05$; ANOVA). Mice of the 1st (control) groups received a balanced, semi-synthetic diet for rodents corresponding to AIN93M [30], with a content of 10% fat by weight, and drinking water purified by reverse osmosis; the 2nd groups received a high-carbohydrate, high-fat diet (HFCD). Part of the starch in HFCD was replaced with fat (a mixture of 1:1 refined vegetable oil and pork lard) to a total fat content equal to 30% by weight of the dry substances of the diet, and drinking water was replaced with a 20% aqueous solution fructose. The 3rd and 4th groups received the same HFCD diet with the addition of Tyr and Trp in the estimated doses of 1250 and 250 mg/kg of bw, respectively. The indicated doses corresponded to a 2-fold increase of both amino acids that were received from the dietary casein and apparently had no large influence on diet palatability, meaning the bitter taste of these amino acids. Mice were kept four individuals to a cage at a temperature of 21 ± 1 °C and a light/dark mode of 12/12 h. The animals were fed diets for 65 days; moreover, food and liquid consumption were determined daily, bw was measured weekly, and the appearance, activity, and behavior were monitored.

4.2. Assessment of Behavioral Responses, Anxiety, and Muscle Tone Indices

Behavioral reactions (locomotor activity and anxiety levels) were studied in the elevated plus maze (EPM) installation (Panlab Harvard Apparatus company, Barcelona, Spain) on the 8th and 59th days of the experiment by the techniques described earlier [15].

4.3. Assessment of Integral and Biochemical Indices

Animals were removed from the experiment on the 66th day by exsanguination from the inferior vena cava under ether anesthesia. Blood was collected in measuring tubes with 0.3 mL of 1% heparin, fixing the dilution of each sample. The liver, spleen, heart, thymus, interscapular brown, and retroperitoneal white fat were collected and weighed on a laboratory balance with an accuracy of ± 0.01 g. Liver tissue samples about 5 mm in diameter were fixed in a 3.7% formaldehyde solution in 0.1M sodium phosphate buffer pH 7.00 ± 0.05, dehydrated in alcohols of increasing concentration, impregnated with xylene, and filled in with a homogenized Histomix® paraffin medium (BioVitrum, St. Petersburg, Russia). Paraffin sections 3–4 μm thick were made on a Microm HM355s microtome (Leica microsystems, Wetzlar, Germany), stained with hematoxylin and eosin using a standard technique, and examined in an AxioImager Z1 microscope (Zeiss, Oberkochen, Germany) with a digital camera at magnification ×400. Plasma biochemical parameters were determined on a Konelab 20i biochemical analyzer (Thermo Fisher Scientific, Waltham, MA, USA). Standard methods such as kinetic methods recommended by the International Federation of Clinical Chemistry and Laboratory Medicine for alanine aminotransferase (AlAT), aspartate aminotransferase (AsAT), creatine phosphokinase (CPK), and total lipase activity were used. Glucoseoxidase enzymatic method for glucose, Trinder reaction with glycerol-3-phosphate oxidase for triglycerides, UV kinetic urease method for urea, Biuret test for proteins, a potentiometric method with ion-selective electrode for calcium, and spectrophotometric method for phosphorus were used also. The erythrocyte glutathione peroxidase (GPO) activity was determined by a direct spectrophotometric endpoint method using reduced glutathione as a substrate with Ellmann's reagent in the presence of sodium azide.

4.4. Cytokines and Adipokines Analysis

Determination of cytokines IL-3, IL-5, IL-10, IL-12p70, IL-17A, MCP-1, and RANTES and peptide hormones—ghrelin, insulin, and leptin—in the plasma of mice was performed by multiplex immunoassay with the basic set of Bio Bio-Plex Pro™ Reagent Kit V supplemented with the following reagents: Bio-Plex Pro™ Mouse Cytokine IL-3 Set, Bio-Plex Pro™ Mouse Cytokine IL-10 Set, Bio-Plex Pro™ Mouse Cytokine IL-12p70 Set, Bio-Plex Pro™ Mouse Cytokine RANTES Set, Bio-Plex Pro™ Mouse Diabetes Set, Bio-Plex Pro™ Mouse Diabetes Insulin Set, and Bio-Plex Pro™ Mouse Diabetes Leptin Set. All are manufactured by Bio-Rad Laboratories, Inc. (Hercules, CA, USA). Studies were run on a Luminex 200 multiplex analyzer (Luminex Corporation, Austin, TX, USA) using xMAP technology using Luminex xPONENT Version 3.1 software.

4.5. Statistical Analysis

Data were processed statistically using the ANOVA multivariate analysis of variance based on dietary intake factors (standard diet/HFCD), animal genotype, and the presence of Tyr and Trp in the diet; Wilcoxon–Mann–Whitney nonparametric test was used for pairwise comparisons; Student's *t*-test was used for pairwise related quantities for daily energy intake; the SPSS 20.0 (IBM Corp., Armonk, NY, USA): software package was used. Differences were considered significant when the probability of accepting the null hypothesis was less than 0.05.

5. Conclusions

Adding amino acids Tyr and Trp to a high-carbohydrate, high-fat diet has a modulating effect on behavioral reactions, signs characterizing the development of alimentary obesity and liver damage, nitrogen metabolism, and the ratio of leptin and ghrelin production, which is most pronounced in DBCB mice that are sensitive to the harmful effects of this diet.

The largest differences in the response of mice of the two lines to the consumption of amino acid supplements related to the effects on the level of locomotor activity and anxiety, liver tissue morphology, and cytokine production. This indicates genotypically determined features of the exchange of aromatic amino acids in mice of two strains, which leads, apparently, to differences in the synthesis and degradation of regulatory biogenic amines, including dopamine and serotonin, as well as some other biologically active and potentially toxic metabolites. Similar genotypic differences that take place in the human population probably present a source of ambiguity in the results of amino acid, as well as other biologically active substances used as supplements in therapeutic and preventive nutrition. Thus, the data obtained indicate the need to apply several qualitatively different in vivo models of obesity and metabolic syndrome for preclinical evaluation of the efficiency of biologically active substances in personalized therapeutic nutrition in patients with disorders related to excess nutrition.

Author Contributions: Conceptualization, I.V.G., S.A.A., V.A.S., and D.B.N.; methodology, V.A.S., I.V.G., N.V.T., and S.A.A.; formal analysis, I.V.G., V.A.S., N.V.T., and S.A.A.; investigation, V.A.S., N.V.T., S.A.A., A.A.S., K.V.M., A.N.T., N.A.R., and I.V.G.; resources, V.A.S., N.V.T., S.A.A., and A.A.S.; data curation, I.V.G. and D.B.N.; writing—original draft preparation, I.V.G., V.A.S., and S.A.A.; writing—review and editing, D.B.N.; visualization, I.V.G. and V.A.S.; supervision, I.V.G. and D.B.N.; project administration, I.V.G.; funding acquisition, I.V.G. All authors have read and agreed to the published version of the manuscript.

Funding: The work was supported by a grant from the Russian Scientific Foundation No. 17-16-01043 "Search for effector units of metabolism regulated by alimentary factors in obesity for the development of innovative specialized food".

Institutional Review Board Statement: The study was conducted according to the guidelines of the Declaration of Helsinki and approved by the Institutional Ethics Committee of the Federal Research Centre of Nutrition and Biotechnology (protocol code # 4, 20 April 2017).

Informed Consent Statement: Not applicable.

Data Availability Statement: The datasets generated during the current study are available from the corresponding author on reasonable request by email.

Acknowledgments: The authors are grateful to Jorge Selada Soto for the biochemical data of blood plasma and to Elena V. Rylina for analyzing the purity of amino acids by HPLC.

Conflicts of Interest: The authors declare no conflict of interest.

References

1. Cheung, W.W.; Mao, P. Recent advances in obesity: Genetics and beyond. *ISRN Endocrinol.* **2012**, *2012*, 536905. [CrossRef]
2. Swinburn, B.A.; Sacks, G.; Hall, K.D.; McPherson, K.; Finegood, D.T.; Moodie, M.L.; Gortmaker, S.L. The global obesity pandemic: Shaped by global drivers and local environments. *Lancet* **2011**, *378*, 804–814. [CrossRef]
3. Bojanowska, E.; Ciosek, J. Can We Selectively Reduce Appetite for Energy-Dense Foods? An Overview of Pharmacological Strategies for Modification of Food Preference Behavior. *Curr. Neuropharmacol.* **2016**, *14*, 118–142. [CrossRef] [PubMed]
4. Rønnestad, I.; Gomes, A.S.; Murashita, K.; Angotzi, R.; Jönsson, E.; Volkoff, H. Appetite-Controlling Endocrine Systems in Teleosts. *Front. Endocrinol.* **2017**, *8*, 73. [CrossRef] [PubMed]
5. Kenny, P.J. Reward mechanisms in obesity: New insights and future directions. *Neuron* **2011**, *69*, 664–679. [CrossRef] [PubMed]
6. Volkow, N.D.; Wise, R.A. How can drug addiction help us understand obesity? *Nat. Neurosci.* **2005**, *8*, 555–560. [CrossRef]
7. Vucetic, Z.; Carlin, J.L.; Totoki, K.; Reyes, T.M. Epigenetic dysregulation of the dopamine system in diet-induced obesity. *J. Neurochem.* **2012**, *120*, 891–898. [CrossRef]
8. Burke, L.K.; Heisler, L.K. 5-hydroxytryptamine medications for the treatment of obesity. *J. Neuroendocrinol.* **2015**, *27*, 389–398. [CrossRef] [PubMed]
9. Gainetdinov, R.R.; Hoener, M.C.; Berry, M.D. Trace Amines and Their Receptors. *Pharmacol. Rev.* **2018**, *70*, 549–620. [CrossRef]
10. Hornykiewicz, O. A brief history of levodopa. *J. Neurol.* **2010**, *257*, S249–S252. [CrossRef]
11. Wurtman, R.J.; Wurtman, J.J. Carbohydrate craving, obesity and brain serotonin. *Appetite* **1986**, *7*, 99–103. [CrossRef]
12. Herrera, C.P.; Smith, K.; Atkinson, F.; Ruell, P.; Chow, C.M.; O'Connor, H.; Brand-Miller, J. High-glycaemic index and glycaemic load meals increase the availability of tryptophan in healthy volunteers. *Br. J. Nutr.* **2011**, *105*, 1601–1606. [CrossRef]
13. Trusov, N.V.; Shipelin, V.A.; Mzhelskaya, K.V.; Shumakova, A.A.; Timonin, A.N.; Riger, N.A.; Apryatin, S.A.; Gmoshinski, I.V. Effect of resveratrol on behavioral, biochemical, and immunological parameters of DBA/2J and tetrahybrid DBCB mice receiving diet with excess fat and fructose. *J. Nutr. Biochem.* **2021**, *88*, 108527. [CrossRef]
14. Allen, J.A.; Varga, J. Eosinophilia-Myalgia Syndrome. In *Encyclopedia of Toxicology*, 3rd ed.; Elsevier: Amsterdam, The Netherlands, 2014; pp. 419–425. [CrossRef]
15. Apryatin, S.A.; Shipelin, V.A.; Sidorova, Y.S.; Petrov, N.A.; Gmoshinskii, I.V.; Nikityuk, D.B. Interspecific Differences in Behavioral Responses and Neuromotorics between Laboratory Rodents Receiving Rations with Easily Digested Carbohydrates. *Bull. Exp. Biol. Med.* **2018**, *165*, 5–9. [CrossRef] [PubMed]
16. Matter, C.M.; Handschin, C. RANTES (regulated on activation, normal T cell expressed and secreted), inflammation, obesity, and the metabolic syndrome. *Circulation* **2007**, *115*, 946–948. [CrossRef] [PubMed]
17. Fujiwara, M.; Iwata, M.; Inoue, T.; Aizawa, Y.; Yoshito, N.; Hayashi, K.; Suzuki, S. Decreased grip strength, muscle pain, and atrophy occur in rats following long-term exposure to excessive repetitive motion. *FEBS Open Bio* **2017**, *7*, 1737–1749. [CrossRef] [PubMed]
18. Botros, M.; Sikaris, K.A. The de ritis ratio: The test of time. *Clin. Biochem. Rev.* **2013**, *34*, 117–130. [PubMed]
19. Nikniaz, L.; Nikniaz, Z.; Tabrizi, J.S.; Sadeghi-Bazargani, H.; Farahbakhsh, M. Is within-normal range liver enzymes associated with metabolic syndrome in adults? *Clin. Res. Hepatol. Gastroenterol.* **2018**, *42*, 92–98. [CrossRef]

20. Esnault, S.; Kelly, E.A. Essential Mechanisms of Differential Activation of Eosinophils by IL-3 Compared to GM-CSF and IL-5. *Crit. Rev. Immunol.* **2016**, *36*, 429–444. [CrossRef]
21. Gregory, B.; Kirchem, A.; Phipps, S.; Gevaert, P.; Pridgeon, C.; Rankin, S.M.; Robinson, D.S. Differential regulation of human eosinophil IL-3, IL-5, and GM-CSF receptor alpha-chain expression by cytokines: IL-3, IL-5, and GM-CSF down-regulate IL-5 receptor alpha expression with loss of IL-5 responsiveness, but up-regulate IL-3 receptor alpha exp. *J. Immunol.* **2003**, *170*, 5359–5366. [CrossRef]
22. Williams, R.L.; Wood, L.G.; Collins, C.E.; Morgan, P.J.; Callister, R. Energy homeostasis and appetite regulating hormones as predictors of weight loss in men and women. *Appetite* **2016**, *101*, 1–7. [CrossRef]
23. Apryatin, S.A.; Shipelin, V.A.; Trusov, N.V.; Mzhelskaya, K.V.; Evstratova, V.S.; Kirbaeva, N.V.; Soto, J.S.; Fesenko, Z.S.; Gainetdinov, R.R.; Gmoshinski, I.V. Comparative analysis of the influence of a high-fat/high-carbohydrate diet on the level of anxiety and neuromotor and cognitive functions in Wistar and DAT-KO rats. *Physiol. Rep.* **2019**, *7*, e13987. [CrossRef] [PubMed]
24. Zhang, L.S.; Davies, S.S. Microbial metabolism of dietary components to bioactive metabolites: Opportunities for new therapeutic interventions. *Genome Med.* **2016**, *8*, 46. [CrossRef] [PubMed]
25. Dou, L.; Jourde-Chiche, N.; Faure, V.; Cerini, C.; Berland, Y.; Dignat-George, F.; Brunet, P. The uremic solute indoxyl sulfate induces oxidative stress in endothelial cells. *J. Thromb. Haemost.* **2007**, *5*, 1302–1308. [CrossRef]
26. Karbownik, M.; Stasiak, M.; Zygmunt, A.; Zasada, K.; Lewiński, A. Protective effects of melatonin and indole-3-propionic acid against lipid peroxidation, caused by potassium bromate in the rat kidney. *Cell Biochem. Funct.* **2006**, *24*, 483–489. [CrossRef]
27. Bercik, P.; Denou, E.; Collins, J.; Jackson, W.; Lu, J.; Jury, J.; Deng, Y.; Blennerhassett, P.; Macri, J.; McCoy, K.D.; et al. The intestinal microbiota affect central levels of brain-derived neurotropic factor and behavior in mice. *Gastroenterology* **2011**, *141*, 599–609.e1–3. [CrossRef] [PubMed]
28. Bartolomeo, M.P.; Maisano, F. Validation of a reversed-phase HPLC method for quantitative amino acid analysis. *J. Biomol. Tech.* **2006**, *17*, 131–137. [PubMed]
29. National Research Council. *Guide for the Care and Use of Laboratory Animals*, 8th ed.; National Academies Press: Washington, DC, USA, 2011.
30. Watson, R.R. *Trace Elements in Laboratory Rodents*; CRC Press: Boca Raton, FL, USA; London, UK, 1996; p. 416.

International Journal of
Molecular Sciences

Article

A Tryptophan-Deficient Diet Induces Gut Microbiota Dysbiosis and Increases Systemic Inflammation in Aged Mice

Ibrahim Yusufu [1,†], Kehong Ding [1,†], Kathryn Smith [2], Umesh D. Wankhade [3,4], Bikash Sahay [5], G. Taylor Patterson [1], Rafal Pacholczyk [6], Satish Adusumilli [7], Mark W. Hamrick [2,8], William D. Hill [9,10], Carlos M. Isales [1,8,*] and Sadanand Fulzele [1,2,8,*]

1. Department of Medicine, Augusta University, Augusta, GA 30912, USA; IYUSUFU@augusta.edu (I.Y.); KDING@augusta.edu (K.D.); GPATTERSON1@augusta.edu (G.T.P.)
2. Department of Cell Biology and Anatomy, Augusta University, Augusta, GA 30912, USA; KATSMITH4@augusta.edu (K.S.); mhamrick@augusta.edu (M.W.H.)
3. Department of Pediatrics, College of Medicine, University of Arkansas for Medical Sciences (UAMS), Little Rock, AR 72202, USA; UWankhade@uams.edu
4. Arkansas Children Nutrition Center, Arkansas Children's Research Institute, Little Rock, AR 72202, USA
5. Department of Infectious Diseases and Immunology, University of Florida, Gainesville, FL 32608, USA; sahayb@ufl.edu
6. Georgia Cancer Center, Augusta University, Augusta, GA 30902, USA; RPACHOLCZYK@augusta.edu
7. Department of Pathology, University of Notre Dame, Notre Dame, IN 46855, USA; sadusumi@nd.edu
8. Institute of Healthy Aging, Augusta University, Augusta, GA 30912, USA
9. Department of Pathology, Medical University of South Carolina, Charleston, SC 29403, USA; hillwi@musc.edu
10. Ralph H Johnson Veterans Affairs Medical Center, Charleston, SC 29403, USA
* Correspondence: cisales@augusta.edu (C.M.I.); sfulzele@augusta.edu (S.F.)
† These authors contributed equally to this work.

Abstract: The gut microflora is a vital component of the gastrointestinal (GI) system that regulates local and systemic immunity, inflammatory response, the digestive system, and overall health. Older people commonly suffer from inadequate nutrition or poor diets, which could potentially alter the gut microbiota. The essential amino acid (AA) tryptophan (TRP) is a vital diet component that plays a critical role in physiological stress responses, neuropsychiatric health, oxidative systems, inflammatory responses, and GI health. The present study investigates the relationship between varied TRP diets, the gut microbiome, and inflammatory responses in an aged mouse model. We fed aged mice either a TRP-deficient (0.1%), TRP-recommended (0.2%), or high-TRP (1.25%) diet for eight weeks and observed changes in the gut bacterial environment and the inflammatory responses via cytokine analysis (IL-1a, IL-6, IL-17A, and IL-27). The mice on the TRP-deficient diets showed changes in their bacterial abundance of Coriobacteriia class, *Acetatifactor* genus, Lachnospiraceae family, *Enterococcus faecalis* species, *Clostridium* sp genus, and *Oscillibacter* genus. Further, these mice showed significant increases in IL-6, IL-17A, and IL-1a and decreased IL-27 levels. These data suggest a direct association between dietary TRP content, the gut microbiota microenvironment, and inflammatory responses in aged mice models.

Keywords: tryptophan; systemic inflammation; dysbiosis; gut; microbiota

1. Introduction

Balanced macronutrients (fats, carbohydrates, and proteins), micronutrients (minerals and trace elements), and vitamins in our diets are important for maintaining healthy physiological systems. However, poor nutrition is commonly observed in older populations and associated with chronic disease conditions, such as impaired digestive health, a decline in cognitive function, and a compromised immune system [1]. Interestingly, studies have shown that diet alone could potentially have therapeutic effects in patients suffering from chronic inflammatory diseases (such as inflammatory bowel disease) [2]. Thus, a better understanding of the relationship between the diet and specific disease processes could have a significant impact on healthy lifespan in the aging population.

Amino acids (AA), the building blocks for protein synthesis, and their many metabolites are particularly crucial for growth, reproduction, immunity, and whole-body homeostasis [3]. Tryptophan (TRP) is one of nine essential amino acids and is responsible for modulating physiological stress responses, neuropsychiatric health, oxidative systems, and the immune system [4]. TRP deficiency is associated with major depressive illness in both males and females, and, in select cases, these depressive symptoms can be reversed by TRP-rich diets [5,6]. Several studies have found that low-TRP diets can lead to growth retardation, impaired reticulocyte function, and fatigue resistance, emphasizing the importance of this amino acid in the human diet [7–9].

Studies have particularly highlighted the importance of TRP and its many metabolites in systemic and local intestinal inflammatory mechanisms. TRP deficiency has been shown to compromise immune response and impair disease resistance in teleost fish [4]. Furthermore, gut health and intestinal immunity require sufficient dietary TRP for efficient immunological response and intestinal homeostasis [10]. TRP metabolites (kynurenines, serotonin, and melatonin) and bacterial TRP metabolites (indole, indolic acid, skatole, and tryptamine) are known to support gut microbiota, microbial metabolism, the host's immune system, and host–microbiota synergy [10]. In fact, disruptions in the gut lactobacillus strains responsible for TRP metabolism can lead to intestinal inflammation and colitis [11].

In this study, we aimed to better define tryptophan's role in the immune response and its effect on the gut microbiome, specifically in relation to age. We fed aged mice TRP-deficient (0.1%), TRP-recommended (0.2%), and high-TRP (1.25%) diets for eight weeks and observed the responses to the gut bacterial environment and the inflammatory responses. We hypothesized that a TRP-deficient diet would exert a systemic pro-inflammatory response and alter gut bacteria homeostasis.

2. Results

2.1. Gut Microbial Taxonomic Analysis

Fecal microbiota analysis was performed on three groups (21 total samples): control diet ($n = 7$), TRP-deficient diet ($n = 7$), and TRP-rich diet ($n = 7$). We analyzed the gut microbiota composition using the 16S rRNA amplicon sequencing of fecal contents and observed distinct differences in the microbial taxa associated with TRP dose variation. Microbiota diversity is typically described in terms of within (i.e., α) and between sample (i.e., β) diversities. α-diversity indices both at the phylum and genus level were not different between the groups of mice that were fed different doses of TRP (Figure 1). Nonmetric multidimensional scaling (NMDS) ordination plots using a Bray–Curtis distancing matrix of β-diversity revealed significant differences at the genus levels (Dose, $p < 0.05$); however, at the phylum level there were no differences due to the TRP content in the diet (Figure 2).

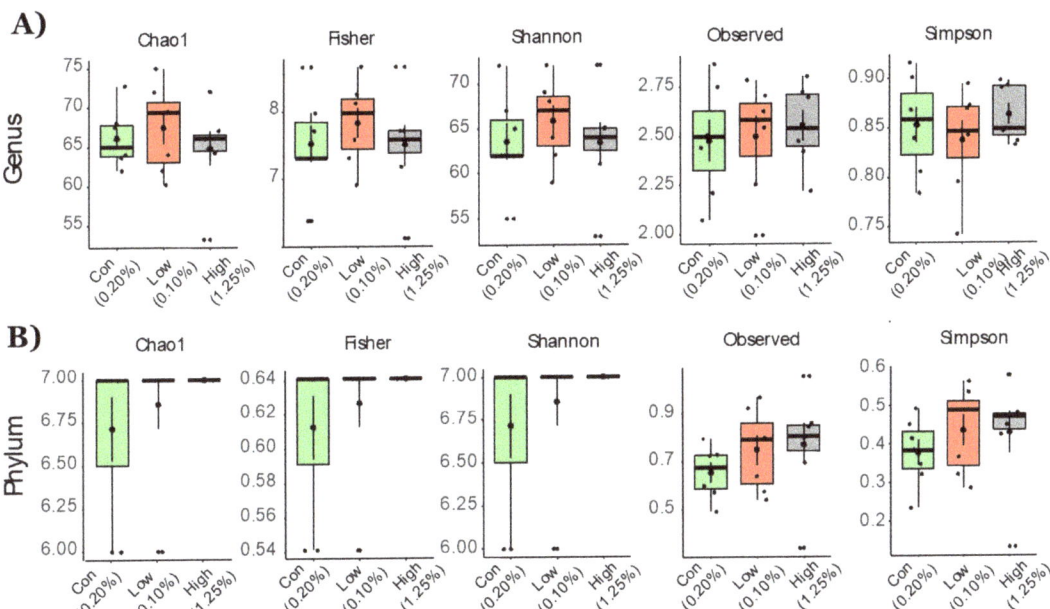

Figure 1. Alpha diversity measurements for aged mice fed with TRP-deficient and TRP-rich diets. α-diversity indices at the (**A**) genus and (**B**) phylum levels. Statistical differences between the group's control diet ($n = 7$), TRP-deficient diet ($n = 7$), and TRP-rich diet ($n = 7$) were determined by Kruskal Wallis one-way ANOVA for doses of TRP. Data are expressed as mean ± SE and all comparisons at all indices were non-significant.

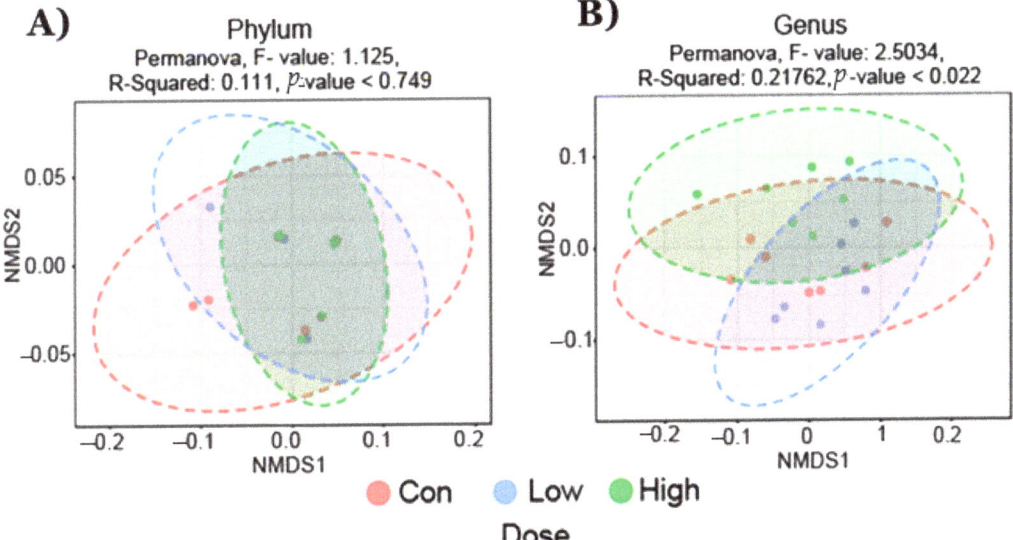

Figure 2. Non-metric multidimensional scaling (NMDS) plot of fecal bacterial community structures in animals fed with control diet ($n = 7$), TRP-deficient diet ($n = 7$), and TRP-rich diet ($n = 7$). Data showed no significant differences at (**A**) the phylum levels but formed distinct clusters at (**B**) the genus level specific to a TRP-deficient diet; (Permanova, F-value; 2.5034, R-squared; 0.21762, p-value < 0.022).

2.2. Effect of TRP Supplementation on Taxonomical Differences

The heat tree analysis leverages the hierarchical structure of taxonomic classifications to quantitatively (using the median abundance) and statistically (using the non-parametric Wilcoxon Rank Sum test) depict taxonomic differences between microbial communities or abundance profiles for a group. The Phylogenetic Heat Tree illustrates the differences in relative bacterial abundance between gut microbiota compositions in mice who were fed a diet with different TRP doses (Figure 3). Red colored nodes denote the enriched bacteria the comparison listed (Figure 3). Mainly Firmicutes and Bacteroidetes dominated gut microbiota composition at the phylum level. Other phyla such as Proteobacteria, Verrucomicrobia, Tenericutes, Deferribacteres, and Actinobacteria comprised the rest (Figure 4). Lower TRP supplementation reduced the Deferribacteres abundance, whereas higher supplementation not only restored but also increased the abundance (Figure 4). A trend for an increase in Proteobacteria abundance was seen in the mice who were fed the low- and high-dose TRP diets. At the genus level, out of 170 genera the Dunn test revealed significant groupwise differences in 21 genera ($p < 0.05$). Notably, Mucispirillum and lachnospiraceae bacteria went down with a low-TRP diet and were restored or increased in abundance on the higher TRP diet (Figure 5). The bacteria Acetatifactor, Enterorhabdus, and Adlercreutzia went up with a low-TRP diet and restored or decreased with a high-TRP diet (Figure 5).

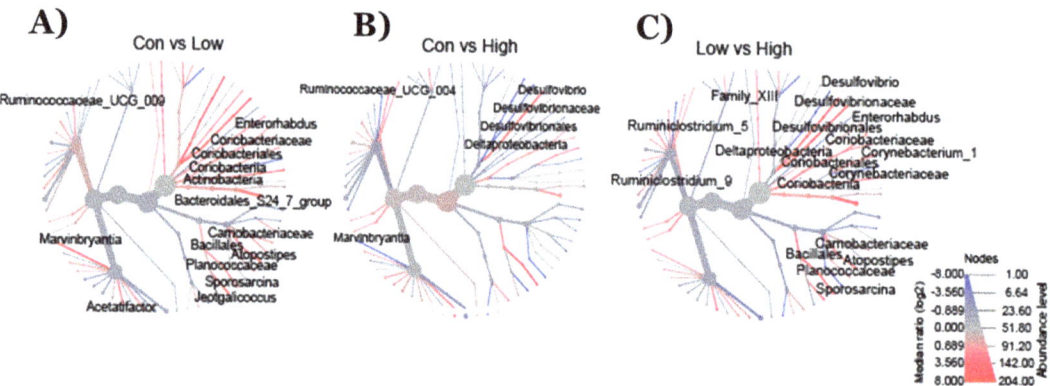

Figure 3. (**A**) Phylogenetic heat tree illustrates the differences in relative bacterial abundance between groups. The data show the changes in the bacterial families in mice fed (**A**) TRP-deficient/low compared to TRP-normal, (**B**) TRP-rich/high compared to TRP-normal, and (**C**) TRP-rich/high compared to TRP-deficient diets. Control diet ($n = 7$), TRP-deficient diet ($n = 7$), and TRP-rich diet ($n = 7$). Red nodes represent more abundant bacterial families, whereas the blue nodes represent less abundant bacterial families.

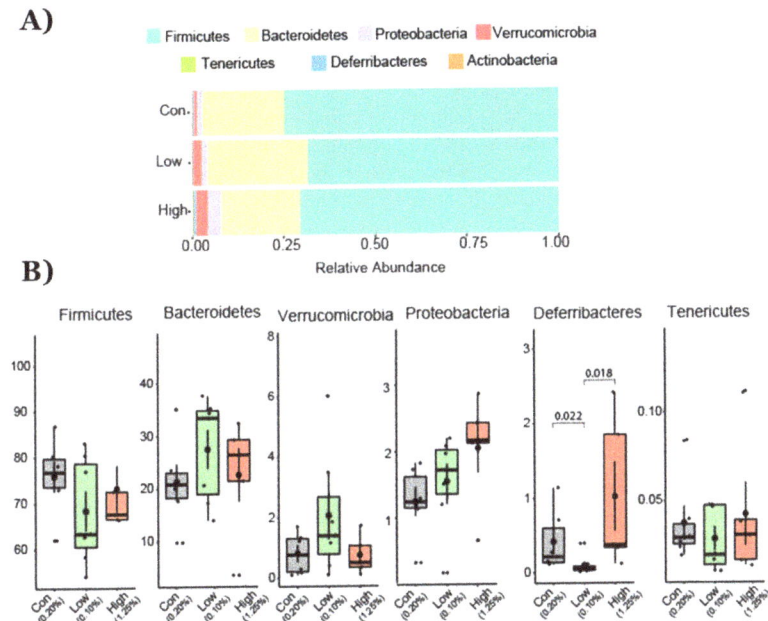

Figure 4. Operational taxonomical unit (OTU) abundance at the phylum level in animals fed with TRP-normal, TRP-deficient, and TRP-rich diets. (**A**) Relative increase in Verrucomicrobia and Bacteroideate and decrease in Firmicutes and Deferribacteres. (**B**) Box and whisker plots depict the operational taxonomic units (OTUs) from different bacterial phyla presented among three groups of mice. p values are shown where the differences were found to be significantly different from each other. Cont (control 0.2% TRP, $n = 7$), def/low (0.1% TRP, $n = 7$), and high/rich (control 1.25% TRP, $n = 7$).

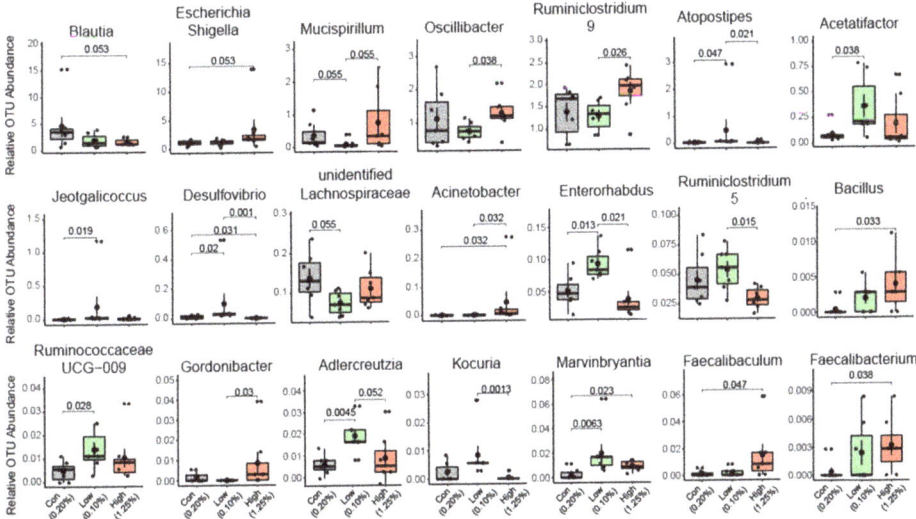

Figure 5. Change in composition of the gut microbiota at the genus level in animals fed with control diet ($n = 7$), TRP-deficient diet ($n = 7$), and TRP-rich diet ($n = 7$). Total of 21 genera were found to be different between at least one comparison (con vs. TRP-def/low, con vs. TRP-high, TRP-def/low vs. TRP high) The data represent OTUs of certain genera found to be different among the groups (data are expressed and mean +/− SE and 1 p-values were calculated using the Kruskal Wallis test).

2.3. Serum Cytokine Analysis

It has been previously reported that an amino acid deficiency can induce systemic inflammation [12]. To assess the effects of the TRP-deficient vs. a TRP-rich diet on the systemic immune profile, we measured the pro-inflammatory cytokines IL-17A and IL-1a and the dual functioning pro- and anti-inflammatory cytokines IL-6 and IL-27 in the serum. We found that IL-6 was significantly (p-value = 0.04) elevated in the TRP-deficient diet, while a trend toward low IL-6 levels was found in the TRP-rich diet (compared to TRP-deficient diet). The pro-inflammatory cytokine IL-1a was significantly up-regulated compared to both the control (p-value = 0.008) and TRP-rich diets (p-value = 0.04). The anti-inflammatory cytokine IL-27 showed a significant (p-value = 0.05) decrease in the TRP-deficient diet compared to the control diet. The pro-inflammatory cytokine IL-17a showed a trend for up-regulation with the TRP-deficient diet compared to both the control and TRP-rich diets (Figure 6). Overall, the results revealed that the animals fed with the TRP-deficient diet had an elevated pro-inflammatory cytokine level compared to those fed a normal and a TRP-rich diet (Figure 6).

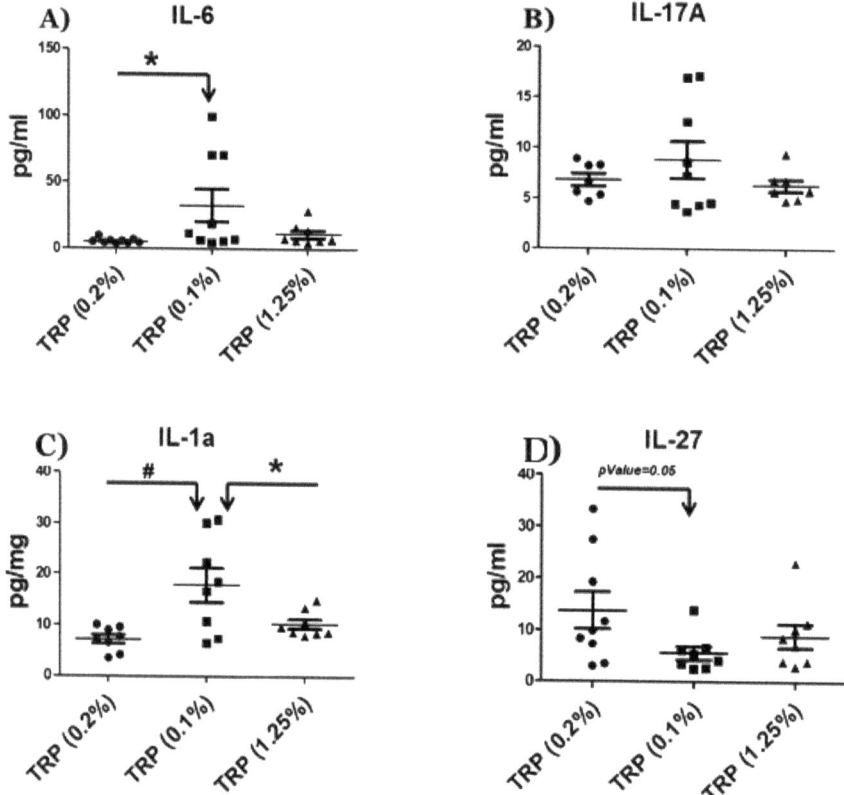

Figure 6. Serum cytokine levels in animals fed with a control diet, a TRP-deficient/low diet, and a TRP-rich diet. Serum was collected at the end of the experiment (week 8), followed by ELISA for immunoreactive (**A**) IL-6, (**B**) IL-17a (**C**) IL-1a, and (**D**) IL-27. Results are means ± SD (n = 7–9/per group). Data were analyzed by ANOVA followed by Bonferroni post hoc test or t-test (* $p < 0.05$, # $p < 0.01$).

3. Discussion

The essential amino acid (AA) Tryptophan (TRP) plays a crucial role in regulating systemic immune responses and mental and gut health. However, an age-dependent relationship between TRP deficiency and gut microbiota health has not yet been reported. The present study explored this potential relationship between the gut microbiota and changes in the inflammatory milieu in aged mice after exposure to either a TRP-deficient (0.1%) or a TRP-rich (1.25%) diet for eight weeks. We found that the low-TRP diet animals exhibited altered bacterial composition in their gut microbiota compared animals fed with control and TRP-rich diets. Moreover, we also found that the low TRP diet-fed animals had an elevated level of systemic inflammation.

Changes in the gut microbiota (dysbiosis) or a reduction in bacterial diversity can result in many intestinal diseases [13,14]. Our study found differences in the bacterial abundance of Coriobacteriia class, *Acetatifactor* genus, Lachnospiraceae family, *Enterococcus faecalis* species, *Clostridium* sp. genus, and oscillibacter genus in animals fed with TRP-deficient diets. Previously, this bacterial abundance/dysbiosis has been reported in a number of pathophysiological conditions [15,16]. *Enterorhabdus* genus is a member of the Actinobacteria phylum and Coriobacteriia class that has been associated with ileocecal mucosal inflammation in mice [1,2]. *Enterorhabdus mucosicola* and *Enterorhabdus caecimuris* species were both isolated from the ileocecal regions of mice suffering from colitis and intestinal inflammation [1,2]. An increased presence of *Enterorhabdus* genus is suggestive of promoting or is associated with mucosal inflammation in the GI tracts of mice. Our study found significant increases in *Enterorhabdus* genus abundance in TRP-deficient mice, suggesting that these animals might be under chronic inflammatory stress.

Our study also found a significantly increased abundance of *Adlercreutzia* genus (Actinobacteria phylum) in the TRP-deficient mouse gut microbiota. Moon et al. (2018) found relatively low abundances of *Adlercreutzia* in diabetic women and increased plasma levels of TRP metabolites (e.g., kynurenine) [17]. Their findings suggest that gut-derived *Adlercreutzia* genus abundances are indirectly associated with increased TRP metabolism. These results are further supported by our finding that *Adlercreutzia* genus is correlated with TRP metabolites. We also noted more than a three-fold increase in *Acetatifactor* bacterium in TRP-deficient mice. Interestingly, *Acetatifactor muris* is associated with intestinal inflammation. Transplanting fecal microbiota rich in *Acetatifactor muris* into healthy wild-type mice induced colonic inflammation [15]. This bacterium's significant increase in the gut microbiota could potentially promote intestinal inflammation and related adverse effects. Although our study found the *Acetatifactor* genus to increase, we found that the Lachnospiraceae family generally decreased significantly. Lachnospiraceae is a family of anaerobic, spore-forming bacteria that produce butyric acid, which is protective against colon cancer development in humans [16,18]. The general decrease in bacteria in the Lachnospiraceae family and increases in the *Acetatifactor* genus strongly indicated a systemic elevation of inflammation in the TRP-deficient group.

It has been previously reported that *Clostridium* sp. metabolizes TRP and generates indoxyl sulfate (IS) and indole-3-propionic acid (IPA) [19]. Indoxyl sulfate is a toxic metabolite that enhances oxidative damage in intestinal epithelial cells and compromises the epithelial layer lining the intestines [20]. Conversely, IPA is a TRP metabolite that protects mice from dextran sodium sulfate (DSS)-induced colitis, suggesting that it has some anti-inflammatory role [21]. Previously, Konopelski et al. reported that rats on a TRP-free diet had a lower concentration of IPA in stool and blood [9]. In our study, we found *Clostridium* sp. to be significantly decreased in TRP-deficient mice. Our data indicated that TRP deficiency might have some protective as well as harmful pro-inflammatory effects via decreased *Clostridium* sp. abundance. *Clostridium* species are also known to regulate the critical neurotransmitter serotonin (5-HT) in the gut [22]. The human colon promotes 5-HT biosynthesis, which regulates many physiological processes, including neurotransmission, mood, sleep, memory, intestinal motility, and digestion [22]. Decreased *Clostridium* sp. may lead to lower levels of 5-HT, leading to depression, impaired digestions, and neuropsychiatric conditions. Similarly, the number of other gut bacterium significantly decreases in TRP-deficient groups such as *Oscillibacter valericigenes, Mucispirillum*, and *Blautia* genus. These gut bacteria play a significant role in maintaining human and animal gut microflora health [5,6,9,23]. Studies have found O. valericigenes to be significantly decreased in Crohn's disease patients' microflora, with elevated inflammation [23]. *Mucispirillum* genus bacterium encodes for proteins that resist the oxidative bursts associated with inflammatory states [5]. These bacterial species express superoxide reductase, catalase, cytochrome c oxidase, and rubrerythrin that utilize diverse reactions to neutralize reactive oxygen species [5,6]. The *Blautia* genus is known for its anti-inflammatory properties. For example, Jenq et al. demonstrated increased abundances of the *Blautia* genus with decreased graft-versus-host-disease mortality and improved survival after allogeneic bone marrow transplantation [9]. Their results suggest some level of anti-inflammatory properties mediated by the *Blautia* genus.

Based on dysbiosis in the gut microbiota of the TRP-deficient diet, we hypothesized that gut microbiota changes might induce the systemic release of cytokines. To investigate this, we measured the predominantly pro-inflammatory cytokines IL-17A and IL-1a and the dual functioning pro- and anti-inflammatory cytokines IL-6 and IL-27 in the serum of animals placed on the TRP-deficient diet. Our data showed significant increases in IL-6, IL-17A, and IL-1a, but a significant decrease in IL-27. It is interesting that the mice fed with a TRP rich diet presented with higher IL-27, a cytokine known to prevent IL-17 transcription. The production of IL-27 along with Aryl hydrocarbon receptor (AhR) activation helps in the generation of regulatory T cells in the gut [24]. IL-27 prevents excessive inflammation by controlling IL-17 responses by the generation of regulatory T cells. It also maintains the gut epithelial barrier, including the enhancement of indoleamine 2,3-dioxygenase (IDO1) expression [25]. The overall effect of TRP deficiency is geared toward a systemic pro-inflammatory state. The elevated level of systemic pro-inflammatory status in a TRP-deficient diet might be due to a combination of different reasons, such as (1) decreased levels of cellular TRP metabolites (Kynurenine pathway metabolites) and bacterial-derived TRP metabolites (indole, indolic acid, and tryptamine), (2) the dysregulation of AhR transcription factor, (3) increased oxidative stress due to the TRP deficiency diet and its metabolites (e.g., IPA), and/or (4) direct changes in the composition of the gut microbiota. The cellular and gut microbiota-derived TRP metabolites are endogenous ligands of AhR [26–28]. A low level of TRP in the diet may decrease endogenous ligands of AhR, leading to the dysregulation of transcription factor (AhR). It has been previously reported that the AhR plays a significant role in maintaining gut and systemic inflammation in humans and rodents [28,29]. Furthermore, a decrease in microbiota-derived TRP metabolites such as Indole-3-propionic acid (IPA) leads to the accumulation of reactive oxygen species (increased oxidative stress). IPA is a potent antioxidant known to play an important role in neuroprotection [30,31], anti-non-alcoholic steatohepatitis [32], and protection against radiation toxicity [33], and reduced the bacterial load in a mouse model of acute *M. tuberculosis* infection [34]. TRP metabolism signaling is complex; the endogenous metabolites of TRP (kynurenine pathway) are known to elevate the inflammatory process [35], whereas gut-derived indoles derivatives reveal an anti-inflammatory effect [21].

A balanced diet is required for minimizing age-associated diseases. The gastrointestinal (GI) system is a critical organ that is commonly compromised in the elderly population. Studies have postulated that gut microbiota's dynamic changes in the aging population could modulate changes in immunity and cognitive functioning, thus contributing to certain diseases [36]. The bacterial gut composition in human and animal models has been shown to play a critical role in regulating intestinal conditions such as colitis and allergic diarrhea via regulatory T-cell (T-reg) modulation [10,11,37]. Most importantly, diet has been directly linked to microbiota composition in humans and rodents [10,11,13,14,36–41]. A further detailed investigation is needed to elucidate the relationships between the TRP metabolites, AhR signaling, and systemic inflammation in age-related pathophysiological conditions.

4. Materials and Methods

4.1. Animal Study

All protocols were conducted by following the guidelines established by the Augusta University Institutional Animal Care and Use Committee (AU-IACUC, Protocol number: 2009-0065). Twenty-month-old male C57BL/6 mice were obtained from the aged rodent colony at the National Institute on Aging. The animals were fed either standard TRP (0.2%), low-TRP (0.1%) or high-TRP (1.25%) diets for eight weeks. TRP concentration was selected based on previously published data [42]. Diets were prepared by Envigo-Teklad (Madison, WI, USA) in consultation with their nutritionist and were isocaloric purified diets that contained all essential amino acids. Fecal samples were collected at the end of 8 weeks, and animals were euthanized using CO_2 overdose followed by thoracotomy according to an AU IACUC-approved protocol.

4.2. Microbial Community Profiling Using 16S rRNA Amplicon Sequencing

The MO BIO PowerSoil DNA Isolation kit (Qiagen, Germantown, MD, USA) was used to isolate genomic DNA samples from the fecal samples of mice. Fecal contents were carefully added to 96-well plates with beads and recommended buffers. The plates were sealed, added to the MO BIO shaker, and shaken horizontally at 20 rpm for 20 min. The isolated genomic DNA was quantified with nanodrop and stored at $-20\ °C$. The DNA samples were shipped to Novogene Corporation, Inc. (Durham, NC, USA), for 16S V3-V4 region amplicon sequencing.

4.3. Bioinformatics Analysis

BIOM files were analyzed using MicrobiomeAnalyst—a web-based tool for the statistical and visual analysis of microbiota data. Reads were initially denoized using filters with a minimum number of five reads in a minimum of one sample required to retain an OTU. Total read counts for the run were 754,355, with 35,921 per sample. Sample reads were normalized and rarefied to the minimum library, then data scaling (total sum scaling) was performed before the final analysis. Minimum read filtering was applied in MicrobiomeAnalyst for alpha and beta diversity calculations but was increased to a minimum of 10% prevalence with a count of 4 for differential abundance analysis. Low variance filter was set at 5% for the inter quartile range. Alpha diversity profiling and significance testing were conducted using one-way ANOVA at the genus and phylum taxonomic level. For beta diversity, permutational MANOVA (PERMANOVA) was used to compute groupwise differences. Differential abundance was calculated univariately. All p-values were calculated using the Kruskal Wallis test and the Dunn test unless stated otherwise. Microbiota figures (Figures 1–4) were prepared using open licensed software R.

4.4. Serum Cytokine Analysis

At the end of study, blood was drawn from animals by cardiac puncture. Levels of selected cytokines were measured in the serum by LEGENDplex Cytokines Detection (BioLegend, San Diego, CA, USA), as described by the manufacturer. The simultaneous quantification of cytokines in mouse sera was performed using the LEGENDplex mouse Inflammation Panel with V-bottom Plate (BioLegend Cat# 740446) according to the manufacturer's instructions. In brief, samples were thawed completely, mixed, and centrifuged to remove particulates prior to use. To achieve measurement accuracy, samples were diluted 2-fold with assay buffer, and standards were mixed with Matrix C (BioLegend) to account for additional components in the serum samples. Standards and samples were plated with capture beads for IL-1α, IL-6, IL-17A, and IL-27 and incubated for 2 h at room temperature on a plate shaker (800 rpm). After washing the plate with wash buffer, detection antibodies were added to each well. The plate was incubated on a shaker for 1 h at room temperature. Finally, without washing, SA-PE was added and incubated for 30 min. Samples were acquired on the CytoFLEX flow cytometer (Beckman Coulter Life Sciences, Indianapolis, IN, USA). Standard curves and protein concentration were calculated using the R package DrLumi [43] installed on R 3.5.2 (https://www.r-project.org/, accessed on 7 May 2021). The limit of detection was calculated as an average of background samples plus $2.5 \times$ SD. Assay and data calculations were performed using the Immune Monitoring Shared Resource (Augusta University).

4.5. Statistical Analysis

The results are shown as means ± standard deviations. GraphPad Prism 5 (La Jolla, CA, USA) was utilized to perform ANOVA with Bonferroni pair-wise comparison or unpaired t-tests as appropriate. A p-value of <0.05 was considered significant.

Author Contributions: Conceptualization, S.F. and C.M.I.; methodology, I.Y., K.D., K.S., B.S., G.T.P., R.P. and S.A.; software, U.D.W., G.T.P.; validation, S.F., C.M.I., M.W.H., and W.D.H.; formal analysis, U.D.W., G.T.P., M.W.H., W.D.H.; investigation, S.F. and C.M.I.; resources, S.A., S.F., C.M.I., M.W.H., and W.D.H.; data curation, I.Y., K.D., K.S., B.S., G.T.P., R.P. and S.A.; writing—original draft preparation, S.F., I.Y. and C.M.I.; writing—review and editing, S.A., S.F., C.M.I., M.W.H., and W.D.H.; visualization, S.F. and C.M.I.; supervision, S.F. and C.M.I.; project administration, S.F. and C.M.I.; funding acquisition, S.A., S.F., C.M.I., M.W.H., and W.D.H. All authors have read and agreed to the published version of the manuscript.

Funding: This publication is based upon work supported in part by the National Institutes of Health AG036675 (National Institute on Aging-AG036675 S.F, W.D.H, M.H, C.I,). The above-mentioned funding did not lead to any conflict of interests regarding the publication of this manuscript.

Institutional Review Board Statement: The study was conducted according to the guidelines of the Declaration of Helsinki, and approved by the Institutional Review Board (or Ethics Committee) of Augusta University (protocol code: 2009-0065 and date of approval: 23 August 2018).

Informed Consent Statement: Not applicable.

Data Availability Statement: The data that support the findings of this study are available from the corresponding author upon reasonable request.

Conflicts of Interest: The authors also declare that there is no other conflict of interest regarding the publication of this manuscript.

Abbreviations

GI	Gastrointestinal
AA	Amino acid
TRP	Tryptophan
IS	Indoxyl sulfate
IPA	IPA; Indole-3-propionic acid
IDO1	IDO1; Indoleamine 2,3-dioxygenase
AhR	AhR; Aryl hydrocarbon receptor

References

1. Shlisky, J.; Bloom, D.E.; Beaudreault, A.R.; Tucker, K.L.; Keller, H.H.; Freund-Levi, Y.; Fielding, R.A.; Cheng, F.W.; Jensen, G.L.; Wu, D.; et al. Nutritional considerations for healthy aging and reduction in age-related chronic disease. *Adv. Nutr.* **2017**, *8*, 17–26. [CrossRef] [PubMed]
2. Kakodkar, S.; Mutlu, E.A. Diet as a Therapeutic option for adult inflammatory bowel disease. *Gastroenterol. Clin. N. Am.* **2017**, *46*, 745–767. [CrossRef] [PubMed]
3. Wu, G. Amino acids: Metabolism, functions, and nutrition. *Amino Acids* **2009**, *37*, 1–17. [CrossRef] [PubMed]
4. Gol, S.; Pena, R.N.; Rothschild, M.F.; Tor, M.; Estany, J. Dietary tryptophan deficiency and its supplementation compromises inflammatory mechanisms and disease resistance in a teleost fish. *Sci. Rep.* **2019**, *9*, 7689. [CrossRef]
5. Benkelfat, C.; Ellenbogen, M.A.; Dean, P.; Palmour, R.M.; Young, S.N. Mood-lowering effect of tryptophan depletion. Enhanced susceptibility in young men at genetic risk for major affective disorders. *Arch. Gen. Psychiatry* **1994**, *51*, 687–697. [CrossRef] [PubMed]
6. Coppen, A.; Eccleston, E.G.; Peet, M. Total and free tryptophan concentration in the plasma of depressive patients. *Lancet* **1973**, *2*, 60–63. [CrossRef]

7. Henderson, L.M.; Koeppe, O.J.; Zimmerman, H.H. Niacin-tryptophan deficiency resulting from amono acid imbalance in non-casein diets. *J. Biol. Chem.* **1953**, *201*, 697–706. [CrossRef]
8. Hori, M.; Fisher, J.M.; Rabinovitz, M. Tryptophan deficiency in rabbit reticulocytes: Polyribosomes during interrupted growth of hemoglobin chains. *Science* **1967**, *155*, 83–84. [CrossRef]
9. Konopelski, P.; Konop, M.; Gawrys-Kopczynska, M.; Podsadni, P.; Szczepanska, A.; Ufnal, M. Indole-3-propionic acid, a tryptophan-derived bacterial metabolite, reduces weight gain in rats. *Nutrients* **2019**, *11*, 591. [CrossRef] [PubMed]
10. Gao, J.; Xu, K.; Liu, H.; Liu, G.; Bai, M.; Peng, C.; Li, T.; Yin, Y. Impact of the gut microbiota on intestinal immunity mediated by tryptophan metabolism. *Front. Cell Infect. Microbiol.* **2018**, *8*, 13. [CrossRef]
11. Lamas, B.; Richard, M.L.; Leducq, V.; Pham, H.P.; Michel, M.L.; Da Costa, G.; Bridonneau, C.; Jegou, S.; Hoffmann, T.W.; Natividad, J.M.; et al. CARD9 impacts colitis by altering gut microbiota metabolism of tryptophan into aryl hydrocarbon receptor ligands. *Nat. Med.* **2016**, *22*, 598–605. [CrossRef] [PubMed]
12. He, F.; Wu, C.; Li, P.; Li, N.; Zhang, D.; Zhu, Q.; Ren, W.; Peng, Y. Functions and Signaling pathways of amino acids in intestinal inflammation. *Biomed. Res. Int.* **2018**, *2018*, 9171905. [CrossRef] [PubMed]
13. Alam, M.T.; Amos, G.C.A.; Murphy, A.R.J.; Murch, S.; Wellington, E.M.H.; Arasaradnam, R.P. Microbial imbalance in inflammatory bowel disease patients at different taxonomic levels. *Gut Pathog.* **2020**, *12*, 1. [CrossRef] [PubMed]
14. Matsuoka, K.; Kanai, T. The gut microbiota and inflammatory bowel disease. *Semin. Immunopathol.* **2015**, *37*, 47–55. [CrossRef] [PubMed]
15. Lee, C.; Hong, S.N.; Paik, N.Y.; Kim, T.J.; Kim, E.R.; Chang, D.K.; Kim, Y.H. CD1d modulates colonic inflammation in NOD2-/-mice by altering the intestinal microbial composition comprising acetatifactor muris. *J. Crohns Colitis* **2019**, *13*, 1081–1091. [CrossRef]
16. Ai, D.; Pan, H.; Li, X.; Gao, Y.; Liu, G.; Xia, L.C. Identifying gut microbiota associated with colorectal cancer using a zero-inflated lognormal model. *Front. Microbiol.* **2019**, *10*, 826. [CrossRef] [PubMed]
17. Moon, J.Y.; Zolnik, C.P.; Wang, Z.; Qiu, Y.; Usyk, M.; Wang, T.; Kizer, J.R.; Landay, A.L.; Kurland, I.J.; Anastos, K.; et al. Gut microbiota and plasma metabolites associated with diabetes in women with, or at high risk for, HIV infection. *EBioMedicine* **2018**, *37*, 392–400. [CrossRef] [PubMed]
18. Meehan, C.J.; Beiko, R.G. A phylogenomic view of ecological specialization in the Lachnospiraceae, a family of digestive tract-associated bacteria. *Genome Biol. Evol.* **2014**, *6*, 703–713. [CrossRef]
19. Wikoff, W.R.; Anfora, A.T.; Liu, J.; Schultz, P.G.; Lesley, S.A.; Peters, E.C.; Siuzdak, G. Metabolomics analysis reveals large effects of gut microflora on mammalian blood metabolites. *Proc. Natl. Acad. Sci. USA* **2009**, *106*, 3698–3703. [CrossRef]
20. Adesso, S.; Ruocco, M.; Rapa, S.F.; Dal Piaz, F.; Di Iorio, B.R.; Popolo, A.; Nishijima, F.; Pinto, A.; Marzocco, S. Effect of indoxyl sulfate on the repair and intactness of intestinal epithelial cells: Role of reactive oxygen species' release. *Int. J. Mol. Sci.* **2019**, *20*, 2280. [CrossRef] [PubMed]
21. Wlodarska, M.; Luo, C.; Kolde, R.; d'Hennezel, E.; Annand, J.W.; Heim, C.E.; Krastel, P.; Schmitt, E.K.; Omar, A.S.; Creasey, E.A.; et al. Indoleacrylic acid produced by commensal peptostreptococcus species suppresses inflammation. *Cell Host Microbe* **2017**, *22*, 25–37. [CrossRef] [PubMed]
22. Yano, J.M.; Yu, K.; Donaldson, G.P.; Shastri, G.G.; Ann, P.; Ma, L.; Nagler, C.R.; Ismagilov, R.F.; Mazmanian, S.K.; Hsiao, E.Y. Indigenous bacteria from the gut microbiota regulate host serotonin biosynthesis. *Cell* **2015**, *161*, 264–276. [CrossRef] [PubMed]
23. Mondot, S.; Kang, S.; Furet, J.P.; Aguirre de Cárcer, D.; McSweeney, C.; Morrison, M.; Marteau, P.; Dore, J.; Leclerc, M. Highlighting new phylogenetic specificities of Crohn's disease microbiota. *Inflamm. Bowel Dis.* **2011**, *17*, 185–192. [CrossRef]
24. Apetoh, L.; Quintana, F.J.; Pot, C.; Joller, N.; Xiao, S.; Kumar, D.; Burns, E.J.; Sherr, D.H.; Weiner, H.L.; Kuchroo, V.K. The aryl hydrocarbon receptor interacts with c-Maf to promote the differentiation of type 1 regulatory T cells induced by IL-27. *Nat. Immunol.* **2010**, *11*, 854–861. [CrossRef] [PubMed]
25. Diegelmann, J.; Olszak, T.; Goke, B.; Blumberg, R.S.; Brand, S. A novel role for interleukin-27 (IL-27) as mediator of intestinal epithelial barrier protection mediated via differential signal transducer and activator of transcription (STAT) protein signaling and induction of antibacterial and anti-inflammatory proteins. *J. Biol. Chem.* **2012**, *287*, 286–298. [CrossRef] [PubMed]
26. Borghi, M.; Puccetti, M.; Pariano, M.; Renga, G.; Stincardini, C.; Ricci, M.; Giovagnoli, S.; Costantini, C.; Romani, L. Tryptophan as a central hub for host/microbial symbiosis. *Int. J. Tryptophan Res.* **2020**, *13*, 1178646920919755. [CrossRef] [PubMed]
27. Hubbard, T.D.; Murray, I.A.; Perdew, G.H. Indole and tryptophan metabolism: Endogenous and dietary routes to ah receptor activation. *Drug Metab. Dispos.* **2015**, *43*, 1522–1535. [CrossRef] [PubMed]
28. Ma, N.; He, T.; Johnston, L.J.; Ma, X. Host-microbiome interactions: The aryl hydrocarbon receptor as a critical node in tryptophan metabolites to brain signaling. *Gut Microbes* **2020**, *11*, 1203–1219. [CrossRef] [PubMed]
29. Iyer, S.S.; Gensollen, T.; Gandhi, A.; Oh, S.F.; Neves, J.F.; Collin, F.; Lavin, R.; Serra, C.; Glickman, J.; de Silva, P.S.A.; et al. dietary and microbial oxazoles induce intestinal inflammation by modulating aryl hydrocarbon receptor responses. *Cell* **2018**, *173*, 1123–1134.e11. [CrossRef] [PubMed]
30. Chyan, Y.J.; Poeggeler, B.; Omar, R.A.; Chain, D.G.; Frangione, B.; Ghiso, J.; Pappolla, M.A. Potent neuroprotective properties against the Alzheimer beta-amyloid by an endogenous melatonin-related indole structure, indole-3-propionic acid. *J. Biol. Chem.* **1999**, *274*, 21937–21942. [CrossRef] [PubMed]
31. Bendheim, P.E.; Poeggeler, B.; Neria, E.; Ziv, V.; Pappolla, M.A.; Chain, D.G. Development of indole-3-propionic acid (OXIGON) for Alzheimer's disease. *J. Mol. Neurosci.* **2002**, *19*, 213–217. [CrossRef] [PubMed]

32. Zhao, Z.H.; Xin, F.Z.; Xue, Y.; Hu, Z.; Han, Y.; Ma, F.; Zhou, D.; Liu, X.L.; Cui, A.; Liu, Z.; et al. Indole-3-propionic acid inhibits gut dysbiosis and endotoxin leakage to attenuate steatohepatitis in rats. *Exp. Mol. Med.* **2019**, *51*, 1–14. [CrossRef] [PubMed]
33. Xiao, H.W.; Cui, M.; Li, Y.; Dong, J.L.; Zhang, S.Q.; Zhu, C.C.; Jiang, M.; Zhu, T.; Wang, B.; Wang, H.C.; et al. Gut microbiota-derived indole 3-propionic acid protects against radiation toxicity via retaining acyl-CoA-binding protein. *Microbiome* **2020**, *8*, 69. [CrossRef] [PubMed]
34. Negatu, D.A.; Liu, J.J.J.; Zimmerman, M.; Kaya, F.; Dartois, V.; Aldrich, C.C.; Gengenbacher, M.; Dick, T. Whole-cell screen of fragment library identifies gut microbiota metabolite indole propionic acid as antitubercular. *Antimicrob. Agents Chemother.* **2018**, *62*, e01571-17. [CrossRef] [PubMed]
35. Sorgdrager, F.J.H.; Naudé, P.J.W.; Kema, I.P.; Nollen, E.A.; Deyn, P.P. Tryptophan metabolism in inflammaging: From biomarker to therapeutic target. *Front. Immunol.* **2019**, *10*, 2565. [CrossRef] [PubMed]
36. Biagi, E.; Franceschi, C.; Rampelli, S.; Severgnini, M.; Ostan, R.; Turroni, S.; Consolandi, C.; Quercia, S.; Scurti, M.; Monti, D.; et al. Gut Microbiota and extreme longevity. *Curr. Biol.* **2016**, *26*, 1480–1485. [CrossRef] [PubMed]
37. Atarashi, K.; Tanoue, T.; Oshima, K.; Suda, W.; Nagano, Y.; Nishikawa, H.; Fukuda, S.; Saito, T.; Narushima, S.; Hase, K.; et al. Treg induction by a rationally selected mixture of Clostridia strains from the human microbiota. *Nature* **2013**, *500*, 232–236. [CrossRef] [PubMed]
38. Murray, I.A.; Patterson, A.D.; Perdew, G.H. Aryl hydrocarbon receptor ligands in cancer: Friend and foe. *Nat. Rev. Cancer* **2014**, *14*, 801–814. [CrossRef]
39. Qin, J.; Li, Y.; Cai, Z.; Li, S.; Zhu, J.; Zhang, F.; Liang, S.; Zhang, W.; Guan, Y.; Shen, D.; et al. A metagenome-wide association study of gut microbiota in type 2 diabetes. *Nature* **2012**, *490*, 55–60. [CrossRef] [PubMed]
40. Wang, Y.; Wiesnoski, D.H.; Helmink, B.A.; Gopalakrishnan, V.; Choi, K.; DuPont, H.L.; Jiang, Z.D.; Abu-Sbeih, H.; Sanchez, C.A.; Chang, C.C.; et al. Fecal microbiota transplantation for refractory immune checkpoint inhibitor-associated colitis. *Nat. Med.* **2018**, *24*, 1804–1808. [CrossRef] [PubMed]
41. Wu, G.D.; Chen, J.; Hoffmann, C.; Bittinger, K.; Chen, Y.Y.; Keilbaugh, S.A.; Bewtra, M.; Knights, D.; Walters, W.A.; Knight, R.; et al. Linking long-term dietary patterns with gut microbial enterotypes. *Science* **2011**, *334*, 105–108. [CrossRef] [PubMed]
42. Refaey, M.E.; McGee-Lawrence, M.E.; Fulzele, S.; Kennedy, E.J.; Bollag, W.B.; Elsalanty, M.; Zhong, Q.; Ding, K.H.; Bendzunas, N.G.; Shi, X.M.; et al. Kynurenine, a Tryptophan metabolite that accumulates with age, induces bone loss. *J. Bone Miner. Res.* **2017**, *32*, 2182–2193. [CrossRef] [PubMed]
43. Sanz, H.; Aponte, J.J.; Harezlak, J.; Dong, Y.; Ayestaran, A.; Nhabomba, A.; Mpina, M.; Maurin, O.R.; Díez-Padrisa, N.; Aguilar, R.; et al. drLumi: An open-source package to manage data, calibrate, and conduct quality control of multiplex bead-based immunoassays data analysis. *PLoS ONE* **2017**, *12*, e0187901. [CrossRef]

Article

Supplementing with L-Tryptophan Increases Medium Protein and Alters Expression of Genes and Proteins Involved in Milk Protein Synthesis and Energy Metabolism in Bovine Mammary Cells

Jay Ronel V. Conejos [1,2,†], Jalil Ghassemi Nejad [1,†], Jung-Eun Kim [1], Jun-Ok Moon [3], Jae-Sung Lee [1] and Hong-Gu Lee [1,*]

1 Department of Animal Science and Technology, Konkuk University, Seoul 05029, Korea; jvconejos@up.edu.ph (J.R.V.C.); jalilgh@konkuk.ac.kr (J.G.N.); sumzzzing@gmail.com (J.-E.K.); jslee78@konkuk.ac.kr (J.-S.L.)
2 Institute of Animal Science, College of Agriculture and Food Sciences, University of the Philippines Los Baños, College Batong Malake, Los Baños, Laguna 4031, Philippines
3 Institute of Integrated Technology, CJ CheilJedang, Suwon 16495, Korea; junok.moon@cj.net
* Correspondence: hglee66@konkuk.ac.kr; Tel.: +82-2-450-0523 or +82-2-457-8367
† These authors contributed equally to this work.

Abstract: The objective of this study was to investigate the effects of supplementing with L-tryptophan (L-Trp) on milk protein synthesis using an immortalized bovine mammary epithelial (MAC-T) cell line. Cells were treated with 0, 0.3, 0.6, 0.9, 1.2, and 1.5 mM of supplemental L-Trp, and the most efficient time for protein synthesis was determined by measuring cell, medium, and total protein at 0, 24, 48, 72, and 96 h. Time and dose tests showed that the 48 h incubation time and a 0.9 mM dose of L-Trp were the optimal values. The mechanism of milk protein synthesis was elucidated through proteomic analysis to identify the metabolic pathway involved. When L-Trp was supplemented, extracellular protein (medium protein) reached its peak at 48 h, whereas intracellular cell protein reached its peak at 96 h with all L-Trp doses. β-casein mRNA gene expression and genes related to milk protein synthesis, such as mammalian target of rapamycin (mTOR) and ribosomal protein 6 (RPS6) genes, were also stimulated ($p < 0.05$). Overall, there were 51 upregulated and 59 downregulated proteins, many of which are involved in protein synthesis. The results of protein pathway analysis showed that L-Trp stimulated glycolysis, the pentose phosphate pathway, and ATP synthesis, which are pathways involved in energy metabolism. Together, these results demonstrate that L-Trp supplementation, particularly at 0.9 mM, is an effective stimulus in β-casein synthesis by stimulating genes, proteins, and pathways related to protein and energy metabolism.

Keywords: L-tryptophan; amino acids; MAC-T cell; proteomics; omics; β-casein; mTOR

1. Introduction

Balancing the profile of essential amino acids (EAAs) can result in higher utilization efficiency of nitrogen, leading to enhanced bovine milk protein synthesis [1,2]. Due to the pivotal importance of milk protein in human health, investigating supplementation with AAs (e.g., methionine, lysine) as the main nutrients that can positively stimulate milk protein synthesis in mammary epithelial cells (MEC) has been a priority of late [2]. Tryptophan (Trp), known as a conditional EAA [3], can be supplemented in the diets of animals and humans when targeting maximal production, cell growth, and proliferation [4–6]. Furthermore, Trp has been widely introduced as a mostly harmless supplement for humans to cope with health issues such as stress and depression. In animals of different species, L-Trp is used for alleviating stress (e.g., heat and cold) [4] by simulating serotonin [7] and melatonin [4,5,8], improving muscle cell growth [6], antioxidation [9], and inducing milk protein, lactose, and unsaturated fatty acids, which can also enhance human health [10,11], as people are the primary consumers of the products. While L-Trp as a supplemental nutrient has been used to induce muscle growth and development in human athletes [12,13] and beef cattle [6], there is a lack of insight into defining the optimum doses for enriching milk quality components to avoid the risk of overdose effects and provide maximum positive effects.

In monogastric cell lines and tissues, the stimulation of protein synthesis through mammalian target of rapamycin (mTOR) signaling is activated by individual AAs, in particular leucine (Leu) [2,14]. However, to what extent the mTOR signaling pathway can be activated directly by changes in EAAs, specifically L-Trp ratios, in bovine MEC remains unknown [2]. Moreover, the underlying mechanisms behind the effects of L-Trp need to be addressed. Before investigating the implication of L-Trp usage in vivo, it is important to define the optimum levels according to several prior in vitro studies in order to validate the benefits while avoiding ineffective high or low levels of supplementation.

In our laboratory, we took steps to examine the effects of L-Trp supplementation on muscle development and gene expression during heat and cold stress, and L-Trp proved to have anti-stress effects by simulating serotonin and melatonin in beef cattle [4–6]. Furthermore, in order to comprehend the mechanisms of L-Trp supplementation in influencing protein synthesis or affecting energy metabolism, metabolic pathways need to be investigated [10,12,13]. Investigating the metabolic pathway related to induced protein synthesis will help to improve our understanding of the reasons behind the phenomena [7,11,15], particularly for the purpose of milk enrichment. However, to date, no provisional and validated study has been conducted to determine what levels of L-Trp in a bovine mammary cell substratum will have the best influence on altered β-casein synthesis and related protein and energy metabolism pathways in vitro. L-Trp may supply nutrients necessary for protein synthesis, and this study may improve our understanding of how it regulates the expression of genes involved in milk protein synthesis in terms of nutrigenomics. Therefore, this study is the first to design supplementation of L-Trp in various doses to test the effects on increased medium protein and alterations in the expression of genes and proteins involved in milk protein synthesis and energy metabolism in bovine mammary cells.

2. Results

In this study, we compared the effects of different levels of L-Trp supplementation at various times on protein synthesis and the expression levels of β-casein, mRNA, and proteins.

2.1. AA Time and Dosage Sampling

Intracellular protein (cell protein) peaked at 96 h with 0 mM concentration of L-Trp (Figure 1a). In terms of extracellular protein (medium protein), when L-Trp was added, most concentrations peaked at 48 h (Figure 1b). Thus, 48 h was the optimal incubation time for the secretion of medium protein by bovine mammary epithelial (MAC-T) cells. The result of the incubation time test suggested that 48 h should be adopted as the incubation time for further tests of L-Trp efficacy for protein synthesis. Tables S1–S3 shows the average relative percentages of cell, medium and total protein, respectively.

To have a clear picture of the distribution or spread of our data, the statistical analysis of heterogeneity (Q) was done on our cell, medium, and total protein (Tables S4–S6). Data show that the replications do not vary from each other or homogeneous ($p < 0.05$).

A confirmatory study was performed to determine the ideal dose of L-Trp at 48 h, which was considered as the optimal cultivation time (Figure 2). Although the value was not significantly different, 0.9 mM showed the numerically highest relative protein quantity. The relative percentage of medium protein level on MAC-T cells supplemented with different levels of L-Trp at 48 h incubation is shown on Table S7. Statistical analysis of heterogeneity (Q) of medium protein (Table S8) of MAC-T cells supplemented with different dosages

(0, 0.3, 0.6, 0.9, 1.2, 1.5 mM) of L-Tryptophan at 48 h incubation time shows that replication samples are homogeneous except for 0.3 mM homogeneous ($p < 0.05$).

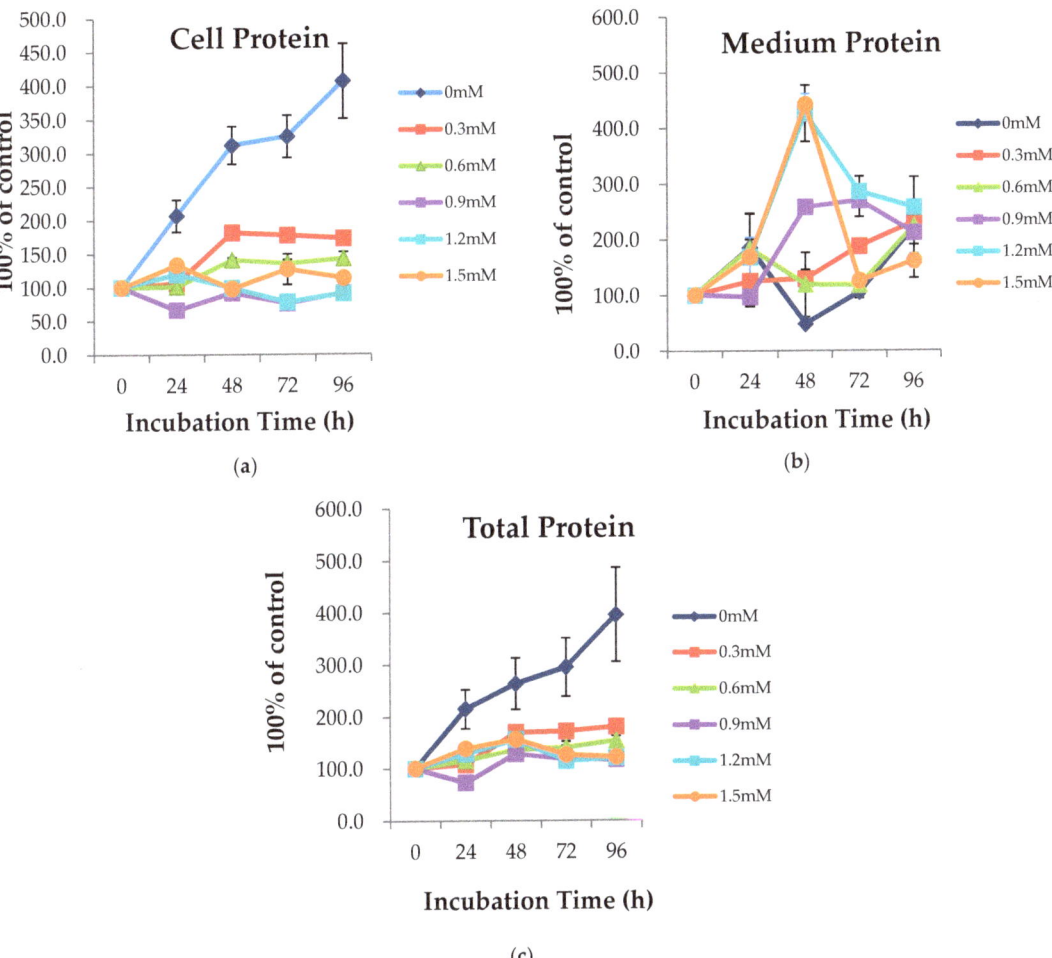

Figure 1. Relative protein content: (**a**) cell protein quantity, (**b**) medium protein quantity, and (**c**) total protein quantity (cell and medium) in bovine mammary epithelial (MAC-T) cells incubated with different levels of L-Trp (0, 0.3, 0.6, 0.9, 1.2, 1.5 mM) for 0, 24, 48, 72, and 96 h. Values are expressed as means ± SE ($n = 6$ per group).

2.2. Real-Time Polymerase Chain Reaction (RT-PCR)

In terms of mRNA relative gene expression, β-casein mRNA expression and genes related to milk protein synthesis, such as *mTOR* and *RPS6*, were also stimulated upon addition of 0.9 mM L-Trp ($p < 0.05$) (Figure 3). On the other hand, there was no effect on *S6K1* and *LDH-B* gene expression upon addition of L-Trp ($p > 0.05$).

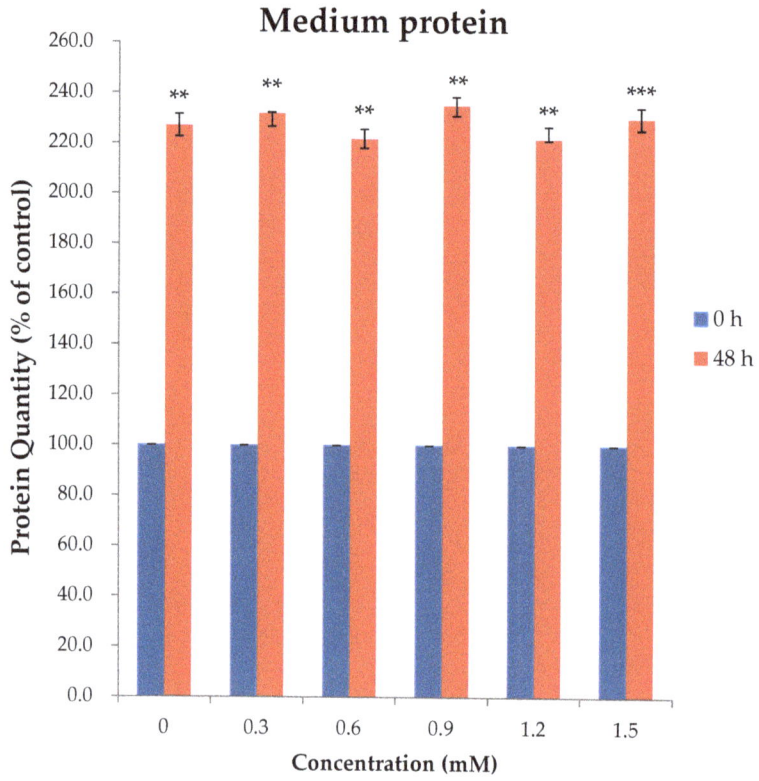

Figure 2. Relative medium protein content in MAC-T cells incubated with different levels of L-Trp (0, 0.3, 0.6, 0.9, 1.2, 1.5 mM) for 48 h. Values are expressed as means ± SE (n = 6 per group). ** < 0.01, *** < 0.001.

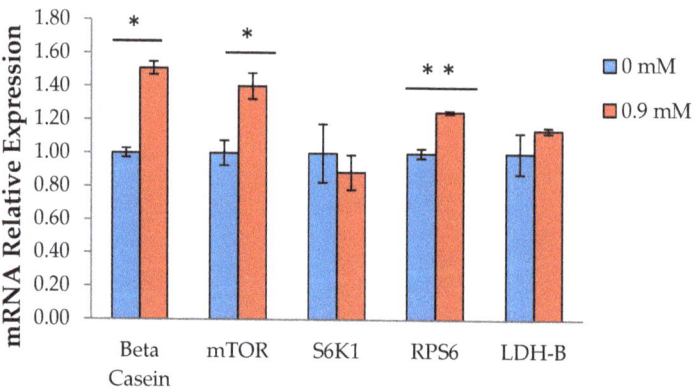

Figure 3. Changes in gene expression levels by addition of 0.9 mM L-Trp to MAC-T cells at 48 h. Analyzed by t-test between 0 and 0.9 mM L-Trp at 48 h: * $p < 0.05$, ** $p < 0.01$.

2.3. Proteome Analysis

In total, the addition of 0.9 mM L-Trp caused upregulation of 51 proteins and downregulation of 59 proteins, many of which are involved in protein synthesis (Table 1). The overall lists of upregulated as well as downregulated proteins in MAC-T cell are found in Tables S9 and S10. The result of protein pathway analysis showed that L-Trp stimulated glycolysis, the pentose phosphate pathway, and ATP synthesis, which are pathways involved in and related to energy metabolism (Table 2). Lastly, to summarize the results, a diagram of the effect of L-Trp supplementation on milk protein synthesis pathway was created (Figure 4).

Table 1. Differentially expressed proteins in MAC-T cells supplemented with L-Trp compared with control.

Detected Proteins		
Number of protein increased		51
Number of protein decreased		59
Selected downregulated and upregulated proteins		▲▼
HSPD1 (60 kDa heat shock protein, mitochondrial)	HSPD1	▲
HSPA1A (Heat shock 70 kDa protein 1A)	HSPA1A	▲
ATP5B (ATP synthase subunit beta, mitochondrial)	ATP5B	▼
EEF1A1 (Elongation factor 1 alpha 1)	EEF1A1	▼
RPSA (Similar to 40S ribosomal protein SA (fragment))	RPSA	▼
ATP synthase subunit alpha, mitochondrial	ATP5A1	▼
RPS18 (40S ribosomal protein S18)	RPS18	▲
EIF4A1 (Eukaryotic initiation factor 4A-I)	EIF4A1	▼
EEF2 (Elongation factor 2)	EEF2	▼
RPS25 (40S ribosomal protein S25)	RPS25	▲
EEF1G (Elongation factor 1-gamma)	EEF1G	▼
RPN2 (Dolichyl-diphosphooligosaccharide–protein glycosyltransferase)	RPN2	▼
ATP1A2 (Sodium/potassium-transporting ATPase subunit alpha-2)	ATP1A2	▼
GPI (Glucose-6-phosphate isomerase)	GPI	▼
RPL11 (60S ribosomal protein L11)	RPL11	▼
RPS2 (40S ribosomal protein S2)	RPS2	▼

▲, Upregulated (>2-fold greater protein expression than control); ▼, downregulated (<0.5-fold greater protein expression than control).

Table 2. Protein and energy metabolism-related pathways stimulated by supplementation of L-Trp compared with control.

Detected Pathways *
Apoptosis signaling pathway
p53 pathway
Glycolysis
Pentose phosphate pathway
ATP synthesis
CCKR signaling map
Endothelin signaling pathway
FGF signaling pathway
Ras pathway
EGF receptor signaling pathway

* Significantly increased protein and energy metabolism-related pathways ($p < 0.05$) compared with control, as determined by the PANTHER online tool for *Bos taurus* (see Methods for data explanation).

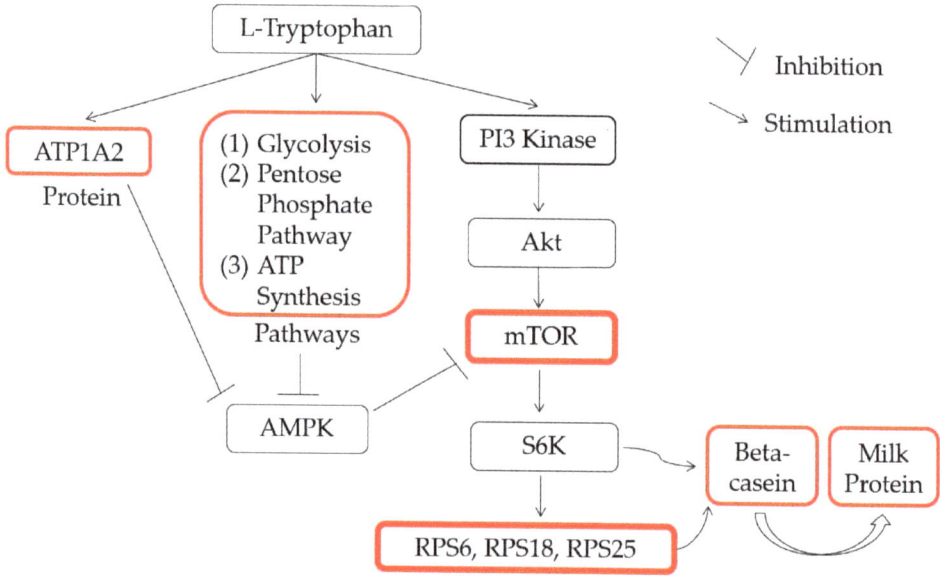

Red boxes are genes, proteins and pathways activated by L-Tryptophan

Figure 4. Diagram of the effect of L-Trp supplementation on milk protein synthesis pathways. ATP1A2, ATP1A2 sodium/potassium-transporting ATPase subunit alpha-2; AMPK, AMP-activated protein kinase; PI3 kinase, phosphoinositide 3-kinase; Akt, protein kinase B; mTOR, mammalian target of rapamycin; S6K, S6 kinase; RPS6, ribosomal protein S6; RPS18, 40S ribosomal protein S18; RPS25, 40S ribosomal protein S25.

3. Discussion

3.1. AA Time and Dosage Sampling

For the optimal time test, most concentrations peaked at 48 h, which was also the optimal incubation time for the secretion of medium protein by MAC-T cells. This is also the time when the secretion of β-casein was at its peak after L-Trp was added. The L-Trp dose test showed that 0.9 mM was the most effective concentration in increasing *CSN2* mRNA expression, indicating that this would be the optimal concentration level. This outcome suggests that 48 h should be selected to test different L-Trp concentrations because of the efficacy in increasing *CSN2* mRNA expression and protein synthesis in MAC-T cells in this time period.

3.2. CSN2 and Protein Synthesis-Related Gene Expression

Supplementation with 0.9 mM L-Trp also increased the relative expression of β-casein. In addition, L-Trp supplementation stimulated the mRNA relative expression of genes related to protein synthesis, especially *mTOR* and *RPS6*. A similar result was observed in a 9 h infusion study, in which a response in RPS6 phosphorylation to the addition of EAA plus glucose was reported [16]. Amino acids (AAs) not only act as substrates for protein synthesis, but also serve as signaling molecules that regulate synthesis [17,18]. The availability of AAs for mammary epithelial cells is of pivotal importance for the regulation of translation, and for the transport rate of AAs as one of the major limiting factors in protein synthesis [19–21]. In another study, a decline in mTOR signaling and fractional synthesis rate (FSR) in bovine mammary acini was observed when media were devoid of total EAA [22]. Hence, L-Trp may not only supply nutrients necessary for protein synthesis, but also regulate the expression of genes involved in milk protein synthesis in terms of nutrigenomics.

Specific AAs can affect translation initiation and elongation rates via two main pathways: the integrated stress response (ISR) and the mammalian target of rapamycin (mTOR) pathways [23]. There is direct evidence that AAs can increase mTOR phosphorylation and/or activity in the case of intact cells [24,25]. Amino acids affect mTOR signaling in bovine mammary epithelial (BME) cells [26], which is associated with increased milk protein synthesis in lactating cattle [16,27]. The stimulation of protein synthesis induced by AA [28] is known to be at least partially mediated by mTOR, a protein kinase present in rapamycin-sensitive mTOR complex 1 [29]. The mechanisms through which AA regulate mTOR stimulation are not yet fully understood, but it has been proposed that several protein factors, such as Ras homolog, class III PI3K Vps34, and Rag GTPases, mediate AA signaling on mTOR [30]. It is worth noting that among protein and energy metabolism-related pathways in this study, the Ras pathway was stimulated upon supplementation of L-Trp.

The mTOR pathway circles around mTOR complex 1 [31]. In mTOR complex 1, mTOR phosphorylates the downstream proteins that monitor the rate of translation initiation and elongation [31,32]. In previous studies, the effect of EAAs on the mTOR pathway was widely demonstrated in splanchnic, muscle, and mammary tissues [33–35]. When activated by AA, mTOR in turn catalyzes phosphorylation of S6K1 and 4EBP1 [36]. In the case of bovine mammary epithelial cells, a study showed that deprivation of all EAAs affected the phosphorylation of mTOR downstream proteins S6K1 and 4E-BP1 and fractional synthesis rates of β-LG [26]. This showed that infusion of EAAs plus glucose reduced the phosphorylation of the IRS-target eIF2 in mammary tissue and increased phosphorylation of the mTOR targets ribosomal S6 kinase 1 (S6K1) and S6 [16]. In addition, in another study, removal of L-Trp reduced S6K1 phosphorylation [37], which is consistent with a prior study in rat liver. Conversely, supplementing with L-Trp did not stimulate S6K1 gene expression, even though this is just expression and not phosphorylation. This is in accordance with the results of another study [38], in which Trp, Phe, and Met addition had no effect on S6K1 phosphorylation.

3.3. Proteomics Analysis

In the proteomics result, supplementation with L-Trp decreased EEF1A1, EEF2, and EEF1G protein expression. Although eEF2 protein is not a direct substrate of mTOR, an inverse relationship between phosphorylation of mTOR and eEF2 has been reported [39]. When eEF2 binds to the ribosome, eEF2 mediates the translocation step of the elongation process. Phosphorylation of eEF2 at Thr56 by eEF2 kinase decreases the affinity of eEF2 for the ribosome [40]. eEF2 phosphorylation, which decreases the elongation rate, is inhibited by the mTOR pathway downstream of kinase S6K1, which results in a positive effect of mTOR on the casein fractional synthesis rate (CFSR) [23], in line with the results obtained in this study. It has been suggested that eEF2 may be a limiting factor in milk protein synthesis [40]. The negative relationship between phosphorylated eEF2 and protein synthesis rates indicates an important role for eEF2 in mammary protein synthesis [37]. Another study also showed an inverse relationship between eEF2 phosphorylation and milk protein yield in dairy cows treated with growth hormone [12]. It has been reported that protein synthesis rates were strongly associated with phosphorylation of eEF2 in mammary tissue slices [37]. The aforementioned pathway review may explain the obtained results.

RPS6 was also negatively correlated with eEF2 [8]. This also agrees with our result that there was increased gene expression of RPS6 but decreased protein expression of eEF2. It has long been known that the addition of a physiological mixture of AAs to hepatocytes will result in strong and rapid phosphorylation of RPS6 [41,42]. RPS6 protein is a component of the 40S ribosomal subunit and one of the endpoints of insulin signaling, and phosphorylation of this protein is needed for the translation of certain mRNA molecules encoding proteins in the protein-translation machinery [30]. The increased expression of RPS6 coincides with the increased mTOR gene expression. This is in accordance with a previous study in which the AA-stimulated phosphorylation of RPS6 was completely inhibited by applying rapamycin, implying that mTOR3 is upstream of RPS6 in the AA signaling pathway [41,42].

It is also especially important to note the decreased expression of eukaryotic initiation factor 4A-I (EIF4A1) protein. Protein synthesis rates primarily depend on translation initiation and elongation rates, which are regulated by several eukaryotic initiation factors (eIFs) and elongation factors (eEFs) [43]. Activation of these proteins occurs through changes in their phosphorylation state, influenced by the addition of AAs and hormones [44]. Accordingly, studies have shown that AAs are likely to regulate eIF and eEF phosphorylation in skeletal muscle and liver cells via the mTOR signaling pathway [33,45].

3.4. Metabolic Pathway Analysis

Protein and energy metabolism-related pathway results showed that glycolysis, the pentose phosphate pathway, and ATP synthesis were stimulated upon supplementation with L-Trp. These energy-related pathways are stimulated in parallel with the increased protein synthesis rates due to the fact that milk protein synthesis is an energy-consuming process [46,47]. The interaction between energy and protein supplementation in the phosphorylation of mTOR in dairy cows infused for 36 h with starch and energy has been reported [48]. In that study, they investigated signaling pathways responsive to casein and starch infusion in primiparous mid-lactation Holstein cows. The results showed that cell signaling molecules involved in the regulation of milk protein synthesis responded differently to the various nutritional stimuli, and the phosphorylation of mTOR was increased in response to starch when casein was infused.

4. Materials and Methods

4.1. AA Dose and Sampling Time

Immortalized mammary epithelial (MAC-T) cells [38] from McGill University, Canada, were grown in 10 cm dishes (TPP, Trasadingen, Switzerland). MAC-T cells were incubated in DMEM/F12 basic medium for 72 h or until the cells reached 90% confluency. The measurement of doubling time of MAC-T cell and the method of determining cells reaching 90% confluency are found in Figures S1 and S2. The cells were then seeded into 6-well U-shaped multiwall plates (BD Falcon™, Franklin Lakes, NJ, USA). They were cultured in DMEM/F12 basic medium (Thermo Scientific, South Logan, UT, USA) supplemented with 10% fetal bovine serum (FBS), 100 units/mL penicillin/streptomycin (Thermo Scientific, South Logan, UT, USA), 5 µg/mL insulin, 1 µg/mL hydrocortisone, and 50 µg/mL gentamycin (Sigma-Aldrich Corp., St. Louis, MO, USA) at 37 °C in a 5% CO_2 incubator [49,50]. When MAC-T cells reached 90% confluence, DMEM/F12 basic medium was replaced with DMEM/F12 lactogenic medium (without FBS) to differentiate MAC-T cells into β-casein (CSN2) secreting cells for 72 h. This medium contained 5 µg/mL bovine insulin, 1 µg/mL hydrocortisone, 100 units/mL penicillin/streptomycin, 50 µg/mL gentamycin, and 5 µg/mL prolactin (Sigma-Aldrich Corp., St. Louis, MO, USA) (Wang et al., 2014, 2015). The complete amino acid profile of lactogenic medium is listed in Table S11. After cell differentiation, a preliminary experiment was performed for the time and dose testing. Cells were treated with 0, 0.3, 0.6, 0.9, 1.2, and 1.5 mM of supplemental L-Trp, and the most efficient time for protein synthesis was determined by measuring cell, medium, and total protein at 0, 24, 48, 72, and 96 h. Then, a confirmatory experiment was performed to determine the ideal dose of L-Trp in the determined optimal cultivation time. Different doses of L-Trp (0, 0.3, 0.6, 0.9, 1.2, and 1.5 mM) were incubated at 48 h, which is the optimal time for protein synthesis. In all experiments, each treatment was replicated 6 times.

4.2. RNA Extraction and cDNA Synthesis

Total RNA was extracted from MAC-T cells using TRIzol® (Life Technologies Corp., Carlsbad, CA, USA). The RNA quality and quantity were measured using a NanoDrop 1000® Spectrophotometer with RNA-40 module (Thermo Fisher Scientific, Wilmington,

DE, USA). RNA integrity number (RIN) was determined using an Agilent 2100 Bioanalyzer (Agilent Technologies Inc., Santa Clara, CA, USA). Then cDNA was prepared using an iScript cDNA synthesis kit (Bio-Rad Laboratories, Inc., Foster City, CA, USA) according to the manufacturer's instructions. After incubating at 25 °C for 5 min, 42 °C for 30 min, and 85 °C for 5 min, the cDNA was quantified using the ssDNA-33 module of the Thermo NanoDrop 1000® Spectrophotometer (Thermo Fisher Scientific, Wilmington, DE, USA).

4.3. Real-Time Polymerase Chain Reaction (RT-PCR)

Real-time PCR (RT-PCR) analysis was performed using a T100™ Thermal Cycler System. ACTB was used as the reference gene. Validated RT-PCR oligonucleotide primer sequences of forward and reverse primers specific for target genes were as follows: *CSN2* forward, 5′-AAATCTGCACCTTCCTCTGC-3′; *CSN2* reverse, 5′-GAACAGGCAG GACTTTGGAC-3′; *ACTB* forward, 5′-GCATGGAATCCTGCGGC-3′; *ACTB* reverse, 5′-GTAGAGGTCCTTGCGGATGT-3′. RT-PCR reactions were performed by initial incubation at 95 °C for 3 min followed by 50 cycles of denaturation at 95 °C for 10 s, annealing at specific temperature for 15 s (bovine CSN2 at 55 °C), and extension at 72 °C for 30 s (Table S12). RT-PCR analysis was conducted using the threshold cycle (2-$\Delta\Delta$CT method) [51] to analyze relative gene expression changes from real-time quantitative PCR experiments. Relative quantification of expression levels of target genes in the treatment group was compared to the untreated group.

4.4. Protein Extraction and Quantification

After incubation in the treatment medium (lactogenic medium plus L-Trp) for 72 h, the culture medium was collected from adherent cells to determine protein quantity. All treatments were done with six replicates. The culture medium was centrifuged at $300\times g$ for 5 min at 4 °C. The supernatant was transferred to a new tube for protein quantification using a Pierce BCA Protein Assay Kit (Pierce Biotechnology, Inc., Rockford, IL, USA) according to the manufacturer's instructions. Cells were washed twice with ice cold $1\times$ PBS and then 200 µL cell lysis buffer containing 10 mM Tris/HCl, pH 8.3, 8 M urea, 5 mM EDTA, 4% CHAPS, and $1\times$ protease inhibitor cocktail (GE Healthcare, Piscataway, NJ, USA) were added. The cell lysates were incubated at room temperature for 30 min and centrifuged at $21,952\times g$ for 30 min at 20 °C. The supernatant was collected and stored at −80 °C until analysis.

4.5. Proteome Analysis

Cellular proteins were extracted using cell lysis buffer containing 20 mM Tris, 10 mM KCl, 1.5 mM $MgCl_2$, 0.5 mM EDTA, 0.1% sodium dodecyl sulfate (SDS), and complete EDTA-free Protease Inhibitor Cocktail (Roche, Basel, Switzerland) after washing twice with ice-cold $1\times$ PBS. Cell lysates were incubated at 4 °C for 30 min and centrifuged at $13,000\times g$ for 10 min at 4 °C [41]. Then, supernatant was collected and stored at −80 °C until analysis. All treatments were replicated three times.

For proteome analysis, 100 µg of cell lysate protein was resuspended in 0.1% SDS in 50 mM triethyl ammonium bicarbonate (TEABC), pH 8.0. Proteins were chemically denatured using 10 mM tris (2-carboxyethyl) phosphine (TCEP) at 60 °C for 30 min and alkylated with 50 mM methyl methanethiosulfonate (MMTS) at room temperature for 30 min in the dark. Proteolytic digestion was conducted using trypsin (protein:trypsin = 50:1, g/g) overnight at 37 °C. Digested peptides were desalted and concentrated, then subjected to liquid chromatography (LC) tandem mass spectrometry (MS/MS) analysis. Total peptides were analyzed by nano-ultra-performance LC–MS/electrospray ionization quadrupole time-of-flight (UPLC–MS/ESI-Q-TOF) (Waters, Manchester, UK). LC peptide separation was performed using the nano Acquity system equipped with a Symmetry C18 5 µm, 5 mm × 300 µm pre-column and CSH C18 1.7 µm, 25 cm × 75 µm analytical column (Waters). Samples were separated using a 3–40% gradient mob Leu phase B (0.1% formic acid in acetonitrile) at a flow rate of 300 nL/min, followed by a 20 min rinse with 90% mob Leu phase B.

Data-dependent analysis (DDA) was performed to obtain two analytical replicates for each of three biological sets. This method is used to read a full MS scan in an m/z range of 400–1600 every 0.5 s and MS/MS scans (m/z range: 100–1990) every 0.5 s for the three most intense ions among the full-scan MS. Protein identification was performed by comparison with the International Protein Index (IPI) bovine database (v. 3.73; 30,403 entries) using the MASCOT search engine v. 2.4 (Matrix Science, Boston, MA, USA), using trypsin as the digestion enzyme, with a parent ion tolerance of 0.2 Da and fragment ion mass tolerance of 0.1 Da. Two missed cleavages were allowed during trypsin digestion. Oxidation (Met) and Methylthio (Cys) were specified as the modification settings. Proteins identified with >95% probability were filtered out. To evaluate the false discovery rate (FDR) of protein identification, data were searched against a combined database of normal and decoy data created by MASCOT. The FDR of all experiments in this study was <1%. The emPAI score of each protein was used to calculate its relative ratio [52]. The emPAI-based abundances in comparison with the actual values were within 63% on average, and this is similar or better than determination of abundance by protein staining [52].

4.6. Statistical Analysis

Statistical analysis (protein quantification data, $n = 6$; proteomics data, $n = 3$; quantitative real-time PCR data, $n = 6$) was conducted for significance testing using SAS v. 9.4 software (SAS Institute, Cary, NC, USA). Data were analyzed using Student's *t*-test. Mean difference was considered statistically significant at $p < 0.05$. The experimental model was

$$Y_{ijk} = \mu + \tau_i + \varepsilon_{ijk},$$

where μ is the grand mean, τi is the L-Trp effect, and ε_{ijk} is the error variability.

Upregulated or downregulated proteins were detected for significance using the semi-quantification relative ratio (≥ 2 or ≤ 0.5), and detected proteins were analyzed using the PANTHER database (http://www.pantherdb.org accessed on 19 August 2020) for pathway analysis (*Bos taurus*).

5. Conclusions

Altogether, 0.9 mM L-Trp supplementation was found to be the optimum dose for stimulating medium protein and β-casein mRNA expression by stimulating the expression of genes related to milk protein synthesis and the increased production of proteins involved in energy metabolism and protein synthesis pathways. Protein and energy metabolism related pathways were also upregulated in the L-Trp treatment, eventually causing increased protein concentration in the MAC-T cell medium supplemented with L-Trp. In conclusion, L-Trp, particularly at 0.9 mM, was effective in increasing protein synthesis in MAC-T cells in vitro by stimulating the genes, proteins, and protein synthesis-related pathways involved in energy and protein synthesis. Also, our findings suggest that L-Trp may not only supply nutrients necessary for protein synthesis, but also regulate the expression of genes involved in milk protein synthesis in terms of nutrigenomics.

Supplementary Materials: The following are available online at https://www.mdpi.com/1422-0067/22/5/2751/s1.

Author Contributions: Conceptualization, H.-G.L., J.R.V.C., and J.-O.M.; methodology, H.-G.L., J.R.V.C., J.-E.K., J.G.N., J.-S.L., and J.-O.M.; software, J.G.N.; formal analysis, J.R.V.C., J.-E.K.; resources, J.-O.M., J.-S.L., and H.-G.L.; data curation, J.G.N., H.-G.L.; writing—original draft preparation, J.R.V.C. and J.G.N.; writing—review & editing, J.R.V.C., J.G.N., and H.-G.L.; supervision, H.-G.L.; project administration, H.-G.L., J.-S.L., and J.-O.M.; funding acquisition, H.-G.L., and J.-O.M. All authors have read and agreed to the published version of the manuscript.

Funding: This research was supported by the Korea Institute of Planning and Evaluation for Technology in Food, Agriculture, Forestry (IPET) through the Agri-Bio Industry Technology Development Program (117030-03-3-HD020).

Institutional Review Board Statement: Not applicable.

Informed Consent Statement: Not applicable.

Data Availability Statement: Data will be available upon request.

Acknowledgments: This research was supported by the Korea Institute of Planning and Evaluation for Technology in Food, Agriculture, Forestry (IPET) through the Agri-Bio Industry Technology Development Program (117030-03-3-HD020). This paper was supported by the KU Research Professor Program of Konkuk University.

Conflicts of Interest: The authors declare no conflict of interest. There was no involvement of the funders in any role pertaining to the choice of research project; design of the study; in the collection, analyses or interpretation of data; in the writing of the manuscript; or in the decision to publish the results.

Abbreviations

CSN2	Beta-casein (β-casein)
ATP1A2	ATP1A2 Sodium/potassium-transporting ATPase subunit alpha-2
AMPK	AMP-activated protein kinase
PI3 Kinase	Phosphoinositide 3-kinase
Akt	Protein kinase B
mTOR	mammalian target of rapamycin
S6K	S6 Kinase
RPS6	Ribosomal protein S6
RPS18	40S ribosomal protein S18
RPS25	40S ribosomal protein S25

References

1. Haque, M.N.; Guinard-Flament, J.; Lamberton, P.; Mustière, C.; Lemosquet, S. Changes in Mammary Metabolism in Response to the Provision of an Ideal Amino Acid Profile at 2 Levels of Metabolizable Protein Supply in Dairy Cows: Consequences on Efficiency. *J. Dairy Sci.* **2015**, *98*, 3951–3968. [CrossRef] [PubMed]
2. Li, S.S.; Loor, J.J.; Liu, H.Y.; Liu, L.; Hosseini, A.; Zhao, W.S.; Liu, J.X. Optimal Ratios of Essential Amino Acids Stimulate β-Casein Synthesis via Activation of the Mammalian Target of Rapamycin Signaling Pathway in MAC-T Cells and Bovine Mammary Tissue Explants. *J. Dairy Sci.* **2017**, *100*, 6676–6688. [CrossRef]
3. Walczak, K.; Langner, E.; Makuch-Kocka, A.; Szelest, M.; Szalast, K.; Marciniak, S.; Plech, T. Effect of Tryptophan-Derived AhR Ligands, Kynurenine, Kynurenic Acid and FICZ, on Proliferation, Cell Cycle Regulation and Cell Death of Melanoma Cells—In Vitro Studies. *Int. J. Mol. Sci.* **2020**, *21*, 7946. [CrossRef]
4. Lee, S.-B.; Lee, K.-W.; Wang, T.; Lee, J.-S.; Jung, U.-S.; Nejad, J.G.; Oh, Y.-K.; Baek, Y.-C.; Kim, K.H.; Lee, H.-G. Intravenous Administration of L-Tryptophan Stimulates Gastrointestinal Hormones and Melatonin Secretions: Study on Beef Cattle. *J. Anim. Sci. Technol.* **2019**, *61*, 239–244. [CrossRef] [PubMed]
5. Lee, S.-B.; Lee, K.-W.; Wang, T.; Lee, J.-S.; Jung, U.-S.; Nejad, J.G.; Oh, Y.-K.; Baek, Y.-C.; Kim, K.H.; Lee, H.-G. Administration of Encapsulated L-Tryptophan Improves Duodenal Starch Digestion and Increases Gastrointestinal Hormones Secretions in Beef Cattle. *Asian Australas. J. Anim. Sci.* **2020**, *33*, 91–99. [CrossRef]
6. Priatno, W.; Jo, Y.-H.; Nejad, J.G.; Lee, J.-S.; Moon, J.-O.; Lee, H.-G. "Dietary Supplementation of L-Tryptophan" Increases Muscle Development, Adipose Tissue Catabolism and Fatty Acid Transportation in the Muscles of Hanwoo Steers. *J. Anim. Sci. Technol.* **2020**, *62*, 595–604. [CrossRef]
7. Barik, S. The Uniqueness of Tryptophan in Biology: Properties, Metabolism, Interactions and Localization in Proteins. *Int. J. Mol. Sci.* **2020**, *21*, 8776. [CrossRef] [PubMed]
8. Mazinani, M.; Naserian, A.; Rude, B.; Tahmasbi, A.; Valizadeh, R. Effects of Feeding Rumen-Protected Amino Acids on the Performance of Feedlot Calves. *J. Adv. Vet. Anim. Res.* **2020**, *7*, 229. [CrossRef]
9. Nayak, B.N.; Buttar, H.S. Evaluation of the Antioxidant Properties of Tryptophan and Its Metabolites in in Vitro Assay. *J. Complement. Integr. Med.* **2016**, *13*, 129–136. [CrossRef]
10. Friedman, M. Analysis, Nutrition, and Health Benefits of Tryptophan. *Int. J. Tryptophan Res.* **2018**, *11*, 117864691880228. [CrossRef]
11. Gostner, J.M.; Geisler, S.; Stonig, M.; Mair, L.; Sperner-Unterweger, B.; Fuchs, D. Tryptophan Metabolism and Related Pathways in Psychoneuroimmunology: The Impact of Nutrition and Lifestyle. *Neuropsychobiology* **2020**, *79*, 89–99. [CrossRef] [PubMed]

12. Strasser, B.; Geiger, D.; Schauer, M.; Gatterer, H.; Burtscher, M.; Fuchs, D. Effects of Exhaustive Aerobic Exercise on Tryptophan-Kynurenine Metabolism in Trained Athletes. *PLoS ONE* **2016**, *11*, e0153617. [CrossRef] [PubMed]
13. Ninomiya, S.; Nakamura, N.; Nakamura, H.; Mizutani, T.; Kaneda, Y.; Yamaguchi, K.; Matsumoto, T.; Kitagawa, J.; Kanemura, N.; Shiraki, M.; et al. Low Levels of Serum Tryptophan Underlie Skeletal Muscle Atrophy. *Nutrients* **2020**, *12*, 978. [CrossRef] [PubMed]
14. Areta, J.L.; Burke, L.M.; Camera, D.M.; West, D.W.D.; Crawshay, S.; Moore, D.R.; Stellingwerff, T.; Phillips, S.M.; Hawley, J.A.; Coffey, V.G. Reduced Resting Skeletal Muscle Protein Synthesis Is Rescued by Resistance Exercise and Protein Ingestion Following Short-Term Energy Deficit. *Am. J. Physiol. Endocrinol. Metab.* **2014**, *306*, E989–E997. [CrossRef] [PubMed]
15. Török, N.; Tanaka, M.; Vécsei, L. Searching for Peripheral Biomarkers in Neurodegenerative Diseases: The Tryptophan-Kynurenine Metabolic Pathway. *Int. J. Mol. Sci.* **2020**, *21*, 9338. [CrossRef] [PubMed]
16. Toerien, C.A.; Trout, D.R.; Cant, J.P. Nutritional Stimulation of Milk Protein Yield of Cows Is Associated with Changes in Phosphorylation of Mammary Eukaryotic Initiation Factor 2 and Ribosomal S6 Kinase 1. *J. Nutr.* **2010**, *140*, 285–292. [CrossRef]
17. Kimball, S.R. Regulation of Global and Specific MRNA Translation by Amino Acids. *J. Nutr.* **2002**, *132*, 883–886. [CrossRef]
18. Meijer, A.J. Amino Acids as Regulators and Components of Nonproteinogenic Pathways. *J. Nutr.* **2003**, *133*, 2057S–2062S. [CrossRef]
19. Baumrucker, C.R. Amino Acid Transport Systems in Bovine Mammary Tissue. *J. Dairy Sci.* **1985**, *68*, 2436–2451. [CrossRef]
20. Reynolds, C.K.; Harmon, D.L.; Cecava, M.J. Absorption and Delivery of Nutrients for Milk Protein Synthesis by Portal-Drained Viscera. *J. Dairy Sci.* **1994**, *77*, 2787–2808. [CrossRef]
21. Shennan, D.B.; Peaker, M. Transport of Milk Constituents by the Mammary Gland. *Physiol. Rev.* **2000**, *80*, 925–951. [CrossRef] [PubMed]
22. Burgos, S.A.; Dai, M.; Cant, J.P. Nutrient Availability and Lactogenic Hormones Regulate Mammary Protein Synthesis through the Mammalian Target of Rapamycin Signaling Pathway. *J. Dairy Sci.* **2010**, *93*, 153–161. [CrossRef] [PubMed]
23. Arriola Apelo, S.I.; Singer, L.M.; Lin, X.Y.; McGilliard, M.L.; St-Pierre, N.R.; Hanigan, M.D. Isoleucine, Leucine, Methionine, and Threonine Effects on Mammalian Target of Rapamycin Signaling in Mammary Tissue. *J. Dairy Sci.* **2014**, *97*, 1047–1056. [CrossRef]
24. Peterson, R.T.; Desai, B.N.; Hardwick, J.S.; Schreiber, S.L. Protein Phosphatase 2A Interacts with the 70-KDa S6 Kinase and Is Activated by Inhibition of FKBP12-Rapamycinassociated Protein. *Proc. Natl. Acad. Sci. USA* **1999**, *96*, 4438–4442. [CrossRef]
25. Navé, B.T.; Ouwens, M.; Withers, D.J.; Alessi, D.R.; Shepherd, P.R. Mammalian Target of Rapamycin Is a Direct Target for Protein Kinase B: Identification of a Convergence Point for Opposing Effects of Insulin and Amino-Acid Deficiency on Protein Translation. *Biochem. J.* **1999**, *344*, 427–431. [CrossRef] [PubMed]
26. Moshel, Y.; Rhoads, R.E.; Barash, I. Role of Amino Acids in Translational Mechanisms Governing Milk Protein Synthesis in Murine and Ruminant Mammary Epithelial Cells. *J. Cell. Biochem.* **2006**, *98*, 685–700. [CrossRef] [PubMed]
27. Hayashi, A.A.; Nones, K.; Roy, N.C.; McNabb, W.C.; Mackenzie, D.S.; Pacheco, D.; McCoard, S. Initiation and Elongation Steps of MRNA Translation Are Involved in the Increase in Milk Protein Yield Caused by Growth Hormone Administration during Lactation. *J. Dairy Sci.* **2009**, *92*, 1889–1899. [CrossRef]
28. Kim, E. Mechanisms of Amino Acid Sensing in MTOR Signaling Pathway. *Nutr. Res. Pract.* **2009**, *3*, 64. [CrossRef]
29. Yang, Q.; Inoki, K.; Kim, E.; Guan, K.-L. TSC1/TSC2 and Rheb Have Different Effects on TORC1 and TORC2 Activity. *Proc. Natl. Acad. Sci. USA* **2006**, *103*, 6811–6816. [CrossRef]
30. Goberdhan, D.C.I.; Ogmundsdóttir, M.H.; Kazi, S.; Reynolds, B.; Visvalingam, S.M.; Wilson, C.; Boyd, C.A.R. Amino Acid Sensing and MTOR Regulation: Inside or Out? *Biochem. Soc. Trans.* **2009**, *37*, 248–252. [CrossRef]
31. Mahoney, S.J.; Dempsey, J.M.; Blenis, J. Cell Signaling in Protein Synthesis Ribosome Biogenesis and Translation Initiation and Elongation. *Prog. Mol. Biol. Transl. Sci.* **2009**, *90*, 53–107. [CrossRef]
32. Dunlop, E.A.; Tee, A.R. Mammalian Target of Rapamycin Complex 1: Signalling Inputs, Substrates and Feedback Mechanisms. *Cell. Signal.* **2009**, *21*, 827–835. [CrossRef] [PubMed]
33. O'Connor, P.M.J.; Kimball, S.R.; Suryawan, A.; Bush, J.A.; Nguyen, H.V.; Jefferson, L.S.; Davis, T.A. Regulation of Neonatal Liver Protein Synthesis by Insulin and Amino Acids in Pigs. *Am. J. Physiol. Endocrinol. Metab.* **2004**, *286*, E994–E1003. [CrossRef] [PubMed]
34. Murgas Torrazza, R.; Suryawan, A.; Gazzaneo, M.C.; Orellana, R.A.; Frank, J.W.; Nguyen, H.V.; Fiorotto, M.L.; El-Kadi, S.; Davis, T.A. Leucine Supplementation of a Low-Protein Meal Increases Skeletal Muscle and Visceral Tissue Protein Synthesis in Neonatal Pigs by Stimulating MTOR-Dependent Translation Initiation. *J. Nutr.* **2010**, *140*, 2145–2152. [CrossRef] [PubMed]
35. Appuhamy, J.A.D.R.N.; Knoebel, N.A.; Nayananjalie, W.A.D.; Escobar, J.; Hanigan, M.D. Isoleucine and Leucine Independently Regulate MTOR Signaling and Protein Synthesis in MAC-T Cells and Bovine Mammary Tissue Slices. *J. Nutr.* **2012**, *142*, 484–491. [CrossRef] [PubMed]
36. Wang, X.; Proud, C.G. The MTOR Pathway in the Control of Protein Synthesis. *Physiology* **2006**, *21*, 362–369. [CrossRef] [PubMed]
37. Appuhamy, J.A.D.R.N.; Bell, A.L.; Nayananjalie, W.A.D.; Escobar, J.; Hanigan, M.D. Essential Amino Acids Regulate Both Initiation and Elongation of MRNA Translation Independent of Insulin in MAC-T Cells and Bovine Mammary Tissue Slices. *J. Nutr.* **2011**, *141*, 1209–1215. [CrossRef] [PubMed]
38. Prizant, R.L.; Barash, I. Negative Effects of the Amino Acids Lys, His, and Thr on S6K1 Phosphorylation in Mammary Epithelial Cells. *J. Cell. Biochem.* **2008**, *105*, 1038–1047. [CrossRef]
39. Wang, X. Regulation of Elongation Factor 2 Kinase by P90RSK1 and P70 S6 Kinase. *EMBO J.* **2001**, *20*, 4370–4379. [CrossRef]

40. Christophersen, C.T.; Karlsen, J.; Nielsen, M.O.; Riis, B. Eukaryotic Elongation Factor-2 (EEF-2) Activity in Bovine Mammary Tissue in Relation to Milk Protein Synthesis. *J. Dairy Res.* **2002**, *69*, 205–212. [CrossRef]
41. Luiken, J.J.F.P.; Blommaart, E.F.C.; Boon, L.; van Woerkom, G.M.; Meijer, A.J. Cell Swelling and the Control of Autophagic Proteolysis in Hepatocytes: Involvement of Phosphorylation of Ribosomal Protein S6? *Biochem. Soc. Trans.* **1994**, *22*, 508–511. [CrossRef] [PubMed]
42. Blommaart, E.F.; Luiken, J.J.; Blommaart, P.J.; van Woerkom, G.M.; Meijer, A.J. Phosphorylation of Ribosomal Protein S6 Is Inhibitory for Autophagy in Isolated Rat Hepatocytes. *J. Biol. Chem.* **1995**, *270*, 2320–2326. [CrossRef] [PubMed]
43. Connors, M.T.; Poppi, D.P.; Cant, J.P. Protein Elongation Rates in Tissues of Growing and Adult Sheep. *J. Anim. Sci.* **2008**, *86*, 2288–2295. [CrossRef] [PubMed]
44. Jackson, R.J.; Hellen, C.U.T.; Pestova, T.V. The Mechanism of Eukaryotic Translation Initiation and Principles of Its Regulation. *Nat. Rev. Mol. Cell Biol.* **2010**, *11*, 113–127. [CrossRef] [PubMed]
45. Orellana, R.A.; Jeyapalan, A.; Escobar, J.; Frank, J.W.; Nguyen, H.V.; Suryawan, A.; Davis, T.A. Amino Acids Augment Muscle Protein Synthesis in Neonatal Pigs during Acute Endotoxemia by Stimulating MTOR-Dependent Translation Initiation. *Am. J. Physiol. Endocrinol. Metab.* **2007**, *293*, E1416–E1425. [CrossRef] [PubMed]
46. Lobley, G. Energy Metabolism Reactions in Ruminant Muscle: Responses to Age, Nutrition and Hormonal Status. *Reprod. Nutr. Dév.* **1990**, *30*, 13–34. [CrossRef]
47. Hanigan, M.D.; Baldwin, R.L. A Mechanistic Model of Mammary Gland Metabolism in the Lactating Cow. *Agric. Syst.* **1994**, *45*, 369–419. [CrossRef]
48. Rius, A.G.; Appuhamy, J.A.D.R.N.; Cyriac, J.; Kirovski, D.; Becvar, O.; Escobar, J.; McGilliard, M.L.; Bequette, B.J.; Akers, R.M.; Hanigan, M.D. Regulation of Protein Synthesis in Mammary Glands of Lactating Dairy Cows by Starch and Amino Acids. *J. Dairy Sci.* **2010**, *93*, 3114–3127. [CrossRef]
49. Wang, T.; Lim, J.-N.; Bok, J.-D.; Kim, J.-H.; Kang, S.-K.; Lee, S.-B.; Hwang, J.-H.; Lee, K.-H.; Kang, H.-S.; Choi, Y.-J.; et al. Association of Protein Expression in Isolated Milk Epithelial Cells and Cis-9, Trans-11 Conjugated Linoleic Acid Proportions in Milk from Dairy Cows: Association of Protein and Cis-9, Trans-11 CLA in Milk. *J. Sci. Food Agric.* **2014**, *94*, 1835–1843. [CrossRef]
50. Wang, T.; Lee, S.B.; Hwang, J.H.; Lim, J.N.; Jung, U.S.; Kim, M.J.; Kang, H.S.; Choi, S.H.; Lee, J.S.; Roh, S.G.; et al. Proteomic Analysis Reveals PGAM1 Altering Cis-9, Trans-11 Conjugated Linoleic Acid Synthesis in Bovine Mammary Gland. *Lipids* **2015**, *50*, 469–481. [CrossRef] [PubMed]
51. Livak, K.J.; Schmittgen, T.D. Analysis of Relative Gene Expression Data Using Real-Time Quantitative PCR and the $2^{-\Delta\Delta CT}$ Method. *Methods* **2001**, *25*, 402–408. [CrossRef] [PubMed]
52. Ishihama, Y.; Oda, Y.; Tabata, T.; Sato, T.; Nagasu, T.; Rappsilber, J.; Mann, M. Exponentially Modified Protein Abundance Index (EmPAI) for Estimation of Absolute Protein Amount in Proteomics by the Number of Sequenced Peptides per Protein. *Mol. Cell. Proteom.* **2005**, *4*, 1265–1272. [CrossRef] [PubMed]

International Journal of
Molecular Sciences

Review

Tryptophan Metabolism via Kynurenine Pathway: Role in Solid Organ Transplantation

Ruta Zulpaite [1,2], Povilas Miknevicius [1,2], Bettina Leber [1], Kestutis Strupas [2], Philipp Stiegler [1,*] and Peter Schemmer [1]

1. General, Visceral and Transplant Surgery, Department of Surgery, Medical University of Graz, Auenbruggerpl. 2, 8036 Graz, Austria; ruta.zulp@gmail.com (R.Z.); povilui.m@gmail.com (P.M.); bettina.leber@medunigraz.at (B.L.); peter.schemmer@medunigraz.at (P.S.)
2. Faculty of Medicine, Vilnius University, M. K. Ciurlionio 21, 03101 Vilnius, Lithuania; kestutis.strupas@santa.lt
* Correspondence: philipp.stiegler@medunigraz.at; Tel.: +43-316-385-84094

Abstract: Solid organ transplantation is a gold standard treatment for patients suffering from an end-stage organ disease. Patient and graft survival have vastly improved during the last couple of decades; however, the field of transplantation still encounters several unique challenges, such as a shortage of transplantable organs and increasing pool of extended criteria donor (ECD) organs, which are extremely prone to ischemia-reperfusion injury (IRI), risk of graft rejection and challenges in immune regulation. Moreover, accurate and specific biomarkers, which can timely predict allograft dysfunction and/or rejection, are lacking. The essential amino acid tryptophan and, especially, its metabolites via the kynurenine pathway has been widely studied as a contributor and a therapeutic target in various diseases, such as neuropsychiatric, autoimmune disorders, allergies, infections and malignancies. The tryptophan-kynurenine pathway has also gained interest in solid organ transplantation and a variety of experimental studies investigating its role both in IRI and immune regulation after allograft implantation was first published. In this review, the current evidence regarding the role of tryptophan and its metabolites in solid organ transplantation is presented, giving insights into molecular mechanisms and into therapeutic and diagnostic/prognostic possibilities.

Keywords: transplantation; ischemia-reperfusion; tolerance; rejection; tryptophan; kynurenine; indoleamine-2,3-dioxygenase

1. Introduction

Solid organ transplantation (Tx) remains the gold standard and the only curative treatment for patients with end-stage organ disease. A substantial improvement in patient and graft survival has been observed during the last decades, mainly due to technical developments and advancements in immunosuppressive regimens [1]. However, there are not sufficient amounts of donor organs to treat all patients on the waiting list for whole organ Tx [2–4]. Therefore, it is inevitable to utilize organs retrieved from extended criteria donors (ECD). These grafts are known to be extremely susceptible to ischemia-reperfusion injury (IRI), often resulting in delayed graft function (DGF) or primary non-function (PNF) after implantation. Therefore, additional efforts to diminish the harm inflicted during organ preservation techniques are necessary [5,6].

The problem of allograft rejection is another important challenge limiting Tx success. The increased variety of effective immunosuppressants resulted in a substantial reduction of acute graft loss rates [7,8]. However, overall graft survival still remains time-limited, mainly due to chronic rejection and subsequent chronic allograft dysfunction [7,8]. The increased risk of organ rejection especially poses a barrier for highly sensitized patients who were already exposed to an antigen or need a repeated Tx [9].

Moreover, specific and sensitive biomarkers to predict organ quality and Tx success are needed in order to expand the potential donor organ pool and transplant the organs safely. Biomarkers for the detection and diagnosis of rejection and/or allograft dysfunction, which would allow a more thorough follow-up and personalized care for recipients at risk, is also of particular interest [10,11] to ameliorate the long-term results after Tx.

Tryptophan is one of the eight essential amino acids not synthesized in the human body and, hence, needs to be supplemented with the nutriment. Among all essential amino acids, the tryptophan concentration is the lowest in the human organism; however, only low concentrations are sufficient for tryptophan-involved processes [12]. Actually, less than 1% of the available tryptophan is used for protein synthesis, while the remaining 99% serves as a precursor of bioactive metabolites, including serotonin, tryptamine, melatonin and kynurenine, as well as the essential coenzyme nicotinamide adenine dinucleotide (NAD+) [13,14]. The role of tryptophan has been extensively studied in a variety of conditions, such as neuropsychiatric disorders [15]; autoimmune and allergic diseases [16,17]; infections [17]; brain tumors [18] and other cancer types, e.g., breast, bladder, colorectal cancers and melanomas [17,19,20]. Tryptophan metabolism via the kynurenine pathway, which is the main route of tryptophan degradation in the human body [14,21,22], has gained particular interest in Tx during the last couple of decades. Since the landmark discovery by Munn et al., who showed that the rejection of an allogeneic fetus in mice is prevented by indoleamine-2,3-dioxygenase (IDO)-mediated tryptophan catabolism [23], enzymes and metabolites of the tryptophan-kynurenine pathway have been widely studied in the field of immune regulation after Tx [17]. In addition, tryptophan and its metabolites also seem to play a role in IRI, thus giving some new direction in the field of organ preservation [24,25]. Most interestingly, inflammation is the basic underlying phenomenon for graft pathology, including IRI and both acute and chronic graft rejections [7,26].

This comprehensive review gives an overview on the current evidence of the tryptophan-kynurenine pathway's significance in Tx, giving both diagnostic/prognostic and therapeutic implications.

2. The Tryptophan-Kynurenine Pathway, Its Enzymes and Metabolites: An Overview

There is evidence that ~90–95% of the overall tryptophan is degraded via the kynurenine pathway, and it is the main route of tryptophan catabolism in the human body [21,22] (Figure 1).

Firstly, tryptophan is converted to N-formyl-L-kynurenine. The reaction is catalyzed by one of three rate-limiting enzymes: tryptophan-2,3-dioxygenase (TDO) and indoleamine-2,3-dioxygenase 1 or 2 (IDO1 or IDO2). All the three enzymes are hemoproteins and use molecular O_2 as a co-substrate, which also allows them to utilize reactive oxygen species (ROS) and regulate the redox balance in the cell [17,27,28]. Unless TDO has been detected in various mouse brain structures, as well as in different tumor types [29,30], this enzyme is mainly expressed in the liver and regulates systemic tryptophan levels in physiological conditions [20,31]. Glucocorticoids, tryptophan itself and heme are the three main regulators of TDO: glucocorticoids induce the synthesis of new TDO apoenzyme, while the substrate tryptophan enhances the conjugation of the apoenzyme with the cofactor heme and stabilizes the whole structure. TDO transcription has also been found to be upregulated by glucagon but inhibited by insulin and adrenaline [32]. Other possible inhibitors described in the literature include reduced forms of nicotinamide adenine dinucleotide phosphate (NAD(P)H), 3-hydroxykynurenine (3-HK) and 3-hydroxyanthranilic acid (3-HAA), as they likely act through the negative feedback mechanism; however, the results obtained by in vitro and in vivo experiments are contradictory [31].

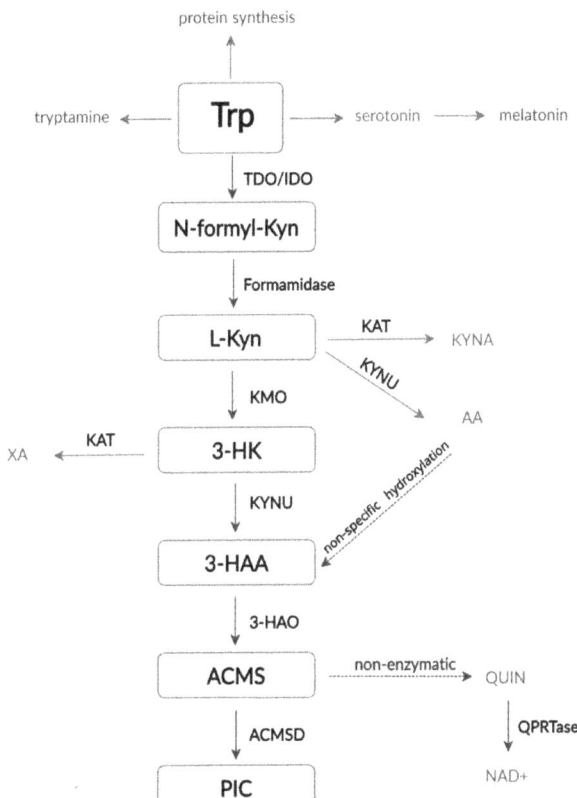

Figure 1. Tryptophan metabolism via the kynurenine pathway. Abbreviations: Trp: tryptophan, N-formyl-Kyn: N-formyl-kynurenine, Kyn: kynurenine, 3-HK: 3-hydroxykynurenine, 3-HAA: 3-hydroxyanthranilic acid, ACMS: 2-amino-3-carboxymuconate-semialdehyde, PIC: picolinic acid, KYNA: kynurenic acid, AA: anthranilic acid, XA: xanturenic acid, QUIN: quinolinic acid, NAD+: nicotinamide adenine dinucleotide, TDO: tryptophan-2,3-dioxygenase, IDO: indoleamine-2,3-dioxygenase, KMO: kynurenine-3-monooxygenase, KYNU: kynureninase, 3-HAO: 3-hydroxyanthranilate-3,4-dioxygenase, ACMSD: amino-carboxy-muconate-semialdehyde-decarboxylase and KAT: kynurenine aminotransferase.

Differently to TDO, IDO is much more widely distributed among tissues and mainly expressed by various immune cells. Interestingly, IDO has a lower capacity but much higher affinity for tryptophan than TDO [33]. Two forms of IDO are currently known: IDO 1 (previously called IDO) and the more recently discovered IDO 2, which is expressed in human liver, kidney and brain [17]. Tryptophan and the cofactor heme both increase IDO activity in a similar way as TDO; however, it was found that high concentrations of tryptophan act as inhibitor of IDO [20,34,35]. Antioxidants [36,37], as well as nitric oxide (NO), are also known to inhibit IDO function [38]. Nevertheless, the principal effectors of IDO activity are pro- and anti-inflammatory cytokines and mediators [17], which will be discussed in more detail below.

In the next step, N-formyl-kynurenine form amidase hydrolyzes N-formyl-L-kynurenine to L-kynurenine, which is subsequently transformed into three alternative metabolites with different properties regarding oxidative stress and organ toxicity: (1) kynurenic acid (KYNA) by kynurenine aminotransferase (KAT), (2) anthranilic acid (AA) by kynureninase (KYNU) and (3) 3-hydroxykynurenine (3-HK) by kynurenine-3-monooxygenase (KMO). The latter metabolite 3-HK can be further converted into xanthurenic acid (XA) by KAT or transformed into 3-hydroxyanthranilic acid (3-HAA) by KYNU [17,20].

The next and the most potent enzyme in the tryptophan-kynurenine pathway is 3-hydroxyanthranilate-3,4-dioxygenase (3-HAO), which rapidly converts 3-HAA into the unstable 2-amino-3-carboxymuconate-semialdehyde (ACMS). The latter is then nonenzymatically converted into quinolinic acid (QUIN) used in NAD+ and NADP+ formation or, alternatively, transformed into picolinic acid (PIC) via an amino-carboxy-muconate-semialdehyde-decarboxylase (ACMS) catalyzed reaction [17,20]. 3-HAA and 3-HK are known generators of free radicals [39]; however, there is evidence that the same metabolites are also able to act as antioxidants, depending on the cell's redox properties [40]. Similarly, PIC and QUIN also enhance oxidative stress by creating free radicals. On the other hand, KYN catalyzation to KYNA by KATs is considered an antioxidant way of tryptophan metabolism. As a result, KMO is located at an important point keeping the balance between anti- and pro-oxidant metabolites [20]. It is important to mention that, due to the high Km (concentration of substrate that permits the enzyme to achieve half of the maximum velocity of its catalyzed reaction) of KAT for its substrates, KAT reactions are only minimally significant, whereas the KMO/KYNU catalyzed branch remains the main one. Nevertheless, KAT can be enhanced by the overload of tryptophan or kynurenine or by the inhibition of KMO activity [41].

Importantly, under physiologic conditions, the tryptophan-kynurenine pathway happens mainly in the liver, where all the enzymes necessary for NAD+ synthesis from tryptophan are present. The extrahepatic way is responsible for only 5–10% of the overall tryptophan degradation in physiologic conditions. However, it becomes much more significant under immunological circumstances, inflammation and oxidative stress. The extrahepatic tryptophan kynurenine pathway does not provide all the necessary enzymes; therefore, its intermediate metabolites and their properties become critical in the pathogenesis and modulation of these conditions [17,20].

3. Role of Tryptophan and Its Metabolites in IRI

IRI is a substantial and unavoidable threat in Tx, determining the early transplant function. Adenosine triphosphate (ATP) depletion, impaired ATPases activity, cellular calcium overload, deterioration of the mitochondrial membrane potential via opened mitochondrial permeability transition pores, the promotion of proapoptotic mechanisms, the generation of ROS after the reintroduction of molecular oxygen, endothelial dysfunction, increased thrombogenicity and the induction of inflammatory responses are examples for the consequences of IRI [26]. Unless the role of tryptophan and its metabolism in IRI has been studied to a much lesser extent compared to immune responses after Tx, several implications have been given in the literature that could lead to further investigations (Table 1).

The histidine-tryptophan-ketoglutarate (HTK) organ preservation solution, which is nowadays considered one of the standard solutions used in Tx, contains 2-mmol/L tryptophan due to its antioxidant capacity and membrane stabilizing potential [42]. On the contrary, several studies have suggested that tryptophan starvation could be a protective factor against IRI. It has been previously demonstrated that short-term dietary restrictions, i.e., reductions of specific food intakes without calorie depletion, increases the resistance to acute stress, including IRI [43,44]. In another experimental study, the total protein or single amino acid L-tryptophan withdrawal from diet protected mice against renal ischemic injury and preserved kidney function. In addition, tryptophan starvation resulted in significantly reduced circulating granulocyte numbers, which was found to be general control nondepressible 2 (GCN2) kinase-dependent [45]. Nevertheless, the same experiments showed that tryptophan withdrawal downregulates the expression of oxidative stress resistance-related genes glutathione synthetase (Gst1/2), catalase, dehydrogenase quinone 1 (Nqo1) and NADPH in the liver [45]. Recently, Eleftheriadis et al. demonstrated a three-fold increase in human renal proximal tubular epithelial cell survival under hypoxic conditions following tryptophan deprivation. They found these effects to be driven by induction of autophagy through the activation of GCN2 kinase and p53-mediated BNIP3L upregulation [46]. Tryptophan deficiency stimulates GCN2 kinase by promoting phosphorylation of the eukaryotic translation initiation factor 2a (eIF2a), which then activates autophagy, while tryptophan supplementation acts the opposite by diminishing activity in the GCN2-eIF2a pathway [47]. In general, autophagy was confirmed to be protective in the situation of acute kidney injury caused by ischemia and reperfusion [48,49]. During the process of autophagy, cytoplasmic fractions are sequestrated within the autophagosomes, which then fuse with lysosomes, where the "captured" material is decomposed. Autophagy is activated in response to various stressors, such as hypoxia or nutrient deprivation, thereby counteracting apoptosis and providing necessary nourishment and time for cells to adapt and survive [46–49]. Moreover, tryptophan starvation-induced autophagy was found to dampen the secretion of proinflammatory cytokines [47]. These interesting results allow speculating that donor kidney preconditioning by tryptophan deprivation could induce the resistance of kidney tissue to hypoxic stress, prevent acute kidney injury and probably decrease the risk of primary nonfunction or delayed graft function after implantation.

IDO has also been investigated concerning its impact in the IRI setting. Transplanted lungs are, compared to other organs, especially prone to IRI due to the vast and constant exposition to environmental oxygen. Reactive oxygen species cause endothelial dysfunction and increase the vascular permeability and extravasation of inflammatory cells, which promotes apoptosis and necrosis processes and causes pulmonary graft injury [50]. As mentioned before, IDO, while catalyzing the reaction of L-tryptophan to L-kynurenine, consumes O_2 and utilizes superoxide anion radicals, thus acting as a powerful antioxidant [27]. Indeed, intravenous nonviral IDO gene delivery to donor rats 24 h before lung Tx significantly protects lung allografts against IRI by reducing the vascular leak and leukocyte extravasation. Furthermore, it preserves a histological structure of the graft and significantly improves the lung function. In vitro experiments showed that enhanced IDO activity in endothelium stabilizes the intracellular redox balance and preserves the mitochondrial structure and function in the context of oxidative stress [28]. On the contrary, Mohib et al. raised concerns regarding the role of IDO in kidney graft protection, showing that, despite the pro-tolerant properties of this enzyme, its activity augments IRI in the kidney [51,52]. They found increased apoptosis in renal tubular cells due to upregulated IDO activity [51], while the inhibition of IDO by 1-metyltryptophan (1-MT) or IDO gene knockout in mice diminished renal IRI [52]. Correspondingly, another study revealed that inhibiting IDO activity by 1-MT changes the transcriptome characteristic to IRI. IRI alone was shown to change 105 coding genes and only three noncoding RNA transcripts. In mice pretreated with 1-MT, only 18 sequences of coding transcripts were altered, while the number of noncoding RNA genes was expanded to 66. The authors speculated that the reduction of IRI affects genes, including those responsible for apoptosis and cell death, and may be related to the reno-protective effects of IDO inhibition [53].

A recent study on normothermic and sub-normothermic machine perfusion for discarded human livers without significant steatosis revealed an increase in tryptophan metabolism via the kynurenine pathway and higher levels of kynurenine and KYNA for organs subjected to normothermic perfusion compared to sub-normothermic perfusion [24]. As it is known that KAT, which converts kynurenine to KYNA, is a temperature-dependent enzyme and KYNA has antioxidant properties [54], decreased KYNA production could be considered as a disadvantage of sub-normothermic perfusion. On the other hand, this finding implicates the therapeutic possibilities, e.g., shunting the kynurenine metabolism from 3-HK to KYNA by suppression of the KMO enzyme [54,55] could be beneficial in sub-normothermic perfusion conditions diminishing IRI. However, further studies using machine perfusion as a platform for tryptophan metabolism regulation and assessing the clinical outcomes of such organs are necessary.

KMO has been found to be expressed in the human kidney proximal tubule's epithelial cells [56]. As explained above, 3-HK is an injurious metabolite of kynurenine, which enhances oxidative stress, causes pathological protein crosslinking and promotes apoptosis, while KYNA protects cells against injury in stress or inflammatory situations [17,20]. Several experimental studies revealed that the pharmacological or transcriptional blockage of KMO prevents acute ischemic kidney injury in multiorgan dysfunctional models [57–60]. Zheng et al. showed that genetically modified mice, lacking a functional KMO gene, were protected against the deleterious effects of kidney IRI. The histological necrotic tubular damage was significantly diminished, and plasma creatinine concentrations were lower compared to wild-type mice. Additionally, the number of apoptotic renal tubular cells and neutrophil infiltration into kidney tissue was significantly lower in KMO knockout mice. These findings underpin the idea of KMO as a potential therapeutic target in the prevention of IRI in the field of Tx [25]. Furthermore, a recent study, investigating glaucoma treatment possibilities, revealed that not only KMO knockout but, also, intravenous or local administration of KYNA itself acts protectively against the IRI in retinal ganglion cells [61], which could also be considered as one of the possibilities for the protection of transplanted organs.

Besides the impact of tryptophan and its metabolites on IRI of solid organ grafts, the role of tryptophan derivatives has recently been under investigation. N-acetyl-L-tryptophan (L-NAT), a ROS scavenger, inhibitor of cytochrome c release from the mitochondria, as well as an antagonist of the neurokinin 1 receptor, which has already been approved for the management of nausea, vomiting, shock and neurodegenerative diseases, was investigated in the context of hepatic IRI. The pretreatment of rats with L-NAT before liver ischemia significantly diminished IRI, prevented morphological changes of hepatocytes and increased their viability. This protective effect was found to be mediated by downregulation of the receptor interacting protein (RIP) 2/caspase1/IL-1b signaling pathway [62]. The same group later demonstrated that L-NAT protects hepatocytes against IRI by the inhibition of excessive autophagy, mitophagy and preservation of the mitochondrial structure and function [63]. Another tryptophan derivative investigated was 5-metoxytryptophan (5-MTP), an endogenous anti-inflammatory endothelial factor synthesized from L-tryptophan catalyzed by tryptophan hydroxylase-1 and hydroxyindole O-methyltransferase. An in vitro 5-MTP treatment protected cardiomyocytes against ROS-induced IRI by preventing cell death, stimulating cell migration and promoting wound healing via cytoskeletal regulations, as well as regulating the intracellular redox state and reducing the endoplasmic reticulum stress [64].

To conclude, unless the current evidence confirms the impact of tryptophan and its metabolites in graft IRI, revealing important implications in possible therapeutic approaches, which potentially would allow to diminish the burden of IRI, it seems that the effect is complex rather than one-sided. Therefore, many questions remain unanswered.

Table 1. Effects of tryptophan and its metabolism via the kynurenine pathway in solid organ IRI in animal models.

Reference	Experimental Model	Treatment/Intervention	Outcomes
Peng et al. [45], 2012	Mice kidney IRI model	Tryptophan deficient diet for 6 days before induction of kidney or liver ischemia.	↓serum creatinine and urea 1 day after bilateral renal ischemia-reperfusion; ↓expression of KIM-1 mRNA; ↓level of acute tubular necrosis in histology; ↓serum levels of ALT, AST, LDH after liver ischemia-reperfusion; ↓P-selectin and IL-6 gene expression; ↓number of circulating neutrophils. Effect dependent on GCN2.
Liu et al. [28], 2007	Rats model of lung Tx after 5 h of warm ischemia	Sleeping beauty transposon mediated hIDO gene delivery to donor animals intravenously 24 h prior to Tx.	↓apoptosis of endothelial cells; ↓leukocyte infiltration; ↑antioxidant capacity; ↓levels of oxidative stress markers (protein carbonyl, MDA); ↓alveolar edema, hemorrhage and formation of focal congestion of lung tissue; preserved mitochondrial structure and function; ↓peak airway pressure, ↑PaO$_2$.
Mohib et al. [52], 2008	Mice kidney IRI model	IDO gene knock-out or IDO inhibition by intraperitoneal injections of 3 mg of 1-MT twice a day for 48 h following reperfusion. Some mice received 1-MT 1 h before ischemia, as well as following reperfusion.	↓serum creatinine and blood urea nitrogen in IDO-knockout mice; ↓blood urea nitrogen but no difference in serum creatinine in 1-MT-treated mice; preserved architecture of kidney tissue; ↓apoptosis and necrosis; ↓neutrophil infiltration in IDO-knockout and 1-MT treated mice.
Merchen et al. [53], 2014	Rats kidney IRI model	IDO inhibition by pretreating rats with 1-MT 140 mg/kg po 1 and 24 h prior to renal ischemia.	IRI alone changed 105 coding genes and 3 noncoding RNA transcripts. In IRI rats pretreated with 1-MT, altered coding transcripts declined to 18 sequences and altered noncoding RNA genes increased to 66.
Zheng et al. [25], 2019	Mice kidney IRI model	KMO gene knockout.	↓plasma creatinine; ↓tubular damage; ↓number of apoptotic cells; ↓neutrophil infiltration; ↓Cxcl 1 and Cxcl2 mRNA levels in kidney tissue.
Wang et al. [62], 2019 Li et al. [63], 2020	Rats liver IRI mode	Intraperitoneal administration of L-NAT (10 mg/kg) 30 min before ischemia.	↓IRI-induced histological changes in hepatocytes; ↓mRNA expression of RIP2, caspase-1 and IL-1b [62]; ↓caspase-1 activity and IL-1b expression; ↓expression of autophagy markers: LC3-II, Beclin1, and ATG-7 and ↑expression of P62; ↓formation of autophagosome; improved morphological and functional changes of mitochondria, maintained the quantity and quality of mtDNA stability; ↓excessive mitophagy [63].

Abbreviations: IRI: ischemia-reperfusion injury, Tx: transplantation, (h)IDO: (human) indoleamine-2,3-dioxygenase, KMO: kynurenine-3-monooxygenase, 1-MT: 1-metyltryptophan, L-NAT: N-acetyl-l-tryptophan, KIM: kidney injury molecule, ALT: alanyl aminotransferase, AST: aspartate aminotransferase, LDH: lactate dehydrogenase, IL: interleukin, GCN2: general control nonderepressible 2 kinase, MDA: malondialdehyde, PaO2: partial oxygen pressure, (m)RNA: (messenger) RNA, Cxcl: reduced chemokine (C-X-C motif) ligand, RIP: receptor interacting protein, LC3-II: microtubule-associated protein 1 light chain 3-II, ATG-7: autophagy-related protein-7 and mtDNA: mitochondrial DNA, ↑: increased, ↓: decreased.

4. The Role of the Tryptophan-Kynurenine Pathway in Immune Regulation after Tx

The immune system plays a key role in Tx. Immunological mechanisms, which normally act as a protective organism response against foreign pathogens, pose a significant challenge to successful Tx. Several types or rejections exist, ranging from hyperacute and acute to chronic, all representing not only different rates of response but, also, different pathophysiological mechanisms in which both cellular and humoral immunity are involved. Despite constantly increasing the knowledge and broadened opportunities in immunosuppressive regimens, the full view of the immune response against allografts is still not fully understood, and the survival of transplanted organs remains time-limited [7].

IDO is probably the most thoroughly investigated tryptophan-kynurenine pathway enzyme, which acts as a natural tolerogenic factor regulating immune responses (Figure 2).

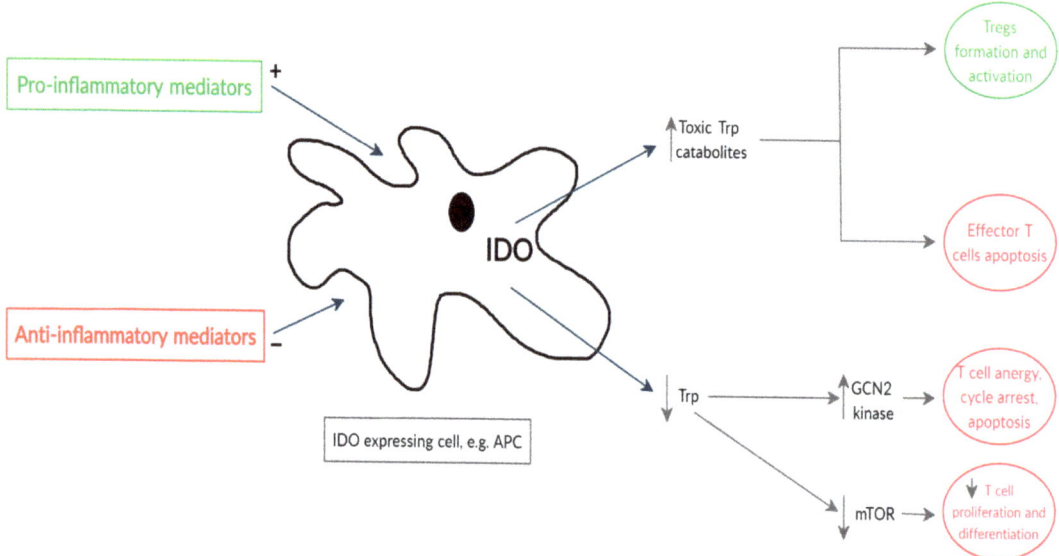

Figure 2. Simplified scheme of IDO activity in immune regulation. IDO is expressed in various cells, such as APCs, fibroblasts, endothelial, epithelial, smooth muscle cells, etc., in response to proinflammatory mediators. Due to IDO activity, decreased levels of tryptophan activate the GCN2 kinase, which leads to T-cell anergy, T-cell cycle arrest and apoptosis. The deactivated mTOR pathway results in suppressed T-cell proliferation and differentiation. Increased levels of toxic tryptophan-kynurenine pathway metabolites induce effector T-cell apoptosis and promote Treg formation and activation. Abbreviations: Trp: tryptophan, IDO: indoleamine-2,3-dioxygenase, APCs: antigen-presenting cells, GCN2: general control nonderepressible 2, mTOR: mammalian target of rapamycin and Tregs–regulatory T cells.

IDO is expressed in several immune cells, such as monocytes, macrophages, dendritic cells (DCs) and microglia, as well as fibroblasts, endothelial cells, epithelial cells and smooth muscle cells [54]. Interferon (IFN)-γ is known as one of the main stimuli that promotes IDO synthesis at the transcriptional level [65]. Other inflammatory mediators capable of enhancing IDO activity—however, mostly to a lesser degree—include IFN-α; IFN-β; tumor necrosis factors (TNF)s; interleukins (e.g., IL-1β, IL-2, IL-6 and IL-27); CD40 ligand; prostaglandins and lipopolysaccharides (LPS) [66–68]. The aforementioned mediators activate specific receptors, such as aryl-hydrocarbon receptor (AhR), Toll-like receptors (TLRs), interferon-β and γ receptors and tumor necrosis factor receptors (TNFR), which subsequently promote intracellular signaling pathways, leading to IDO upregulation. Similarly, anti-inflammatory cytokines, such as IL-4, IL-10 and tumor growth factor (TGF)-β, were found to diminish the expression of IDO, confirming that the IDO status directly depends on the balance of pro- or anti-inflammatory responses [20,68]. The production of IDO was also found to be promoted by the engagement of CD80 (B7.1) and CD86 (B7.1) located on the surface of DCs with cytotoxic T-lymphocyte-associated antigen 4 (CTLA4) or CTLA4-Ig fusion protein [69,70]. The fusion of CD28 situated on T cells with CD80/86 on DCs is necessary for naïve T-cell activation. Endogenous CTLA4 expressed in activated conventional T cells and T-regulatory cells (Tregs) or exogenous CTLA4Ig (used as immunotherapy drug) are known to counterbalance CD28 in T cells due to a higher affinity for B7 molecules, resulting in the deactivation of the immune response [70–73]. It has been demonstrated that IDO inhibitor 1-MT abrogates a positive CTLA4Ig effect on the prolonged survival of mice pancreatic islet allografts [70]. However, recent data from a multicenter sub-study of the phase II trial showed that Betalacept, a second-generation CTLA4-Ig fusion protein, was not able to induce detectable IDO activity in human liver

Tx recipients. Similarly, Betalacept did not have any effect on IDO expression in human dendritic cells in vitro [74].

As mentioned before (compare Figure 1), IDO, as the first enzyme in the tryptophan-kynurenine pathway, determines the local levels of tryptophan and tryptophan-kynurenine pathway metabolites. Tryptophan depletion increases the amount of free transfer RNA in T cells, which activates the GCN2 stress kinase pathway, leading to T-cell anergy and cell cycle arrest at the mid-G1 phase, thus sensitizing T cells to apoptosis [17,20,46,75]. Additionally, IDO significantly reduces the macrophage activity through the induction of apoptosis via the GCN2 kinase pathway and inhibition of inducible nitric oxide synthase (iNOS) expression, and this effect is also more related to tryptophan starvation rather than increased concentrations of tryptophan metabolites [76]. Moreover, it has been found that decreased levels of tryptophan may inhibit the mammalian target of the rapamycin (mTOR) pathway, resulting in the subsequential blockade of the translation process in T cells [77]. However, not all studies have confirmed that mTOR reacts to tryptophan depletion [46,75]. The theory of "tryptophan starvation" was also supported by in vitro experiments that showed that T-effector cells and macrophages may be reactivated in IDO-overexpressing cultured cells by adding an excessive amount of tryptophan [75,76].

On the other hand, active tryptophan metabolites kynurenine, 3-HK and 3-HAA, generated following IDO and other tryptophan-kynurenine pathway enzyme activity, promote oxidative stress and have a direct cytotoxic effect on T-effector cells, as well as suppress their proliferation by inducing apoptosis. Moreover, by inducing the apoptosis of T-helper 1 (Th1) cells, kynurenine metabolites are able to shift the balance between Th1 and Th2 towards Th2 and stimulate the formation of Tregs [20,78]. Kynurenine metabolites are also agonists of AhR, which plays an important role in Th17 cell differentiation and promotes the generation of Tregs [79–81]. Among the tryptophan metabolites produced downstream in the kynurenine pathway, 3-HAA appears to be the most potent immune regulator [82–84]. It has been previously revealed that 3-HAA induces the T-cell apoptosis activating PDK1 kinase [85] and caspase 8 [86]. Its dimerized form, cinnabarinic acid, produced in oxidant conditions was found to activate AhR as well [87]. A recent study also showed that 3-HAA engages nuclear coactivator 7 (NCOA7), expressed in dendritic cells, which increases the kynurenine-driven transcriptional activity of AhR. This combined mechanism of 3-HAA and kynurenine is important for the DC-mediated induction of Treg cells [88].

Even though the impact of IDO and the tryptophan-kynurenine pathway on cellular immune mechanisms has been widely investigated, much less is known about the humoral alloimmune response, which plays a role in Tx as well [89]. It has been reported that the B-cell-intrinsic promotion of IDO 1 via activation by Toll-like receptor ligands or B-cell receptor crosslinking is a key mechanism, reducing the production of antibodies against T-cell-independent antigens [90]. Another study, however, showed that IDO 2 expression in B cells increases the humoral autoimmunity by supporting the cross-interaction between reactive T and B cells [91]. An in vitro model by Sounidaki et al. recently demonstrated that the IDO inhibitor 1-metyltryptophan (1-MT) enhanced the humoral alloimmune response, demonstrating that IDO is also a possible inhibitor of humoral rejection mechanisms. However, the lack of effects from the AhR inhibitor and GCN2 kinase activator in these experiments suggests that molecular pathways that are responsible for the IDO effect on humoral alloimmunity differ from those participating in cellular alloimmunity inhibition, demanding further investigation [92].

Despite similar enzymatic function, in the field of immune regulation, TDO has been studied in a much lesser extent than IDO. Nevertheless, at the beginning of the current century, the TDO role in murine fetal tolerance was proposed [93,94]. An experimental study by Schmidt el al. revealed that TDO, similarly to IDO, is also capable of inhibiting the growth of bacteria, as well as restricting alloantigen-induced T-cell activation and inhibiting IFN-γ production. These effects were inhibited by additional tryptophan supplementation [95]. Another aspect, suggesting that TDO participates in the regulation of immune responses after Tx, is its promotion by glucocorticoids, which are included in classical immunosuppressive regimens [32]. Additionally, TDO has been detected in different tumors, such as melanomas, hepatocellular or bladder carcinomas, and its significance in tumor immune escape processes has been investigated [17,96].

KMO, as an enzyme standing at a cross-point of the "pro- and anti-inflammatory" or "oxidant and antioxidant" branches of the tryptophan-kynurenine pathway, has also gained interest in the research of immune regulation. The inhibition of KMO results in increased levels of the tolerogenic and anti-inflammatory metabolite KYNA, which diminish inflammation by activating AhR, reducing TNF expression in monocytes, IL-4 secretion by natural killer (NK) cells and IL-23 formation in DCs [17,20].

Due to the aforementioned immunomodulatory properties and the natural capability to downregulate allogeneic immune responses, the tryptophan catabolism via the kynurenine pathway has gained interest as an attractive therapeutic target in Tx. The in vivo studies investigating the role of enzymes and metabolites of the tryptophan-kynurenine pathway in immune regulation in solid organ Tx are summarized in Table 2 and discussed in detail below.

4.1. Heart

The impact of IDO gene delivery via an adenoviral vector directly into cardiac allografts has been investigated in several animal experimental models [97,98]. Li et al. reported that the intracoronary administration, as well as intramyocardial injection, of an adenoviral vector encoding for IDO cDNA (Ad-IDO) resulted in significantly prolonged cardiac graft survival in rats. Similarly, IDO gene transfer, together with a short course of low-dose Cyclosporine A (CsA), was more efficient than CsA alone. Interestingly, almost all the cells expressing IDO were cardiomyocytes. Histological examinations also revealed less cardiac infiltration by monocytes, macrophages and T cells (CD4+ and CD8α+), which indicates an attenuated rejection process [97]. These findings were confirmed by Yu et al., showing that intramyocardial Ad-IDO injection prolonged the cardiac allograft survival in mice [98]. Both studies reported significantly reduced mRNA transcript levels for proinflammatory cytokines and chemokines in Ad-IDO grafts [97,98]. However, despite the delayed rejection, Yu et al. showed a similar cellular infiltration in Ad-IDO-treated grafts as in control grafts, which emphasizes a probable capability of IDO to modify the function rather than number of alloreactive T cells. Ad-IDO grafts also showed a significantly higher proportion of Tregs [98]. As IDO-transduced DCs demonstrated the ability to inhibit allo-specific T-cell proliferation in vitro [97], IDO-overexpressing DC injection into the recipient before cardiac Tx has also been investigated [99–101]. Decreased INFγ and increased IL-10 intra-graft transcription-induced CD4+-cell apoptosis, greatly diminished histopathology changes and significantly prolonged graft survival have been reported after pre-Tx IDO-overexpressing DC transfusion. Interestingly, IDO was also mainly expressed in cardiomyocytes [99]. Ly et al. used a combined recipient preconditioning with a pre-Tx infusion of Ad-IDO-transfected DCs and a post-Tx infusion of CD40L mAb, which also acts via IDO upregulation [100]. They reported significantly lower serum INF-γ and IL-2 expression, an increased apoptosis of peripheral CD3+ T cells and decreased creatine kinase (CK) and lactate dehydrogenase (LDH) levels, referring to the improvement of allograft function after cardiac Tx. More significant effects were achieved by using the combination therapy of IDO-overexpressing DCs and CD40L mAb compared to IDO-overexpressing DCs alone [100]. The combination treatment with IDO-transfected DCs and tryptophan metabolites was significantly more effective in allograft rejection delay and survival prolongation in comparison to each of these strategies separately [101]. Among tryptophan metabolites, contrary to AA and QUIN, 3-HAA, 3-HK and L-kynurenine showed the best ability to inhibit the proliferation of unstimulated spleen-derived T cells by the induction of apoptosis in vitro. A single 3-HAA and allogeneic bone marrow-derived DC injection of recipient rats resulted in significantly prolonged cardiac graft survival and lowered the pathological grade of rejection, while a 3-HAA injection alone had only minimal protective effects. The most important finding of this experiment was that 3-HAA-suppressed T cells could not be restimulated by donor-specific DCs [83]. Another two studies by He et al. showed that IDO-overexpressing bone marrow mesenchymal stem cells [102] or exosomes derived from IDO-overexpressing bone marrow mesenchymal stem cells [103] injected intravenously 48 h after allogeneic heart Tx of rats similarly resulted in significantly improved graft function. This was demonstrated by determination of the ejection fraction and left ventricular fractional shortening, as well as decreased graft damage and infiltration by inflammatory cells compared to mycophenolate mofetil-treated or untreated animals. The authors reported significantly lower serum levels of IL-1α, IL-4, IL-1β, IL-2, IFN-γ and IL-18 and increased levels of IL-10, TGFβ1, TGFβ2 and TGFβ3 in both models. The flow cytometry of isolated cells of rat spleens showed significantly lower expressions of CD40, CD86, CD80, MHC-II, CD45RA and CD45RA+CD45RB and a higher expression of CD274 and a higher proportion of Tregs than in the control groups [102,103]. A proteomic analysis demonstrated that exosomes secreted by IDO-overexpressing mesenchymal cells contained significantly upregulated immunoregulatory protein FHL-1 [103]. Importantly, the tolerogenic state, achieved by direct IDO gene transfer to cardiac cells or by IDO-overexpressing DC infusion, is allograft-specific and not only limited to the local milieu but is rather systemic, as shown by the increased survival of secondary implanted skin allografts that were genetically identical to implanted hearts [98,101].

4.2. Lungs

The survival rates after lung Tx remain significantly lower compared to other solid organ Tx, mainly due to the more active immune response to lung allografts [104]. Lung interstitial and peribronchial tissue are well-supplied with antigen-presenting cells (APCs), such as DCs, monocytes and macrophages, that can strongly stimulate or suppress the immune response. There is evidence that IDO is significantly upregulated in lung tissue during microbial infections after allogeneic hematopoietic stem cell Tx and under other inflammatory conditions [105–107]. Importantly, it has been found that IDO is induced and the tryptophan-kynurenine pathway is promoted not only in antigen-presenting but, also, in nonimmune pulmonary cells, such as epithelial cells, which is important in the local protection of lung tissue from collateral damage [105–107]. Therefore, among a bunch of therapeutic strategies, targeting the tryptophan-kynurenine pathway has also been investigated in several in vivo animal models of lung Tx. Swanson et al. were the first to report that systemic Ad-IDO vector instillation into the donor's organism before lung Tx results in an abrogated delayed-type immune response and lowers the histopathological grade of allograft rejection [108]. In another study, the intratracheal nonviral delivery of IDO preserved the allograft function, which was reflected by a significantly reduced peak airway pressure and increased PaO$_2$ levels in IDO-transfected grafts. In vitro experiments revealed that the protective effect of IDO is achieved not only through T-cell inhibition but, also, through the enhancement of the local antioxidant defense system when the graft faces the burden of oxidants [109]. The possible therapeutic role of IDO was also assessed in a model of chronic lung allograft injury using IDO gene delivery via the "Sleeping beauty" transposon, which, differently to most viral vectors, is capable of promoting a long-lasting expression of the desired gene in target tissues. IDO-positive lung cells inhibited the TGFβ-mediated proliferation of fibroblasts and excessive accumulation of collagenous tissue, which resulted in reduced graft fibrosis, preserved bronchus-alveolar architecture and significantly improved the function of the preconditioned lung allograft [110]. Further investigating the mechanism of the tryptophan metabolism-induced tolerance of lung allografts, the same group reported that, despite IDO overexpression, the number of allograft-infiltrating CD8 T cells is still higher than in normal lungs or isografts, revealing that IDO effects in immune modulation may be incomplete. Nevertheless, a significant reduction of T-cell cytotoxicity by disturbance of the granule perforin and granzyme exocytosis was demonstrated. T cells from treated allografts produced significantly less IL-2 and TNF-α, while the INF-γ production remained obvious. Interestingly, the impaired function of complex I of the electron transport chain in CD8+ cell mitochondria was revealed, while complexes II-IV were still intact. This may be the reason for the survival of CD8+ cells in a high-IDO environment [111]. In another study, 3-HAA delivery seven days after lung Tx showed similar immune-protective effects as donor preconditioning by IDO gene transfer. Moreover, comparable results in GCN2 gene knockout mice were demonstrated, strengthening the theory that the upregulation of tryptophan metabolites may be more critical in immune regulation than tryptophan depletion. The study found that high IDO/3-HAA levels inhibit T cells due to the impairment of calcium signaling via the T-cell receptor (TCR)/Ca2+ signaling pathway [84]. Finally, in recent experiments, the IDO gene was transferred into tissue-engineered lung allografts, which were constructed from decellularized rat lungs, differentiating medium and rat bone marrow mesenchymal stem cells. IDO gene transfer into such composites allowed to achieve tolerogenic status by the reduction of inflammatory cytokines levels and upregulation of regulatory T cells. Thus, IDO overexpression may be considered as one of the possible methods allowing to go one step further in setting up a nonimmunogenic lung tissue construct and to overcome the problem of graft shortage [112].

4.3. Liver

It is well-known that, unlike other solid organs, fully allogeneic liver grafts are accepted spontaneously in most mice strains combinations. Miki et al. were the first investigators who demonstrated that, unless IDO mRNA is not expressed in the mouse liver in physiological conditions, it is induced after allogeneic liver Tx, and the inhibition of IDO activity and tryptophan metabolism leads to the rejection of otherwise spontaneously accepted liver grafts [113]. The IDO gene was found to be significantly upregulated in allogeneic spontaneously accepted liver tissue, while naïve and syngeneic livers express IDO mRNA. Moreover, IDO was expressed only in liver antigen-presenting cells such as dendritic or Kupffer cells, while hepatocytes lacked an IDO signal [114,115]. Several studies suggested that TDO, which is a liver-specific tryptophan-kynurenine pathway enzyme, may also play a role in immune response regulation after allogeneic liver Tx [95,116]. However, in studies by Lin et al., TDO mRNA was found to be downregulated in allogeneic livers, implicating the probability that TDO does not participate in immune modulation, but rather, the depletion of this enzyme induces hepatocyte damage [114,115]. The reinfusion of IFN-γ-treated dendritic cells, showing vastly enhanced IDO mRNA expression, into liver-transplanted rats resulted in attenuated allogeneic liver graft rejection. The tryptophan levels decreased during the first week after Tx, followed by a continuous increase from day seven [117]. Contrary to these findings, Laurence et al., by inhibiting IDO activity by means of the IDO inhibitor 1-metyltryptophan (1-MT), in allogeneic liver-transplanted rats for the first week after Tx, did not cause acute rejection [118]. The same group confirmed their findings in another experiment with preconditioned liver donor rats with an IDO-containing recombinant adeno-associated virus vector; this intervention did not result in the prevention of liver allograft rejection signs, despite the confirmed IDO activity in vivo [119]. Another interesting animal allogeneic liver Tx experiment was conducted using N-(3′,4′-dimethoxycinnamonyl) anthranilic acid (3,4-DAA), which was injected intraperitoneally into recipient rats immediately after Tx. A synthetic derivative of the natural tryptophan-kynurenine pathway branch metabolite anthranilic acid, 3,4-DAA, has been approved in Japan as an antiallergic drug and is known for its ability to increase IDO expression. This study demonstrated that 3,4-DAA treatment resulted in attenuated liver allograft injury through the decreased TNF-α and IFN-γ, as well as increased IL-10 inflammatory signaling, while the IDO inhibitor 1-MT abrogated this effect. Significantly increased L-kynurenine levels were observed in 3,4-DAA-treated rats [120].

4.4. Kidneys

Cook et al. were the first to demonstrate the immunomodulatory effects of reduced local tryptophan concentrations and increased tryptophan metabolites in renal allografts.

They suggested that the early reduction of cellular rejection in allografts is associated with the TGF-β-mediated mechanism, while long-term tolerance is achieved via the IDO-induced regulation of reactive T cells [121]. Oppositely, the pre-Tx injection of immature IDO knockout dendritic cells to renal allograft recipient mice overturned the tolerogenic effects of normal immature dendritic cells. This resulted in a partial reverse of the reduction of T-cell proliferation, decreased proportions of CD4+CD25+Foxp3+ cells and increased cellular infiltrations in grafts, as well as upregulated IgG production, IL-2 and IFN-γ expression and, most importantly, deteriorated allograft survival and function, as confirmed by elevated serum creatinine levels [122]. Renal tubular epithelial cells, representing 75% of the renal parenchymal cells, are the main targets of the T-cell-mediated immune response. There is evidence that these cells have immunomodulatory capacities similar to mesenchymal stem cells or antigen-presenting cells and express IDO when stimulated by IFN-γ and TNF-α [51]. Tubular epithelial cells were found to have the ability to inhibit CD4+ and CD8+ T-cell proliferation. However, the inhibition of IDO by 1-MT only partially restored the activity of effector T cells. Importantly, it has been revealed that the tubular epithelial cell immunosuppressive ability is cell–cell contact-dependent; therefore, they are not able to influence T cells, which infiltrate the renal interstitial compartment and are not directly exposed to epithelial cells [123]. Mesenchymal stem cells are known for their natural ability to restrain the proliferation and activation of T, B and natural killer cells, as well as to inhibit inflammation. IDO, produced by mesenchymal stem cells, has been proven to be crucial for the promotion of Treg generation and improvement of graft survival in rat kidney Tx models [124]. Unless the injection of high doses of mesenchymal stem cells revealed positive effects in preventing acute allograft rejection and tolerance induction [125], high doses of these cells may increase the risk for other complications like infections, thrombotic microangiopathy or organ infarction [126]. He et al. injected IDO-expressing mesenchymal stem cells into renal allograft recipients, resulting in significantly prolonged graft survival and enhanced donor-specific tolerance by inducing the production and improving the function of antigen-specific Tregs in a rabbit model. Importantly, low doses of IDO-mesenchymal stem cells were sufficient to achieve tolerogenic effects [127].

The group of Vavrincova-Yaghi investigated the role of IDO gene therapy in kidney allograft preconditioning after organ retrieval. IDO transgene delivery directly into the kidney graft resulted in a significantly attenuated increase in plasma creatinine and improved allograft function, as well as reduced levels of kidney injury markers KIM-1 and alpha smooth muscle actin (α-SMA). Significantly reduced inflammation and upregulated Treg cell markers were found in IDO-preconditioned kidneys, which was accompanied by preserved tubular morphology and reduced interstitial pre-fibrosis. Despite the locally increased transgene expression in the renal interstitium, no significant difference in the systemic kynurenine/tryptophan ratio was found in IDO-transfected rats, suggesting that the effect of IDO overexpression remained local [128]. Interestingly, human IDO mRNA was absent in IDO-preconditioned transplanted kidneys at 12 weeks after Tx, which shows again that adenoviral transgene expression is limited in time. However, several clinical improvements, like the gaining of body weight, increase of systolic blood pressure or prevention of proteinuria, were still observed within the 12-week period in IDO-treated animals [129]. It has been previously found that tryptophan metabolism to kynurenines induced by IDO expressed in endothelium leads to the relaxation of arterial vessels via the adenylate and guanylate cyclase pathways during inflammatory conditions and sepsis [130]. Vavrincova-Yaghi et al. found reduced angiotensin-converting enzyme mRNA expression in IDO-preconditioned renal grafts, which could be another favorable mechanism through which IDO may reduce arterial blood pressure [129]. Moreover, kidney graft preconditioning with IDO in a long time period protected against transplant vasculopathy, which is one of the most important pathophysiologic elements in chronic transplant rejection, as proliferated neointima and narrowed vascular lumen lead to hypoperfusion, subsequent graft fibrosis and chronic transplant dysfunction [129]. A murine model of heterotopic aortic Tx revealed that the daily administration of various doses of antioxidant sodium sulfite blocked tryptophan breakdown through the kynurenine pathway and resulted in significant vasculopathy, as well as increased levels of α-SMA, vascular cell adhesion molecule (VCAM)-1 and P-selectin. These findings proved the role of tryptophan metabolism in chronic transplant rejection and highlighted the challenges associated with antioxidant treatments, as they apparently may have ambivalent consequences [37].

In a recent porcine kidney allo-Tx study, the allogenic immune response significantly promoted the IDO gene and proinflammatory cytokines expression and, additionally, suppressed another tryptophan kynurenine pathway enzyme KMO [131]. As 3-HK, a KMO-catalyzed reaction product, is known to be a more potent immunosuppressive catabolite than L-kynurenine [83], its loss due to KMO suppression may explain the inability of elevated allograft IDO activity to fully inhibit rejection [131] and give some implications about KMO or KAT as another potential therapeutic target of immune modulation.

4.5. Pancreas

Unfortunately, there are no studies, examining the immunoregulatory properties of tryptophan-kynurenine pathway enzymes and metabolites in the setting of whole pancreas Tx so far. Nevertheless, these questions have been investigated in several experimental animal models of pancreatic islet Tx. Alexander et al. found IDO gene transfer via the adenoviral vector to pancreatic cells resulting in significantly prolonged graft survival, increased amounts of insulin-positive β cell mass and the inhibition of T-cell proliferation [132]. Avoiding the potential harmful effects of the adenovirus on pancreatic islets [133,134], Jalili et al. created a three-dimensional islet graft by embedding allogeneic mouse islets and IDO-transduced fibroblasts within a collagen gel matrix. Interestingly, IDO was still expressed in the graft, whereas T-cell infiltration was not detected 40 days after Tx. The inclusion of IDO-overexpressing fibroblasts into this composite model also promoted Treg upregulation in graft-draining lymph nodes, generated anti-inflammatory cytokine profiles, inhibited the production of donor specific alloantibodies and significantly increased the number of insulin-producing cells. However, this effect was only observed until the fifth week after graft implantation, proving that IDO expression and its mediated immunosuppression was transient [135]. By using a lentiviral IDO vector combined with a protease-resistant composite scaffold, the immunosuppression could be prolonged for two weeks. However, a gradual decline of IDO expression and the respective immunosuppression remained inevitable. The authors emphasized that immune stability is maintained until a sufficient amount of Tregs is present [136]. The same group proposed a pancreatic islets xenogeneic Tx model. They implanted a collagen matrix-containing rat islet and adenoviral-transduced IDO-expressing mouse fibroblasts into recipient mice. The results indicated that IDO suppresses macrophage and T-cell infiltration into the graft, as well as inhibits iNOS expression by macrophages, which is extremely important in xenogeneic Tx, as macrophage activity is mainly responsible for xenograft rejection. Interestingly, in vitro experiments showed that a tryptophan deficiency, rather than increased levels of tryptophan metabolites, was responsible for macrophage viability reduction and apoptosis induction [76]. Another study investigated, in a systemic approach, the effects of intraperitoneal-injected IDO-expressing fibroblasts in recipients. This treatment delayed the rejection of pancreatic islet allografts culminating in a part of recipients, even reaching immune tolerance and leading to the assumption that the systemic delivery of IDO-overexpressing fibroblasts may be more efficient than local applications, probably due to the possibility to inject more cells [137].

4.6. Small Bowel

One of the most challenging features of intestinal allografts is the extremely high immunogenicity compared to other organ grafts; proper immune regulation is still a major obstacle. A murine small bowel Tx model revealed that a recipient treatment with IDO-transfected DCs, intravenous 3-HAA or a combination of these approaches results in a significantly prolonged graft survival, reduced inflammation and diminished morphological graft distortion. Interestingly, IDO-DCs seem to have stronger immunosuppressive effects than 3-HAA. Despite that in vitro 3-HAA enhanced the immunosuppressive effect of IDO-DCs when used in combination, in vivo, no significant improvement in allograft survival was observed compared to the IDO-DC treatment alone. Despite the potential therapeutic effect of IDO in intestinal Tx, the increased risk of intestinal tumors due to IDO activity should be taken into account [82].

To sum it up, targeting tryptophan metabolism via the kynurenine pathway seems an attractive strategy for immune regulation in solid organ Tx via mechanisms decreasing the inflammatory changes, such as IRI and both acute and chronic rejection. While human recipient preconditioning or post-Tx treatment with IDO gene therapy might still raise concerns, especially due to unknown long-term systemic effects and their related dangers, intensifying the investigation on novel organ preservation techniques, such as normothermic or sub-normothermic machine perfusion, will potentially provide an attractive platform for gene therapy by direct IDO gene transfer to graft cells or IDO-transfected APC delivery directly to the organ. Nevertheless, the current evidence is limited to in vitro experiments and animal models, revealing that there are still several challenges to overcome. Most of the studies only succeeded in delaying the acute rejection, but absolute allograft tolerance was usually not achieved, and allograft survival remained time-limited. This means that the use of more stable vectors in gene therapy, multiple dosage recipient treatment after Tx or a combination with conventional immunosuppressive therapy may be necessary. There is still also a lack of evidence regarding the role of tryptophan and its metabolites in the regulation of humoral alloimmunity, which is also an important player in the organism's general immune response against allografts [89,92]. The current experimental evidence still lacks information about the IDO gene therapy effects on long-term graft conditions and functions, as well as systemic consequences. For example, the creation of a tolerogenic environment by enhancing the tryptophan breakdown carries a risk of the development and expansion of malignant cells, which could potentially be one of the biggest drawbacks of IDO gene therapy [17]. Therefore, further investigation is still necessary to answer the remaining questions regarding the effectiveness and safety of targeting the tryptophan-kynurenine pathway, especially using gene therapy, for post-Tx immune regulation.

Table 2. Immunomodulatory effects of targeting the tryptophan-kynurenine pathway in animal models of solid organ transplantation (Tx).

Reference	Experimental Model	Treatment/Intervention	Outcomes
		Heart	
Li et al. [97], 2007	Rats model of heart Tx	IDO gene transfer into the donor heart via adenoviral vector by an intracoronary infusion or intracardiac injections of vector-containing (10^{10} PFU) solution immediately before Tx.	↓mRNA levels of IFN-γ, TNF-α, TGF-β, IL-1β; ↓graft infiltration with monocytes, macrophages, T effector cells; ↑graft survival.
Yu et al. [98], 2008	Mice model of heart Tx	IDO gene transfer into the donor heart via adenoviral vector by intracardiac injections of vector-containing (10^{10} PFU) solution immediately after Tx.	↓mRNA levels for IL-2, IL-17, IFN-γ; ↑proportion of Tregs, no difference in leucocyte infiltration; delayed rejection, ↑graft survival.
Dai et al. [83], 2009	Rats model of heart Tx	Single injection of allogeneic bone marrow dendritic cells+3-HAA for recipient animals 7 days before receiving the graft.	↓mRNA levels for IL-2, IL-17, IFN-γ; ↑proportion of Tregs, no difference in leucocyte infiltration; delayed rejection, ↑graft survival.
Li et al. [99], 2016	Mice model of heart Tx	10^6 donor's DCs transfected with IDO gene via adenoviral vector, infused intravenously 3- and 1- day before Tx.	↓mRNA levels of IFN-γ, ↑ IL-10; ↑ CD4+ T cells apoptosis; ↓histopathological changes; ↑ graft survival.

Table 2. Cont.

Reference	Experimental Model	Treatment/Intervention	Outcomes
Lv et al. [100], 2018	Mice model of heart Tx	10^6 donor's DCs transfected with IDO gene via adenoviral vector, infused intravenously 5- and 3- days before Tx, combined with 250-μg/d CD40L mAb infused intravenously at 0, 1, 2 and 4 days after Tx.	↓serum levels of INF-γ and IL-2; ↑apoptosis of peripheral CD3+ T cells; ↓serum CK and LDH levels. More significant effect of combined therapy than IDO+ DCs alone.
He et al. [103], 2018	Rats model of heart Tx	IDO 1-overexpressing (lentivirus-transfected) bone marrow mesenchymal stem cells exosomes injected intravenously (1 mL; 20 mg/mL) 48 h after heart Tx.	↑graft function (↑ejection fraction and left ventricular fractional shortening on days 2, 4 and 7); ↓graft infiltration by inflammatory cells; ↓serum levels of IL-1α, IL-4, IL-1β, IL-2, IFN-γ, IL-18; ↑levels of IL-10, TGFβ1, TGFβ2 and TGFβ3; ↓expression of CD40, CD86, CD80, MHC-II, CD45RA and CD45RA+CD45RB and ↑expression of CD274; ↑proportion of Tregs in spleen; ↑immunoregulatory protein FHL-1. miR-540-3p was the most highly upregulated microRNA, and miR-338-5p was the most highly downregulated microRNA in IDO+ exosomes.
He et al. [102], 2020	Rats model of heart Tx	10^6 donor bone marrow mesenchymal stem cells transfected with IDO via the lentivirus injected intravenously 48 h after heart Tx.	↑graft function (↑ejection fraction and left ventricular fractional shortening on days 2, 4 and 7 post-Tx; ↓graft infiltration by inflammatory cells; ↓hemorrhage, edema, and myocardial damage; ↓serum levels of IL-1, IL-4, IL-2, IFN-γ, IL-18 and ↑ IL-10, TGFβ1, TGFβ2 and TGFβ3. ↓expression of CD40, CD86, CD80, MHCII, and CD45RA+CD45RB, and ↑ CD274; ↑Tregs.
Li et al. [101], 2020	Mice model of heart Tx	Donors DCs transfected with IDO gene via adenoviral vector combined + tryptophan catabolic products infused intravenously to recipient 3 days before Tx.	↓IL-2, IFN-γ and TFN-α, ↑IL-10 mRNA and protein expression; ↑CD4+ T cells apoptosis; ↓histopathological injury level; ↑allograft survival. Effect more significant with combined therapy than either of therapies alone.
Lungs			
Swanson et al. [108], 2004	Rats model of lung Tx	IDO gene transfer into the donor lungs via adenoviral vector by instillation of vector-containing (10^7 PFU) solution 24 h before Tx.	↓delayed-type hypersensitivity responses to donor antigens, ↓graft rejection histopathological grade.
Liu et al. [109], 2006	Rats model of lung Tx	hIDO gene transfer via PEI carrier. 0.2 mL of transfection solution containing 20 mcg of plasmid DNA delivered to donor rat lung via an intratracheal catheter 24 h before Tx.	↓peak airway pressure, ↑PaO$_2$; ↓level of acute cellular rejection and preserved graft architecture in histopathology; ↓CD3 and MPO-positive cells infiltration; ↓necrosis and apoptosis of lung cells, ↓ intracellular ROS formation.
Liu et al. [110], 2006	Rats model of lung Tx	Sleeping beauty transposon mediated hIDO gene delivery (50 mcg of plasmid DNA) to donor rat lung via an intratracheal catheter 24 h before Tx.	↓peak airway pressure, ↑PaO$_2$; inhibition of TGFβ mediated proliferation of fibroblasts; ↓graft fibrosis; preserved bronchus-alveolar architecture.
Liu et al. [111], 2009	Rats model of lung Tx	Sleeping beauty transposon mediated hIDO gene delivery. 450 mcg of plasmid DNA delivered to donors intravenously 24 h before Tx.	↓acute cellular rejection grade and graft injury; ↓peak airway pressure, ↑PaO$_2$. Impaired function of complex I of the electron transport chain in mitochondria, inhibited cytotoxic function of lung infiltrating T cells; ↓production of IL-2 and TNF-α, but remaining production of IFN-γ (in CD8+ T cells isolated from lung allografts).

Table 2. Cont.

Reference	Experimental Model	Treatment/Intervention	Outcomes
Iken et al. [84], 2012	Mice model of lung Tx	hIDO gene transferred to donor lung via non-viral PEI carrier; Intraperitoneal injections of 250–350 mg/kg 3-HAA for 7 days.	↓acute cellular reduction grade and graft injury; ↓peak airway pressure, ↑PaO$_2$; ↑graft survival due to T cell inhibition: (a) Impaired TCR activation through the interruption of intracellular calcium (Ca21) and of the TCR/Ca21 signaling pathway; (b) ↓levels of proinflammatory cytokines and chemokines (evident ↓IL-2, IL-4 IL-6, IL-5, IL-13, 50% ↓IFN-γ, TNF-α, IL-12, no change of IL-10, IL-17) in allografts; Maintenance of naïve T cells rather than the generation of effector memory T cells.
Ebrahimi et al. [112], 2016	Rats model of engineered lung tissue Tx	IDO transduction to the engineered Lung Tissue via IDO expressing lentivirus.	↓histopathological score of acute rejection, ↓TNF-α and IFN-γ gene expression, ↑level of FOXP3 expression (↑Tregs), ↓RANTES.
Liver			
Laurence et al. [119], 2009	Rats model of liver Tx	IDO gene transfer via recombinant adeno-associated virus vector. Donors pretreated with 10^{13} Vg of the recombinant virus (rAAV2/8-LSP1-rIDO) by infusion into the portal vein 2 weeks prior to Tx.	Despite confirmed in vivo IDO activity, rAAV2/8-LSP1-rIDO failed to prevent liver allograft rejection (recipient and graft survival, histological features did not differ between groups).
Sun et al. [120], 2011	Rats model of liver Tx	200 mg·kg^{-1}·day^{-1} of 3,4-DAA injected intraperitoneally immediately after surgery.	↓serum ALT; ↓level of injury in histopathology; ↓TNF-α, IFN-γ, ↑IL-10 mRNA expression.
Kidney			
Vavrincova-Yaghi et al. [128], 2011	Rats model of kidney Tx	IDO gene transfer to donor kidneys via RGD-modified adenovirus. Solution with 4×10^{11} viral particles infused into renal artery of retrieved kidney and incubated for 20 min in 4 °C saline.	↓increase in plasma creatinine; ↓the interstitial infiltration of CD8+ T cells and macrophages; ↓expression of SMA-α and KIM-1 mRNA; ↓expression of IL-2, IL-17, TGF-β mRNA, ↑levels of foxp3 mRNA (↑Tregs).
He et al. [127], 2015	Rabbits model of kidney Tx	Mesenchymal stem cells transfected with IDO gene via recombinant lentivirus. IDO-MSCs (2×10^6 cells/kg) injected intravenously to recipient mice after Tx.	Induced donor specific allograft tolerance via ↓of CD4+CD25− T-cells, ↑of CD4+CD25+ Foxp3+ (Tregs), ↑ antigen-specific immune-suppressive functions of CD4+CD25+ Tregs, ↑CTLA-4 expression by CD4+CD25+ Tregs, stimulation of Treg cells to secrete IL-10 and TGF-β1; ↑graft survival, ↓acute rejection signs (↓serum creatinine levels, preserved normal histological graft structure).
Vavrincova-Yaghi et al. [129], 2016	Rats model of kidney Tx	IDO gene transfer to donor kidneys via RGD-modified adenovirus. Solution with 4×10^{11} viral particles infused into renal artery of retrieved kidney and incubated for 20 min in 4 °C saline.	In a long-time period (12 weeks) treatment protected against development of transplant vasculopathy, rise of systolic blood pressure and proteinuria; plasma creatinine did not reduce significantly; ↑expression of TGF-β and foxp3 mRNA; ↓expression of ACE mRNA.
Pancreas			
Alexander et al. [132], 2002	Mice model of pancreatic islets Tx	IDO gene transfected into pancreatic islets via recombinant adenovirus.	↑graft survival, ↓proliferation of T cells, ↑remaining insulin producing cells.

Table 2. Cont.

Reference	Experimental Model	Treatment/Intervention	Outcomes
Jalili et al. [135], 2010	Mice model of pancreatic islets Tx	Composite three-dimensional islet grafts engineered by embedding allogeneic mouse islets and adenoviral-transduced IDO-expressing syngeneic fibroblasts within collagen gel matrix.	↓CD4+ and CD8+ effector T-cells infiltration at the graft site, ↑number of Treg cells in graft-draining lymph nodes; ↓gene expression of proinflammatory cytokines (IL-2, IL-17, CXCL9 and CXCL10), ↑gene expression of anti-inflammatory cytokines (IL-4, IL-10); delayed allo-specific antibody production, ↑viability of insulin secreting cells. Immunosuppressive effect limited to 5 weeks post-Tx.
Poormasjedi-Meibod et al. [76], 2013	Mice model of pancreatic islets xenoTx	Composite three-dimensional islet grafts engineered by embedding allogeneic rat islets and adenoviral-transduced IDO-expressing syngeneic mouse fibroblasts within the collagen gel matrix.	Well-preserved islet morphology, ↑number of insulin and glucagon expressing β cells, ↓number of infiltrating macrophages and CD3+ T cells, ↓iNOS expression.
Hosseini-Tabatabaei et al. [136], 2015	Mice model of pancreatic islets Tx	Composite three-dimensional islet grafts engineered by embedding allogeneic rat islets and lentivirus-transduced IDO-expressing syngeneic mouse fibroblasts within protease-resistant scaffold.	↑islet allograft survival (until 7 weeks); ↑population of foxp3+ Tregs at the graft site and graft-draining lymph nodes, ↓T-cell infiltration. Better-preserved functional β cell mass.
Small bowel			
Xie et al. [82], 2015	Mice model of small bowel Tx	IDO gene transfected to DCs via adenoviral vector. Three treatment groups: (A) 2×10^6 IDO gene-transfected DCs injected intravenously immediately after Tx; (B) 120-mg/kg 3-HAA injected intravenously 7 days before Tx; (C) Both.	All treatment types ↓effector T cells and ↑Foxp3+ Tregs, ↓plasma pro-inflammatory cytokines (IFN-γ, IL-2) and ↑anti-inflammatory cytokines (IL-10, TGFβ), preserved histological graft structure. All strategies ↑graft survival compared to control, however IDO-DCs were more effective than 3-HAA. No significant difference in graft survival when used both.

Abbreviations: Tx: transplantation, (h)IDO: (human) indoleamine-2,3-dioxygenase, PFU: plaque-forming unit, IFN: interferon, TNF: tumor necrosis factor, TGF: tumor growth factor, IL: interleukin, 3-HAA: 3-hydroxyanthranilic acid, CD: cluster of differentiation, DCs: dendritic cells, CK: creatine kinase, LDH: lactate dehydrogenase, MHC: major histocompatibility complex, PEI: polymer polyethyleneimine, MPO: myeloperoxidase, TCR: T-cell receptor, foxp3: Forkhead box P3, RANTES: Regulated upon Activation, Normal T Cell Expressed and Presumably Secreted, Tregs: regulatory T cells, RNA: ribonucleic acid, DNA: deoxyribonucleic acid, ALT: alanine aminotransferase, 3,4-DAA: 3,4,-dimethoxycinnamoyl anthranilic acid, CTLA: cytotoxic T-lymphocyte-associated protein, ACE: angiotensin-converting enzyme, ROS: reactive oxygen species, PaO$_2$: partial oxygen pressure, KIM: kidney injury molecule, SMA: smooth muscle actin and iNOS: inducible nitric oxide synthase, ↑: increased, ↓: reduced.

5. Diagnostic and Prognostic Role of Tryptophan and Its Metabolites in Solid Organ Tx

The assessment of graft quality, prediction of the early- and long-term graft outcomes and risk of post-Tx mortality of recipients is another important aspect in the field of solid organ Tx. The early detection of specific biomarkers of rejection, delayed graft function or primary nonfunction are also essential to identify patients at risk and put more effort on the follow-up and timely treatment in these cases. Tryptophan and its metabolites have been investigated as attractive diagnostic and/or prognostic tools in solid organ Tx (Table 3).

There is evidence that about half of recipients experience inflammatory conditions after kidney Tx, such as cellular or humoral graft rejections, bacterial, viral or mycotic infections, which increase the level of proinflammatory mediators and induce tryptophan catabolism by IDO [138]. Correspondingly, it has been suggested that low IDO activity and stable low serum levels of tryptophan catabolite kynurenine after kidney Tx are related to superior graft survival, as an immune balance has been reached [139]. Lahdou et al. [140] and Brandacher et al. [141], investigating adult kidney recipients, reported that the serum kynurenine/tryptophan ratio in the recipients without acute rejection drops to a level comparable with healthy non-transplanted subjects as early as in the first three weeks [141] or six months [140] after graft implantation. Interestingly, another group showed that stable pediatric kidney Tx recipients (no acute graft rejection and no infection) had significantly higher urine kynurenine/tryptophan ratios within the first month post-Tx compared to healthy children. This suggests that a slightly increased tryptophan catabolism may probably reflect the maintenance of the stable immunological/anti-inflammatory state after Tx [142]. Holmes et al. were the first to report that serum kynurenine levels significantly increase at five–seven days before biopsy-proved kidney allograft rejections, as well as in cases of viral or bacterial infections [143]. Brandacher et al. confirmed these results, showing that the serum and urine kynurenine/tryptophan ratios and kynurenine alone permit an accurate diagnosis of acute graft rejection. Importantly, changes of kynurenine/tryptophan were observed as early as one day after Tx, much earlier than the episode of rejection, and suggests a predictive role of tryptophan catabolism in transplanted patients. However, no significant changes in either levels of tryptophan, kynurenine or the kynurenine/tryptophan ratio were associated with infection episodes in the observed cohort [141]. Lahdou et al. demonstrated that recipients experiencing acute kidney graft rejection had significantly higher plasma kynurenine levels already prior to Tx compared to patients who had favorable early graft outcomes [140]. Although these early observations were limited to small cohorts [140,141,143], later larger retrospective and prospective studies confirmed rather than denied these results. In a large retrospective study, Kaden et al. demonstrated that kidney recipients with immediately functioning grafts show a significant decrease of serum kynurenine levels to almost normal values already at day five after Tx, and this normal course in kynurenine dynamics is changed by inflammatory events that activate IDO. The increased levels of kynurenine in the context of acute graft rejection in this study was dependent on the severity of rejection (the lowest increase in steroid-sensitive rejection and the highest one in vascular rejection). Moreover, the successful treatment of rejection resulted in decreased plasma kynurenine levels [144]. In a prospective study by Vavrincova-Yaghi et al., a rapidly increasing serum kynurenine/tryptophan ratio as early as the first day after Tx was associated with acute graft rejection [145]. Kim et al. detected five urinary biomarkers related to T-cell mediated kidney allograft rejection by using a metabolomics approach: a positive association was found with guanidoacetic acid, methylimidazoleacetic acid and dopamine and a negative one with 4-guanidinobutyric acid and L-tryptophan. Together, these metabolites predicted acute T-cell-mediated rejection with an accuracy of 87% for the training set; however, in the validation set, the accuracy decreased to 62.5% [146]. In another targeted metabolomics study that investigated urinary metabolites, kynurenine was found as one of the top 10 metabolites able to identify acute T-cell-mediated rejection in pediatric kidney Tx recipients [147]. On the contrary, Dharnidharka et al. did not observe such diagnostic and prognostic effects of urine tryptophan, kynurenine or the kynurenine/tryptophan ratio. However, they found that the increased serum kynurenine/tryptophan ratio, but not tryptophan or kynurenine alone, reflects episodes of acute rejection in children after renal Tx [148]. Interestingly, another metabolomics study, investigating metabolic changes in acute renal allograft rejection, reported converse results. Serum tryptophan was decreased and kynurenine increased in the group of patients experiencing no acute graft rejection in comparison to recipients who did. The authors explained this finding as a reflection of the increased IDO activity and its probable tolerogenic role for relieving acute rejection [149].

The role of kynurenine in the diagnosis of infectious events of kidney Tx recipients has also been demonstrated. In the case of cytomegalovirus (CMV) infection, elevated serum kynurenine levels were found even in asymptomatic patients [144]. Sadeghi et al. similarly showed that plasma levels of two tryptophan metabolites, kynurenine and quinolinic acid, correlated with the severity of CMV infection in kidney Tx recipients [150]. Moreover, patients experiencing pneumonia and/or sepsis showed significantly increased serum kynurenine levels already five days prior to diagnosis and treatment initiation compared to the conventional marker C-reactive protein (CRP), which increased only immediately before the start of therapy. Furthermore, the increase of kynurenine despite the ongoing antibiotics was associated with lethal endpoints [144]. Dharnidharka et al. also reported that the urine kynurenine/tryptophan ratio was significantly increased prior to major infection events in pediatric kidney transplant patients, even though the tryptophan metabolite levels in urine did not reflect acute rejection in this study [148]. Nevertheless, it is important to emphasize that kynurenine may not be suitable in the prediction or diagnosis of acute rejection or major infections in patients who are already dialyzed, because dialysis itself leads to elevated kynurenine levels [144,151].

Several attempts to use tryptophan and its metabolites as diagnostic or prognostic tools to predict graft function in a short-term or long-term period after Tx have been reported. A metabolomics study that enrolled [42] patients with or without acute kidney injury after renal Tx revealed the serum tryptophan and symmetric dimethylarginine levels to have the most significant negative correlations with blood urea nitrogen, serum creatinine and uric acid, being the most accurate markers for graft injury. When combined, these two markers reached an area under the curve of 0.901 with a sensitivity of 0.889 and a specificity of 0.831 for the diagnosis of acute kidney injury. These results implicate that decreased tryptophan levels may serve as a highly potential biomarker for graft injury [152]. Another group investigated the metabolomics of 40 human kidney allografts, looking for abnormalities in failing transplants. They found significantly reduced serum tryptophan levels in patients with low GFR after renal Tx. The decline in serum tryptophan was already detected in patients with slightly diminished GFR and became apparent in the group with the lowest GFR (21 ± 39 mL/min), revealing a dose response trend of tryptophan reduction. Similarly, a tryptophan decline in urine confirms that lower tryptophan levels reflect accelerated catabolism rather than urinary loss [11]. Vavrincova-Yaghi et al. focused on the prediction of chronic kidney allograft dysfunction and long-term kidney outcome. In general, the kynurenine/tryptophan ratio was significantly decreased at two years after Tx compared to two weeks and six months, thus reflecting the pattern of tryptophan metabolism changes. Serum and urine tryptophan and kynurenine levels, as well as the serum tryptophan/kynurenine ratio at two weeks and six months after renal Tx, were associated with two-year kidney graft function, as reflected by serum creatinine levels and albuminuria. Importantly, serum kynurenine at six months was found to be the only independent predicting factor for two-year serum creatinine. Interestingly, an association between the early levels of tryptophan or its metabolites with a histopathological level of graft damage was not found [145]. Another large prospective cohort study of 561 recipients with a more than one-year functioning renal graft revealed that baseline serum kynurenine, 3-HK and the 3-HK/kynurenine ratio were strongly associated with graft failure in long-term follow-up and kidney graft function. Out of all of these parameters, 3-HK was best-associated with long-term kidney graft outcomes. Serum 3-HK and the 3HK/kynurenine ratio were also linked to mortality, while serum kynurenine and the kynurenine/tryptophan ratio did not show such an association. However, the same parameters in urine did not show any such independent associations with long-term outcomes, suggesting that serum metabolites of the tryptophan-kynurenine pathway may be more favorable predictors in terms of kidney function [153].

Despite the fact that most of the predictive/diagnostic studies investigating tryptophan metabolism were conducted in the setting of kidney Tx, similar results were demonstrated in other transplant recipients. Meloni et al. investigated the plasma kynurenine/tryptophan levels in 26 patients with clinically stable lung grafts for >36 months after Tx and 64 patients who presented with bronchiolitis obliterans syndrome of various grades. A significantly increased kynurenine/tryptophan ratio was found in recipients with bronchiolitis obliterans syndrome in comparison to patients with stable grafts, supporting the idea that an increased IDO activity and tryptophan catabolism rather reflects chronic rejection than the protective/tolerogenic state of the immune system [154]. In a retrospective cohort study, Oweira et al. found significantly increased pre-Tx tryptophan, kynurenine, KYNA and quinolinic acid levels in patients who died due to sepsis or graft failure during the first year after liver Tx compared to those who survived more than one year. Together with serum bilirubin at post-Tx day five and phenylalanine at days five and 10, the pre-Tx serum kynurenine and kynurenic acid levels correlated significantly with one-year mortality after liver Tx [155].

In addition, tryptophan and its metabolites were also investigated as potential biomarkers of the organ quality and severity of graft IRI in livers. A metabolomics analysis of 37 human liver grafts revealed nearly two-fold increased levels of tryptophan and kynurenine in donation after circulatory death (DCD) grafts (in comparison with donation after brain-stem death (DBD) grafts) in the cold ischemia phase. All grafts were stored in tryptophan-free organ preservation solution, suggesting that tryptophan metabolism is probably responsible for the organ quality and may be used as a biomarker. Two primary nonfunctioning allografts in this cohort also had significantly higher levels of tryptophan and kynurenine in the cold phase. However, significant differences regarding tryptophan and its metabolites were not observed during reperfusion [10]. A metabolomics study investigating micro-dialysates of human liver grafts during the complex process of IRI revealed that kynurenine might potentially serve as a biomarker of aggravated ischemic injury. A nearly three-fold increased expression of kynurenine was found in DCD (donation after circulatory death) grafts at the end of cold ischemia compared to DBD (donation after brain-stem death) grafts. Moreover, the kynurenine levels gradually increased during reperfusion. Cold ischemia-phase kynurenine levels were also linked to graft outcome. Grafts that later developed a primary nonfunction or initially poor function had four-times higher kynurenine levels at 10 min of cold ischemia compared to immediately functioning ones [156]. There is an increased interest in the detection of biomarkers in the organ machine perfusion setting, which would allow determining the graft quality and predicting outcomes after implantation. As already mentioned before, a recent study found differences in tryptophan metabolism during discarded human liver normothermic machine perfusion compared to sub-normothermic machine perfusion. An increased tryptophan metabolism, as well as higher kynurenine and KYNA levels, have been found in livers subjected to normothermic compared to sub-normothermic machine perfusion. As grafts were not implanted, no associations with the later graft function were possible. However, this study suggests that measurements of tryptophan catabolites in perfusate could be used for graft quality assessment or the prediction of outcomes [24]. A recent experimental study using a mouse model raised concerns about the capability of tryptophan metabolism markers to reflect graft IRI. The authors found that metabolic profiles in acute cellular rejection and IRI after kidney Tx differ. Kynurenine was significantly increased during acute allograft rejection but not in IRI. This finding sheds some light on the potential discriminative property of tryptophan metabolites in distinguishing reasons for graft dysfunction [157].

The increased tryptophan metabolism seems to be paradox. On the one hand, it should work as an endogenous process inducing immune suppression and preventing graft rejection. On the other hand, studies have shown that increased kynurenine/tryptophan ratios and/or decreased tryptophan and/or increased levels of tryptophan metabolites reflect the graft rejection or injury. It raises a question if this IDO-dependent pathway is sufficient to achieve a tolerogenic state after solid organ Tx. Importantly, it should be taken into account that observational studies reveal correlations and associations between metabolite and graft functions or the event of rejection rather than causal relationships.

Most of the current evidence is available in the setting of kidney Tx, while sufficient data about the diagnostic role of tryptophan catabolism in other organ Tx settings is scarce. However, there is still a question whether tryptophan and its metabolites might be considered as markers in solid organ Tx, as the results of the studies are inconclusive. This may not only be explained by inconsistent types and sizes of study populations, endpoints of interest and varying follow-up periods but, also, by different measurement techniques. In general, the reliable and accurate quantification of tryptophan and kynurenine pathway metabolites in plasma, urine or organ perfusate might be challenging, especially in daily clinical practice, mainly due to their lability, low physiological levels and interference of other endogenous compounds in biological fluids [158]. Extreme concentration differences among tryptophan and its metabolites (e.g., plasma concentrations of tryptophan, kynurenine and 3HK in healthy individuals are about 60 µmol/L, 2 µmol/L and <0.13 µmol/L, respectively) [159] may hamper the simultaneous quantification of these analytes. Determination of the tryptophan-kynurenine pathway compounds mainly relies on chromatographic methods such as high-performance liquid chromatography (HPLC) coupled to ultraviolet (UV), fluorescent or electrochemical detection, as well as liquid or gas chromatography with mass spectrometric detection (LC-MS or GC-MS, respectively). However, specific drawbacks of all these techniques should be taken into account [155,158]. Although the high-performance LC with UV detection is easily accessible in clinical practice, its low sensitivity and specificity in tryptophan-kynurenine metabolite quantification due to the interference of other biological compounds, as well as the problems in the simultaneous determination of multiple metabolites, are major drawbacks of the method. Selectivity may be improved with fluorescence detection; however, not all kynurenines are fluorescent, making quantification without derivatization sometimes impossible. Electrochemical detection high sensitivity is an advantage for the detection of metabolites in low concentrations. However, its level of selectivity and reproducibility may be not satisfactory [158]. The most suitable and reliable chromatographic method for the simultaneous multi-compound analysis of tryptophan and kynurenine pathway metabolites is LC-MS, which allows the accurate quantification of analytes in a wider range of concentrations. However, due to the necessity of costly equipment, expensive isotope-labeled internal standards and precise sample preparation process, its utility in daily clinical practice is limited as well [11,146,147,149,152,158]. Nuclear magnetic resonance (NMR) is another promising technique that allows not only the rapid and precise quantification of various biochemical analytes in biological fluids or tissues but, also, the noninvasive in vivo analysis of metabolism in spectroscopic images and have also been successfully used to determine the levels of tryptophan-kynurenine pathway compounds [11,160]. Nevertheless, this method also requires sophisticated equipment, which currently limits its use mainly to the scientific setting.

Nevertheless, tryptophan metabolism via the kynurenine pathway remains an attractive potential indicator of graft quality and post-Tx function, as well as a reflector of the immune state. Therefore, we encourage the establishment of robust, accurate and universal detection techniques that could be available not only in scientific but, also, in clinical settings. Additionally, further studies validating tryptophan metabolites as noninvasive biomarkers in the field of Tx are needed.

Table 3. Tryptophan and its metabolite association with post-Tx outcomes in human studies.

Reference	Study Design	Population	Measurement	Follow-Up	Graft Function	AR	Infection	Recipient Mortality
Holmes et al. [143], 1992	Retrospective	32 kidney Tx recipients	Post-Tx serum Trp and Kyn	3 weeks post-Tx	DGF: Trp↓, Kyn↑; SCr: Trp↓, Kyn↔	Trp↔, but ↓if failed AR treatment or concomitant infection; Kyn↑	CMV, EBV, pneumonia, UTI: Trp↓, Kyn↑	n.d.
Brandacher et al. [141], 2007	Prospective	43 kidney Tx recipients	Serum and urine Trp and Kyn 1, 8, 15 and 21 days post-Tx and at the time of AR	3 weeks post-Tx	n.d.	Serum Trp↓, Kyn↑, Kyn/Trp↓, Urine Kyn/Trp↓	H. simplex (n = 4), UTI (n = 1): serum Trp↔, Kyn↔, Kyn/Trp↔; sepsis (n = 1): serum Trp↔, Kyn↔, Kyn/Trp↔	n.d.
Lahdou et al. [140], 2010	Retrospective	210 first kidney Tx recipients	Pre-Tx plasma Trp, Kyn, post-Tx plasma Trp, Kyn, Kyn/Trp *available for 10 AR+ patients (median 7 days post-Tx) and 24 AR- patients (median 11 days post-Tx)	6 months post-Tx	n.d.	Pre-Tx Trp↑ (sens. 61%, spec. 71%), Kyn↑ (sens. 64%, spec. 71%); Post-Tx Trp↔, Kyn↑ (sens. 80%, spec. 79%), Kyn/Trp↑ (sens. 70%, spec. 79%)	n.d.	n.d.
Kaden et al. [144], 2015	Retrospective	355 kidney Tx recipients	Post-Tx serum Kyn	3 weeks post-Tx	DGF and PNF: slowed decrease or increase of serum Kyn	Kyn↑ (level of increase depends strongly on AR severity in non-dialyzed patients)	CMV: Kyn↑ Pneumonia: Kyn↑ Sepsis: Kyn↑	n.d.
Vavrincova-Yaghi et al. [145], 2015	Prospective	48 kidney Tx recipients	Serum and urine Trp, Kyn, Kyn/Trp at 2 weeks, 6 months and 2 years post-Tx	24 months	SCr at 2 years: serum Kyn/Trp at 6 months↑; Kyn at 6 months↑ (AUC 0.76); CrCl at 2 years: urine Trp at 2 weeks↓ (AUC 0.44). Albuminuria at 2 years: Urine Trp↓; Kyn↑ at 2 weeks, Kyn/Trp↑ at 6 months	n.d.	n.d.	n.d.
Kim et al. [146], 2019	Cross-sectional	385 kidney Tx recipients	Urine metabolomic analysis (LC-MS)	n.d.	n.d.	TCMR: Trp↓; Panel of 5 top metabolites, including Trp showed AUC 0.926, acc. 87.0% (training set) and 62.5% (validation set)	n.d.	n.d.

115

Table 3. Cont.

Reference	Study Design	Population	Measurement	Follow-Up	Outcome Association			
					Graft Function	AR	Infection	Recipient Mortality
Blydt-Hansen et al. [147], 2014	Cross-sectional	277 biopsy-paired urine samples from 57 pediatric kidney Tx recipients	Urine metabolomic analysis (LC-MS)	n.d.	n.d.	TCMR: Kyn↑; TMCR discriminant score of 10 top metabolites, including Kyn showed AUC 0.892	n.d.	n.d.
Dharnidharka et al. [148], 2013	Prospective	29 pediatric kidney Tx recipients	Serum and urine Trp, Kyn, Kyn/Trp	12 months after Tx	n.d.	AR within the next 30 days from the test: serum Kyn/Trp↑; urine Kyn/Trp↔; serum and urine Trp and Kyn↔	MIE (CMV, EBV, BKV, Tx pyelonephritis, fever with bacteriemia): serum and urine Kyn/Trp↔; serum and urine Trp and Kyn↔	n.d.
Zhao et al. [149], 2014	Cross-sectional	27 primary kidney Tx recipients	Serum metabolomics analysis (LC-MS) pre-Tx and 7 days post-Tx	n.d.	n.d.	In non-rejecting patients: Trp↓; Kyn↑; Kyn/Trp↑	n.d.	n.d.
Sadeghi et al. [150], 2012	Cross-sectional	86 kidney Tx recipients	Plasma Trp, Kyn, Quin, Kyn/Trp, Quin/Trp	n.d.	n.d.	n.d.	CMV: Trp↔; Kyn↑ (AUC 0.82), Kyn/Trp↑ (AUC 0.83) Quin↑ (AUC 0.85), Quin/Trp (sens. 83%, spec. 74%) (correlate with the severity); BKV: Trp, Kyn, Quin, Kyn/Trp, Quin/Trp↔	n.d.
Zhang et al. [152], 2018	Retrospective	42 kidney Tx recipients	Plasma metabolomics (25 amino acids) analysis (UHPLC-MS/MS)	n.d.	AKI (↑ of SCr of > 0.3 mg/dL or ↑ ≥50% over baseline): Trp↓ (AUC 0.78); Trp+SDMA↓ (AUC 0.901)	n.d.	n.d.	n.d.
Bassi et al. [11], 2017	Cross-sectional	40 kidney Tx recipients at least 6 months after Tx	Ex vivo (serum and urine) and in vivo metabolomics (LC-MS/MS, FIA-MS/MS (n = 40), 2D COSY with 3D-image transformation (n = 15)	n.d.	GFR 21–39 mL/min: serum Trp↓	n.d.	n.d.	n.d.

Table 3. Cont.

Reference	Study Design	Population	Measurement	Follow-Up	Graft Function	Outcome Association AR	Infection	Recipient Mortality
de Vries et al. [153], 2017	Prospective	561 stable kidney Tx recipients with functioning graft for at least 1 year	Serum and urine Trp, Kyn, 3-HK, Kyn/Trp, 3-HK, Kyn/Trp, 3-HK/Kyn	Median 7.0 [6.2–7.5] years	Graft failure, SCr, proteinuria: serum Trp↓, Kyn, 3-HK, Kyn/Trp, 3-HK/Trp↑; no significant changes in urine; GFR: serum Trp↑, Kyn, 3-HK, Kyn/Trp, 3-HK/Trp↓	n.d.	n.d.	Serum Trp↔; Kyn; 3-HK; Kyn/Trp; 3-HK/Kyn↑; no significant changes in urine
Meloni et al. [154], 2009	Cross-sectional	90 lung Tx recipients	Plasma Trp, Kyn, Kyn/Trp	n.d.	n.d.	BOS: Trp↔, Kyn↑, Kyn/Trp↑	n.d.	n.d.
Oweira et al. [155], 2018	Retrospective	89 liver Tx recipients	Pre-Tx and early post-Tx serum Trp, Kyn, Quin, KYNA	1 year	EAD: pre-Tx Kyn↑ (AUC 0.64); post-Tx day 3 Kyn↑ (AUC 0.69); day 5 Kyn↑ (AUC 0.74); day 10 Kyn↑ (AUC 0.77)	n.d.	n.d.	1-year mortality: pre-Tx Kyn↑ (AUC 0.77), KYNA↑ (AUC 0.74), Quin↑ (AUC 0.72), Trp↑ (AUC 0.72); Post-Tx day 1 KYNA↑ (AUC 0.73); day 5 KYNA↑ (AUC 0.71); Kyn↑ (AUC 0.73); Pre-Tx Kyn↑ associated with 1-year mortality in univariate analysis
Perera et al. [156], 2014	Prospective	40 liver Tx recipients	Metabolomic analysis (CEAD) of liver micro-dialysis samples: pre-Tx (bench micro-dialysis) and every 6 h for 48 h post-Tx	7 days post Tx	PNF/IPF: pre-Tx (bench) Kyn↑; Kyn↑ in DCD vs. DBD grafts at the end of cold ischemia	n.d.	n.d.	n.d.

Abbreviations: ↑: positive association, ↓: negative association, ↔: no association, n.d.: no data, sens.: sensitivity, spec: specificity, acc.: accuracy, AUC: area under the curve, Tx: transplantation, Trp: tryptophan, Kyn: kynurenine, Quin: quinolinic acid, KYNA: kynurenic acid, SDMA: symmetric dimethylarginine, LC-MS: liquid chromatography-tandem mass spectrometry, UHPLC–MS/MS: ultra-high performance liquid chromatography-tandem mass spectrometry, FIA-MS/MS: flow injection analysis-tandem mass spectrometry, 2D COSY: two-dimensional correlated spectroscopy, CEAD: coulometric electrochemical array detection, DGF: delayed graft function, PNF: primary nonfunction, EAD: early allograft dysfunction, IPF: initial poor function, SCr: serum creatinine, GFR: glomerular filtration rate, CrCl: creatinine clearance, AR: acute rejection, TCMR: T-cell-mediated rejection, BOS: bronchiolitis obliterans syndrome, MIE: major infection event, UTI: urinary tract infection, CMV: cytomegalovirus, EBV: Ebstein Barr virus, DCD: donation after circulatory death, DBD: donation after brain-stem death and BKV: BK virus.

6. Conclusions

It seems that the tryptophan metabolism via the kynurenine pathway may have multiple clinical effects in solid organ Tx, depending on the timepoint and the dominant pathophysiological mechanism (IRI or immune response against allograft), as well as the prevalent enzymes and metabolites of this metabolic pathway. However, this evidence grants some interesting insights into therapeutic possibilities by offering several attractive pharmacological and/or genetic modification targets, probably paving the way for protection from deleterious pathological effects in solid organ Tx and the improvement of short- and long-term outcomes. Moreover, metabolites of the tryptophan-kynurenine pathway may potentially serve as diagnostic and prognostic tools that allow improving and personalizing the care of transplanted patients. Further studies investigating this topic and answering the remaining questions are crucial.

Author Contributions: P.S. (Philipp Stiegler) was responsible for the study concept, design and critical revision of the drafted manuscript. R.Z., P.M., P.S. (Peter Schemmer), B.L. and K.S. were responsible for the literature review, interpretation and drafting of the manuscript. All authors have read and agreed to the published version of the manuscript.

Funding: This research received no external funding.

Conflicts of Interest: The authors declare no conflict of interest.

References

1. Black, C.K.; Termanini, K.M.; Aguirre, O.; Hawksworth, J.S.; Sosin, M. Solid Organ Transplantation in the 21st Century. *Ann. Transl. Med.* **2018**, *6*, 409. [CrossRef]
2. Kittleson, M.M.; Kobashigawa, J.A. Cardiac Transplantation: Current Outcomes and Contemporary Controversies. *JACC Heart Fail.* **2017**, *5*, 857–868. [CrossRef]
3. Annual Reports. Available online: https://www.eurotransplant.org/statistics/annual-report/ (accessed on 29 November 2020).
4. Organ Specific Reports. Available online: https://statistics-and-reports/organ-specific-reports/ (accessed on 29 November 2020).
5. Jochmans, I.; Brat, A.; Davies, L.; Hofker, H.S.; van de Leemkolk, F.E.M.; Leuvenink, H.G.D.; Knight, S.R.; Pirenne, J.; Ploeg, R.J. COMPARE Trial Collaboration and Consortium for Organ Preservation in Europe (COPE). Oxygenated versus Standard Cold Perfusion Preservation in Kidney Transplantation (COMPARE): A Randomised, Double-Blind, Paired, Phase 3 Trial. *Lancet* **2020**, *396*, 1653–1662. [CrossRef]
6. Nasralla, D.; Coussios, C.C.; Mergental, H.; Akhtar, M.Z.; Butler, A.J.; Ceresa, C.D.L.; Chiocchia, V.; Dutton, S.J.; García-Valdecasas, J.C.; Heaton, N.; et al. Consortium for Organ Preservation in Europe. A Randomized Trial of Normothermic Preservation in Liver Transplantation. *Nature* **2018**, *557*, 50–56. [CrossRef] [PubMed]
7. Moreau, A.; Varey, E.; Anegon, I.; Cuturi, M.-C. Effector Mechanisms of Rejection. *Cold Spring Harb. Perspect. Med.* **2013**, *3*, a015461. [CrossRef]
8. Charlton, M.; Levitsky, J.; Aqel, B.; O'Grady, J.; Hemibach, J.; Rinella, M.; Fung, J.; Ghabril, M.; Thomason, R.; Burra, P.; et al. International Liver Transplantation Society Consensus Statement on Immunosuppression in Liver Transplant Recipients. *Transplantation* **2018**, *102*, 727–743. [CrossRef] [PubMed]
9. Abbas, K.; Zafar, M.N.; Aziz, T.; Musharraf, W.; Naqvi, A.; Rizvi, A. Renal Transplantation in Highly Sensitized Patients: SIUT Experience. *Transplantation* **2018**, *102*, S618. [CrossRef]
10. Hrydziuszko, O.; Perera, M.T.P.R.; Laing, R.; Kirwan, J.; Silva, M.A.; Richards, D.A.; Murphy, N.; Mirza, D.F.; Viant, M.R. Mass Spectrometry Based Metabolomics Comparison of Liver Grafts from Donors after Circulatory Death (DCD) and Donors after Brain Death (DBD) Used in Human Orthotopic Liver Transplantation. *PLoS ONE* **2016**, *11*, e0165884. [CrossRef]
11. Bassi, R.; Niewczas, M.A.; Biancone, L.; Bussolino, S.; Merugumala, S.; Tezza, S.; D'Addio, F.; Ben Nasr, M.; Valderrama-Vasquez, A.; Usuelli, V.; et al. Metabolomic Profiling in Individuals with a Failing Kidney Allograft. *PLoS ONE* **2017**, *12*, e0169077. [CrossRef]
12. Richard, D.M.; Dawes, M.A.; Mathias, C.W.; Acheson, A.; Hill-Kapturczak, N.; Dougherty, D.M. L-Tryptophan: Basic Metabolic Functions, Behavioral Research and Therapeutic Indications. *Int. J. Tryptophan Res.* **2009**, *2*, 45–60. [CrossRef]
13. Rodriguez Cetina Biefer, H.; Vasudevan, A.; Elkhal, A. Aspects of Tryptophan and Nicotinamide Adenine Dinucleotide in Immunity: A New Twist in an Old Tale. *Int. J. Tryptophan Res.* **2017**, *10*, 1178646917713491. [CrossRef]
14. Höglund, E.; Øverli, Ø.; Winberg, S. Tryptophan Metabolic Pathways and Brain Serotonergic Activity: A Comparative Review. *Front. Endocrinol.* **2019**, *10*, 158.
15. Gostner, J.M.; Geisler, S.; Stonig, M.; Mair, L.; Sperner-Unterweger, B.; Fuchs, D. Tryptophan Metabolism and Related Pathways in PsychoneuroImmunol.ogy: The Impact of Nutrition and Lifestyle. *Neuropsychobiology* **2020**, *79*, 89–99. [CrossRef]
16. Opitz, C.A.; Wick, W.; Steinman, L.; Platten, M. Tryptophan Degradation in Autoimmune Diseases. *Cell. Mol. Life Sci.* **2007**, *64*, 2542–2563. [CrossRef] [PubMed]

17. Boros, F.A.; Vécsei, L. Immunomodulatory Effects of Genetic Alterations Affecting the Kynurenine Pathway. *Front. Immunol.* **2019**, *10*, 2570. [CrossRef]
18. Zhai, L.; Lauing, K.L.; Chang, A.L.; Dey, M.; Qian, J.; Cheng, Y.; Lesniak, M.S.; Wainwright, D.A. The Role of IDO in Brain Tumor Immunotherapy. *J. Neuro-Oncol.* **2015**, *123*, 395–403. [CrossRef] [PubMed]
19. Platten, M.; Nollen, E.A.A.; Röhrig, U.F.; Fallarino, F.; Opitz, C.A. Tryptophan Metabolism as a Common Therapeutic Target in Cancer, Neurodegeneration and Beyond. *Nat. Rev. Drug Discov.* **2019**, *18*, 379–401. [PubMed]
20. Badawy, A.A.-B. Kynurenine Pathway of Tryptophan Metabolism: Regulatory and Functional Aspects. *Int. J. Tryptophan Res.* **2017**, *10*, 1178646917691938. [CrossRef] [PubMed]
21. Oxenkrug, G.F. Genetic and Hormonal Regulation of Tryptophan–Kynurenine Metabolism. *Ann. N. Y. Acad. Sci.* **2007**, *1122*, 35–49. [CrossRef]
22. Badawy, A.A.-B. Tryptophan Metabolism, Disposition and Utilization in Pregnancy. *Biosci. Rep.* **2015**, *35*, e00261.
23. Munn, D.H.; Zhou, M.; Attwood, J.T.; Bondarev, I.; Conway, S.J.; Marshall, B.; Brown, C.; Mellor, A.L. Prevention of Allogeneic Fetal Rejection by Tryptophan Catabolism. *Science* **1998**, *281*, 1191–1193. [CrossRef] [PubMed]
24. Zhang, A.; Carroll, C.; Raigani, S.; Karimian, N.; Huang, V.; Nagpal, S.; Beijert, I.; Porte, R.J.; Yarmush, M.; Uygun, K.; et al. Tryptophan Metabolism via the Kynurenine Pathway: Implications for Graft Optimization during Machine Perfusion. *J. Clin. Med.* **2020**, *9*, 1864. [CrossRef] [PubMed]
25. Zheng, X.; Zhang, A.; Binnie, M.; McGuire, K.; Webster, S.P.; Hughes, J.; Howie, S.E.M.; Mole, D.J. Kynurenine 3-Monooxygenase Is a Critical Regulator of Renal Ischemia-Reperfusion Injury. *Exp. Mol. Med.* **2019**, *51*, 1–14. [CrossRef] [PubMed]
26. Kalogeris, T.; Baines, C.P.; Krenz, M.; Korthuis, R.J. Cell Biology of Ischemia/Reperfusion Injury. *Int. Rev. Cell Mol. Biol.* **2012**, *298*, 229–317.
27. Thomas, S.R.; Stocker, R. Redox Reactions Related to Indoleamine 2,3-Dioxygenase and Tryptophan Metabolism along the Kynurenine Pathway. *Redox Rep.* **1999**, *4*, 199–220.
28. Liu, H.; Liu, L.; Visner, G.A. Nonviral Gene Delivery with Indoleamine 2,3-Dioxygenase Targeting Pulmonary Endothelium Protects against Ischemia-Reperfusion Injury. *Am. J. Transpl.* **2007**, *7*, 2291–2300. [CrossRef]
29. Kanai, M.; Nakamura, T.; Funakoshi, H. Identification and Characterization of Novel Variants of the Tryptophan 2,3-Dioxygenase Gene: Differential Regulation in the Mouse Nervous System during Development. *Neurosci. Res.* **2009**, *64*, 111–117. [CrossRef]
30. Opitz, C.A.; Litzenburger, U.M.; Sahm, F.; Ott, M.; Tritschler, I.; Trump, S.; Schumacher, T.; Jestaedt, L.; Schrenk, D.; Weller, M.; et al. An Endogenous Tumour-Promoting Ligand of the Human Aryl Hydrocarbon Receptor. *Nature* **2011**, *478*, 197–203. [CrossRef]
31. Badawy, A.A.-B.; Bano, S. Tryptophan Metabolism in Rat Liver After Administration of Tryptophan, Kynurenine Metabolites, and Kynureninase Inhibitors. *Int. J. Tryptophan Res.* **2016**, *9*, 51–65.
32. Nakamura, T.; Niimi, S.; Nawa, K.; Noda, C.; Ichihara, A.; Takagi, Y.; Anai, M.; Sakaki, Y. Multihormonal Regulation of Transcription of the Tryptophan 2,3-Dioxygenase Gene in Primary Cultures of Adult Rat Hepatocytes with Special Reference to the Presence of a Transcriptional Protein Mediating the Action of Glucocorticoids. *J. Biol. Chem.* **1987**, *262*, 727–733. [CrossRef]
33. Badawy, A.A.-B. Tryptophan Availability for Kynurenine Pathway Metabolism across the Life Span: Control Mechanisms and Focus on Aging, Exercise, Diet and Nutritional Supplements. *Neuropharmacology* **2017**, *112*, 248–263.
34. Ozaki, Y.; Reinhard, J.F.; Nichol, C.A. Cofactor Activity of Dihydroflavin Mononucleotide and Tetrahydrobiopterin for Murine Epididymal Indoleamine 2,3-Dioxygenase. *Biochem. Biophys. Res. Commun.* **1986**, *137*, 1106–1111. [CrossRef]
35. Cook, J.S.; Pogson, C.I.; Smith, S.A. Indoleamine 2,3-Dioxygenase. A New, Rapid, Sensitive Radiometric Assay and Its Application to the Study of the Enzyme in Rat Tissues. *Biochem. J.* **1980**, *189*, 461–466.
36. Thomas, S.R.; Salahifar, H.; Mashima, R.; Hunt, N.H.; Richardson, D.R.; Stocker, R. AntiOxid.ants Inhibit Indoleamine 2,3-Dioxygenase in IFN-Gamma-Activated Human Macrophages: Posttranslational Regulation by Pyrrolidine Dithiocarbamate. *J. Immunol.* **2001**, *166*, 6332–6340. [CrossRef] [PubMed]
37. Sucher, R.; Hautz, T.; Mohr, E.; Mackowitz, M.; Mellitzer, V.; Steger, C.; Cardini, B.; Resch, T.; Margreiter, C.; Schneeberger, S.; et al. Sodium Sulfite Exacerbates Allograft Vasculopathy and Affects Tryptophan Breakdown in Murine Heterotopic Aortic Transplantation. *Oxid. Med. Cell. Longev.* **2019**, *2019*, 8461048. [CrossRef] [PubMed]
38. Hucke, C.; MacKenzie, C.R.; Adjogble, K.D.Z.; Takikawa, O.; Däubener, W. Nitric Oxid.e-Mediated Regulation of Gamma Int.erferon-Induced Bacteriostasis: Inhibition and Degradation of Human Indoleamine 2,3-Dioxygenase. *Infect. Immun.* **2004**, *72*, 2723–2730. [CrossRef]
39. Okuda, S.; Nishiyama, N.; Saito, H.; Katsuki, H. 3-Hydroxykynurenine, an Endogenous Oxid.ative Stress Generator, Causes Neuronal Cell Death with Apoptotic Features and Region Selectivity. *J. Neurochem.* **1998**, *70*, 299–307. [CrossRef]
40. Zhuravlev, A.V.; Zakharov, G.A.; Shchegolev, B.F.; Savvateeva-Popova, E.V. AntiOxid.ant Properties of Kynurenines: Density Functional Theory Calculations. *PLoS Comput. Biol.* **2016**, *12*, e1005213.
41. Bender, D.A. Biochemistry of Tryptophan in Health and Disease. *Mol. Asp. Med.* **1983**, *6*, 101–197. [CrossRef]
42. Mohr, A.; Brockmann, J.G.; Becker, F. HTK-N: Modified Histidine-Tryptophan-Ketoglutarate Solution-A Promising New Tool in Solid Organ Preservation. *Int. J. Mol. Sci.* **2020**, *21*, 6468. [CrossRef] [PubMed]
43. Mitchell, J.R.; Verweij, M.; Brand, K.; van de Ven, M.; Goemaere, N.; van den Engel, S.; Chu, T.; Forrer, F.; Müller, C.; de Jong, M.; et al. Short-Term Dietary Restriction and Fasting Precondition against Ischemia Reperfusion Injury in Mice. *Aging Cell* **2010**, *9*, 40–53. [PubMed]

44. Ahmet, I.; Wan, R.; Mattson, M.P.; Lakatta, E.G.; Talan, M. Cardioprotection by Int.ermittent Fasting in Rats. *Circulation* **2005**, *112*, 3115–3121. [CrossRef]
45. Peng, W.; Robertson, L.; Gallinetti, J.; Mejia, P.; Vose, S.; Charlip, A.; Chu, T.; Mitchell, J.R. Surgical Stress Resistance Induced by Single Amino Acid Deprivation Requires Gcn2 in Mice. *Sci. Transl. Med.* **2012**, *4*, 118ra11. [CrossRef]
46. Eleftheriadis, T.; Pissas, G.; Sounidaki, M.; Antoniadis, N.; Antoniadi, G.; Liakopoulos, V.; Stefanidis, I. Preconditioning of Primary Human Renal Proximal Tubular Epithelial Cells without Tryptophan Increases Survival under Hypoxia by Inducing Autophagy. *Int. Urol. Nephrol.* **2017**, *49*, 1297–1307. [CrossRef] [PubMed]
47. Fougeray, S.; Mami, I.; Bertho, G.; Beaune, P.; Thervet, E.; Pallet, N. Tryptophan Depletion and the Kinase GCN2 Mediate IFN-γ-Induced Autophagy. *J. Immunol.* **2012**, *189*, 2954–2964. [CrossRef] [PubMed]
48. Guan, X.; Qian, Y.; Shen, Y.; Zhang, L.; Du, Y.; Dai, H.; Qian, J.; Yan, Y. Autophagy Protects Renal Tubular Cells against Ischemia/Reperfusion Injury in a Time-Dependent Manner. *Cell. Physiol. Biochem.* **2015**, *36*, 285–298.
49. Liu, S.; Hartleben, B.; Kretz, O.; Wiech, T.; Igarashi, P.; Mizushima, N.; Walz, G.; Huber, T.B. Autophagy Plays a Critical Role in Kidney Tubule Maintenance, Aging and Ischemia-Reperfusion Injury. *Autophagy* **2012**, *8*, 826–837. [CrossRef]
50. Kozower, B.D.; Christofidou-Solomidou, M.; Sweitzer, T.D.; Muro, S.; Buerk, D.G.; Solomides, C.C.; Albelda, S.M.; Patterson, G.A.; Muzykantov, V.R. Immunotargeting of Catalase to the Pulmonary Endothelium Alleviates Oxidative Stress and Reduces Acute Lung Transplantation Injury. *Nat. Biotechnol.* **2003**, *21*, 392–398. [CrossRef]
51. Mohib, K.; Guan, Q.; Diao, H.; Du, C.; Jevnikar, A.M. Proapoptotic Activity of Indoleamine 2,3-Dioxygenase Expressed in Renal Tubular Epithelial Cells. *Am. J. Physiol. Ren. Physiol.* **2007**, *293*, F801–F812. [CrossRef] [PubMed]
52. Mohib, K.; Wang, S.; Guan, Q.; Mellor, A.L.; Sun, H.; Du, C.; Jevnikar, A.M. Indoleamine 2,3-Dioxygenase Expression Promotes Renal Ischemia-Reperfusion Injury. *Am. J. Physiol. Ren. Physiol.* **2008**, *295*, F226–F234. [CrossRef]
53. Merchen, T.D.; Boesen, E.I.; Gardner, J.R.; Harbarger, R.; Kitamura, E.; Mellor, A.; Pollock, D.M.; Ghaffari, A.; Podolsky, R.; Nahman, N.S. Indoleamine 2,3-Dioxygenase Inhibition Alters the Non-Coding RNA Transcriptome Following Renal Ischemia-Reperfusion Injury. *Transpl. Immunol.* **2014**, *30*, 140–144. [CrossRef] [PubMed]
54. Wang, Q.; Liu, D.; Song, P.; Zou, M.-H. Tryptophan-Kynurenine Pathway Is Dysregulated in Inflammation, and Immune Activation. *Front. Biosci.* **2015**, *20*, 1116–1143.
55. Giorgini, F.; Huang, S.-Y.; Sathyasaikumar, K.V.; Notarangelo, F.M.; Thomas, M.A.R.; Tararina, M.; Wu, H.-Q.; Schwarcz, R.; Muchowski, P.J. Targeted Deletion of Kynurenine 3-Monooxygenase in Mice: A New Tool for Studying Kynurenine Pathway Metabolism in Periphery and Brain. *J. Biol. Chem.* **2013**, *288*, 36554–36566. [CrossRef]
56. Thul, P.J.; Lindskog, C. The Human Protein Atlas: A Spatial Map of the Human Proteome. *Protein Sci.* **2018**, *27*, 233–244. [CrossRef] [PubMed]
57. Mole, D.J.; Webster, S.P.; Uings, I.; Zheng, X.; Binnie, M.; Wilson, K.; Hutchinson, J.P.; Mirguet, O.; Walker, A.; Beaufils, B.; et al. Kynurenine-3-Monooxygenase Inhibition Prevents Multiple Organ Failure in Rodent Models of Acute Pancreatitis. *Nat. Med.* **2016**, *22*, 202–209.
58. Hutchinson, J.P.; Rowland, P.; Taylor, M.R.D.; Christodoulou, E.M.; Haslam, C.; Hobbs, C.I.; Holmes, D.S.; Homes, P.; Liddle, J.; Mole, D.J.; et al. Structural and Mechanistic Basis of Differentiated Inhibitors of the Acute Pancreatitis Target Kynurenine-3-Monooxygenase. *Nat. Commun.* **2017**, *8*, 15827.
59. Liddle, J.; Beaufils, B.; Binnie, M.; Bouillot, A.; Denis, A.A.; Hann, M.M.; Haslam, C.P.; Holmes, D.S.; Hutchinson, J.P.; Kranz, M.; et al. The Discovery of Potent and Selective Kynurenine 3-Monooxygenase Inhibitors for the Treatment of Acute Pancreatitis. *Bioorg. Med. Chem. Lett.* **2017**, *27*, 2023–2028. [CrossRef]
60. Walker, A.L.; Ancellin, N.; Beaufils, B.; Bergeal, M.; Binnie, M.; Bouillot, A.; Clapham, D.; Denis, A.; Haslam, C.P.; Holmes, D.S.; et al. Development of a Series of Kynurenine 3-Monooxygenase Inhibitors Leading to a Clinical Candidate for the Treatment of Acute Pancreatitis. *J. Med. Chem.* **2017**, *60*, 3383–3404. [CrossRef]
61. Nahomi, R.B.; Nam, M.-H.; Rankenberg, J.; Rakete, S.; Houck, J.A.; Johnson, G.C.; Stankowska, D.L.; Pantcheva, M.B.; MacLean, P.S.; Nagaraj, R.H. Kynurenic Acid Protects Against Ischemia/Reperfusion-Induced Retinal Ganglion Cell Death in Mice. *Int. J. Mol. Sci.* **2020**, *21*, 1795. [CrossRef]
62. Wang, J.; Yu, S.; Li, J.; Li, H.; Jiang, H.; Xiao, P.; Pan, Y.; Zheng, J.; Yu, L.; Jiang, J. Protective Role of N-Acetyl-l-Tryptophan against Hepatic Ischemia-Reperfusion Injury via the RIP2/Caspase-1/IL-1β Signaling Pathway. *Pharm. Biol.* **2019**, *57*, 385–391.
63. Li, H.; Pan, Y.; Wu, H.; Yu, S.; Wang, J.; Zheng, J.; Wang, C.; Li, J.; Jiang, J. Inhibition of Excessive Mitophagy by N-Acetyl-L-Tryptophan Confers Hepatoprotection against Ischemia-Reperfusion Injury in Rats. *Peer J.* **2020**, *8*, e8665. [CrossRef]
64. Chou, H.-C.; Chan, H.-L. 5-Methoxytryptophan-Dependent Protection of Cardiomyocytes from Heart Ischemia Reperfusion Injury. *Arch. Biochem. Biophys.* **2014**, *543*, 15–22. [CrossRef]
65. Mándi, Y.; Vécsei, L. The Kynurenine System and Immunoregulation. *J. Neural. Transm.* **2012**, *119*, 197–209. [CrossRef] [PubMed]
66. Baban, B.; Hansen, A.M.; Chandler, P.R.; Manlapat, A.; Bingaman, A.; Kahler, D.J.; Munn, D.H.; Mellor, A.L. A Minor Population of Splenic Dendritic Cells Expressing CD19 Mediates IDO-Dependent T Cell Suppression via Type I IFN Signaling Following B7 Ligation. *Int. Immunol.* **2005**, *17*, 909–919. [CrossRef]
67. Carbotti, G.; Barisione, G.; Airoldi, I.; Mezzanzanica, D.; Bagnoli, M.; Ferrero, S.; Petretto, A.; Fabbi, M.; Ferrini, S. IL-27 Induces the Expression of IDO and PD-L1 in Human Cancer Cells. *Oncotarget* **2015**, *6*, 43267–43280. [CrossRef] [PubMed]
68. Yanagawa, Y.; Iwabuchi, K.; Onoé, K. Co-Operative Action of Int.erleukin-10 and Int.erferon-Gamma to Regulate Dendritic Cell Functions. *Immunology* **2009**, *127*, 345–353. [CrossRef]

69. Munn, D.H.; Sharma, M.D.; Mellor, A.L. Ligation of B7-1/B7-2 by Human CD4+ T Cells Triggers Indoleamine 2,3-Dioxygenase Activity in Dendritic Cells. *J. Immunol.* **2004**, *172*, 4100–4110.
70. Grohmann, U.; Orabona, C.; Fallarino, F.; Vacca, C.; Calcinaro, F.; Falorni, A.; Candeloro, P.; Belladonna, M.L.; Bianchi, R.; Fioretti, M.C.; et al. CTLA-4-Ig Regulates Tryptophan Catabolism in Vivo. *Nat. Immunol.* **2002**, *3*, 1097–1101. [CrossRef] [PubMed]
71. Walker, L.S.K.; Sansom, D.M. The Emerging Role of CTLA4 as a Cell-Extrinsic Regulator of T Cell Responses. *Nat. Rev. Immunol.* **2011**, *11*, 852–863.
72. Sucher, R.; Fischler, K.; Oberhuber, R.; Kronberger, I.; Margreiter, C.; Ollinger, R.; Schneeberger, S.; Fuchs, D.; Werner, E.R.; Watschinger, K.; et al. IDO and Regulatory T Cell Support Are Critical for Cytotoxic T Lymphocyte-Associated Ag-4 Ig-Mediated Long-Term Solid Organ Allograft Survival. *J. Immunol.* **2012**, *188*, 37–46. [CrossRef]
73. Furuzawa-Carballeda, J.; Lima, G.; Uribe-Uribe, N.; Avila-Casado, C.; Mancilla, E.; Morales-Buenrostro, L.E.; Pérez-Garrido, J.; Pérez, M.; Cárdenas, G.; Llorente, L.; et al. High Levels of IDO-Expressing CD16+ Peripheral Cells, and Tregs in Graft Biopsies from Kidney Transplant Recipients under Belatacept Treatment. *Transpl. Proc.* **2010**, *42*, 3489–3496. [CrossRef]
74. Bigenzahn, S.; Juergens, B.; Mahr, B.; Pratschke, J.; Koenigsrainer, A.; Becker, T.; Fuchs, D.; Brandacher, G.; Kainz, A.; Muehlbacher, F.; et al. No Augmentation of Indoleamine 2,3-Dioxygenase (IDO) Activity through Belatacept Treatment in Liver Transplant Recipients. *Clin. Exp. Immunol.* **2018**, *192*, 233–241. [CrossRef]
75. Laing, A.G.; Fanelli, G.; Ramirez-Valdez, A.; Lechler, R.I.; Lombardi, G.; Sharpe, P.T. Mesenchymal Stem Cells Inhibit T-Cell Function through Conserved Induction of Cellular Stress. *PLoS ONE* **2019**, *14*, e0213170. [CrossRef] [PubMed]
76. Poormasjedi-Meibod, M.-S.; Jalili, R.B.; Hosseini-Tabatabaei, A.; Hartwell, R.; Ghahary, A. Immuno-Regulatory Function of Indoleamine 2,3 Dioxygenase through Modulation of Innate Immune Responses. *PLoS ONE* **2013**, *8*, e71044. [CrossRef]
77. Metz, R.; Rust, S.; Duhadaway, J.B.; Mautino, M.R.; Munn, D.H.; Vahanian, N.N.; Link, C.J.; Prendergast, G.C. IDO Inhibits a Tryptophan Sufficiency Signal That Stimulates MTOR: A Novel IDO Effector Pathway Targeted by D-1-Methyl-Tryptophan. *Oncoimmunology* **2012**, *1*, 1460–1468. [CrossRef] [PubMed]
78. Badawy, A.A.-B.; Namboodiri, A.M.A.; Moffett, J.R. The End of the Road for the Tryptophan Depletion Concept in Pregnancy and Infection. *Clin. Sci.* **2016**, *130*, 1327–1333. [CrossRef] [PubMed]
79. Nguyen, N.T.; Nakahama, T.; Le, D.H.; Van Son, L.; Chu, H.H.; Kishimoto, T. Aryl Hydrocarbon Receptor and Kynurenine: Recent Advances in Autoimmune Disease Research. *Front. Immunol.* **2014**, *5*, 551. [CrossRef]
80. Mezrich, J.D.; Fechner, J.H.; Zhang, X.; Johnson, B.P.; Burlingham, W.J.; Bradfield, C.A. An Int.eraction between Kynurenine and the Aryl Hydrocarbon Receptor Can Generate Regulatory T Cells. *J. Immunol.* **2010**, *185*, 3190–3198.
81. Eleftheriadis, T.; Pissas, G.; Antoniadi, G.; Liakopoulos, V.; Tsogka, K.; Sounidaki, M.; Stefanidis, I. Differential Effects of the Two Amino Acid Sensing Systems, the GCN2 Kinase and the MTOR Complex 1, on Primary Human Alloreactive CD4+ T-Cells. *Int. J. Mol. Med.* **2016**, *37*, 1412–1420. [CrossRef]
82. Xie, F.T.; Cao, J.S.; Zhao, J.; Yu, Y.; Qi, F.; Dai, X.C. IDO Expressing Dendritic Cells Suppress Allograft Rejection of Small Bowel Transplantation in Mice by Expansion of Foxp3+ Regulatory T Cells. *Transpl. Immunol.* **2015**, *33*, 69–77. [CrossRef]
83. Dai, X.; Zhu, B.T. Suppression of T-Cell Response and Prolongation of Allograft Survival in a Rat Model by Tryptophan Catabolites. *Eur. J. Pharm.* **2009**, *606*, 225–232. [CrossRef]
84. Iken, K.; Liu, K.; Liu, H.; Bizargity, P.; Wang, L.; Hancock, W.W.; Visner, G.A. Indoleamine 2,3-Dioxygenase and Metabolites Protect Murine Lung Allografts and Impair the Calcium Mobilization of T Cells. *Am. J. Respir. Cell Mol. Biol.* **2012**, *47*, 405–416. [CrossRef]
85. Hayashi, T.; Mo, J.-H.; Gong, X.; Rossetto, C.; Jang, A.; Beck, L.; Elliott, G.I.; Kufareva, I.; Abagyan, R.; Broide, D.H.; et al. 3-Hydroxyanthranilic Acid Inhibits PDK1 Activation and Suppresses Experimental Asthma by Inducing T Cell Apoptosis. *Proc. Natl. Acad. Sci. USA* **2007**, *104*, 18619–18624.
86. Fallarino, F.; Grohmann, U.; Vacca, C.; Bianchi, R.; Orabona, C.; Spreca, A.; Fioretti, M.C.; Puccetti, P. T Cell Apoptosis by Tryptophan Catabolism. *Cell Death Differ.* **2002**, *9*, 1069–1077. [CrossRef]
87. Lowe, M.M.; Mold, J.E.; Kanwar, B.; Huang, Y.; Louie, A.; Pollastri, M.P.; Wang, C.; Patel, G.; Franks, D.G.; Schlezinger, J.; et al. Identification of Cinnabarinic Acid as a Novel Endogenous Aryl Hydrocarbon Receptor Ligand That Drives IL-22 Production. *PLoS ONE* **2014**, *9*, e87877.
88. Gargaro, M.; Vacca, C.; Massari, S.; Scalisi, G.; Manni, G.; Mondanelli, G.; Mazza, E.M.C.; Bicciato, S.; Pallotta, M.T.; Orabona, C.; et al. Engagement of Nuclear Coactivator 7 by 3-Hydroxyanthranilic Acid Enhances Activation of Aryl Hydrocarbon Receptor in Immunoregulatory Dendritic Cells. *Front. Immunol.* **2019**, *10*, 1973. [CrossRef]
89. Loupy, A.; Lefaucheur, C. Antibody-Mediated Rejection of Solid-Organ Allografts. *N. Engl. J. Med.* **2018**, *379*, 1150–1160. [CrossRef] [PubMed]
90. Shinde, R.; Shimoda, M.; Chaudhary, K.; Liu, H.; Mohamed, E.; Bradley, J.; Kandala, S.; Li, X.; Liu, K.; McGaha, T.L. B Cell-Intrinsic IDO1 Regulates Humoral Immunity to T Cell-Independent Antigens. *J. Immunol.* **2015**, *195*, 2374–2382. [PubMed]
91. Merlo, L.M.F.; DuHadaway, J.B.; Grabler, S.; Prendergast, G.C.; Muller, A.J.; Mandik-Nayak, L. IDO2 Modulates T Cell-Dependent Autoimmune Responses through a B Cell-Int.rinsic Mechanism. *J. Immunol.* **2016**, *196*, 4487–4497. [CrossRef]
92. Sounidaki, M.; Pissas, G.; Eleftheriadis, T.; Antoniadi, G.; Golfinopoulos, S.; Liakopoulos, V.; Stefanidis, I. Indoleamine 2,3-Dioxygenase Suppresses Humoral Alloimmunity via Pathways That Different to Those Associated with Its Effects on T Cells. *Biomed. Rep.* **2019**, *1*, 1–5. [CrossRef]

93. Suzuki, S.; Toné, S.; Takikawa, O.; Kubo, T.; Kohno, I.; Minatogawa, Y. Expression of Indoleamine 2,3-Dioxygenase and Tryptophan 2,3-Dioxygenase in Early Concepti. *Biochem. J.* **2001**, *355*, 425–429. [CrossRef] [PubMed]
94. Tatsumi, K.; Higuchi, T.; Fujiwara, H.; Nakayama, T.; Egawa, H.; Itoh, K.; Fujii, S.; Fujita, J. Induction of Tryptophan 2,3-Dioxygenase in the Mouse Endometrium during Implantation. *Biochem. Biophys. Res. Commun.* **2000**, *274*, 166–170. [CrossRef]
95. Schmidt, S.K.; Müller, A.; Heseler, K.; Woite, C.; Spekker, K.; MacKenzie, C.R.; Däubener, W. Antimicrobial and Immunoregulatory Properties of Human Tryptophan 2,3-Dioxygenase. *Eur. J. Immunol.* **2009**, *39*, 2755–2764. [CrossRef] [PubMed]
96. Pilotte, L.; Larrieu, P.; Stroobant, V.; Colau, D.; Dolusic, E.; Frédérick, R.; De Plaen, E.; Uyttenhove, C.; Wouters, J.; Masereel, B.; et al. Reversal of Tumoral Immune Resistance by Inhibition of Tryptophan 2,3-Dioxygenase. *Proc. Natl. Acad. Sci. USA* **2012**, *109*, 2497–2502. [CrossRef] [PubMed]
97. Li, J.; Meinhardt, A.; Roehrich, M.-E.; Golshayan, D.; Dudler, J.; Pagnotta, M.; Trucco, M.; Vassalli, G. Indoleamine 2,3-Dioxygenase Gene Transfer Prolongs Cardiac Allograft Survival. *Am. J. Physiol. Heart Circ. Physiol.* **2007**, *293*, H3415–H3423. [CrossRef]
98. Yu, G.; Dai, H.; Chen, J.; Duan, L.; Gong, M.; Liu, L.; Xiong, P.; Wang, C.-Y.; Fang, M.; Gong, F. Gene Delivery of Indoleamine 2,3-Dioxygenase Prolongs Cardiac Allograft Survival by Shaping the Types of T-Cell Responses. *J. Gene Med.* **2008**, *10*, 754–761. [CrossRef] [PubMed]
99. Li, C.; Liu, T.; Zhao, N.; Zhu, L.; Wang, P.; Dai, X. Dendritic Cells Transfected with Indoleamine 2,3-Dioxygenase Gene Suppressed Acute Rejection of Cardiac Allograft. *Int. Immunopharmacol.* **2016**, *36*, 31–38. [CrossRef] [PubMed]
100. Lv, Y.; Pang, X.; Jia, P.-Y.; Jia, D.-L. Combined Therapy of Infusion of DC from Rats with Higher Expression of IDO and CD40L on Rejection Post Heart Transplantation. *Eur. Rev. Med. Pharm. Sci.* **2018**, *22*, 7977–7984.
101. Li, C.; Sun, Z.; Yuan, F.; Zhao, Z.; Zhang, J.; Zhang, B.; Li, H.; Liu, T.; Dai, X. Mechanism of Indoleamine 2, 3-Dioxygenase Inhibiting Cardiac Allograft Rejection in Mice. *J. Cell. Mol. Med.* **2020**, *24*, 3438–3448. [CrossRef]
102. He, J.-G.; Li, B.-B.; Zhou, L.; Yan, D.; Xie, Q.-L.; Zhao, W. Indoleamine 2,3-Dioxgenase-Transfected Mesenchymal Stem Cells Suppress Heart Allograft Rejection by Increasing the Production and Activity of Dendritic Cells and Regulatory T Cells. *J. Investig. Med.* **2020**, *68*, 728–737. [CrossRef]
103. He, J.-G.; Xie, Q.-L.; Li, B.-B.; Zhou, L.; Yan, D. Exosomes Derived from IDO1-Overexpressing Rat Bone Marrow Mesenchymal Stem Cells Promote Immunotolerance of Cardiac Allografts. *Cell Transpl.* **2018**, *27*, 1657–1683. [CrossRef] [PubMed]
104. Chambers, D.C.; Cherikh, W.S.; Goldfarb, S.B.; Hayes, D.; Kucheryavaya, A.Y.; Toll, A.E.; Khush, K.K.; Levvey, B.J.; Meiser, B.; Rossano, J.W.; et al. International Society for Heart and Lung Transplantation. The International Thoracic Organ Transplant Registry of the International Society for Heart and Lung Transplantation: Thirty-Fifth Adult Lung and Heart-Lung Transplant Report-2018; Focus Theme: Multiorgan Transplantation. *J. Heart Lung Transpl.* **2018**, *37*, 1169–1183.
105. Hayashi, T.; Beck, L.; Rossetto, C.; Gong, X.; Takikawa, O.; Takabayashi, K.; Broide, D.H.; Carson, D.A.; Raz, E. Inhibition of Experimental Asthma by Indoleamine 2,3-Dioxygenase. *J. Clin. Investig.* **2004**, *114*, 270–279. [CrossRef] [PubMed]
106. Desvignes, L.; Ernst, J.D. Interferon-Gamma-Responsive Nonhematopoietic Cells Regulate the Immune Response to *Mycobacterium Tuberculosis*. *Immunity* **2009**, *31*, 974–985. [CrossRef] [PubMed]
107. Lee, S.M.; Park, H.Y.; Suh, Y.S.; Yoon, E.H.; Kim, J.; Jang, W.H.; Lee, W.S.; Park, S.G.; Choi, I.W.; Choi, I.; et al. Inhibition of acute lethal pulmonary inflammation by the IDO-AhR pathway. *Proc. Natl. Acad. Sci. USA* **2017**, *114*, E5881–E5890. [CrossRef]
108. Swanson, K.A.; Zheng, Y.; Heidler, K.M.; Mizobuchi, T.; Wilkes, D.S. CDllc+ Cells Modulate Pulmonary Immune Responses by Production of Indoleamine 2,3-Dioxygenase. *Am. J. Respir Cell Mol. Biol.* **2004**, *30*, 311–318. [CrossRef]
109. Liu, H.; Liu, L.; Fletcher, B.S.; Visner, G.A. Novel Action of Indoleamine 2,3-Dioxygenase Attenuating Acute Lung Allograft Injury. *Am. J. Respir Crit Care Med.* **2006**, *173*, 566–572. [CrossRef]
110. Liu, H.; Liu, L.; Fletcher, B.S.; Visner, G.A. Sleeping Beauty-Based Gene Therapy with Indoleamine 2,3-Dioxygenase Inhibits Lung Allograft Fibrosis. *Faseb. J.* **2006**, *20*, 2384–2386. [CrossRef]
111. Liu, H.; Liu, L.; Liu, K.; Bizargity, P.; Hancock, W.W.; Visner, G.A. Reduced Cytotoxic Function of Effector CD8+ T Cells Is Responsible for Indoleamine 2,3-Dioxygenase-Dependent Immune Suppression. *J. Immunol.* **2009**, *183*, 1022–1031. [CrossRef]
112. Ebrahimi, A.; Kardar, G.A.; Teimoori-Toolabi, L.; Toolabi, L.; Ghanbari, H.; Sadroddiny, E. Inducible Expression of Indoleamine 2,3-Dioxygenase Attenuates Acute Rejection of Tissue-Engineered Lung Allografts in Rats. *Gene* **2016**, *576*, 412–420. [CrossRef]
113. Miki, T.; Sun, H.; Lee, Y.; Tandin, A.; Kovscek, A.M.; Subbotin, V.; Fung, J.J.; Valdivia, L.A. Blockade of Tryptophan Catabolism Prevents Spontaneous Tolerogenicity of Liver Allografts. *Transpl. Proc.* **2001**, *33*, 129–130. [CrossRef]
114. Lin, Y.-C.; Chen, C.-L.; Nakano, T.; Goto, S.; Kao, Y.-H.; Hsu, L.-W.; Lai, C.-Y.; Jawan, B.; Cheng, Y.-F.; Tateno, C.; et al. Immunological Role of Indoleamine 2,3-Dioxygenase in Rat Liver Allograft Rejection and Tolerance. *J. Gastroenterol. Hepatol.* **2008**, *23*, e243–e250. [CrossRef]
115. Lin, Y.C.; Goto, S.; Tateno, C.; Nakano, T.; Cheng, Y.F.; Jawan, B.; Kao, Y.H.; Hsu, L.W.; Lai, C.Y.; Yoshizato, K.; et al. Induction of Indoleamine 2,3-Dioxygenase in Livers Following Hepatectomy Prolongs Survival of Allogeneic Hepatocytes after Transplantation. *Transpl. Proc.* **2008**, *40*, 2706–2708. [CrossRef]
116. Benseler, V.; McCaughan, G.W.; Schlitt, H.J.; Bishop, G.A.; Bowen, D.G.; Bertolino, P. The Liver: A Special Case in Transplantation Tolerance. In *Seminars in Liver Disease*; Thieme Medical Publishers, Inc.: New York, NY, USA, 2007; Volume 27, pp. 194–213.
117. Sun, X.; Gong, Z.; Wang, Z.; Li, T.; Zhang, J.; Sun, H.; Liu, S.; Huang, L.; Huang, C.; Peng, Z. IDO-Competent-DCs Induced by IFN-γ Attenuate Acute Rejection in Rat Liver Transplantation. *J. Clin. Immunol.* **2012**, *32*, 837–847. [CrossRef]

118. Laurence, J.M.; Wang, C.; Park, E.T.; Buchanan, A.; Clouston, A.; Allen, R.D.M.; Mccaughan, G.W.; Bishop, G.A.; Sharland, A.F. Blocking Indoleamine Dioxygenase Activity Early after Rat Liver Transplantation Prevents Long-Term Survival but Does Not Cause Acute Rejection. *Transplantation* **2008**, *85*, 1357–1361. [CrossRef] [PubMed]
119. Laurence, J.M.; Wang, C.; Zheng, M.; Cunningham, S.; Earl, J.; Tay, S.S.; Allen, R.D.M.; McCaughan, G.W.; Alexander, I.E.; Bishop, G.A.; et al. Overexpression of Indoleamine Dioxygenase in Rat Liver Allografts Using a High-Efficiency Adeno-Associated Virus Vector Does Not Prevent Acute Rejection. *Liver Transplant.* **2009**, *15*, 233–241. [CrossRef] [PubMed]
120. Sun, Q.-F.; Ding, J.-G.; Sheng, J.-F.; Zhu, M.-H.; Li, J.-J.; Sheng, Z.-K.; Tang, X.-F. Novel Action of 3,4-DAA Ameliorating Acute Liver Allograft Injury. *Cell Biochem. Funct.* **2011**, *29*, 673–678. [CrossRef]
121. Cook, C.H.; Bickerstaff, A.A.; Wang, J.-J.; Nadasdy, T.; Della Pelle, P.; Colvin, R.B.; Orosz, C.G. Spontaneous Renal Allograft Acceptance Associated with "Regulatory" Dendritic Cells and IDO. *J. Immunol.* **2008**, *180*, 3103–3112. [CrossRef]
122. Na, N.; Luo, Y.; Zhao, D.; Yang, S.; Hong, L.; Li, H.; Miao, B.; Qiu, J. Prolongation of Kidney Allograft Survival Regulated by Indoleamine 2, 3-Dioxygenase in Immature Dendritic Cells Generated from Recipient Type Bone Marrow Progenitors. *Mol. Immunol.* **2016**, *79*, 22–31. [CrossRef]
123. Demmers, M.W.H.J.; Korevaar, S.S.; Roemeling-van Rhijn, M.; van den Bosch, T.P.P.; Hoogduijn, M.J.; Betjes, M.G.H.; Weimar, W.; Baan, C.C.; Rowshani, A.T. Human Renal Tubular Epithelial Cells Suppress Alloreactive T Cell Proliferation. *Clin. Exp. Immunol.* **2015**, *179*, 509–519. [CrossRef]
124. Ge, W.; Jiang, J.; Arp, J.; Liu, W.; Garcia, B.; Wang, H. Regulatory T-Cell Generation and Kidney Allograft Tolerance Induced by Mesenchymal Stem Cells Associated with Indoleamine 2,3-Dioxygenase Expression. *Transplantation* **2010**, *90*, 1312–1320. [CrossRef] [PubMed]
125. Casiraghi, F.; Noris, M.; Remuzzi, G. Immunomodulatory Effects of Mesenchymal Stromal Cells in Solid Organ Transplantation. *Curr. Opin. Organ. Transpl.* **2010**, *15*, 731–737. [CrossRef] [PubMed]
126. Tan, J.; Wu, W.; Xu, X.; Liao, L.; Zheng, F.; Messinger, S.; Sun, X.; Chen, J.; Yang, S.; Cai, J.; et al. Induction Therapy with Autologous Mesenchymal Stem Cells in Living-Related Kidney Transplants: A Randomized Controlled Trial. *JAMA* **2012**, *307*, 1169–1177. [CrossRef]
127. He, Y.; Zhou, S.; Liu, H.; Shen, B.; Zhao, H.; Peng, K.; Wu, X. Indoleamine 2, 3-Dioxgenase Transfected Mesenchymal Stem Cells Induce Kidney Allograft Tolerance by Increasing the Production and Function of Regulatory T Cells. *Transplantation* **2015**, *99*, 1829–1838. [CrossRef]
128. Vavrincova-Yaghi, D.; Deelman, L.E.; Goor, H.; Seelen, M.; Kema, I.P.; Smit-van Oosten, A.; Zeeuw, D.; Henning, R.H.; Sandovici, M. Gene Therapy with Adenovirus-Delivered Indoleamine 2,3-Dioxygenase Improves Renal Function and Morphology Following Allogeneic Kidney Transplantation in Rat. *J. Gene Med.* **2011**, *13*, 373–381. [CrossRef] [PubMed]
129. Vavrincova-Yaghi, D.; Deelman, L.E.; van Goor, H.; Seelen, M.A.; Vavrinec, P.; Kema, I.P.; Gomolcak, P.; Benigni, A.; Henning, R.H.; Sandovici, M. Local Gene Therapy with Indoleamine 2,3-Dioxygenase Protects against Development of Transplant Vasculopathy in Chronic Kidney Transplant Dysfunction. *Gene* **2016**, *23*, 797–806. [CrossRef]
130. Hofmann, F. Ido Brings down the Pressure in Systemic Inflammation. *Nat. Med.* **2010**, *16*, 265–267. [CrossRef] [PubMed]
131. Wang, Y.; Merchen, T.D.; Fang, X.; Lassiter, R.; Ho, C.-S.; Jajosky, R.; Kleven, D.; Thompson, T.; Mohamed, E.; Yu, M.; et al. Regulation of Indoleamine 2,3 Dioxygenase and Its Role in a Porcine Model of Acute Kidney Allograft Rejection. *J. Investig. Med.* **2018**, *66*, 1109–1117. [CrossRef]
132. Alexander, A.M.; Crawford, M.; Bertera, S.; Rudert, W.A.; Takikawa, O.; Robbins, P.D.; Trucco, M. Indoleamine 2,3-Dioxygenase Expression in Transplanted NOD Islets Prolongs Graft Survival after Adoptive Transfer of Diabetogenic Splenocytes. *Diabetes* **2002**, *51*, 356–365.
133. Jalili, R.B.; Forouzandeh, F.; Moeenrezakhanlou, A.; Rayat, G.R.; Rajotte, R.V.; Uludag, H.; Ghahary, A. Mouse Pancreatic Islets Are Resistant to Indoleamine 2,3 Dioxygenase-Induced General Control Nonderepressible-2 Kinase Stress Pathway and MaInt.ain Normal Viability and Function. *Am. J. Pathol.* **2009**, *174*, 196–205. [CrossRef]
134. Jalili, R.B.; Rayat, G.R.; Rajotte, R.V.; Ghahary, A. Suppression of Islet Allogeneic Immune Response by Indoleamine 2,3 Dioxygenase-Expressing Fibroblasts. *J. Cell. Physiol.* **2007**, *213*, 137–143. [CrossRef] [PubMed]
135. Jalili, R.B.; Forouzandeh, F.; Rezakhanlou, A.M.; Hartwell, R.; Medina, A.; Warnock, G.L.; Larijani, B.; Ghahary, A. Local Expression of Indoleamine 2,3 Dioxygenase in Syngeneic Fibroblasts Significantly Prolongs Survival of an Engineered Three-Dimensional Islet Allograft. *Diabetes* **2010**, *59*, 2219–2227. [CrossRef] [PubMed]
136. Hosseini-Tabatabaei, A.; Jalili, R.B.; Khosravi-Maharlooei, M.; Hartwell, R.; Kilani, R.T.; Zhang, Y.; Ghahary, A. Immunoprotection and Functional Improvement of Allogeneic Islets in Diabetic Mice, Using a Stable Indoleamine 2,3-Dioxygenase Producing Scaffold. *Transplantation* **2015**, *99*, 1341–1348. [CrossRef]
137. Khosravi-Maharlooei, M.; Pakyari, M.; Jalili, R.B.; Kilani, R.T.; Ghahary, A. Int.raperitoneal Injection of IDO-Expressing Dermal Fibroblasts Improves the Allograft Survival. *Clin. Immunol.* **2017**, *174*, 1–9. [CrossRef] [PubMed]
138. Ingelsten, M.; Gustafsson, K.; Oltean, M.; Karlsson-Parra, A.; Olausson, M.; Haraldsson, B.; Nyström, J. Is Indoleamine 2,3-Dioxygenase Important for Graft Acceptance in Highly Sensitized Patients after Combined Auxiliary Liver-Kidney Transplantation? *Transplantation* **2009**, *88*, 911–919. [CrossRef]
139. Kaden, J.; May, G.; Völp, A.; Wesslau, C. Factors Impacting Short and Long-Term Kidney Graft Survival: Modification by Single Int.ra-Operative -High-Dose Induction with ATG-Fresenius. *Ann. Transpl.* **2011**, *16*, 81–91. [CrossRef] [PubMed]

140. Lahdou, I.; Sadeghi, M.; Daniel, V.; Schenk, M.; Renner, F.; Weimer, R.; Löb, S.; Schmidt, J.; Mehrabi, A.; Schnitzler, P.; et al. Increased Pretransplantation Plasma Kynurenine Levels Do Not Protect from but Predict Acute Kidney Allograft Rejection. *Hum. Immunol.* **2010**, *71*, 1067–1072. [CrossRef]
141. Brandacher, G.; Cakar, F.; Winkler, C.; Schneeberger, S.; Obrist, P.; Bösmüller, C.; Werner-Felmayer, G.; Werner, E.R.; Bonatti, H.; Margreiter, R.; et al. Non-Invasive Monitoring of Kidney Allograft Rejection through IDO Metabolism Evaluation. *Kidney Int.* **2007**, *71*, 60–67. [CrossRef]
142. Al Khasawneh, E.; Gupta, S.; Tuli, S.Y.; Shahlaee, A.H.; Garrett, T.J.; Schechtman, K.B.; Dharnidharka, V.R. Stable Pediatric Kidney Transplant Recipients Run Higher Urine Indoleamine 2, 3 Dioxygenase (IDO) Levels than Healthy Children. *Pediatric Transpl.* **2014**, *18*, 254–257. [CrossRef]
143. Holmes, E.W.; Russell, P.M.; Kinzler, G.J.; Reckard, C.R.; Flanigan, R.C.; Thompson, K.D.; Bermes, E.W. Oxid.ative Tryptophan Metabolism in Renal Allograft Recipients: Increased Kynurenine Synthesis Is Associated with Inflammation and OKT3 Therapy. *Cytokine* **1992**, *4*, 205–213.
144. Kaden, J.; Abendroth, D.; Völp, A.; Marzinzig, M. Dynamics and Diagnostic Relevance of Kynurenine Serum Level after Kidney Transplantation. *Ann. Transpl.* **2015**, *20*, 327–337.
145. Vavrincova-Yaghi, D.; Seelen, M.A.; Kema, I.P.; Deelman, L.E.; van der Heuvel, M.C.; Breukelman, H.; Van den Eynde, B.J.; Henning, R.H.; van Goor, H.; Sandovici, M. Early Posttransplant Tryptophan Metabolism Predicts Long-Term Outcome of Human Kidney Transplantation. *Transplantation* **2015**, *99*, e97–e104. [CrossRef] [PubMed]
146. Kim, S.-Y.; Kim, B.K.; Gwon, M.-R.; Seong, S.J.; Ohk, B.; Kang, W.Y.; Lee, H.W.; Jung, H.-Y.; Cho, J.-H.; Chung, B.H.; et al. Urinary Metabolomic Profiling for Noninvasive Diagnosis of Acute T Cell-Mediated Rejection after Kidney Transplantation. *J. Chromatogr. B Anal. Technol. Biomed. Life Sci.* **2019**, *1118–1119*, 157–163.
147. Blydt-Hansen, T.D.; Sharma, A.; Gibson, I.W.; Mandal, R.; Wishart, D.S. Urinary Metabolomics for Noninvasive Detection of Borderline and Acute T Cell-Mediated Rejection in Children after Kidney Transplantation. *Am. J. Transpl.* **2014**, *14*, 2339–2349. [CrossRef] [PubMed]
148. Dharnidharka, V.R.; Al Khasawneh, E.; Gupta, S.; Shuster, J.J.; Theriaque, D.W.; Shahlaee, A.H.; Garrett, T.J. Verification of Association of Elevated Serum IDO Enzyme Activity with Acute Rejection and Low CD4-ATP Levels with Infection. *Transplantation* **2013**, *96*, 567–572. [CrossRef]
149. Zhao, X.; Chen, J.; Ye, L.; Xu, G. Serum Metabolomics Study of the Acute Graft Rejection in Human Renal Transplantation Based on Liquid Chromatography-Mass Spectrometry. *J. Proteome Res.* **2014**, *13*, 2659–2667. [CrossRef]
150. Sadeghi, M.; Lahdou, I.; Daniel, V.; Schnitzler, P.; Fusch, G.; Schefold, J.C.; Zeier, M.; Iancu, M.; Opelz, G.; Terness, P. Strong Association of Phenylalanine and Tryptophan Metabolites with Activated Cytomegalovirus Infection in Kidney Transplant Recipients. *Hum. Immunol.* **2012**, *73*, 186–192. [CrossRef]
151. Yilmaz, N.; Ustundag, Y.; Kivrak, S.; Kahvecioglu, S.; Celik, H.; Kivrak, I.; Huysal, K. Serum Indoleamine 2,3 Dioxygenase and Tryptophan and Kynurenine Ratio Using the UPLC-MS/MS Method, in Patients Undergoing Peritoneal Dialysis, Hemodialysis, and Kidney Transplantation. *Ren. Fail.* **2016**, *38*, 1300–1309.
152. Zhang, F.; Wang, Q.; Xia, T.; Fu, S.; Tao, X.; Wen, Y.; Chan, S.; Gao, S.; Xiong, X.; Chen, W. Diagnostic Value of Plasma Tryptophan and Symmetric Dimethylarginine Levels for Acute Kidney Injury among Tacrolimus-Treated Kidney Transplant Patients by Targeted Metabolomics Analysis. *Sci. Rep.* **2018**, *8*, 14688. [CrossRef]
153. de Vries, L.V.; Minović, I.; Franssen, C.F.M.; van Faassen, M.; Sanders, J.-S.F.; Berger, S.P.; Navis, G.; Kema, I.P.; Bakker, S.J.L. The Tryptophan/Kynurenine Pathway, Systemic Inflammation, and Long-Term Outcome after Kidney Transplantation. *Am. J. Physiol. Ren. Physiol.* **2017**, *313*, F475–F486. [CrossRef]
154. Meloni, F.; Giuliano, S.; Solari, N.; Draghi, P.; Miserere, S.; Bardoni, A.M.; Salvini, R.; Bini, F.; Fietta, A.M. Indoleamine 2,3-Dioxygenase in Lung Allograft Tolerance. *J. Heart Lung Transpl.* **2009**, *28*, 1185–1192. [CrossRef]
155. Oweira, H.; Lahdou, I.; Opelz, G.; Daniel, V.; Terness, P.; Schmidt, J.; Mehrabi, A.; Fusch, G.; Schefold, J.; Zidan, A.; et al. Association of Pre- and Early Post-Transplant Serum Amino Acids and Metabolites of Amino Acids and Liver Transplant Outcome. *Transpl. Immunol.* **2018**, *46*, 42–48. [CrossRef] [PubMed]
156. Perera, M.T.P.R.; Higdon, R.; Richards, D.A.; Silva, M.A.; Murphy, N.; Kolker, E.; Mirza, D.F. Biomarker Differences between Cadaveric Grafts Used in Human Orthotopic Liver Transplantation as Identified by Coulometric Electrochemical Array Detection (CEAD) Metabolomics. *OMICS* **2014**, *18*, 767–777. [CrossRef] [PubMed]
157. Beier, U.H.; Hartung, E.A.; Concors, S.; Hernandez, P.T.; Wang, Z.; Perry, C.; Baur, J.A.; Denburg, M.R.; Hancock, W.W.; Gade, T.P.; et al. Tissue Metabolic Profiling Shows That Saccharopine Accumulates during Renal Ischemic-Reperfusion Injury, While Kynurenine and Itaconate Accumulate in Renal Allograft Rejection. *Metabolomics* **2020**, *16*, 65. [CrossRef] [PubMed]
158. Sadok, I.; Gamian, A.; Staniszewska, M.M. Chromatographic analysis of tryptophan metabolites. *J. Sep. Sci.* **2017**, *40*, 3020–3045. [PubMed]
159. de Jong, W.H.A.; Smit, R.; Bakker, S.J.L.; de Vries, E.G.E.; Kema, I.P. Plasma tryptophan, kynurenine and 3-hydroxykynurenine measurement using automated online solid-phase extraction HPLC-tandem mass spectrometry. *J. Chromatogr. B Anal. Technol. Biomed. Life Sci.* **2009**, *877*, 603–609. [CrossRef] [PubMed]
160. Tombari, R.J.; Saunders, C.M.; Wu, C.Y.; Dunlap, L.E.; Tantillo, D.J.; Olson, D.E. Ex Vivo Analysis of Tryptophan Metabolism Using 19F NMR. *Acs. Chem. Biol.* **2019**, *14*, 1866–1873. [CrossRef]

Article

Strong Dependence between Tryptophan-Related Fluorescence of Urine and Malignant Melanoma

Anna Birková [1,†], Marcela Valko-Rokytovská [2,*,†], Beáta Hubková [1,*], Marianna Zábavníková [3] and Mária Mareková [1]

1 Department of Medical and Clinical Biochemistry, Faculty of Medicine, Pavol Jozef Šafárik University in Košice, Tr. SNP 1, 040 11 Košice, Slovakia; anna.birkova@upjs.sk (A.B.); maria.marekova@upjs.sk (M.M.)
2 Department of Chemistry, Biochemistry and Biophysics, University of Veterinary Medicine and Pharmacy in Košice, Komenského 73, 041 81 Košice, Slovakia
3 KOREKTCHIR s.r.o., Zborovská 7, 040 01 Košice, Slovakia; marianna.zabavnikova@gmail.com
* Correspondence: marcela.valko-rokytovska@gmail.com (M.V.-R.); beata.hubkova@upjs.sk (B.H.)
† These authors contributed equally to this work.

Abstract: Urine autofluorescence at 295 nm is significantly higher in patients with malignant melanoma at each clinical stage compared to the healthy group. The largest difference is in the early-stages and without metastases. With increasing stage, the autofluorescence at 295 nm decreases. There is also a significant negative correlation between autofluorescence and Clark classification. Based on our results, it is assumed that the way malignant melanoma grows also affects urinary autofluorescence.

Keywords: malignant melanoma; urine; autofluorescence

Citation: Birková, A.; Valko-Rokytovská, M.; Hubková, B.; Zábavníková, M.; Mareková, M. Strong Dependence between Tryptophan-Related Fluorescence of Urine and Malignant Melanoma. *Int. J. Mol. Sci.* **2021**, *22*, 1884. https://doi.org/10.3390/ijms22041884

Academic Editor: Burkhard Poeggeler

Received: 23 December 2020
Accepted: 11 February 2021
Published: 13 February 2021

Publisher's Note: MDPI stays neutral with regard to jurisdictional claims in published maps and institutional affiliations.

Copyright: © 2021 by the authors. Licensee MDPI, Basel, Switzerland. This article is an open access article distributed under the terms and conditions of the Creative Commons Attribution (CC BY) license (https://creativecommons.org/licenses/by/4.0/).

1. Introduction

Human urine is a complex biological fluid containing a variety of both endogenous and exogenous chemical compounds excreted by the body. The optical properties of the biological system obtained by fluorescence analysis reflect some of the physicochemical properties of the metabolites. Urine is a multicomponent mixture of different fluorophores and nonfluorescent metabolites [1–4].

In modern medicine, the use of urine as one of the biological fluids is very widespread because it does not require invasive sampling. Under physiological conditions, the range of urine output is 800 to 2000 mL per day, providing a sufficient amount of urine for sampling. There are more than 1700 metabolites identified and quantified in the human urine, and other almost 3000 expected or identified, but not quantified metabolites, listed in the Urine Metabolome Database with their structures and links to their known health and disease associations [4]. Urine analysis is advantageous over other biological matrices because the metabolites present, mostly those representing the final breakdown products of foods and beverages, technological additives, drugs, environmental contaminants, and even endogenous waste metabolites and bacterial byproducts, are more stable in the urine. Urine requires minimal pretreatment of the sample and reflect the physiological/pathological state of the biological system [5].

In recent years, progress has been made in the use of fluorescence spectroscopic techniques for the diagnosis of selected oncological diseases, such as ovarian cancer [2,6,7], pancreatic cancer [8], breast cancer [9,10], bladder cancer [11].

The incidence of skin cancer—melanoma—is increasing worldwide, and it represents 3% of all skin cancers but 65% of skin cancer deaths [12]. Melanoma is currently the fifth and sixth most common solid malignancy diagnosed both in men and women [13]. Malignant melanoma is a neoplasm derived from specialized melanin-producing cells called melanocytes or cells that develop from melanocytes. Melanoma spreads through the lymphatic system and blood and can therefore metastasize to any organ in the body. For this reason, early detection is very important. Malignant melanoma is a tumor with a high production of melanin. Precursors of melanin and their metabolites play an important role in melanogenesis, but in a broader sense, melanogenesis also refers to the process of melanosome formation and transfer to the keratinocytes. While melanins are polymers, the whole process of their synthesis, storage and transport takes place in the organelles called melanosomes. Before generating a sufficient amount of melanins to be seen by light microscopy, the melanosomes are known as pre-melanosomes.

Melanins are synthesized from the precursor amino acid L-tyrosine. It is hydroxylated by tyrosinase in the presence of dioxygen to L-dihydroxyphenylalanine (L-DOPA) and consequently to dopaquinone (melanogenic pathway, Raper–Mason pathway). In the following step, dopaquinone is metabolized to dopachrome by rapid cyclization. Dopachrome can be decarboxylated to 5,6-dihydroxyindole (DHI) or tautomerized to 5,6-dihydroxyindole-2-carboxylic acid (DHICA). Intermediates DHI and DHICA are oxidized to form eumelanin, with brown and black subtype [14].

Red-orange pheomelanins are synthesized after conjugation with cysteine and the subsequent oxidation of the intermediate. Neuromelanin, which has an approved role in aging and in the development of Parkinson's disease, can be formed both from eumelanin and from pheomelanin intermediates in dopamine and norepinephrine neurons [15].

Some studies have shown that, in addition to tyrosine, tryptophan and some of its metabolites are also involved in melanin biosynthesis [16–18]. Serotonin and its metabolite 5-hydroxyindole-3-acetic acid (5-HIAA) occurring in urine act as specific tumor markers, and increased production of 5-HIAA have been documented in human epidermal keratinocytes and melanoma cells [17].

Many of these metabolites have native fluorescent properties and are part of biological samples such as tissue, blood, and urine. Metabolism of cancerous tissues differs from healthy ones, which also affects the composition of natural fluorophores in body biofluids.

The use of urinary fluorescence analysis offers the possibility to detect urinary metabolites potentially associated with the neoplastic process, which could provide a new direction in the current search for predictive and prognostic markers [2,11,19,20]. In our previous studies, we have confirmed the role of fluorescence spectroscopy as a useful diagnostic tool with high-efficiency in ovarian cancer [7] as well as the feasibility of synovial fluid fluorescence fingerprinting to identify disease-specific profiles of synovial fluid metabolites [21].

In this study, we focused on the analysis of changes in urine autofluorescence from patients with malignant melanoma in comparison to healthy subjects as a tool of non-invasive melanoma detection.

2. Results

2.1. Fluorescence Measurements in Control and Malignant Melanoma Group

Statistically significant differences were found in two wide spectral regions at region 254–348 nm (with the most significant difference at 289–302 nm, $p = 8.4 \times 10^{-13}$) and at region 451–470 nm (with the most significant difference at 455–458 nm, $p = 0.025$) (Figure 1, Table 1, Figure 2).

2.2. Confrontation of Fluorescence Analysis and Histological Findings in Malignant Melanoma Group

In the subsequent analysis, we focused mainly on the area around the wavelength 295 nm, as in this region was found the most significant difference between the control and melanoma group ($p = 8.4 \times 10^{-13}$).

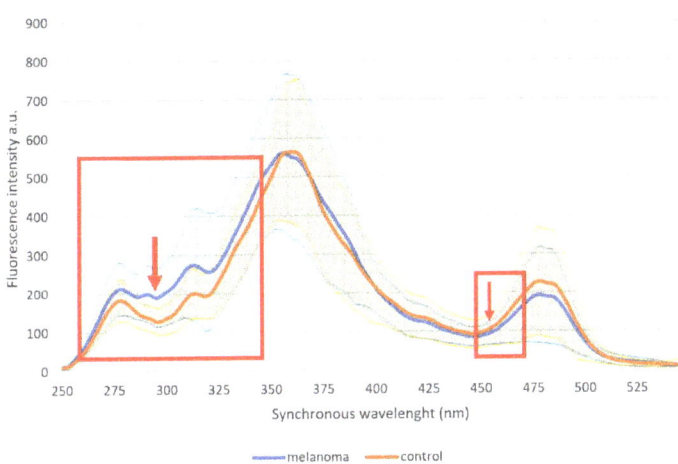

Figure 1. Average fluorescence profiles of urine in the control group and whole malignant melanoma group. Rectangles indicate wavelength areas with significant differences; arrows indicate wavelengths with the largest statistical difference.

Table 1. Comparison of fluorescence intensity at various wavelengths through whole measured spectrum (in 25 nm steps) in healthy and control group. For evaluation of differences, Student's t-test was used. Values of fluorescence intensity are expressed as mean ± SD.

λ (nm)	Group		p-Value of Student's t-Test
	Control	Malignant Melanoma	
250	9.8 ± 2.6	9.6 ± 5.1	0.74
275	171.4 ± 51.2	199.7 ± 68.8	0.00069
300	134.8 ± 40.7	204.4 ± 85.7	3.5×10^{-12}
325	228.1 ± 78.9	273.1 ± 146.2	0.0056
350	492.9 ± 128.5	527.5 ± 175.9	0.098
375	417.4 ± 169.2	440.8 ± 201.3	0.35
400	216.1 ± 95.7	215.0 ± 68.8	0.92
425	132.1 ± 53.0	121.8 ± 42.2	0.11
450	94.7 ± 37.8	86.7 ± 24.7	0.059
475	220.6 ± 134.5	187.4 ± 117.1	0.052
500	85.0 ± 64.4	74.5 ± 41.6	0.14
525	14.9 ± 8.7	17.0 ± 12.3	0.15
550	9.4 ± 15.3	8.2 ± 4.2	0.42

When analyzing according to the type of the malignant melanoma, there were strong significant differences between control group and nodular-type malignant melanoma group ($N = 26$; $p = 1.64 \times 10^{-6}$), superficial-spreading-type group ($N = 54$; $p = 5.63 \times 10^{-8}$), and weak difference in melanoma in situ group ($N = 6$; $p = 0.01$), but not between control group and acral lentiginous malignant melanoma and lentigo maligna group ($N = 7$; $p = 0.077$) or nevoid melanoma ($N = 3$; $p = 0.1$; Table 2, Figure 3).

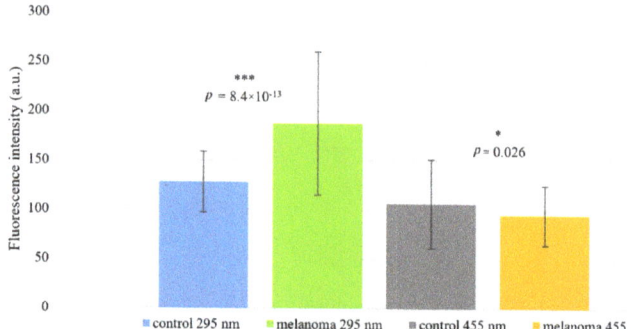

Figure 2. Fluorescence intensities of the control and malignant melanoma groups at 295 nm and 455 nm. Statistically significant differences are related to the control group. Values are expressed as mean ± SD. * indicates $p < 0.05$ and *** indicates $p < 0.001$.

Table 2. Statistical analysis of types of malignant melanoma compared to the control group. For evaluation of differences, Student's *t*-test was used. Values of fluorescence intensity are expressed as mean ± SD.

	Control	Nodular Type	Superficial Spreading Type	Acral Lentiginous Melanoma + Lentigo Maligna	Melanoma In Situ	Nevoid Type
N	119	26	54	7	6	3
Fluorescence intensity Mean ± SD	127.9 ± 31.0	164.3 ± 40.6	199.5 ± 82.1	181.5 ± 65.9	208.9 ± 58.7	154.3 ± 29.8
t-test		1×10^{-5}	5.6×10^{-8}	0.077	0.011	0.097

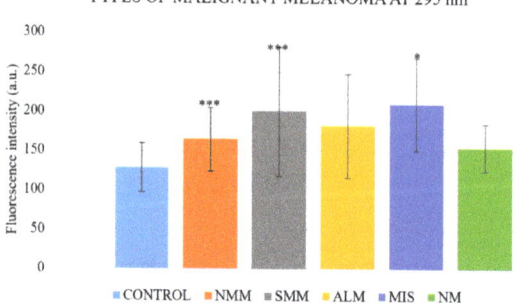

Figure 3. Fluorescence intensity in different types of malignant melanoma. NMM—nodular type malignant melanoma, SMM—superficial spreading-type melanoma, ALM—acral lentiginous malignant melanoma and lentigo maligna group, MIS—melanoma in situ, NM—nevoid melanoma. Statistically significant differences are related to the control group. Values are expressed as mean ± SD. * indicates $p < 0.05$ and *** indicates $p < 0.001$.

Within the melanoma group, there were found significant differences between fluorescence intensity in nodular-type malignant melanoma group and superficial-spreading-type melanoma ($p = 0.037$), and nodular-type malignant melanoma group and melanoma in situ ($p = 0.016$). There was not a significant difference ($p = 0.52$) in fluorescence intensity of melanoma group at 295 nm between those who had in the histological finding of ulceration or not, nor those with positive or negative HMB-45 and MelanA marker. MelanA and HMB-45 are commonly determined differentiation antigens in malignant melanoma, and their loss is relatively common, especially in progressive disease with metastatic lesions. When evaluating the effect of metastases on fluorescence analysis, it shows a significant difference ($p = 0.013$) in fluorescence intensity at 295 nm between those with ($N = 34$, 169.2 ± 94.1) or without ($N = 71$, 201.8 ± 83.1) metastases (Figure 4).

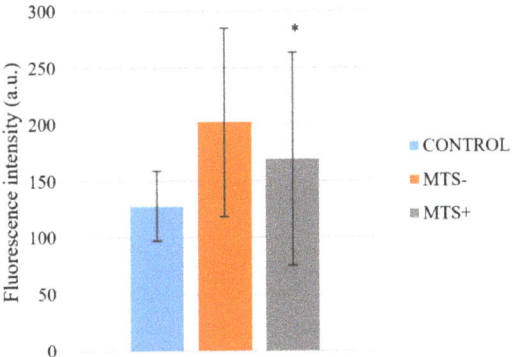

Figure 4. Fluorescence intensity in patients with malignant melanoma with (MTS+) or without metastases (MTS-) compared to the control. Statistically significant difference is between MTS+ and MTS- group. Values are expressed as mean ± SD. * indicates $p < 0.05$.

Analysis of clinical stage relation to fluorescence shows significant difference between fluorescence at 295 nm of the control group and particular stage (0-IV), but with different significance (0–IV: $p = 0.013, 0.000014, 0.000033, 0.0002$ and 0.026, respectively, Figure 5). The highest fluorescence intensity was at stage 0 ($N = 6$, 218.1 ± 58.4) and decreases with increasing clinical stage (I: $N = 42$, 201.2 ± 95.5, II: $N = 24$, 179.4 ± 47.2, III: $N = 23$, 173.2 ± 48.3, IV: $N = 10$, 173.0 ± 53.4).

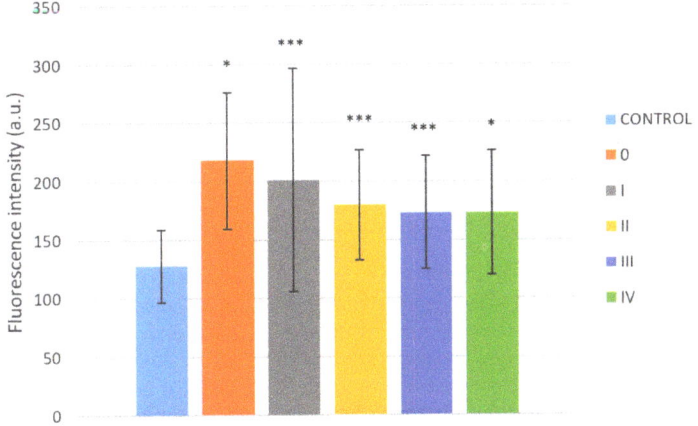

Figure 5. Fluorescence intensity in various clinical stages. Statistically significant difference is related to the control group. Values are expressed as mean ± SD. * indicates $p < 0.05$ and *** indicates $p < 0.001$.

A significant negative correlation was found ($r = -0.20$, $p = 0.041$) between fluorescence intensity and clinical stage. There was also a significant negative correlation of fluorescence intensity ($r = -0.23$, $p = 0.027$) with Clark level of invasion (Figure 6), but not with Breslow thickness ($r = -0.14$, $p = 0.177$, Figure 7).

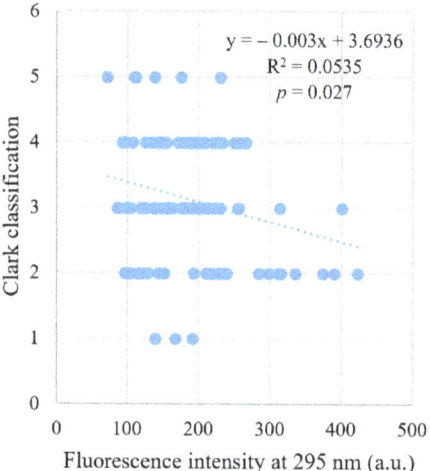

Figure 6. Correlation between fluorescence intensity and Clark scale.

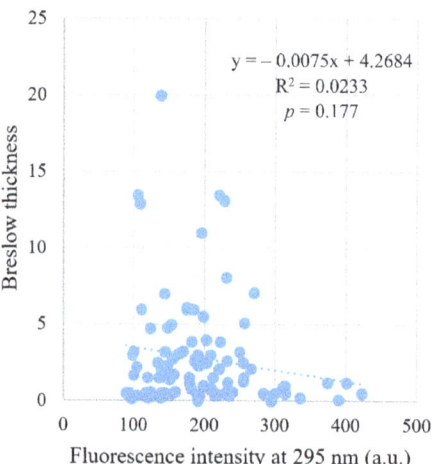

Figure 7. Correlation between fluorescence intensity and Breslow thickness.

3. Discussion

Malignant melanoma is a type of cancer with high metastatic potential, and early detection is crucial for the later assessment of prognosis and survival of patients. In our study, we have found that the urine of patients with malignant melanoma has higher fluorescence at a wide range of wavelengths, with the most significant difference at 295 nm, and the early-stages have higher fluorescence than the later ones. There exists very little information on the use of native fluorescence in malignant melanoma. In vivo diagnostics of skin, lesions have been published only a few times in the past. Chwirot et al. [22] tested the sensitivity of fluorescence detection of melanoma compared to other pigmented lesions. The tested wavelengths showed at ex/em 366/475 nm 82.7% sensitivity and 59.9% specificity. The authors concluded that the origin of the observed autofluorescence is not fully understood, and a plausible hypothesis may be that the spatial distributions of the autofluorescence observed for the skin (suggested fluorophores could be collagen in combination with elastin and desmosine or keratin) surrounding the pigmented lesions may result from responses of the host cells reacting to the presence of the lesions. In 2006, Borisova reported the high distinguishing potential of a fluorescence signal measured from skin lesions in vivo by a fiber-optic spectrophotometer. The most significant difference was found at similar wavelengths ex/em 337/480–490 nm, and the author reported a possible distinction between benign nevi and malignant melanoma [23]. The lower fluorescence of the reported area was interpreted by the author as a consequence of the hypervascularization of malignant neoplasia and thus the hemoglobin quenching effect. Lower fluorescence around 450 nm in melanoma patient's urine was also detected in this study, but we can exclude the effect of hemoglobin, as the urine samples were checked on the presence of the blood using strip tests and were blood negative. On the other hand, in vivo diagnostics of pigmented lesions can also be dangerous, especially upon excitation of the light in the UV area [24,25]. It appears to be a safer out-of-body autofluorescence examination or measurement of the biological fluids. Recently, was published a paper in which the authors focused on the measurement of analysis of the urine in malignant melanoma patients after removing NADH fluorescence by glutathione reductase derivation, and some results are also related to underived urine autofluorescence [26]. They focused on the range of 300–500 nm and described significantly lower emission at 460 nm in melanoma patients compared to the control group. The lower fluorescence emission at 460 nm in underived urines of melanoma patients correlates with our lower fluorescence at 450 nm, which differs slightly from the control group. An additional common finding of the compared study is that there was no difference in urine fluorescence in MelanA positive or negative patients, but when comparing positive or negative ulceration, there was a difference, which is not in correlation with the result of our study with $p = 0.76$. The divergence in this result probably stems from the fact that the study of Špaková et al. recruited 56 people (46 melanoma patients and 10 controls), while our study included more than 220 people (105 melanoma patients and 119 controls).

In this study, we focused more on the wavelength at 295 nm as there was a very significant difference between the healthy control group and the melanoma group ($p = 8.4 \times 10^{-13}$). Fluorescence is higher in all clinical stages of malignant melanoma, but the difference at early stages, even in the in situ stages of melanoma, with a Breslow thickness equal to zero, is very interesting. The clinical stage correlates slightly negatively with fluorescence, and this is also in agreement with a larger difference compared to the control group and patients without metastases than with metastases. The Breslow thickness is not related, but the Clark invasion is related to fluorescence, again with a negative correlation. Based on results, the type of malignant melanoma also plays a role in influencing fluorescence, but regardless of the positivity or negativity of the markers Melan A and HMB-45. The fluorescence area around 295 nm is typical for the fluorescence of proteins and is close to the fluorescence of amino acids tryptophan and phenylalanine and their metabolites and derivatives [27,28]. Malignant melanoma is associated with many molecules that could hypothetically fluoresce in this area. Recently, Belter et al. reported a list of potential biochemical serum markers that could be useful in diagnosing or assessing the prognosis of melanoma cancer patients. Among these markers are large molecules such as many proteins with functions such as enzymes, antigens, growth factors, inflammatory markers, e.g., lactate dehydrogenase, tyrosinase, cyclooxygenase 2, many matrix metalloproteinases, indoleamine-2,3-dioxygenase, tissue inhibitor of metalloproteinase, vascular endothelial growth factor, osteopontin, C-reactive protein, etc. [29], but also small molecules that are precursors of eumelanin and pheomelanin, such as 5-S cysteinyldopa, 6-hydroxy-5-methoxyindole-2-carboxylic acid, 5,6-dihydroxyindole, 5,6-dihydroxyindole-2-carboxylic acid [30,31]. Other melanoma-related derivatives of amino acids phenylalanine and tryptophan are also described, such as vanilmandelic acid, homovanilic acid, 5-hydroxyindole-3-acetic acid and indoxyl sulfate [18]. Many of these mentioned small molecules fluoresce naturally and are also present in urine. The amino acids phenylalanine and tryptophan and their metabolites have similar fluorescent properties in the characteristic fluorescence zone of 250–350 nm. The major fluorophore in this fluorescent area is tryptophan, and its derivative, indoxyl sulfate, is also present. Tryptophan is the most abundant component among the three fluorescent amino acid components of proteins. The contribution of phenylalanine to the intrinsic fluorescence of the protein is negligible due to its low absorption and very low quantum yield. Tyrosine has a quantum yield similar to tryptophan, but the tryptophan indole group is considered to be the dominant source of UV absorption at 280 nm and emission at 350 nm in proteins [32]. It is also possible that some proteins can contribute to increased fluorescence at 295 nm. The already mentioned amino acids and many of their metabolites with fluorescent characteristics are naturally present in the urine, and their presence is associated with several different diseases [3,7,20,33]. Matrix metalloproteinases are present in the urine in some cancer types, and their combination is cancer-specific [34]; some are specific for melanoma [35]. Based on our ongoing research, we believe that the increased urine fluorescence at 295 nm is related to the increased tryptophan concentration, which can be derived from free tryptophan as well as from tryptophan-rich protein degradation. This unpublished research pointed to a significant positive correlation between urinary tryptophan concentration and fluorescence at 295 nm ($r = 0.24$, $p = 0.041$; data not published).

4. Materials and Methods

4.1. Composition of the Study Group

The study group consisted of 105 patients with malignant melanoma with clinical stage 0 to IV; average age 56 ± 15.47 years; 52 men (49.5%) and 53 women (50.5%). The Control group consisted of 119 healthy controls (average age 40.4 ± 11 years; 75 men (63%) and 44 women (27%). Patients were recruited during hospitalization at the Department of Plastic and Reconstructive Surgery UPJŠ LF in Košice. The diagnosis of malignant melanoma was confirmed histologically, and the melanoma staging was based on the eighth edition of the American Joint Committee on Cancer (AJCC) staging system [36] that uses the following key information for assigning Tumor-Node-Metastasis (TNM) classifications: Breslow thickness of the tumor, ulceration presence, depth invasion by Clark scale and the presence of nodal or distant metastases. The healthy control group was chosen by random assignment, with the following criteria: the absence of any oncological disease, negative oncological family anamnesis and any serious disease (Table 3).

Table 3. Descriptive statistics of the study group.

	N	Age in Years		
		Minimum	Maximum	Mean
Healthy Control	119	22	65	40.4 ± 11
Malignant Melanoma	105	17	87	56.0 ± 15.5
Clinical Stage 0	6	17	64	45.2 ± 20.2
Clinical Stage I	42	17	87	52.7 ± 17.6
Clinical Stage II	24	43	81	59.5 ± 11.1
Clinical Stage III	23	29	85	59.7 ± 14.0
Clinical Stage IV	10	34	78	58.1 ± 13.3

Written informed consent was obtained from all patients prior to sample collection. All clinical investigations were conducted in accordance with the Declaration of Helsinki, and the study was approved by the Ethics Committee of the University of P. J. Šafárik in Košice, Medical Faculty (20 N/2016).

The study participants were not following any special diet and were asked not to take vitamin supplements prior to urine collection. Other medicaments and drugs prescribed by the general practitioner were retained. The use of drugs by probands was sporadic and statistically unevaluable, so we neglected the effect of the drugs in the present study.

4.2. Urine Samples

Urine samples were obtained from patients with malignant melanoma immediately following admission to the hospital. Participants of healthy controls were given a morning appointment and asked to fast at least 8 h before the sample collection. Samples were taken under standard conditions as first morning urines. Leukocytes, nitrite, pH, specific gravity, protein, glucose, ketones, urobilinogen, bilirubin, blood were evaluated semiquantitative using the 10 parameter urine strip test Dekaphan Leuco (Erba Lachema, Brno, Czech Republic). The average values of the pH and the specific gravity in the healthy control group were 5.76 ± 0.62, and 1021.5 ± 8.2, respectively. Ketones were detected in 6 members of the healthy control group, in an amount of 1.5 mmol/L. Leukocytes, nitrite, protein, glucose, urobilinogen, bilirubin and blood were negative. The result of the listed urine parameters in the malignant melanoma patients' group was not significantly different. The average pH and specific gravity values were 5.82 ± 0.50 and 1015.5 ± 10.5, respectively. Subsequently, the samples were stored at $-27\ °C$. After thawing and centrifugation at 5000 rpm for 10 min at laboratory temperature (Centrifuge Boeco S8, Boeco Germany, Hamburg, Germany), urine samples were analyzed.

4.3. Instrumentation and Statistical Analysis

The autofluorescence of urine samples was measured at room temperature, using a Luminescence Spectrophotometer PerkinElmer LS55 (PerkinElmer, Waltham, Massachusetts, USA) in 1 cm quartz cuvettes (Helmut Fischer, Sindelfingen, Germany). From every sample, we measured 12 synchronous spectra with $\Delta\lambda$ 30 nm in the range 250–550, step 0.5 nm, excitation/emission slits 5/5 nm, and a scan speed of 1200 nm/min. The measurements were performed, and urine fluorescence profiles were constructed as previously published [2].

Statistical analysis was performed with SPSS Statistics software version 22 (IBM, Armonk, New York, USA). The F-test was used to assess whether the standard deviations within the groups differ from each other. A Student's unpaired t-test was performed to determine the differences in fluorescence intensity within the control group and various descriptive parameters in malignant melanoma patients. A Student's unpaired t-test was also used to analyze the differences in fluorescence intensities between the types of malignant melanoma compared to the control group. Pearson's correlation analysis was performed to demonstrate the strength and direction of the linear relationship between fluorescence intensity and clinical stage, Clark scale or Breslow thickness. Statistical significance was assigned for p-value < 0.05. Values of fluorescence intensity were expressed as mean $\pm$ SD.

5. Conclusions

Fluorescence spectrometry has become a standard tool in many areas of research and is a universal alternative technique for studies using radioactive labels. A large number of fluorescence techniques and methods have been developed to study biological processes. Autofluorescence of biological systems, such as urine, has facilitated the development and review of several approaches without the necessary labeling. Significantly higher autofluorescence of the urine at 295 nm in melanoma patients may be useful background for further studies revealing metabolic changes in malignant melanoma, but may also serve as a potential diagnostic marker, as the early-stages show even more pronounced changes in tryptophan-related fluorescence intensity than the later ones.

Author Contributions: Conceptualization, A.B., M.V.-R. and B.H.; methodology, A.B., M.V.-R.; software, A.B., M.V.-R.; validation, A.B., M.V.-R. and B.H.; formal analysis, A.B., M.V.-R. and B.H.; investigation, M.Z.; resources, M.Z.; data curation, A.B.; writing—original draft preparation, A.B., M.V.-R. and B.H.; writing—review and editing, A.B., M.V.-R. and B.H.; visualization, A.B., M.V.-R. and B.H.; supervision, A.B., M.V.-R. and B.H.; project administration, M.M.; funding acquisition, M.M. All authors have read and agreed to the published version of the manuscript.

Funding: This research was funded by Scientific Grant Agency of the Ministry of Education, Science, Research and Sport of the Slovak Republic and Slovak Academy of Sciences VEGA, grant number 1/0372/17.

Institutional Review Board Statement: The study was conducted according to the guidelines of the Declaration of Helsinki, and approved by the Ethics Committee of the University of P. J. Šafárik in Košice, Medical Faculty (20N/2016).

Informed Consent Statement: Informed consent was obtained from all subjects involved in the study.

Data Availability Statement: The data presented in this study are available on request from the corresponding authors. The data are not publicly available due to ethical restrictions.

Conflicts of Interest: The authors declare no conflict of interest.

References

1. Valko-Rokytovská, M.; Hubková, B.; Birková, A.; Mareková, M. Factors Affecting Fluorescence Analysis of Diagnostically Important Urinary Metabolites—Influence of Mixture Composition. *Spectrosc. Lett.* **2014**, *48*, 227–233. [CrossRef]
2. Birková, A.; Oboril, J.; Kréta, R.; Čižmárová, B.; Hubková, B.; Šteffeková, Z.; Genči, J.; Paralič, J.; Mareková, M. Human fluorescent profile of urine as a simple tool of mining in data from autofluorescence spectroscopy. *Biomed. Signal Process. Control.* **2020**, *56*, 101693. [CrossRef]
3. Masilamani, V.; Vijmasi, T.; Al Salhi, M.; Govindaraj, K.; Vijaya-Raghavan, A.P.; Antonisamy, B. Cancer detection by native fluorescence of urine. *J. Biomed. Opt.* **2010**, *15*, 057003. [CrossRef] [PubMed]
4. Bouatra, S.; Aziat, F.; Mandal, R.; Guo, A.C.; Wilson, M.R.; Knox, C.; Bjorndahl, T.C.; Krishnamurthy, R.; Saleem, F.; Liu, P.; et al. The Human Urine Metabolome. *PLoS ONE* **2013**, *8*, e73076. [CrossRef] [PubMed]
5. Valko-Rokytovská, M.; Očenáš, P.; Salayová, A.; Kostecká, Z. New Developed UHPLC Method for Selected Urine Metabolites. *J. Chromatogr. Sep. Tech.* **2018**, *9*, 2. [CrossRef]
6. Zvarik, M.; Martinicky, D.; Hunakova, L.; Lajdova, I.; Sikurova, L. Fluorescence characteristics of human urine from normal individuals and ovarian cancer patients. *Neoplasma* **2013**, *60*, 533–537. [CrossRef]
7. Birkova, A.; Grešová, A.; Steffekova, Z.; Kraus, V.; Ostró, A.; Toth, R.; Mareková, M. Changes in urine autofluorescence in ovarian cancer patients. *Neoplasma* **2014**, *61*, 724–731. [CrossRef]
8. Lwin, T.M.; Hoffman, R.M.; Bouvet, M. The development of fluorescence guided surgery for pancreatic cancer: From bench to clinic. *Expert Rev. Anticancer. Ther.* **2018**, *18*, 651–662. [CrossRef]
9. Veys, I.; Pop, C.; Barbieux, R.; Moreau, M.; Noterman, D.; De Neubourg, F.; Nogaret, J.; Liberale, G.; Larsimont, D.; Bourgeois, P. ICG Fluorescence Imaging as a New Tool for Optimization of Pathological Evaluation in Breast Cancer Tumors after Neo-adjuvant Chemotherapy. *PLoS ONE* **2018**, *13*, e0197857. [CrossRef]
10. Sordillo, L.A.; Sordillo, P.P.; Budansky, Y.; Pu, Y.; Alfano, R.R. Optical Spectral Fingerprints of Tissues from Patients with Different Breast Cancer Histologies Using a Novel Fluorescence Spectroscopic Device. *Technol. Cancer Res. Treat.* **2013**, *12*, 455–461. [CrossRef]
11. Kollarik, B.; Zvarik, M.; Bujdak, P.; Weibl, P.; Rybar, L.; Sikurova, L.; Hunakova, L. Urinary Fluorescence Analysis in Diagnosis of Bladder Cancer. *Neoplasma* **2018**, *65*, 234–241. [CrossRef] [PubMed]
12. Valko-Rokytovska, M.; Bruchatá, K.; Šimková, J.; Milkovicova, M.; Kostecka, Z. Current trends in the treatment of malignant melanoma. *Neoplasma* **2016**, *63*, 333–341. [CrossRef]
13. Levine, S.M.; Shapiro, R.L. Surgical Treatment of Malignant Melanoma. *Dermatol. Clin.* **2012**, *30*, 487–501. [CrossRef]
14. Slominski, A.; Zmijewski, M.A.; Pawelek, J.M. L-tyrosine and L-dihydroxyphenylalanine as hormone-like regulators of melanocyte functions. *Pigment Cell Melanoma Res.* **2012**, *25*, 14–27. [CrossRef]
15. Zucca, F.A.; Segura-Aguilar, J.; Ferrari, E.; Muñoz, P.; Paris, I.; Sulzer, D.; Sarna, T.; Casella, L.; Zecca, L. Interactions of iron, dopamine and neuromelanin pathways in brain aging and Parkinson's disease. *Prog. Neurobiol.* **2017**, *155*, 96–119. [CrossRef]
16. Slominski, A.T.; Pisarchik, A.; Semak, I.; Sweatman, T.; Wortsman, J.; Szczesniewski, A.; Slugocki, G.; McNulty, J.; Kauser, S.; Tobin, D.J.; et al. Serotoninergic and melatoninergic systems are fully expressed in human skin. *FASEB J.* **2002**, *16*, 896–898. [CrossRef] [PubMed]
17. Lee, H.; Park, M.; Kim, S.; Choo, H.P.; Lee, A.; Lee, C. Serotonin induces melanogenesis via serotonin receptor 2A. *Br. J. Dermatol.* **2011**, *165*, 1344–1348. [CrossRef] [PubMed]
18. Valko-Rokytovská, M.; Hubková, B.; Birková, A.; Mašlanková, J.; Stupák, M.; Zábavníková, M.; Čižmárová, B.; Mareková, M. Specific Urinary Metabolites in Malignant Melanoma. *Medicina* **2019**, *55*, 145. [CrossRef]
19. Zhang, L.; Pu, Y.; Jianpeng, X.; Pratavieira, S.; Xu, B.; Achilefu, S.; Alfano, R.R. Tryptophan as the Fingerprint for Distinguishing Aggressiveness among Breast Cancer Cell Lines Using Native Fluorescence Spectroscopy Aggressiveness among Breast Cancer Cell Lines Using. *J. Biomed. Opt.* **2014**, *19*. [CrossRef] [PubMed]

20. Rajasekaran, R.; Aruna, P.R.; Koteeswaran, D.; Padmanabhan, L.; Muthuvelu, K.; Rai, R.R.; Thamilkumar, P.; Krishna, C.M.; Ganesan, S. Characterization and Diagnosis of Cancer by Native Fluorescence Spectroscopy of Human Urine. *Photochem. Photobiol.* **2012**, *89*, 483–491. [CrossRef]
21. Bilská, K.; Šteffeková, Z.; Birková, A.; Mareková, M.; Ledecký, V.; Hluchý, M.; Kisková, T. The use of native fluorescence analysis of synovial fluid in the diagnosis of medial compartment disease in medium- and large-breed dogs. *J. Veter- Diagn. Investig.* **2016**, *28*, 332–337. [CrossRef] [PubMed]
22. Chwirot, B.W.; Sypniewska, N.; Michniewicz, Z.; Redzinski, J.; Chwirot, S.; Kurzawski, G.; Ruka, W. Fluorescence In Situ Detection of Human Cutaneous Melanoma: Study of Diagnostic Parameters of the Method. *J. Investig. Dermatol.* **2001**, *117*, 1449–1451. [CrossRef]
23. Borisova, E. Fluorescence detection improves malignant melanoma diagnosis. *SPIE Newsroom* **2006**. [CrossRef]
24. Elwood, J.M.; Williamson, C.; Stapleton, P.J. Malignant melanoma in relation to moles, pigmentation, and exposure to fluorescent and other lighting sources. *Br. J. Cancer* **1986**, *53*, 65–74. [CrossRef]
25. Beral, V.; Evans, S.; Shaw, H.; Milton, G. Malignant melanoma and exposure to fluorescent lighting at work. *Lancet* **1982**, *320*, 290–293. [CrossRef]
26. Špaková, I.; Dubayová, K.; Nagyová, V.; Mareková, M. Fluorescence Biomarkers of Malignant Melanoma Detectable in Urine. *Open Chem.* **2020**, *18*, 898–910. [CrossRef]
27. Lakowicz, J.R. *Principles of Fluorescence Spectroscopy*, 2nd ed.; Springer: New York, NY, USA, 1999; ISBN 978-1-4757-3063-0.
28. Khatun, M.; Jana, G.C.; Nayim, S.; Das, S.; Patra, A.; Dhal, A.; Hossain, M. Hydrophobic ring substitution on 9-O position of berberine act as a selective fluorescent sensor for the recognition of bovine serum albumin. *Microchem. J.* **2020**, *153*, 104453. [CrossRef]
29. Belter, B.; Haase-Kohn, C.; Pietzsch, J. Biomarkers in Malignant Melanoma: Recent Trends and Critical Perspective. In *Cutaneous Melanoma: Etiology and Therapy*; Codon Publications: Singapore, 2017; pp. 39–56.
30. Wakamatsu, K.; Fukushima, S.; Minagawa, A.; Omodaka, T.; Hida, T.; Hatta, N.; Takata, M.; Uhara, H.; Okuyama, R.; Ihn, H. Significance of 5-S-Cysteinyldopa as a Marker for Melanoma. *Int. J. Mol. Sci.* **2020**, *21*, 432. [CrossRef]
31. Hara, H.; Walsh, N.; Yamada, K.; Jimbow, K. High Plasma Level of a Eumelanin Precursor, 6-Hydroxy-5-Methoxyindole-2-Carboxylic Acid as a Prognostic Marker for Malignant Melanoma. *J. Investig. Dermatol.* **1994**, *102*, 501–505. [CrossRef]
32. Ghisaidoobe, A.B.T.; Chung, S.J. Intrinsic Tryptophan Fluorescence in the Detection and Analysis of Proteins: A Focus on Förster Resonance Energy Transfer Techniques. *Int. J. Mol. Sci.* **2014**, *15*, 22518–22538. [CrossRef]
33. Funai, K.; Honzawa, K.; Suzuki, M.; Momiki, S.; Asai, K.; Kasamatsu, N.; Kawase, A.; Shinke, T.; Okada, H.; Nishizawa, S.; et al. Urinary fluorescent metabolite O-aminohippuric acid is a useful biomarker for lung cancer detection. *Metabolomics* **2020**, *16*, 1–8. [CrossRef]
34. Roy, R.; Louis, G.; Loughlin, K.R.; Wiederschain, D.; Kilroy, M.; Lamb, C.C.; Zurakowski, D.; Moses, M.A. Tumor-Specific Urinary MMP Fingerprinting: Identification of High Molecular Weight Urinary MMP Species. *Clin Cancer Res.* **2008**, *14*, 6610–6617. [CrossRef]
35. Giricz, O.; Lauer, J.L.; Fields, G.B. Variability in melanoma metalloproteinase expression profiling. *J. Biomol. Tech. JBT* **2010**, *21*, 194–204.
36. Keung, E.Z.; Gershenwald, J.E. The eight edition American Joint Committee on Cancer (AJCC) melanoma staging system: Implications for melanoma treatment and care. *Expert Rev. Anticancer Ther.* **2018**, *18*, 775–784. [CrossRef]

Review

Tryptophan Metabolites and Aryl Hydrocarbon Receptor in Severe Acute Respiratory Syndrome, Coronavirus-2 (SARS-CoV-2) Pathophysiology

George Anderson [1], Annalucia Carbone [2] and Gianluigi Mazzoccoli [2,*]

1. CRC Scotland & London, Eccleston Square, London SW1V 1PX, UK; anderson.george@rocketmail.com
2. Department of Medical Sciences, Division of Internal Medicine and Chronobiology Laboratory, Fondazione IRCCS "Casa Sollievo della Sofferenza", 71013 San Giovanni Rotondo, Italy; annalucia.carbone@gmail.com
* Correspondence: g.mazzoccoli@operapadrepio.it

Abstract: The metabolism of tryptophan is intimately associated with the differential regulation of diverse physiological processes, including in the regulation of responses to severe acute respiratory syndrome, coronavirus-2 (SARS-CoV-2) infection that underpins the COVID-19 pandemic. Two important products of tryptophan metabolism, viz kynurenine and interleukin (IL)4-inducible1 (IL41)-driven indole 3 pyruvate (I3P), activate the aryl hydrocarbon receptor (AhR), thereby altering the nature of immune responses to SARS-CoV-2 infection. AhR activation dysregulates the initial pro-inflammatory cytokines production driven by neutrophils, macrophages, and mast cells, whilst AhR activation suppresses the endogenous antiviral responses of natural killer cells and CD8+ T cells. Such immune responses become further dysregulated by the increased and prolonged pro-inflammatory cytokine suppression of pineal melatonin production coupled to increased gut dysbiosis and gut permeability. The suppression of pineal melatonin and gut microbiome-derived butyrate, coupled to an increase in circulating lipopolysaccharide (LPS) further dysregulates the immune response. The AhR mediates its effects via alterations in the regulation of mitochondrial function in immune cells. The increased risk of severe/fatal SARS-CoV-2 infection by high risk conditions, such as elderly age, obesity, and diabetes are mediated by these conditions having expression levels of melatonin, AhR, butyrate, and LPS that are closer to those driven by SARS-CoV-2 infection. This has a number of future research and treatment implications, including the utilization of melatonin and nutraceuticals that inhibit the AhR, including the polyphenols, epigallocatechin gallate (EGCG), and resveratrol.

Keywords: tryptophan; aryl hydrocarbon receptor; severe acute respiratory syndrome; SARS-CoV-2; COVID-19

1. Introduction

There is a growing appreciation that tryptophan and its metabolites are crucial aspects of the severe acute respiratory syndrome, coronavirus-2 (SARS-CoV-2) pathophysiology that has driven the COVID-19 pandemic. SARS-CoV-2 severity and fatality are strongly driven by advanced age as well as pre-existing co-morbidities, such as obesity and type 2 diabetes (T2D), as well as by stress-associated conditions, including racial discrimination stress [1]. It is widely accepted that it is alterations in the immune response that drive an individual's susceptibility to severe SARS-CoV-2 infection. This is supported by the World Health Organization (WHO), which recently admitted that their recommended treatments, viz Remdesivir, Lopinavir, β-interferon (IFN), and Hydroxychloroquine proved of little utility, whilst treatments more directly targeting the immune response, such as Dexamethasone, were more beneficial [2].

Tryptophan and its metabolites, including kynurenine and indole-3-pyruvate (I3P), can be differentially regulated over the course of SARS-CoV-2 infection, which is strongly determined by the influence of pre-existing high-risk conditions, such as obesity/T2D and ageing, on the immune response [1]. The influence of the tryptophan metabolites, kynurenine and I3P, are mediated via the activation of the aryl hydrocarbon receptor (AhR). The AhR can have a number of complex effects, that may be partly ligand-dependent as well as from the complex interactions of the AhR with metabolic and circadian processes [3].

AhR activation can dysregulate the immune response, contributing to an increase in the initial pro-inflammatory cytokine wave/storm to SARS-CoV-2. It is the heightened and prolonged initial 'cytokine storm' that underpins SARS-CoV-2 driven severity and fatality [4]. Increased AhR activation contributes to this heightened and prolonged 'cytokine storm'. In the normal course of a viral infection, such initial macrophage, neutrophil, and mast cell-driven inflammatory processes are eventually superseded by the endogenous anti-viral cells, especially natural killer (NK) cells and CD8+ T cells [1]. NK cells and CD8+ T cells target and kill viral infected cells by mechanisms similar to how these cells kill and remove cancer cells. AhR activation on NK cells and CD8+ T cells leads to a state of 'exhaustion', which prevents or suppresses their capacity to eliminate virus-infected cells or cancer cells [5,6]. Consequently, the differential regulation of tryptophan metabolites to increase AhR ligands, such as kynurenine and I3P, are important determinants of SARS-CoV-2 infection severity and fatality.

The present article reviews data on the role of tryptophan and tryptophan metabolites in driving SARS-CoV-2 infection severity via AhR activation in immune cells. The driving of tryptophan to kynurenine and I3P decreases the availability of tryptophan for serotonin synthesis. As such, SARS-CoV-2 driven processes are linked to a decrease in serotonin and serotonin-derived N-acetylserotonin (NAS) and melatonin. The inhibition of the melatonergic pathway may therefore be an intimate aspects of SARS-CoV-2 pathophysiology, which is proposed to arise from alterations in immune cell metabolism [1] (Figure 1).

2. Tryptophan Metabolites and the Aryl Hydrocarbon Receptor

Although tryptophan is classically associated with being the necessary precursor for serotonin synthesis, the majority (95%) of the body's tryptophan is converted to kynurenine by indoleamine 2,3-dioxygenase (IDO) and tryptophan 2,3-dioxygenase (TDO). An increase in pro-inflammatory cytokines, including IL-1β, IL-6, IL-18, and tumor necrosis factor (TNF), but especially IFNγ, increases IDO (and/or TDO in some cells) to further raise kynurenine levels and suppress serotonin, NAS, and melatonin levels [7]. TDO may also be increased by stress-associated cortisol and hypothalamus–pituitary–adrenal (HPA) axis activation [8]. As such, many medical conditions that are associated with an increase in pro-inflammatory cytokine or stress-induced dysregulation of the HPA axis will have increased IDO/kynurenine/AhR activation as an aspect of their pathophysiology, as evident in cancers [9], Alzheimer's disease [10], depression [11], and arthritis [12], as well as in many other medical conditions. Consequently, the 'cytokine storm' that occurs to many viral infections, including influenza and SARS-CoV-2, will be associated with the induction of IDO/TDO/kynurenine/AhR activation and therefore wills alterations in the immune response and decreases in the serotonergic and melatonergic pathway. This is supported by data showing circulating kynurenine to be significantly increased in SARS-CoV-2 infection, in correlation with levels of severity [13].

However, there is another route whereby alterations in tryptophan metabolism may induce ligands that activate the AhR. Many bacterial and viral infections have been shown to increase interleukin-4-induced 1 (IL4I1), especially in macrophages [14,15], with IL4I1 leading to the induction of other AhR ligands, especially via I3P production. This would suggest that SARS-CoV-2 infection may be mediating some of its effects via the upregulation of IDO/kynurenine-independent AhR ligands. This is given some support by data showing that SARS-CoV-2 can increases AhR activation in an IDO-independent manner [16]. IL4I1 can be upregulated by a number of other immune cells, including dendritic cells, CD4+ T cells $\gamma\delta$ ($\gamma\delta$T-cells), and B-lymphocytes [17,18], although to a lesser extent than in macrophages. The IL4I1 induction in macrophages does not seem to have any suppressive impact on macrophage activation [17]. Rather, the effects of macrophage and dendritic cell IL4I1 and I3P induction are on the regulation of T and B lymphocytes [19], especially the suppression of CD3+, CD4+, and CD8+ T cells [18]. To date, there is no data on IL4I1 and I3P effects on NK cells, although it is clear that AhR activation is the major driver of 'exhaustion' and suppression of NK cells in the tumor microenvironment [3,6].

I3P and some of its derivates activate the AhR [20,21]. I3P can also generate indole-3-acetaldehyde (I3A), another AhR ligand, with both I3P and I3A being able to rearrange to the classical endogenous AhR ligand, 6-formylindolo(3,2-b)carbazole (FICZ), as well as the FICZ oxidation product, indolo(3,2-b)carbazole-6-carboxylic acid (CICZ) [22]. As such, the tryptophan metabolite pathway may be a rich source for AhR ligands that will be variably induced under different cellular conditions. It is also of note that the gut microbiome is an important source of I3P and I3A, with AhR activation in the gut being important to the maintenance of the gut barrier [23]. Future research will have to clarify as to whether I3P and/or its metabolites are AhR agonists.

3. Aryl Hydrocarbon Receptor

Classically, the humans AhR has been thought to function as a xenobiotic chemical sensor, and may still be referred to as the dioxin receptor. AhR activation aromatic (aryl) hydrocarbons is how the AhR derived its name. The AhR mediates many of its diverse and important effects via AhR activation induction of cytochrome P450 (CYP) metabolizing enzymes, including CYP1A, CYP1B1, and CYP1A2. CYP1A1 is important in the metabolism and regulation of estradiol, and therefore, in hormonal regulation, whilst CYP1B1 can O-demethylate melatonin to its immediate precursor, NAS, thereby driving alterations in the NAS/melatonin ratio, with some dramatic and contrasting consequences [24].

The AhR can be activated by a growing array of endogenous (e.g., FICZ), induced (e.g., kynurenine), and exogenous (e.g., air pollutants) ligands, highlighting its frequent involvement under a wide array of diverse circumstances. The AhR is highly expressed in the placenta and immune cells, with a wide array of diverse developmental effects. As indicated by AhR activation maintaining the gut barrier under challenge [23], the AhR has many beneficial effects, as well as detrimental effects when the production of AhR ligands becomes dysregulated. AhR activation leads to the induction of its own repressor, the AhR repressor (AHRR). The AhR is also differentially expressed over the circadian rhythm, indicating its involvement in wider systemic processes. It should also be noted that the AhR can be expressed in the mitochondrial membrane, although any direct effects on mitochondria regulation are still to be investigated [25].

The AhR is classified as one of a group of 'basic helix-loop-helix' transcription factors. Typically, the AhR is bound to, and inactivated by, a variety of chaperones in the cytoplasm. Upon ligand-binding, the AhR dissociates from these chaperones and nuclear translocates. In the nucleus, the AhR dimerizes with the AhR nuclear translocator (ARNT), where upon it then regulates numerous genes, including those expressing the xenobiotic-responsive element (XRE).

4. Gut Dysbiosis and Permeability in COVID-19

There is a growing appreciation of the role of the gut microbiome and gut permeability in the regulation of a host of diverse medical conditions. Such an array of consequences partly arises from two important process, viz: gut dysbiosis-associated decrease in the short-chain fatty acid, butyrate; and increased gut permeability-associated transfer of lipopolysaccharide (LPS) into the circulation. A decrease in butyrate and increase in circulating LPS has significant impacts on a wide array of cells, especially on the immune response.

Suppressed butyrate levels are evident over the course of SARS-CoV-2 infection [26], indicating a loss of butyrate's histone deacetylase (HDAC) inhibitory activity as an aspect of SARS-CoV-2 infection. As butyrate and HDAC inhibition are important regulators of the immune response, with butyrate increasing the cytotoxicity and levels of NK cells in response to cancers [27], it would seem clear that such alterations in the gut microbiome short-chain fatty acid production are relevant to SARS-CoV-2 pathophysiology.

Hypertension is one of the high-risk conditions for severe/fatal SARS-CoV-2 infection, with a decrease in butyrate contributing to alterations in the gut–lung axis that contribute to hypertension-linked SARS-CoV-2 responses [28]. These authors indicate that butyrate's inhibition of the high-mobility group box (HMGB)1 may be a crucial aspect of the protection afforded by butyrate, with decreased butyrate increasing the risk of SARS-CoV-2 symptom severity in people with hypertension [28]. However, as decreased butyrate production is an aspect of many medical conditions, including obesity and T2D [29], the optimization of butyrate production or supplementation with sodium butyrate is likely to afford protection in many of the high-risk medical conditions associated with increased SARS-CoV-2 severity/fatality [29]. Butyrate has also been found to increase a number of genes, which are regarded as anti-viral genes, thereby having wider anti-viral efficacy, including melatonin [30].

It should also be noted that gut-derived butyrate helps to prevent gut permeability, with effects that are partly mediated via an increase in mitochondrial function and associated induction of the melatonergic pathway [30]. Butyrate, like pineal melatonin, may ultimately act via the disinhibition of the pyruvate dehydrogenase complex (PDC), leading to an increased conversion of pyruvate to acetyl-CoA, thereby increasing ATP production from the tricarboxylic acid (TCA) cycle and oxidative phosphorylation (OXPHOS) as well as providing acetyl-CoA as a necessary co-factor for the first enzyme in the melatonergic pathway, arylalkylamine N-acetyltransferase (AANAT) [29]. As such, the effects of butyrate will be modulated by variations in the levels of tryptophan driven to serotonin production, given that serotonin is a necessary precursor for activation of the melatonergic pathway. Such effects of butyrate link gut dysbiosis and decreased butyrate production to gut permeability and increased circulating LPS levels.

By elevating circulating LPS, increased gut permeability modulates the patterning of the immune response via the activation of toll-like receptor (TLR)4 on different immune cells. Most viruses make their first host contact with mucosal surfaces, where bacterial microbials will already be well-established. Consequently, bacteria–virus interactions are an integral aspect of most viral infections [31]. However, an increase in circulating LPS, or other TLR4 agonists, such as gut-derived HMGB1 [32], can modulate the immune response to viral infection, including potentiating influenza virus lethality via effects in dendritic cells [33].

Clearly, alterations in the gut may act to regulate SARS-CoV-2 infection via variations in butyrate, LPS, and HMGB1. However, the gut and gut microbiome are also important to the uptake and metabolism of tryptophan, with the gut also being an important source for I3P, and therefore for AhR activation. In contrast to the many negative consequences of AhR activation in SARS-CoV-2 infection and in many other medical conditions, gut AhR activation helps to seal the gut barrier [23].

Overall, the gut is an important hub for many of the physiological factors and high-risk medical conditions associated with SARS-CoV-2 severity and fatality. Other factors known to regulate tryptophan metabolism, immunity, and SARS-CoV-2 infection may be acting, at least partly, via the gut. One such factor is vitamin D.

5. Vitamin D and COVID-19

A growing body of data shows vitamin D to afford protection against SARS-CoV-2 infection severity and fatality [34,35] ARS-CoV-2 infection outcome [36].

However, not everyone is convinced by the data indicating a role for vitamin D in SARS-CoV-2 infection severity/fatality [37], especially as lower vitamin D levels may interact with other non-measured and/or difficult to modify factors, such as age, obesity, ethnicity [38], and racial discrimination stress [1], in modulating immune responses to SARS-CoV-2 infection. It is long appreciated that vitamin D upregulates the levels and cytotoxicity of NK cells [39], with NK cells being important drivers of the endogenous anti-viral response to SARS-CoV-2 infection. Low vitamin D levels significantly correlate with reduced NK cell numbers and cytotoxicity in ICU and non-ICU SARS-CoV-2 infected patients with pneumonia [40]. Vitamin D also acts to suppress the heightened pro-inflammatory macrophage and myeloid derived suppressor cell contribution to the initial 'cytokine storm' during SARS-CoV-2 infection [41].

Vitamin D can also regulate mitochondrial function, including mitochondria ROS production and complexes II and IV, as shown in a variety of different cell types [42]. The effects of vitamin D on mitochondrial function, mitochondrial ROS, and levels of oxidative stress have recently been proposed to underpin the utility of vitamin D in the regulation of the SARS-CoV-2 infection [43]. It is also important to note the vitamin D receptor is also expressed in mitochondria [44], where it acts to regulate mitochondrial function and ROS production [45].

Importantly, vitamin D is a significant regulator of tryptophan metabolism. Murine data shows vitamin D to increase tryptophan hydroxylase (TPH)2, leading to a dramatic increase in neuronal serotonin production, whilst decreasing adipocyte leptin production [46]. Serotonin is a necessary precursor for the melatonergic pathway, and therefore in the regulation of metabolism and immune cell activation/deactivation [47], whilst heightened leptin levels correlate with SARS-CoV-2 infection severity on overweight patients [48]. As leptin influences the structural organization of NK cells [49], alterations in the regulation of leptin may be co-ordinated with wider regulation of the tryptophan/serotonergic/melatonergic pathway. Vitamin D also increases IDO in dendritic cells, thereby increasing the immune-suppressive levels of T-regulator (Treg) cells [50]. Alterations in Treg function may be an important aspect of immune dysregulation to severe SARS-CoV-2 infection [51].

As noted, vitamin D may also act via the gut, with vitamin D suppressing gut dysbiosis and increasing gut bacteria diversity [52]. Consequently, many of the effects of vitamin D on immune, mitochondria, epigenetic, and melatonergic pathway regulation may be mediated via effects in the gut. The prevention of gut permeability by vitamin D [53], may therefore be mediated via the upregulation of butyrate, whilst butyrate also acts to upregulate vitamin D receptor signaling [54]. Vitamin D may also suppress pro-inflammatory cytokine production [55], indicating that it may decrease pro-inflammatory cytokine-induced IDO, thereby impacting on the kynurenine/tryptophan ratio and inflammation driven AhR activation. However, as well as in dendritic cells, vitamin D can also upregulate the AhR in some cells [55]. This requires further investigation in different cell types, particularly as data in keratinocytes indicate that the hydroxyl-derivative of vitamin D3, the CYP11A1-derived 20,23(OH)$_2$D3, is an AhR ligand [56]. The effects of vitamin D on AhR activation may be confounded by data showing that the vitamin D receptor can directly bind to an everted repeat (ER) 8 motif in the human CYP1A1 promoter [57]. CYP1A1 is typically used as an indicant of AhR activation, whilst it is CYP1B1 that seems responsible for the significant effects of AhR activation on mitochondrial and immune function [3]. HDAC inhibitors, such as butyrate, can suppress CYP11A1 [58], suggesting that variations in butyrate may act to regulate the induction of CYP11A1-derived 20,23(OH)$_2$D3 and its activation of the AhR. The effects of butyrate, and other short-chain fatty acids, on CYP1B1 is context and cell dependent [59]. The interactions of the gut microbiome short-chain fatty acid, butyrate, acetate and propionate, via HDAC inhibition, on AhR-linked inductions will be important to determine in different cell types.

Such data indicate the complex interactions that vitamin D can have with other processes relevant to SARS-CoV-2 infection. For example, an increase in stress, including racial discrimination stress, may increase gut permeability and gut dysbiosis, with any decrease in butyrate attenuating butyrate's potentiation of the vitamin D receptor. As such, as well as directly regulating gut permeability, vitamin D effects will be influenced by other factors, such as stressors, that increase gut dysbiosis/permeability.

The complexity of vitamin D receptor effects are further increased by data showing that melatonin binds to the vitamin D receptor and increases vitamin D-driven transcription [60]. This suggests that variations in pineal and local melatonin production may act to regulate vitamin D effects more directly, via melatonin interactions with the vitamin D receptor. It should be noted that melatonin may interact with, and regulate the activity of, a wide array of different receptors [61], considerably complicating its effects. Such data would indicate that the decrease in circadian, pineal melatonin with age may contribute to age-linked SARS-CoV-2 infection severity via alterations in melatonin's regulation of the vitamin D receptor, and other receptors, as well as from the melatonin/Bmal1/SIRT1/SIRT3/PDC/acetyl-CoA/metabolism pathway. It will also be important to determine as to whether variations in mitochondrial and/or cytoplasmic melatonin production, as influenced by AhR/CYP1B1, act to modulate the mitochondrial vitamin D receptor. This could suggest differential effects of the mitochondrial vitamin D receptor under conditions of altered mitochondrial function that is co-ordinated with variations in the AhR regulation of the melatonergic pathway. As such, the variations in mitochondrial melatonergic pathway may then act to differential regulate the effects of vitamin D. Likewise variations in the circadian, pineal melatonin production may modulate the vitamin D receptor effects over the circadian rhythm.

6. Circadian Rhythm and COVID-19

The above factors, viz the AhR, vitamin D, the gut microbiome, and especially pineal melatonin, are all intimately linked to the circadian rhythm. Although the circadian rhythm and circadian genes have long been associated with the regulation of viral infections [62,63], there is a relative paucity of data on the role of the circadian rhythm in the SARS-CoV-2 infection. Sleep-wake disruption is a SARS-CoV-2 infection severity risk factor, including for people with dementia and diabetes [64,65]. It is of note that elderly age is the major risk factor for severity/fatality in the COVID-19 pandemic. People over 80 years of age show a 10-fold decrease in the levels of pineal melatonin production, suggesting that the loss of pineal melatonin over ageing may contribute to the association of SARS-CoV-2 infection severity with age. Ageing associated factors may contribute to the loss of pineal melatonin, including via increased levels of pro-inflammatory cytokines, amyloid-β and circulating LPS, all of which may act to inhibit pineal melatonin production. The loss of pineal melatonin is important, as melatonin at night acts to reset the cells of the immune system, thereby better optimizing immune cell function during daytime, when the immune system is more likely to be under challenge.

Immunosenescence is widely used to explain the catastrophic changes that can arise in dementia and the general inability of the elderly to resist immune challenge [66]. Recent work has indicated that the loss of pineal melatonin may underpin immunosenescence via the lost/suppressed night-time resetting of immune cell metabolism [29]. All immune cells require the upregulation of glycolysis, coupled to maintained OXPHOS, in order to become activated. By shifting cells from glycolytic metabolism to OXPHOS, pineal melatonin dampens any lingering pro-inflammatory activity in the immune system, whilst optimizing immune cells for daytime challenge. This is achieved by melatonin's induction of the circadian gene, Bmal1, and SIRT1, leading to a melatonin/Bmal1/SIRT1/SIRT3/PDC/acetyl-CoA pathway, whereby acetyl-CoA not only increases ATP from the TCA cycle and OXPHOS, but also allows for the activation of the mitochondrial and cytoplasmic melatonergic pathway. The latter arises from acetyl-CoA being a necessary co-substrate for the initial enzyme in the melatonergic pathway, viz AANAT. Acetyl-CoA also dampens COX2-driven inflammatory activity by the acetylation and inhibition of COX2, as occurs with aspirin. Consequently, the loss of pineal melatonin will contribute to, if not drive, the changes arising in immunosenescence.

The efficacy of gut microbiome-derived butyrate is mediated via optimized mitochondrial function, arising from an increased PDC and associated melatonergic pathway activation [30]. As such, the effects of the gut microbiome-derived butyrate are intimately linked to the circadian alterations driven by pineal melatonin. Gut dysbiosis and associated increased gut permeability, by raising levels of the TLR4 ligands, LPS and HMGB1, will suppress pineal melatonin production [67]. Consequently, ageing-associated gut dysbiosis/permeability will contribute to the inhibition of pineal melatonin's optimization and resetting of immune cell metabolism. As noted, the effects of vitamin D at the vitamin D receptor may be significantly determined by variations in pineal and local melatonin production [60]. The other major circadian factor highlighted above, the AhR, is in negative reciprocal interactions with melatonin [68].

Overall, such data highlights the important interactions of COVID-19 regulatory factors, such as the AhR, vitamin D, butyrate, amyloid-β, and LPS, with pineal melatonin and the circadian rhythm regulation of the immune system.

7. Integrating Tryptophan Metabolism into COVID-19 Pathophysiology

As indicated throughout, variations in the regulation of tryptophan and its metabolites are intimately linked to key processes underpinning SARS-CoV-2 pathophysiology. The raised levels of pro-inflammatory cytokines during the initial 'cytokine' storm will drive the upregulation of IDO and its conversion of tryptophan to kynurenine. This has two important consequences, viz increasing AhR activation and decreasing the availability of tryptophan for the serotonergic and melatonergic pathways. As SARS-CoV-2 may increase AhR ligands independently of IDO induction, it is proposed that SARS-CoV-2, like other viral infections, may be associated with the upregulation of IL4I1, and thereby with the driving of tryptophan to the production of other AhR ligands, viz I3P and I3A [16]. See Figure 1.

Increased AhR activation will dysregulate the immune response, contributing to a heightened pro-inflammatory phase during the initial 'cytokine storm', whilst also suppressing the endogenous antiviral cell responses of NK cells and CD8+ T cells. Consequently, there is prolongation of heightened pro-inflammatory activity in the initial phase of infection, at least in part driven by the suppressed antiviral responses of NK cells and CD8+ T cells, which would typically control the immune response around 7 days after initial infection. It has been recognized that this dysregulated immune response is the major driver of SARS-CoV-2 severity and fatality.

SARS-CoV-2 severity risk factors, including elderly age, obesity, and T2D, prime for an altered immune response by a number of mechanisms. These conditions are all associated with an increase in AhR levels and activity as well as an increase in pro-inflammatory cytokines/IDO/kynurenine, leading to AhR activation. Diet is known to regulate IL4I1 [69], suggesting that dietary factors contributing to obesity and T2D may be acting in part via the upregulation of IL4I1 and I3P.

As well as regulating the immune response, AhR activation primes platelets for activation, coagulation, and thrombin formation, as does TLR4 activation that may arise from gut permeability-derived LPS and HMGB1. Gut dysbiosis/permeability and AhR activation are therefore important contributors to the association of activated platelets to SARS-CoV-2 fatality. The gut is intimately associated with the regulation of tryptophan and its metabolites. Many of the beneficial effects of butyrate are mediated by its upregulation of the melatonergic pathway, highlighting that the driving of tryptophan to the production of AhR ligands and away from serotonin, NAS, and melatonin production, will attenuate the beneficial effects of butyrate. This may be of particular relevance in immune cells, where butyrate's induction of PDC/acetyl-CoA/OXPHOS/TCA cycle/melatonergic pathway are important to a more optimized antiviral immune response by NK cells and CD8+ T cells.

The important role of the circadian rhythm may also be similarly modulated by variations in tryptophan regulation. Recent work would indicate that the importance of pineal melatonin to immune cell 'resetting' over the circadian rhythm may be mediated by the activation of the Bmal1/SIRT1/SIRT3/PDC/acetyl-CoA/OXPHOS/TCA cycle/melatonergic pathway [3]. The driving of tryptophan away from serotonin production will attenuate pineal melatonin's influence on the metabolism of immune cells. Importantly, AhR activation arising from the induction of kynurenine and I3P will increase AhR-induced CYP1B1, thereby backward converting melatonin to NAS, whilst the ability of AhR activation to suppress 14-3-3 [70] can prevent the activation of the AANAT and the melatonergic pathway. 14-3-3 is necessary for the stabilization of AANAT, indicating that AhR suppression of 14-3-3 may inhibit mitochondrial and cytoplasmic melatonin production. As the release and autocrine effects of melatonin are required to switch immune cells from an M1-like pro-inflammatory phenotype to an M2-like, anti-inflammatory, pro-phagocytic phenotype [47], the induction of tryptophan metabolites can significantly alter the nature of the immune response. As such, as well as modulating circadian regulation of the immune system, an increase in tryptophan metabolite-driven AhR activation can change the nature of the activation–deactivation processes in individual immune cells.

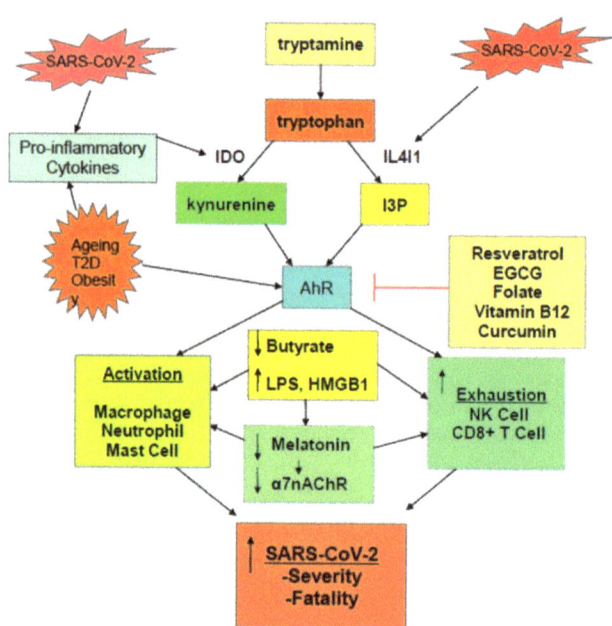

Figure 1. Scheme showing how the SARS-CoV-2 virus, and pre-existent high-risk medical conditions, shift tryptophan metabolism to increase AhR ligands. The activation of the AhR alters the nature of the initial 'cytokine storm' and suppresses the endogenous antiviral responses of NK cells and CD8+ T cells, leading to a prolonged activation of macrophages, neutrophils, and mast cells, as evident in severe SARS-CoV-2 infection. The driving to tryptophan to kynurenine and I3P, along with the elevated pro-inflammatory cytokines of the 'cytokine storm', also suppresses pineal melatonin production and therefore the induction of the α7nAChR by melatonin, thereby contributing lost vagal dampening of immune activity and raising sympathetic nervous system activation. The elevation in pro-inflammatory cytokines also increases gut dysbiosis/permeability, leading to a decrease in butyrate and raising LPS levels, further contributing to metabolic dysregulation of patterned immune responses to SARS-CoV-2 infection. A number of readily available nutraceuticals, such as resveratrol, EGCG, folate, vitamin B12, and curcumin, by AhR inhibition, may act to modulate how many processes influence SARS-CoV-2 pathophysiology.

There is a growing appreciation of the pathophysiological and treatment relevance of vitamin D in the SARS-CoV-2 infection. By regulating serotonin levels [46], vitamin D will modulate the serotonin-melatonergic pathway, whilst the suppression of pro-inflammatory cytokine production by vitamin D [55] will attenuate cytokine-induced IDO and the pro-inflammatory route for activation of the AhR by kynurenine. As noted, melatonin can bind the vitamin D receptor, thereby potentiating some vitamin D-driven transcription [60]. The suggests that attenuation of the tryptophan/serotonin/melatonin pathway by IDO, TDO, and IL4I1 can suppress vitamin D-driven transcription. As such, the effects of vitamin D in the SARS-CoV-2 infection may be intimately linked to alterations in tryptophan metabolism.

8. Future Research

Does SARS-CoV-2 increase IL4I1, and therefore the production of the AhR ligands, I3P and I3A?

Preclinical data show that prenatal/early postnatal exposure to AhR ligands leads to a reprogramming of the CD4+ and CD8+ T cell responses to viral infection, via the long-term maintenance of alterations in DNA methylation [71,72]. This is also supported by human data [73]. This could suggest that the upregulation of tryptophan-derived ligands, including kynurenine and I3P, in early development will modulate cytolytic cell responses to later infection. The relevance of AhR regulation of tryptophan metabolites in such processes, including via CYP1B1 regulation of the melatonergic pathway, will be important to determine. As many high-risk medical conditions for severe/fatal SARS-CoV-2 infection, including obesity and T2D [74], can have an early developmental etiology, the developmental overlaps of these conditions with alterations in responses to viral infection in later life will be important to determine.

Recent data show that even small elevations in plasma glucose associate with SARS-CoV-2 infection severity/fatality [75], indicating a role regulation of the gut microbiome within the context of the gut–liver and gut–brain axes [76], in modulating SARS-CoV-2 severity. The roles of the tryptophan metabolites, kynurenine and I3P, in the activation of the AhR in driving such glucose dysregulation [77], in concert with AhR-mediated dysregulation of the immune response, will be important to determine.

As to how the tryptophan metabolites interact with the alpha 7 nicotinic acetylcholine receptor (α7nAChR) in SARS-CoV-2 pathophysiology will be important to determine. The conversion of kynurenine to kynurenic acid (KYNA) may contribute to KYNA antagonism of the α7nAChR, as well as N-methyl-d-aspartate (NMDA) receptor antagonism. The α7nAChR can significantly inhibit pulmonary viral infections via effects in pulmonary epithelial cells as well as in the regulation of the immune response, reviewed in [29]. Importantly, the α7nAChR is regulated by melatonin [78], suggesting that the suppression of pineal, and perhaps local, melatonin in SARS-CoV-2 infection may contribute to a decrease in α7nAChR activation in pulmonary epithelial cells, and the immune system α7nAChR level and activity in SARS-CoV-2 infection will contribute to inflammation during the initial 'cytokine storm' [29,79]. The regulation of tryptophan metabolism may then directly modulate the α7nAChR via KYNA, which also activates the AhR, as well as via melatonin.

The α7nAChR is an important mediator of vagal nerve activity, and therefore in the sympathetic/parasympathetic balance of the autonomic nervous system. Many of the high-risk conditions associated with SARS-CoV-2 infection severity/fatality have heightened levels of sympathetic nervous system activity [80,81], which may be contributed by increased KYNA and decreased melatonin that suppresses α7nAChR activity and levels, respectively. The effects of KYNA and decreased melatonin on the α7nAChR may therefore be an aspect of high-risk medical conditions per se, as well as being driven by SARS-CoV-2 infection via tryptophan metabolite regulation. Clearly, disentangling the effects of α7nAChR activation from research focused on promoting anti-cigarette smoking [82] will be important for future research. It is also important that this is looked at in human cells, given the unique human duplicant dupα7 (CHRFAM7A), which negatively regulates the α7nAChR, and which can be independently regulated by different factors, including LPS, reviewed in [29]. As such, measurement of the role duplicant dupα7 (CHRFAM7A) in different experimental protocols will be necessary in order to clarify α7nAChR effects, if any.

The presence of pulmonary embolism in COVID-19 patients is high, with one Spanish study of consecutive hospitalized patients showing 35.6% of patients having a pulmonary embolism [83]. The role of I3P and kynurenine, via AhR activation, in priming platelets for elevated activity, coagulation, and thrombin production in severe SARS-CoV-2 infection [84], will be important to determine.

Given the paucity of SARS-CoV-2 data pertaining to many of the processes highlighted above, it is clear that many future research directions may be indicated. For example, the well-proven role of vitamin D in the regulation of immune responses may interact with the regulation of serotonin and melatonin production, as indicated above. The interactions of vitamin D with experiences of racial discrimination in modulating SARS-CoV-2 pathophysiology will be important to determine, including in tropical countries such as Brazil, where heightened levels of racial discrimination stress are evident. As such, vitamin D levels in equatorial countries may be in interaction with other important processes that occlude a simple interpretation of how vitamin D variations may interact with SARS-CoV-2 susceptibility and severity. As stressors, including possibly racial discrimination stress, can increase gut permeability and dysregulate the gut microbiome, whilst vitamin D also regulates gut permeability, it is clear that indicants of how vitamin D may modulate SARS-CoV-2 infection susceptibility and severity will need the recording of wider body systems. The acquisition of such data should provide a more detailed basis as to how tryptophan metabolites and the AhR interact with wider body systems.

9. Treatment Implications

Targeting aspects of tryptophan metabolism may have important prophylactic and treatment implications for the management of SARS-CoV-2 infection, particularly as the emergence of new variants would suggest that this virus is likely to have impacts for years, if not decades. Clearly, the suppression of infectivity and symptomatology by nutraceuticals is preferable to treatment with pharmaceuticals, given the lack of any efficacy of the four WHO recommended pharmaceuticals at the beginning of the COVID-19 pandemic [2].

10. Prophylactic Treatment

A number of studies have shown that people taking melatonin have up to a 64% decrease in the likelihood of getting SARS-CoV-2 infection, or have had no conscious symptoms [85]. The optimization of night-time melatonin production by 2–10 mg melatonin, about 20 min before bed-time is a readily achievable intervention. As melatonin is a natural product with no relevant side-effects and is widely used by millions of people worldwide, the utilization of melatonin as a prophylactic would not require extensive safety trials. The dramatic loss of pineal melatonin in the elderly and the proven utility of melatonin in the management of dementia, obesity, T2D would indicate that melatonin may have wider clinical benefits for people with high-risk conditions for SARS-CoV-2 severity/fatality.

Likewise, utilizing vitamin D supplements will have immune and gut microbiome/barrier integrity benefits as well as suppressing the pro-inflammatory cytokine response that drives tryptophan to the production of AhR ligands.

A number of nutraceuticals afford antagonism at the AhR, including green tea and its polyphenol, epigallocatechin gallate (EGCG). EGCG has also been shown to decrease SARS-CoV-2 entry into cells in vitro, with over two-times the efficacy of WHO recommended, Lopinavir [86,87].

Other AhR antagonists include resveratrol, vitamin B12, folate, and curcumin, all of which have wider health benefits, including via the regulation of immune function [16].

As melatonin may mediate some of its effects via increased α7nAChR, with the α7nAChR activation having possible prophylactic benefits in pulmonary epithelial cells, as in other viral infections [84], nicotine and/or pharmaceutical α7nAChR agonists, may have prophylactic utility. Data shows nicotine to have significant impacts on pulmonary epithelial cells, which could indicate some relevant impacts on SARS-CoV-2 infection [82], although it is highly likely that in vivo the effects of nAChR agonism will be in both immune cells and pulmonary epithelial cell interactions [88].

Pre- and pro-biotics, as well as dietary alterations, can slowly improve the α-diversity of gut microbiome, thereby increasing butyrate production. However, the utilization of the nutraceutical, sodium butyrate, will achieve optimal butyrate levels more quickly, whilst also increasing the production of butyrate-producing gut bacteria.

As with melatonin, such nutraceutical supplements will have wider benefits on the pathophysiology of SARS-CoV-2 high risk medical conditions, such as obesity and T2D.

11. Treating SARS-CoV-2 Infection

There is a growing appreciation of the potential clinical utility of melatonin in the treatment of SARS-CoV-2 infection. This has been indicated by data showing that patients needing intubation for SARS-CoV-2 infection are more likely to survive if they have been taking melatonin [89]. Such data highlight how the prophylactic use of melatonin, if unsuccessful in preventing infection, may still prove beneficial by decreasing the severity of any subsequent SARS-CoV-2 infection.

The results of an ongoing pilot study, looking at the utility of intravenous melatonin at 5–8 mg/kg body weight, in the treatment of severe SARS-CoV-2 infection [90], should clarify as to whether melatonin has treatment utility in the management of SARS-CoV-2 infection. Given the well-proven anti-inflammatory effects of melatonin, and its utilization by the body over the course of evolution to reset immune cells over the circadian rhythm as well as switching immune cells from and M1-like to an M2-like anti-inflammatory phenotype via its autocrine effects [47], it is likely that such high doses of melatonin will modulate the course of severe SARS-CoV-2 infection. Unlike dexamethasone, melatonin does not totally suppress the antiviral NK cells and CD8+ T cells, but rather seems to increase their cytotoxicity and antiviral efficacy, it may be that melatonin will have more utility than the limited efficacy of dexamethasone. Future research should clarify this in the near future, as well as to whether melatonin would have any impact on mitochondrial AhR regulation of mitochondrial transcription [25,91].

Melatonin also has negative reciprocal interactions with the AhR, suggesting that it may modulate some of the consequences of tryptophan metabolism dysregulation. However, it may be that more direct inhibition of the AhR over the course of SARS-CoV-2 infection will prove useful. For example, the dampening effects of melatonin on the initial 'cytokine storm' may be paired with AhR antagonism, e.g., via EGCG or resveratrol supplementation, in order to prevent the AhR antagonism/exhaustion in NK cells and CD8+ T cells. If of utility, the temporal utilization and dose of such AhR antagonists would have to be ascertained.

The utility of sodium butyrate, via its HDAC inhibitory capacity, induction of PDC and the melatonergic pathway [30], as well as its ability to increase the levels and cytotoxicity of NK cells requires investigation [27], including as to route and dose of administration.

12. Conclusions

Accumulating data on the pathophysiological changes driven by SARS-CoV-2 infection indicate an important role for variations in the metabolism of tryptophan. Pro-inflammatory cytokine induction of IDO, leads to kynurenine activation of the AhR, which can significantly suppress the endogenous antiviral responses of NK cells and CD8+ T cells, whilst dysregulating the initial pro-inflammatory cytokine responses of macrophages, neutrophils, and mast cells. Such processes are relevant in many other medical conditions, including as to how age associates with an increased susceptibility to a range of diverse medical conditions. As data show SARS-CoV-2 to have IDO/kynurenine-independent activation of the AhR, it is proposed that the SARS-CoV-2 virus, like some other viruses, activates IL4I1, thereby increasing the production of other AhR, such as I3P. As such, SARS-CoV-2 viral infection may have a number of routes to activate the AhR. The utilization of tryptophan to increase kynurenine in SARS-CoV-2 infection decreases the production of serotonin, and therefore suppresses the availability of serotonin as a necessary precursor for the melatonergic pathway. The cytokine suppression of pineal melatonin and the AhR-induced CYP1B1 suppression of local melatonin, leads to suboptimal melatonin regulation of mitochondrial metabolism, which underpins the dysregulated immune response to SARS-CoV-2 infection. This is further confounded by pro-inflammatory cytokine induced gut dysbiosis/permeability that suppresses butyrate and increases circulating LPS. Such a SARS-CoV-2 dysregulated immune responses, aided by pre-existent changes in high risk medical conditions, provide readily achievable treatment targets, including melatonin and

AhR antagonists, such as EGCG, resveratrol, and curcumin, both as prophylactic and in the course of treatment of severe SARS-CoV-2 infection.

Author Contributions: G.A. conceived and wrote the article; G.M. and A.C. reviewed the scientific literature and contributed to article reviewing and figure drawing. All authors have read and agreed to the published version of the manuscript.

Funding: This research was funded by the "5 × 1000" voluntary contribution and by a grant from the Italian Ministry of Health (Ricerca Corrente 2020-2021) to GM. The APC was funded by the Italian Ministry of Health (Ricerca Corrente 2020) to GM.

Conflicts of Interest: The authors declare that there are no conflict of interest with respect to the authorship and/or publication of this article.

Abbreviations

α7nAChR	alpha 7 nicotinic acetylcholine receptor
AANAT	aralkylamine N-acetyltransferase
AhR	aryl hydrocarbon receptor
ASMT	acetylserotonin methyltransferase
CYP	cytochrome P450
dupα7	duplicating alpha 7nicotinic receptor
EGCG	epigallocatechin gallate
FICZ	6-formylindolo(3,2-b)carbazole
HDAC	histone deacetylase
HMGB	high-mobility group box
HPA	hypothalamic-pituitary-adrenal
I3A	indole-3-acetaldehyde
I3P	indole-3-pyruvate
IDO	indoleamine 2,3-dioxygenase
IFN	interferon
IL	interleukin
IL4I1	interleukin 4-inducible 1
KYNA	kynurenic acid
LPS	lipopolysaccharide
nAChR	nicotinic acetylcholine receptor
NAS	N-acetylserotonin
NK	natural killer cell
OXPHOS	oxidative phosphorylation
PDC	pyruvate dehydrogenase complex
SARS-CoV-2	severe acute respiratory disease, coronavirus 2
T2D	type II diabetes
TCA	tricarboxylic acid
TDO	tryptophan 2,3-dioxygenase
Th	T-helper
TLR	toll-like receptor
TNF	tumor necrosis factor

References

1. Anderson, G.; Carbone, A.; Mazzoccoli, G. Aryl Hydrocarbon Receptor Role in Co-Ordinating SARS-CoV-2 Entry and Symptomatology: Linking Cytotoxicity Changes in COVID-19 and Cancers; Modulation by Racial Discrimination Stress. *Biology* **2020**, *9*, 249. [CrossRef] [PubMed]
2. WHO Solidarity Trial Consortium; Pan, H.; Peto, R.; Henao-Restrepo, A.M.; Preziosi, M.P.; Sathiyamoorthy, V.; Abdool Karim, Q.; Alejandria, M.M.; Hernández García, C.; Kieny, M.P.; et al. Repurposed Antiviral Drugs for Covid-19—Interim WHO Solidarity Trial Results. *N. Engl. J. Med.* **2020**. [CrossRef]
3. Anderson, G. Tumour Microenvironment: Roles of the Aryl Hydrocarbon Receptor, O-GlcNAcylation, Acetyl-CoA and Melatonergic Pathway in Regulating Dynamic Metabolic Interactions across Cell Types—Tumour Microenvironment and Metabolism. *Int. J. Mol. Sci.* **2020**, *22*, 141. [CrossRef] [PubMed]
4. Iqubal, A.; Hoda, F.; Najmi, A.K.; Haque, S.E. Macrophage Activation and Cytokine Release Syndrome in COVID-19: Current Updates and Analysis of Repurposed and Investigational Anti-Cytokine Drugs. *Drug Res.* **2021**. [CrossRef]

5. Li, M.; Guo, W.; Dong, Y.; Wang, X.; Dai, D.; Liu, X.; Wu, Y.; Li, M.; Zhang, W.; Zhou, H.; et al. Elevated Exhaustion Levels of NK and CD8+ T Cells as Indicators for Progression and Prognosis of COVID-19 Disease. *Front. Immunol.* **2020**, *11*, 580237. [CrossRef]
6. Liu, Y.; Liang, X.; Dong, W.; Fang, Y.; Lv, J.; Zhang, T.; Fiskesund, R.; Xie, J.; Liu, J.; Yin, X.; et al. Tumor-Repopulating Cells Induce PD-1 Expression in CD8+ T Cells by Transferring Kynurenine and AhR Activation. *Cancer Cell* **2018**, *33*, 480–494. [CrossRef]
7. Anderson, G. Local Melatonin Regulates Inflammation Resolution: A Common Factor in Neurodegenerative, Psychiatric and Systemic Inflammatory Disorders. *CNS Neurol. Disord. Drug Targets* **2014**, *13*, 817–827. [CrossRef]
8. Morris, G.; Anderson, G.; Maes, M. Hypothalamic-Pituitary-Adrenal Hypofunction in Myalgic Encephalomyelitis (ME)/Chronic Fatigue Syndrome (CFS) as a Consequence of Activated Immune-Inflammatory and Oxidative and Nitrosative Pathways. *Mol. Neurobiol.* **2016**, *54*, 6806–6819. [CrossRef]
9. Le Naour, J.; Galluzzi, L.; Zitvogel, L.; Kroemer, G.; Vacchelli, E. Trial watch: IDO inhibitors in cancer therapy. *OncoImmunology* **2020**, *9*, 1777625. [CrossRef]
10. Souza, L.C.; Jesse, C.R.; Antunes, M.S.; Ruff, J.R.; Espinosa, D.D.O.; Gomes, N.S.; Donato, F.; Giacomeli, R.; Boeira, S.P. Indoleamine-2,3-dioxygenase mediates neurobehavioral alterations induced by an intracerebroventricular injection of amyloid-β1-42 peptide in mice. *Brain Behav. Immun.* **2016**, *56*, 363–377. [CrossRef] [PubMed]
11. Filho, A.J.M.C.; Lima, C.N.C.; Vasconcelos, S.M.M.; De Lucena, D.F.; Maes, M.; Macedo, D.S. IDO chronic immune activation and tryptophan metabolic pathway: A potential pathophysiological link between depression and obesity. *Prog. Neuro-Psychopharmacol. Biol. Psychiatry* **2018**, *80*, 234–249. [CrossRef]
12. Williams, R.O. Exploitation of the IDO Pathway in the Therapy of Rheumatoid Arthritis. *Int. J. Tryptophan Res.* **2013**, *6*, 67–73. [CrossRef] [PubMed]
13. Thomas, T.; Stefanoni, D.; Reisz, J.A.; Nemkov, T.; Bertolone, L.; Francis, R.O.; Hudson, K.E.; Zimring, J.C.; Hansen, K.C.; Hod, E.A.; et al. COVID-19 infection alters kynurenine and fatty acid metabolism, correlating with IL-6 levels and renal status. *JCI Insight* **2020**, *5*, 140327. [CrossRef]
14. Feng, M.; Xie, T.; Li, Y.; Zhang, N.; Lu, Q.; Zhou, Y.; Shi, M.; Sun, J.; Zhang, X. A balanced game: Chicken macrophage response to ALV-J infection. *Vet. Res.* **2019**, *50*, 20. [CrossRef] [PubMed]
15. Hu, X.; Chen, S.; Jia, C.; Xue, S.; Dou, C.; Dai, Z.; Xu, H.; Sun, Z.; Geng, T.; Cui, H. Gene expression profile and long non-coding RNA analysis, using RNA-Seq, in chicken embryonic fibroblast cells infected by avian leukosis virus. *J. Arch. Virol.* **2017**, *163*, 639–647. [CrossRef] [PubMed]
16. Turski, W.A.; Wnorowski, A.; Turski, G.N.; Turski, C.A.; Turski, L. AhR and IDO1 in pathogenesis of Covid-19 and the "Systemic AhR Activation Syndrome:" a translational review and therapeutic perspectives. *Restor. Neurol. Neurosci.* **2020**, *38*, 343–354. [CrossRef]
17. Elsheimer-Matulova, M.; Polansky, O.; Seidlerova, Z.; Varmuzova, K.; Štěpánová, H.; Fedr, R.; Rychlik, I. Interleukin 4 inducible 1 gene (IL4I1) is induced in chicken phagocytes by Salmonella Enteritidis infection. *Vet. Res.* **2020**, *51*, 1–8. [CrossRef] [PubMed]
18. Boulland, M.-L.; Marquet, J.; Molinier-Frenkel, V.; Möller, P.; Guiter, C.; Lasoudris, F.; Copie-Bergman, C.; Baia, M.; Gaulard, P.; Leroy, K.; et al. Human IL4I1 is a secreted l-phenylalanine oxidase expressed by mature dendritic cells that inhibits T-lymphocyte proliferation. *Blood* **2007**, *110*, 220–227. [CrossRef]
19. Aubatin, A.; Sako, N.; Decrouy, X.; Donnadieu, E.; Molinier-Frenkel, V.; Castellano, F. IL4-induced gene 1 is secreted at the immune synapse and modulates TCR activation independently of its enzymatic activity. *Eur. J. Immunol.* **2018**, *48*, 106–119. [CrossRef]
20. Chowdhury, G.; Dostalek, M.; Hsu, E.L.; Nguyen, L.P.; Stec, D.F.; Bradfield, C.A.; Guengerich, F.P. Structural Identification of Diindole Agonists of the Aryl Hydrocarbon Receptor Derived from Degradation of Indole-3-pyruvic Acid. *Chem. Res. Toxicol.* **2009**, *22*, 1905–1912. [CrossRef]
21. Aoki, R.; Aoki-Yoshida, A.; Suzuki, C.; Takayama, Y. Indole-3-Pyruvic Acid, an Aryl Hydrocarbon Receptor Activator, Suppresses Experimental Colitis in Mice. *J. Immunol.* **2018**, *201*, 3683–3693. [CrossRef]
22. Smirnova, A.; Wincent, E.; Bergander, L.V.; Alsberg, T.; Bergman, J.; Rannug, A.; Rannug, U. Evidence for New Light-Independent Pathways for Generation of the Endogenous Aryl Hydrocarbon Receptor Agonist FICZ. *Chem. Res. Toxicol.* **2015**, *29*, 75–86. [CrossRef] [PubMed]
23. Scott, S.A.; Fu, J.; Chang, P.V. Microbial tryptophan metabolites regulate gut barrier function via the aryl hydrocarbon receptor. *Proc. Natl. Acad. Sci.* **2020**, *117*, 19376–19387. [CrossRef] [PubMed]
24. Anderson, G.; Mazzoccoli, G. Left Ventricular Hypertrophy: Roles of Mitochondria CYP1B1 and Melatonergic Pathways in Co-Ordinating Wider Pathophysiology. *Int. J. Mol. Sci.* **2019**, *20*, 4068. [CrossRef] [PubMed]
25. Anderson, G.; Maes, M. Interactions of Tryptophan and Its Catabolites with Melatonin and the Alpha 7 Nicotinic Receptor in Central Nervous System and Psychiatric Disorders: Role of the Aryl Hydrocarbon Receptor and Direct Mitochondria Regulation. *Int. J. Tryptophan Res.* **2017**, *10*, 1178646917691738. [CrossRef]
26. Zuo, T.; Zhang, F.; Lui, G.C.; Yeoh, Y.K.; Li, A.Y.; Zhan, H.; Wan, Y.; Chung, A.C.; Cheung, C.P.; Chen, N.; et al. Alterations in Gut Microbiota of Patients With COVID-19 During Time of Hospitalization. *Gastroenterology* **2020**, *159*, 944–955. [CrossRef] [PubMed]
27. Ma, X.; Zhou, Z.; Zhang, X.; Fan, M.; Hong, Y.; Feng, Y.; Dong, Q.-H.; Diao, H.; Wang, G. Sodium butyrate modulates gut microbiota and immune response in colorectal cancer liver metastatic mice. *Cell Biol. Toxicol.* **2020**, *36*, 509–515. [CrossRef] [PubMed]
28. Li, J.; Richards, E.M.; Handberg, E.M.; Pepine, C.J.; Raizada, M.K. Butyrate Regulates COVID-19–Relevant Genes in Gut Epithelial Organoids from Normotensive Rats. *Hypertension* **2021**, *77*, 13–16. [CrossRef] [PubMed]

29. Anderson, G.; Reiter, R.J. Melatonin: Roles in influenza, Covid-19, and other viral infections. *Rev. Med. Virol.* **2020**, *30*, e2109. [CrossRef]
30. Jin, C.J.; Engstler, A.J.; Sellmann, C.; Ziegenhardt, D.; Landmann, M.; Kanuri, G.; Lounis, H.; Schröder, M.; Vetter, W.; Bergheim, I. Sodium butyrate protects mice from the development of the early signs of non-alcoholic fatty liver disease: Role of melatonin and lipid peroxidation. *Br. J. Nutr.* **2016**, *23*, 1–12. [CrossRef]
31. Shi, Z.; Gewirtz, A.T. Together Forever: Bacterial–Viral Interactions in Infection and Immunity. *Viruses* **2018**, *10*, 122. [CrossRef]
32. Chen, Y.; Sun, H.; Bai, Y.; Zhi, F. Gut dysbiosis-derived exosomes trigger hepatic steatosis by transiting HMGB1 from intestinal to liver in mice. *Biochem. Biophys. Res. Commun.* **2019**, *509*, 767–772. [CrossRef]
33. Perrin-Cocon, L.; Aublin-Gex, A.; Sestito, S.E.; Shirey, K.A.; Patel, M.C.; André, P.; Blanco, J.C.; Vogel, S.N.; Peri, F.; Lotteau, V. TLR4 antagonist FP7 inhibits LPS-induced cytokine production and glycolytic reprogramming in dendritic cells, and protects mice from lethal influenza infection. *Sci. Rep.* **2017**, *7*, 40791. [CrossRef] [PubMed]
34. Bennouar, S.; Cherif, A.B.; Kessira, A.; Bennouar, D.-E.; Abdi, S. Vitamin D Deficiency and Low Serum Calcium as Predictors of Poor Prognosis in Patients with Severe COVID-19. *J. Am. Coll. Nutr.* **2021**, 1–11. [CrossRef]
35. Katz, J.; Yue, S.; Xue, W. Increased risk for COVID-19 in patients with vitamin D deficiency. *Nutrition* **2020**, *84*, 111106. [CrossRef]
36. Liu, N.; Sun, J.; Wang, X.; Zhang, T.; Zhao, M.; Li, H. Low vitamin D status is associated with coronavirus disease 2019 outcomes: A systematic review and meta-analysis. *Int. J. Infect. Dis.* **2021**, *104*, 58–64. [CrossRef] [PubMed]
37. Rubin, R. Sorting Out Whether Vitamin D Deficiency Raises COVID-19 Risk. *JAMA* **2021**, *325*, 329. [CrossRef] [PubMed]
38. Clark, E.C.; Masoli, J.; Warren, F.C.; Soothill, J.; Campbell, J.L. Vitamin D and COVID-19 in older age: Evidence versus expectations. *Br. J. Gen. Pr.* **2021**, *71*, 10–11. [CrossRef]
39. Balogh, G.; De Boland, A.R.; Boland, R.; Barja, P. Effect of 1,25(OH)2-Vitamin D3 on the Activation of Natural Killer Cells: Role of Protein Kinase C and Extracellular Calcium. *Exp. Mol. Pathol.* **1999**, *67*, 63–74. [CrossRef]
40. Vassiliou, A.G.; Jahaj, E.; Pratikaki, M.; Keskinidou, C.; Detsika, M.; Grigoriou, E.; Psarra, K.; Orfanos, S.E.; Tsirogianni, A.; Dimopoulou, I.; et al. Vitamin D deficiency correlates with a reduced number of natural killer cells in intensive care unit (ICU) and non-ICU patients with COVID-19 pneumonia. *Hell. J. Cardiol.* **2020**. [CrossRef]
41. Kloc, M.; Ghobrial, R.M.; Lipińska-Opałka, A.; Wawrzyniak, A.; Zdanowski, R.; Kalicki, B.; Kubiak, J.Z. Effects of vitamin D on macrophages and myeloid-derived suppressor cells (MDSCs) hyperinflammatory response in the lungs of COVID-19 patients. *Cell Immunol.* **2021**, *360*, 104259. [CrossRef] [PubMed]
42. Consiglio, M.; Destefanis, M.; Morena, D.; Foglizzo, V.; Forneris, M.; Pescarmona, G.; Silvagno, F. The Vitamin D Receptor Inhibits the Respiratory Chain, Contributing to the Metabolic Switch that Is Essential for Cancer Cell Proliferation. *PLoS ONE* **2014**, *9*, e115816. [CrossRef]
43. Heras, N.D.L.; Giménez, V.M.M.; Ferder, L.; Manucha, W.; Lahera, V. Implications of Oxidative Stress and Potential Role of Mitochondrial Dysfunction in COVID-19: Therapeutic Effects of Vitamin D. *Antioxidants* **2020**, *9*, 897. [CrossRef]
44. Silvagno, F.; De Vivo, E.; Attanasio, A.; Gallo, V.; Mazzucco, G.; Pescarmona, G. Mitochondrial localization of vitamin D receptor in human platelets and differentiatedmegakaryocytes. *PLoS ONE* **2010**, *5*, e8670. [CrossRef] [PubMed]
45. Ricca, C.; Aillon, A.; Bergandi, L.; Alotto, D.; Castagnoli, C.; Silvagno, F. Vitamin D Receptor Is Necessary for Mitochondrial Function and Cell Health. *Int. J. Mol. Sci.* **2018**, *19*, 1672. [CrossRef] [PubMed]
46. Kaneko, I.; Sabir, M.S.; Dussik, C.M.; Whitfield, G.K.; Karrys, A.; Hsieh, J.-C.; Haussler, M.R.; Meyer, M.B.; Pike, J.W.; Jurutka, P.W. 1,25-Dihydroxyvitamin D regulates expression of the tryptophan hydroxylase 2 and leptin genes: Implication for behavioral influences of vitamin D. *FASEB J.* **2015**, *29*, 4023–4035. [CrossRef]
47. Muxel, S.M.; Pires-Lapa, M.A.; Monteiro, A.W.A.; Cecon, E.; Tamura, E.K.; Floeter-Winter, L.M.; Markus, R.P. NF-κB Drives the Synthesis of Melatonin in RAW 264.7 Macrophages by Inducing the Transcription of the Arylalkylamine-N-Acetyltransferase (AA-NAT) Gene. *PLoS ONE* **2012**, *7*, e52010. [CrossRef]
48. Wang, J.; Xu, Y.; Zhang, X.; Wang, S.; Peng, Z.; Guo, J.; Jiang, H.; Liu, J.; Xie, Y.; Wang, J.; et al. Leptin correlates with monocytes activation and severe condition in COVID-19 patients. *J. Leukoc. Biol.* **2021**. [CrossRef]
49. Oswald, J.; Büttner, M.; Jasinski-Bergner, S.; Jacobs, R.; Rosenstock, P.; Kielstein, H. Leptin affects filopodia and cofilin in NK-92 cells in a dose- and time-dependent manner. *Eur. J. Histochem.* **2018**, *62*, 2848. [CrossRef]
50. Correale, J.; Ysrraelit, M.C.; Gaitán, M.I. Immunomodulatory effects of Vitamin D in multiple sclerosis. *Brain* **2009**, *132*, 1146–1160. [CrossRef] [PubMed]
51. Kalfaoglu, B.; Almeida-Santos, J.; Tye, C.A.; Satou, Y.; Ono, M. T-Cell Hyperactivation and Paralysis in Severe COVID-19 Infection Revealed by Single-Cell Analysis. *Front. Immunol.* **2020**, *11*, 589380. [CrossRef]
52. Singh, P.; Rawat, A.; Alwakeel, M.; Sharif, E.; Al Khodor, S. The potential role of vitamin D supplementation as a gut microbiota modifier in healthy individuals. *Sci. Rep.* **2020**, *10*, 1–14. [CrossRef] [PubMed]
53. Lee, P.; Hsieh, Y.; Huo, T.; Yang, U.; Lin, C.; Li, C.; Huang, Y.; Hou, M.; Lin, H.; Lee, K. Active Vitamin D 3 Treatment Attenuated Bacterial Translocation via Improving Intestinal Barriers in Cirrhotic Rats. *Mol. Nutr. Food Res.* **2020**, e2000937. [CrossRef]
54. Battistini, C.; Ballan, R.; Herkenhoff, M.E.; Saad, S.M.I.; Sun, J. Vitamin D Modulates Intestinal Microbiota in Inflammatory Bowel Diseases. *Int. J. Mol. Sci.* **2020**, *22*, 362. [CrossRef] [PubMed]
55. Li, H.; Li, W.; Wang, Q. 1,25-dihydroxyvitamin D3 suppresses lipopolysaccharide-induced interleukin-6 production through aryl hydrocarbon receptor/nuclear factor-κB signaling in oral epithelial cells. *BMC Oral Health* **2019**, *19*, 236. [CrossRef]

56. Slominski, A.T.; Kim, T.-K.; Janjetovic, Z.; Brożyna, A.A.; Zmijewski, M.A.; Xu, H.; Sutter, T.R.; Tuckey, R.C.; Jetten, A.M.; Crossman, D. Differential and Overlapping Effects of 20,23(OH)$_2$D$_3$ and 1,25(OH)$_2$D$_3$ on Gene Expression in Human Epidermal Keratinocytes: Identification of AhR as an Alternative Receptor for 20,23(OH)$_2$D$_3$. *Int. J. Mol. Sci.* **2018**, *19*, 3072. [CrossRef] [PubMed]
57. Matsunawa, M.; Akagi, D.; Uno, S.; Endo-Umeda, K.; Yamada, S.; Ikeda, K.; Makishima, M. Vitamin D Receptor Activation Enhances Benzo[a]pyrene Metabolism via CYP1A1 Expression in Macrophages. *Drug Metab. Dispos.* **2012**, *40*, 2059–2066. [CrossRef] [PubMed]
58. Chen, W.-Y.; Weng, J.-H.; Huang, C.-C.; Chung, B.-C. Histone Deacetylase Inhibitors Reduce Steroidogenesis through SCF-Mediated Ubiquitination and Degradation of Steroidogenic Factor 1 (NR5A1). *Mol. Cell. Biol.* **2007**, *27*, 7284–7290. [CrossRef]
59. Jin, U.-H.; Cheng, Y.; Park, H.; Davidson, L.A.; Callaway, E.S.; Chapkin, R.S.; Jayaraman, A.; Asante, A.; Allred, C.; Weaver, E.A.; et al. Short Chain Fatty Acids Enhance Aryl Hydrocarbon (Ah) Responsiveness in Mouse Colonocytes and Caco-2 Human Colon Cancer Cells. *Sci. Rep.* **2017**, *7*, 1–12. [CrossRef]
60. Fang, N.; Hu, C.; Sun, W.; Xu, Y.; Gu, Y.; Wu, L.; Peng, Q.; Reiter, R.J.; Liu, L.-F. Identification of a novel melatonin-binding nuclear receptor: Vitamin D receptor. *J. Pineal Res.* **2019**, *68*, 12618. [CrossRef] [PubMed]
61. Williams, W.R. Dampening of neurotransmitter action: Molecular similarity within the melatonin structure. *Endocr. Regul.* **2018**, *52*, 199–207. [CrossRef] [PubMed]
62. Borrmann, H.; McKeating, J.A.; Zhuang, X. The Circadian Clock and Viral Infections. *J. Biol. Rhythm.* **2020**. [CrossRef] [PubMed]
63. Squibb, R.L. Interrelationship of Lighting Regimen and Virus Infection to Diurnal Rhythms in Liver Components Associated with Protein Metabolism in the Chick. *J. Nutr.* **1968**, *95*, 357–362. [CrossRef]
64. Barone, M.T.U.; Ngongo, B.; Menna-Barreto, L. Sleep-wake cycle impairment adding on the risk for COVID-19 severity in people with diabetes. *Sleep Sci.* **2020**, *13*, 191–194. [CrossRef] [PubMed]
65. Roccaro, I.; Smirni, D. Fiat Lux: The Light Became Therapy. An Overview on the Bright Light Therapy in Alzheimer's Disease Sleep Disorders. *J. Alzheimer's Dis.* **2020**, *77*, 113–125. [CrossRef] [PubMed]
66. Zhao, Y.; Zhan, J.-K.; Liu, Y. A Perspective on Roles Played by Immunosenescence in the Pathobiology of Alzheimer's Disease. *Aging Dis.* **2020**, *11*, 1594–1607. [CrossRef]
67. Barbosa-Lima, L.E.; Muxel, S.M.; Kinker, G.S.; Carvalho-Sousa, C.E.; Cruz-Machado, S.D.S.; Markus, R.P.; Fernandes, P.A.C.M.; Cruz-Machado, S.S. STAT1-NFκB crosstalk triggered by interferon gamma regulates noradrenaline-induced pineal hormonal production. *J. Pineal Res.* **2019**, *67*, e12599. [CrossRef]
68. Chang, T.K.H.; Chen, J.; Yang, G.; Yeung, E.Y.H. Inhibition of procarcinogen-bioactivating human CYP1A1, CYP1A2 and CYP1B1 enzymes by melatonin. *J. Pineal Res.* **2010**, *48*, 55–64. [CrossRef]
69. Arpon, A.; Riezu-Boj, J.I.; Milagro, F.I.; Marti, A.; Razquin, C.; Martinez-Gonzalez, M.A.; Corella, D.; Estruch, R.; Casas, R.; Fito, M.; et al. Adherence to Mediterranean diet is associated with methylation changes in inflammation-related genes in peripheral blood cells. *J. Physiol. Biochem.* **2016**, *73*, 445–455. [CrossRef]
70. Zhao, Y.; Fu, Y.; Sun, Y.; Zou, M.; Peng, X. Transcriptional Regulation of gga-miR-451 by AhR:Arnt in Mycoplasma gallisepticum (HS Strain) Infection. *Int. J. Mol. Sci.* **2019**, *20*, 3087. [CrossRef]
71. Burke, C.G.; Myers, J.R.; Post, C.M.; Boulé, L.A.; Lawrence, B.P. DNA Methylation Patterns in CD4+ T Cells of Naïve and Influenza A Virus-Infected Mice Developmentally Exposed to an Aryl Hydrocarbon Receptor Ligand. *Environ. Health Perspect.* **2021**, *129*, 017007. [CrossRef]
72. Winans, B.; Nagari, A.; Chae, M.; Post, C.M.; Ko, C.-I.; Puga, A.; Kraus, W.L.; Lawrence, B.P. Linking the Aryl Hydrocarbon Receptor with Altered DNA Methylation Patterns and Developmentally Induced Aberrant Antiviral CD8+ T Cell Responses. *J. Immunol.* **2015**, *194*, 4446–4457. [CrossRef]
73. Su, K.-Y.; Li, M.-C.; Lee, N.-W.; Ho, B.-C.; Cheng, C.-L.; Chuang, Y.-C.; Yu, S.-L.; Guo, Y.L. Perinatal polychlorinated biphenyls and polychlorinated dibenzofurans exposure are associated with DNA methylation changes lasting to early adulthood: Findings from Yucheng second generation. *Environ. Res.* **2019**, *170*, 481–486. [CrossRef]
74. Rughani, A.; Friedman, J.E.; Tryggestad, J.B. Type 2 Diabetes in Youth: The Role of Early Life Exposures. *Curr. Diabetes Rep.* **2020**, *20*, 1–11. [CrossRef]
75. Kesavadev, J.; Misra, A.; Saboo, B.; Aravind, S.; Hussain, A.; Czupryniak, L.; Raz, I. Blood glucose levels should be considered as a new vital sign indicative of prognosis during hospitalization. *Diabetes Metab. Syndr. Clin. Res. Rev.* **2021**, *15*, 221–227. [CrossRef]
76. Schertzer, J.D.; Lam, T.K. Peripheral and central regulation of insulin by the intestine and microbiome. *Am. J. Physiol. Endocrinol. Metab.* **2021**, *320*, 234–239. [CrossRef] [PubMed]
77. Roh, E.; Kwak, S.H.; Jung, H.S.; Cho, Y.M.; Pak, Y.K.; Park, K.S.; Kim, S.Y.; Lee, H.K. Serum aryl hydrocarbon receptor ligand activity is associated with insulin resistance and resulting type 2 diabetes. *Acta Diabetol.* **2014**, *52*, 489–495. [CrossRef] [PubMed]
78. Markus, R.P.; Silva, C.L.M.; Gil Franco, D.; Barbosa, E.M.; Ferreira, Z.S. Is modulation of nicotinic acetylcholine receptors by melatonin relevant for therapy with cholinergic drugs? *Pharmacol. Ther.* **2010**, *126*, 251–262. [CrossRef]
79. Booz, G.W.; Altara, R.; Eid, A.H.; Wehbe, Z.; Fares, S.; Zaraket, H.; Habeichi, N.J.; Zouein, F.A. Macrophage responses associated with COVID-19: A pharmacological perspective. *Eur. J. Pharmacol.* **2020**, *887*, 173547. [CrossRef]
80. Rastovic, M.; Srdić-Galić, B.; Barak, O.; Stokić, E.; Polovina, S. AGING, HEART RATE VARIABILITY AND METABOLIC IMPACT OF OBESITY. *Acta Clin. Croat.* **2019**, *58*, 430–438. [CrossRef] [PubMed]

81. Del Rio, R.; Marcus, N.J.; Inestrosa, N.C. Potential Role of Autonomic Dysfunction in Covid-19 Morbidity and Mortality. *Front. Physiol.* **2020**, *11*, 561749. [CrossRef]
82. Lupacchini, L.; Maggi, F.; Tomino, C.; De Dominicis, C.; Mollinari, C.; Fini, M.; Bonassi, S.; Merlo, D.; Russo, P. Nicotine Changes Airway Epithelial Phenotype and May Increase the SARS-COV-2 Infection Severity. *Molecules* **2020**, *26*, 101. [CrossRef]
83. García-Ortega, A.; Oscullo, G.; Calvillo, P.; López-Reyes, R.; Méndez, R.; Gómez-Olivas, J.D.; Bekki, A.; Fonfría, C.; Trilles-Olaso, L.; Zaldívar, E.; et al. Incidence, risk factors, and thrombotic load of pulmonary embolism in patients hospitalized for COVID-19 infection. *J. Infect.* **2021**. [CrossRef]
84. Anderson, G.; Reiter, R.J. COVID-19 pathophysiology: Interactions of gut microbiome, melatonin, vitamin D, stress, kynurenine and the alpha 7 nicotinic receptor: Treatment implications. *Melatonin Res.* **2020**, *3*, 322–345. [CrossRef]
85. Zhou, Y.; Hou, Y.; Shen, J.; Mehra, R.; Kallianpur, A.; Culver, D.A.; Gack, M.U.; Farha, S.; Zein, J.; Comhair, S.; et al. A network medicine approach to investigation and population-based validation of disease manifestations and drug repurposing for COVID-19. *PLoS Biol.* **2020**, *18*, e3000970. [CrossRef]
86. Sharma, S.; Deep, S. In-silico drug repurposing for targeting SARS-CoV-2 main protease (Mpro). *J. Biomol. Struct. Dyn.* **2020**, 1–8. [CrossRef]
87. Bhardwaj, V.K.; Singh, R.; Sharma, J.; Rajendran, V.; Purohit, R.; Kumar, S. Identification of bioactive molecules from tea plant as SARS-CoV-2 main protease inhibitors. *J. Biomol. Struct. Dyn.* **2020**, 1–10. [CrossRef] [PubMed]
88. Dratcu, L.; Boland, X. Does Nicotine Prevent Cytokine Storms in COVID-19? *Cureus* **2020**, *12*, 11220. [CrossRef] [PubMed]
89. Ramlall, V.; Zucker, J.; Tatonetti, N. Melatonin is significantly associated with survival of intubated COVID-19 patients. *medRxiv* **2020**. [CrossRef]
90. Acuña-Castroviejo, D.; Escames, G.; Figueira, J.C.; De La Oliva, P.; Borobia, A.M.; Acuña-Fernández, C. Clinical trial to test the efficacy of melatonin in COVID-19. *J. Pineal Res.* **2020**, *69*, e12683. [CrossRef]
91. Kwon, Y.J.; Shin, S.; Chun, Y.J. Biological roles of cytochrome P450 1A1, 1A2, and 1B1 enzymes. *Arch. Pharm. Res.* **2021**. [CrossRef] [PubMed]

Review

A New Insight into the Potential Role of Tryptophan-Derived AhR Ligands in Skin Physiological and Pathological Processes

Monika Szelest [†,‡], Katarzyna Walczak [*,†] and Tomasz Plech

Department of Pharmacology, Medical University of Lublin, Chodźki 4a, 20-093 Lublin, Poland; m.wlodarczyk214@gmail.com (M.S.); tomasz.plech@umlub.pl (T.P.)
* Correspondence: katarzyna.walczak@umlub.pl; Tel.: +48-81-448-6774
† These authors contributed equally to this work.
‡ A Volunteer in the Department of Pharmacology, Medical University of Lublin.

Abstract: The aryl hydrocarbon receptor (AhR) plays a crucial role in environmental responses and xenobiotic metabolism, as it controls the transcription profiles of several genes in a ligand-specific and cell-type-specific manner. Various barrier tissues, including skin, display the expression of AhR. Recent studies revealed multiple roles of AhR in skin physiology and disease, including melanogenesis, inflammation and cancer. Tryptophan metabolites are distinguished among the groups of natural and synthetic AhR ligands, and these include kynurenine, kynurenic acid and 6-formylindolo[3,2-b]carbazole (FICZ). Tryptophan derivatives can affect and regulate a variety of signaling pathways. Thus, the interest in how these substances influence physiological and pathological processes in the skin is expanding rapidly. The widespread presence of these substances and potential continuous exposure of the skin to their biological effects indicate the important role of AhR and its ligands in the prevention, pathogenesis and progression of skin diseases. In this review, we summarize the current knowledge of AhR in skin physiology. Moreover, we discuss the role of AhR in skin pathological processes, including inflammatory skin diseases, pigmentation disorders and cancer. Finally, the impact of FICZ, kynurenic acid, and kynurenine on physiological and pathological processes in the skin is considered. However, the mechanisms of how AhR regulates skin function require further investigation.

Keywords: aryl hydrocarbon receptor; tryptophan; kynurenine; FICZ; skin; kynurenic acid; atopic dermatitis; psoriasis; melanoma

1. Introduction

The aryl hydrocarbon receptor (AhR) is expressed in various tissues characterized by a rapid growth rate, including skin [1]. Gene expression analysis revealed that AhR activation enhances or suppresses the expression of several genes, thus influencing the gene expression profile [2]. Previous studies revealed the crucial role of AhR in several physiological and pathological processes in the skin. Among the groups of natural and synthetic AhR ligands is the group of tryptophan derivatives [1,3]. Some of them, including kynurenine, kynurenic acid, and 6-formylindolo[3,2-b]carbazole (FICZ), have been previously recognized as ligands of this receptor. However, recently discovered biological properties of these substances, their widespread presence, and potential continuous exposure may suggest the important role of tryptophan-derived AhR ligands in many physiological and pathological processes in the skin [4]. Unfortunately, the role of AhR itself and the biological effect of the tryptophan-derived ligands in the prevention, pathogenesis, and progression of skin diseases are not fully understood to date.

2. Aryl Hydrocarbon Receptor (AhR)

AhR is a transcription factor from the evolutionarily old family of a basic helix-loop-helix/Per-ARNT-Sim (bHLH-PAS) transcription regulators, acting in a DNA sequence-specific manner. The bHLH motif contains two domains that are responsible for DNA sequence binding and protein dimerization [1]. Several low-molecular-weight chemical compounds activate the cytosolic AhR after entering cells via diffusion [3,5]. Air pollution compounds [3], endogenous amino acid derivatives [6–8], some food components (e.g., indoles, polyphenols, glucosinolates) [9–11], and some yeast and bacterial metabolites [12] are considered AhR ligands. AhR has only one binding pocket, whose amino acid composition determines ligand binding strength [1]; however, AhR may also be activated by a number of stress factors and some substances that might not fit into the binding pocket (e.g., hypoxia and oxidized low-density lipoproteins) [13].

The type of AhR ligand determines the level of activation and the spectrum of genes transcribed [14–16]. An increased AHR expression is observed in the placenta, liver, lungs, intestines, and skin, which are barrier tissues or play an important role in metabolism. The lowest AHR expression is reported in the brain, kidneys, and skeletal muscles [2]. Recent studies revealed additional functions of AhR in the body, including control of liver and vascular development, intestinal immunity, hematopoiesis, and perinatal growth [17–22]. Moreover, AhR signaling may be associated with stem cell proliferation and carcinogenesis [2].

After ligand binding, AhR dissociates from its chaperones (e.g., proto-oncogene tyrosine-protein kinase c-Src, heat shock protein 90 (HSP90), p23, and the hepatitis B virus X-associated protein 2 (XAP2)) and undergoes conformational changes, resulting in the exposition of the nuclear translocation signal and induction of AhR transport into the nucleus (Figure 1A) [23].

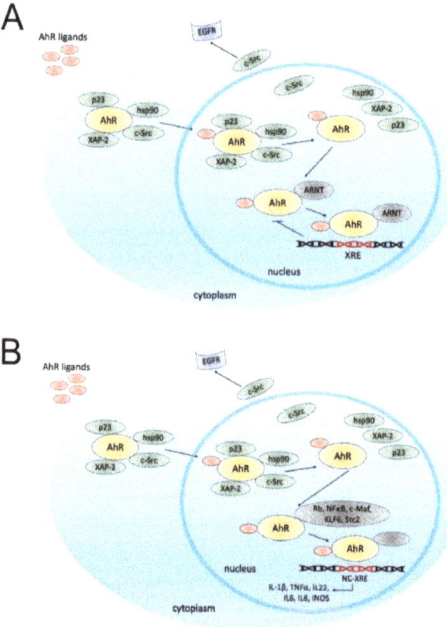

Figure 1. Schematic overview of aryl hydrocarbon receptor (AhR) signaling pathways: canonical (**A**) and noncanonical (**B**). In physiological conditions, AhR is localized in the cytosol and complexed with specific proteins, such as the hepatitis B virus X-associated protein 2 (XAP-2), heat shock protein 90 (HSP90), c-Src and p23. Upon ligand binding, AhR changes its conformation and is translocated to the nucleus, where it dimerizes with AhR nuclear transporter (ARNT) (**A**) or other partners, such as transcription factors (e.g., Kruppel-like factor 6 (KLF6)). (**B**). Dissociated c-Src interacts with the epidermal growth factor receptor (EGFR). To date, several different types of crosstalk between AhR and other proteins have been described.

For instance, AhR interaction with the hyperphosphorylated form of the retinoblastoma protein (Rb) results in growth arrest at the G1/S phase of the cell cycle [24]. AhR signaling may also promote nuclear factor kappa-light-chain-enhancer of activated B cell (NF-κB) activation via RelA and RelB interaction [25–27]. Moreover, AhR signaling is associated with the activity and function of the estrogen receptor [28]. The AhR/ARNT complex binds to the xenobiotic-responsive element (XRE) and induces the transcription of AhR-responsive genes (e.g., *CYP1A1*). On the other hand, AhR ligation promotes the transcription of its inhibitor—the AhR repressor (AhRR). AhRR forms a heterodimer with ARNT and competes with AhR/ARNT to bind to the XRE sequence, inhibiting AhR-induced transcription. However, previous studies suggest that AhR repression may not occur solely by inhibition of the DNA binding site and AhR/ARNT complex formation [29,30]. Moreover, Wilson et al. indicated that AhR–KLF6 complex formation may be involved in cell cycle regulation [31]. AhR and KLF6 proteins form a heterodimer that recognizes novel nonconsensus XRE (NC-XRE), highlighting a distinction from the XRE-dependent AhR signaling mechanism. This noncanonical signaling pathway may influence cell cycle regulation, as it controls the expression of the cyclin-dependent kinase inhibitor p21Waf1/Cip1 [31]. Jackson et al. revealed that 2,3,7,8-tetrachlorodibenzo-p-dioxin (TCDD)-mediated p21Waf1/Cip1 activation is associated with disrupted liver regeneration [32]. Therefore, while the KLF6-mediated noncanonical AhR signaling pathway might suppress tumor growth by regulating p21Waf1/Cip1 expression, carcinogenic AhR agonists might activate the canonical AhR signaling pathway and promote tumorigenesis. As outlined above, AhR might influence cell survival by various mechanisms. However, AhR might also interact with different genes that have a similar binding pattern, such as *STC2* gene, encoding a glycoprotein responsible for the regulation of endoplasmic reticulum stress [33]. Vogel et al. revealed that AhR forms a complex with NF-κB subunit RelB. The NF-κB-RelB-binding site is targeted by AhR and promotes the expression of chemokine genes, such as *BAFF*, *BLC*, and *IRF3* [34]. Furthermore, Ge et al. identified Rb as an AhR dimerization partner, suggesting its role in cell cycle arrest [35]. Recently, Huang et al. described a novel NC-XRE in the promoter of the gene encoding plasminogen activator inhibitor-1 (*PAI-1*) that might be targeted by a distinct protein complex [36]. However, further investigations are needed to determine the contribution of canonical and noncanonical AhR signaling pathways to cell homeostasis. The scheme is based on previously reported data [2,3,14].

Chaperone c-Src disconnected from AhR initiates internalization and nuclear translocation of the epidermal growth factor receptor (EGFR) and activation of mitogen-activated protein kinases (MAPK) signaling cascades, which are involved in cell proliferation, migration, and angiogenesis [37]. In the nucleus, AhR dimerizes with the AhR nuclear transporter (ARNT), another member of the bHLH-PAS family [38]. Genes possessing AhR binding sites (xenobiotic response elements, XRE) in their promoters are transcribed when bound to an AhR/ARNT dimer. After the interaction with XRE, AhR is transported back into the cytosol and degraded. The crosstalk between AhR and other signaling pathways may modify the effects of AhR and its ligands' interaction (Figure 1B) [1].

The wide spectrum of genes interacting with AhR within XRE indicates that AhR signaling is specific to the type of cell, tissue, or prevailing conditions (Table 1) [15,16].

Table 1. The effect of AhR on selected cellular processes.

Alterations in Cellular Functions		Biological Effect	Type of Cell/ Mouse Model	Reference
Cell metabolism		• AhR stimulates the expression of enzymes involved in drug metabolism (e.g., CYP1A1, CYP1A2, and CYP1B1).	*AHR*-deficient mice	[39]
Cell proliferation	Inhibition	• AhR-dependent mechanisms induce the expression of genes for the CDK inhibitory factors p27 ^{Kip1} and p21 $^{Waf1/Cip1}$ leading to an inhibition of CDKs and Rb inactivation.	5L cells	[40]
			HUVEC	[41]
		• AhR binds directly to hypophosphorylated Rb and prevents its phosphorylation by cyclin-dependent kinases (CDKs).	LNCaP cells	[42]
			BP8 5L HEK293	[43]
		• AhR interacts directly with E2F, promoting the expression of S-phase-specific genes.	Hepa-1c1c7 MCF-7	[44]
		• *AHR* silencing inhibits cell cycle progression and proliferation.	HepG2	[45]
	Stimulation	• *AHR* silencing stimulates cell cycle and proliferation of cells.	MCF-7	[45]
		• AhR acts as potent transcriptional coactivator of E2F1-dependent transcription.	A549	[46]
		• AhR induce the expression of SOS1, accelerating cell proliferation.	HepG2	[47]
Cell migration *		• AhR promotes the increased formation of cytoskeleton stress fibers and reduction of lamellipodia formation, and decreases migration of fibroblasts in *AHR* knockdown mice.	T-FGM- *AHR*−/− myofibroblasts	[48]
		• TCDD-induced AhR activity promotes cell motility and cytoskeleton remodeling.	MCF-7 HepG2	[49]
		• Flavin, an AhR agonist, induces the inhibition of breast cancer cell growth and motility.	MDA-MB-231 T47D	[50]
		• Omeprazole, an AhR agonist, decreases breast cancer cell invasion and suppress metastasis.	MDA-MB-231	[51]
		• *AHR* knockout reduces cell migration due to heregulin signaling activation in breast cancer cells displaying HER2 overexpression.	MCF-7	[52]
		• Hyperactivation of AhR accelerates cell migration of oral squamous cell carcinoma cells, while AhR inhibition reduces migration of these cells.	HSC-3 CAL27	[53]

Table 1. Cont.

Alterations in Cellular Functions	Biological Effect	Type of Cell/ Mouse Model	Reference
Regulation of Signaling Pathways and Nuclear Receptors			
NF-κB signaling pathway	• TCDD-mediated AhR activation stimulates the transcription of inflammatory genes within the NF-κB signaling pathway, e.g., IL 8.	U937 macrophages	[54]
	• AhR-mediated *IL 17A* and *CCL20* transcriptional activation is dependent on RelB activity.	B6 mice	[25]
Nuclear factor-erythroid 2-related factor-2 (Nrf2) signaling pathway	• AhR promotes the expression of antioxidant enzymes such as glutathione S-transferases and NAD(P)H quinone dehydrogenase 1 (NQO1).	NHEK	[55,56]
Calcium-dependent signaling pathways	• AhR ligands TCDD and polycyclic aromatic hydrocarbons (PAH) are able to produce rapid and sustained calcium influx.	Hepa-1	[57]
	• AhR is involved in stimulating the 3′5-cyclic adenosine monophosphate (cAMP), protein kinase C (PKC), and protein kinase A (PKA) activity, promoting the inflammatory response to TCDD.	3T3-L1	[58,59]
Hypoxia-induced factor (HIF)	• Under hypoxic conditions, HIF1α binds to ARNT, limiting its bioavailability for AhR and inhibiting AhR transcriptional response.	HepG2 HaCaT	[60]
	• Low O_2 conditions stabilize HIF1α and HIF2A, the absence of which impairs the expression of the HIF-targeted gene encoding filaggrin; thus, keratinocyte terminal differentiation and epidermal barrier formation are impaired.	HEK Krt14-Cre+ mice	[61]
	• TCDD-mediated AhR activation improperly expresses R-Spondin1, which mediates through LRP6 to activate the Wnt/β-catenin signaling; Activation of Wnt/β-catenin results in the stabilization of β-catenin, which in turn causes the misexpression of various Wnt target genes, resulting in the inhibition of tissue regeneration.	Zebrafish caudal fin regeneration model	[62]
Estrogen and retinoid receptors	• AhR ligand TCDD stimulates the expression of a gene product that inhibits estrogen receptor α (ERα)-dependent induction of transcription.	BG1	[63]
	• CYP1A/1B induction increases estrogen catabolism.	MCF-7	Reviewed in [28,64]

* Although AhR activity may influence cell migration and invasion, the ability of the AhR to drive tumor growth is mostly tissue specific. AhR—aryl hydrocarbon receptor; CDK—cyclin-dependent kinase; Rb—retinoblastoma protein; E2F—a group of transcription factors, which are downstream effectors of Rb; SOS1—son of sevenless 1; TCDD—2,3,7,8-tetrachlorodibenzo-p-dioxin; IL17A—interleukin 17A; CCL20—chemokine (C-C motif) ligand 20; NQO1—NAD(P)H quinone dehydrogenase 1; PAH—polycyclic aromatic hydrocarbons; cAMP—3′5′-cyclic adenosine monophosphate; PKC—protein kinase C; PKA—protein kinase A; ARNT—aryl hydrocarbon receptor nuclear translocator; LPR6—LDL receptor-related protein 6; ERα—estrogen receptor α; NF-κB—nuclear factor kappa-light-chain-enhancer of activated B cells; HIF—hypoxia-induced factor; Nrf2—nuclear factor-erythroid 2-related factor-2.

The transcription of genes encoding xenobiotic-metabolizing enzymes (*CYP1A1*, *CYP1A2*, and *CYP1B1*), genes responsible for cell differentiation and regulation of the cell cycle, and genes coordinating the immune response is dependent, even partially, on AhR activity [65,66]. Moreover, induction of *CYP1A1* expression allows for the degradation of some AhR ligands, including FICZ. During this process, a large amount of reactive oxygen species (ROS) is produced. ROS activity affects cell metabolism, leading to DNA damage and expression of various cytokines [65,67].

Interestingly, microarray studies indicated a ligand-specific differences in AhR-induced gene expression profile [68]. To date, many cellular metabolites and xenobiotic compounds were defined as AhR agonists [2,67].

The mechanism of downregulation of AhR signaling is still unclear. It is based on the activity of the negative AhR regulator, the AhR repressor (AhRR). Upon AhR activation, *AHRR* expression is induced. The AhRR forms a heterodimer with ARNT and competes with the AhR/ARNT complex for the XRE binding site. This feedback loop consequently inhibits AhR transcriptional activity [69]. On the other hand, Evans et al. reported that AhRR-mediated AhR inhibition is not the cause of ARNT sequestration [29]. A distinct mechanism of AhRR action was proposed, indicating that AhR inhibition occurs through protein–protein interaction [29]. Therefore, AhR repression does not occur solely by inhibition of the DNA binding site and AhR/ARNT complex formation.

3. The Role of AhR in Skin Physiology

Skin, the largest organ of the human body, is a protective barrier against harmful environmental factors. The maintenance of body fluid balance and a constant temperature depends on the proper condition and function of the skin. A battery of receptors and nerve endings present in the skin enable a reaction to various stimuli and communication with the surrounding environment [70].

The skin has a layered structure, consisting of (from the outside): epidermis, formed mainly by keratinocytes; dermis, created mainly by fibroblasts; and subcutaneous tissue. Among skin cells, there are also Langerhans cells (LCs), melanocytes, sebocytes, and immune cells (mast cells, CD8+ T cells, and dendritic cells (DCs)) [70]. AhR is observed in all skin cells, but particular cell types differ in its expression level [2].

The skin is exposed to biological, physical, mechanical, and chemical factors. Interestingly, AhR signaling appears to play an important role in maintaining skin homeostasis as it participates in many processes such as metabolism of environmental toxins, maintaining redox balance in the cell, response to ultraviolet (UV) radiation, melanogenesis, regulation of immunological processes, and functioning of the epidermal barrier [2].

AhR/ARNT signaling initiates the activation of the OVO-like 1 (OVOL1) transcription factor, which subsequently enhances the expression of filaggrin (FLG) and loricrin (LOR), proteins specific to fully differentiated keratinocytes and corneocytes [71]. Thus, the activation of this pathway contributes to accelerating the final differentiation of the epidermis and formation of epidermal barrier.

The role of the skin in immune processes is based on protecting the host from pathogens while suppressing excessive inflammation. High levels of AhR in the skin cells may be associated with an AhR-mediated immune response. AhR signaling is essential for the maturation of LCs and its capacity to present antigens, as demonstrated in studies on *AHR*-null mice [72]. Interestingly, inflammatory skin lesions were observed in mice with permanently active AhR in keratinocytes [73]. High *AHR* expression was previously reported in Th17 cells. Moreover, IL-22 secretion by these lymphocytes depends on AhR activation [74]. In summary, AhR deficiency or alterations within AhR activity may disrupt the immune response or impair the development and function of the epidermal barrier [75].

4. AhR and Skin Pathological Processes

In addition to the prominent and well-documented role of AhR in skin homeostasis, this receptor is also involved in many pathological processes within the skin through alterations in AhR-controlled signaling pathways. Moreover, it may be associated with exposure to toxic AhR ligands present in air pollution. Disorders whose pathomechanism is associated with AhR function in the skin include, among others, chloracne, hyperpigmentation, and vitiligo, as well as inflammatory diseases such as psoriasis or atopic dermatitis [66].

Skin diseases may be related to air pollution. The most common air pollutants with high affinity for AhR include the following: 2,3,7,8-tetrachlorodibenzo-p-dioxin (TCDD), benzo[a]pyrene (BaP), polychlorinated dibenzofurans (PCDFs), polychlorinated dibenzo-p-dioxins (PCDDs), and polychlorinated biphenyls (PCBs) [2,66]. Although exposure to high doses of toxic AhR ligands is relatively rare and accidental, even low doses of these compounds have previously caused skin irritation or worsened the symptoms of diseases [76]. Chloracne and hyperpigmentation are the most frequently mentioned among skin diseases in people exposed to high doses of air pollution components [77–80]. AhR may also be activated due to chronic exposure to PM2.5, which is made up of dioxin derivatives [76].

Activation of the AhR signaling pathway in epidermal keratinocytes is sufficient to initiate inflammatory skin lesions [73]. Tauchi et al. suggest that the activation of the AhR signaling pathway and the expression of AhR target genes are the main mechanisms of inflammatory skin disorders induced by PAH [73]. Thus, blocking of AhR signals that induce transcription of selected genes may be a potential therapeutic target in the treatment of some skin diseases.

A Double Agent: The Role of AhR in Oxidative Stress

Oxygen molecules do not always undergo full four-electron reduction, which leads to the formation of unstable ROS. In physiological conditions, ROS are formed during biochemical reactions and are characterized by high reactivity. Moreover, ROS are produced as a result of exposure to environmental stress, such as UV radiation or ionizing radiation. Xenobiotics and air pollutants may also increase ROS formation. The balance between the rate of ROS formation and the activity of antioxidants produced by the cells determines the biological response to ROS. The consequences of increased cell exposure to ROS include the following: a decrease in adenosine triphosphate (ATP) levels, lipid peroxidation, cell membrane depolarization, morphological changes in cell surfaces, and DNA damage. However, the biological activity of ROS is not limited to adverse effects, as at a physiological concentration, they play an important role in cell homeostasis by regulation of proliferation, apoptosis, and migration [81].

The interaction of TCDD with AhR enhances the expression of cytochrome P450 family members such as CYP1A1, CYP1A2, and CYP1B1. Due to the stable structure of TCDD, these enzymes are not able to metabolize TCDD dioxin effectively. Furthermore, excessive CYP1A1 activity, resulting from the constant interaction between TCDD and AhR, induces the generation of ROS. Increasing oxidative stress may cause oxidation of fatty acids in skin cell membranes, structural proteins (mainly collagen), and enzymatic proteins [14]. CYP1A1-induced excessive production of ROS may indirectly affect cell metabolism due to direct activation of multiple signaling pathways. Moreover, an interaction of ROS with various molecules such as NF-κB, c-Jun oncoprotein, or Rb may affect the cell cycle [82,83].

Activation of the AhR/CYP1A1 signaling pathway also contributes to increased production of inflammatory mediators, including interleukin 1 (IL-1), IL-6, and IL-8. Furthermore, BaP exposure is associated with an increased expression of *CYP1A1*, and IL-8, ROS production. This phenomenon may underlie the inflammatory skin diseases in tobacco smokers, as BaP is a component of tobacco smoke [84,85].

On the other hand, AhR activity induces the expression of nuclear factor erythroid 2-related factor 2 (Nrf2), a transcription factor with antioxidant properties. Upon AhR-mediated activation, Nrf2 increases the expression of antioxidant enzymes such as glutathione S-transferases and NAD(P)H quinone dehydrogenase 1 (NQO1) [56]. Some AhR ligands are more active in promoting the antioxidant response (Table 2).

Table 2. The AhR signaling pathway mediates antioxidative signals in response to different substances, e.g., herbal medicines and flavonoids.

Substance	Outcome	Cell Type	References
Ketoconazole	• Activation of antioxidative Nrf2 and NQO1 pathways; Anti-inflammatory effect mediated by TNF-α inhibition; Inhibition of BaP-mediated ROS and IL-8 production (cytoprotective effect); • AhR activation without ROS production.	NHEK	[55]
Bidens pilosa	• Inhibition of TNF-α and BaP-mediated ROS production; Activation of antioxidative Nrf2 and NQO1 pathways.	Human dermal endothelial cells	[86]
Epigallocatechin gallate	• Activation of antioxidative Nrf2 and NQO1 pathways; Downregulation of AhR and CYP1A1 expression.	Primary vascular endothelial cells	[87]
Quercitrin	• Inhibition of UVB-mediated ROS production; Reduction of UVB-mediated oxidative DNA damage.	JB6 cells	[88]
Quercetin, kaempferol	• Reduction of BaP-mediated increased expression of Nrf2; • Counteraction of BaP-mediated suppression of AhRR.	Caco2	[89]
Cinnamaldehyde	• Inhibition of AhR activation; • Activation of antioxidative Nrf2 and NQO1 pathways.	HaCaT	[11]
Cynaropicrin (*Cynara scolymus*)	• Activation of antioxidative Nrf2 and NQO1 pathways; Inhibition of ROS production in cells after exposure on UVB radiation; • Inhibition of proinflammatory cytokine (IL-6 and TNF-α) production in cells after exposure on UVB radiation.	NHEK	[10]
Opuntia ficus indica	• Activation of antioxidative Nrf2 and NQO1 pathways; Induction of *FLG* and *LOR* expression.	HNEK	[9]
Hesperetin	• Inhibition of AhR transactivation; Inhibition of AhR downstream gene expression (*CYP1A1*, *CYP1A2*, and *CYP1B1*).	MCF-7	[90]
Quercetin, resveratrol, curcumin	• Induction of CYP1A1 in an indirect manner by inhibiting the metabolic turnover of FICZ.	HaCaT	[91]

AhR—aryl hydrocarbon receptor; Nrf2—nuclear factor-erythroid 2-related factor-2; NQO1—NAD(P)H quinone dehydrogenase 1; TNF-α—tumor necrosis factor alpha; BaP—benzo[a]pyrene; ROS—reactive oxygen species; AhRR—aryl hydrocarbon receptor repressor; FLG—filaggrin; FICZ—6-formylindolo[3,2-b]carbazole.

Modulation of the activity of various proteins, including downstream AhR targets, such as CYP1A1 or Nrf2 via the AhR/ARNT pathway, determines the redox balance of the cells [56].

In contrast to TCDD, coal tar induces Nrf2 nuclear translocation and follows the induction of *NQO1* expression, thereby triggering an antioxidant signal pathway that neutralizes the negative effect of ROS in keratinocytes [92]. Activation of this pathway may be a clue suggesting the lack of toxicity and carcinogenicity of coal tar used in the treatment of psoriasis [93]. On the other hand, chronic exposure to TCDD results in growing immunotoxicity, thereby increasing the risk of cancer [65]. Although the mechanism of coal tar activity is not fully understood, a comprehensive study in a large group of patients with psoriasis and eczema has not indicated a relationship between the use of coal tar and an increased risk of skin cancers [93]. In summary, both ROS production and antioxidative response resulting from AhR activation depend on the AhR ligand type.

5. Role of AhR in Inflammatory Skin Diseases

5.1. Atopic Dermatitis

Atopic dermatitis (AD) is a heterogeneous skin disease accompanied by eczema, Th2-deviated inflammation, and chronic itching. Due to the reduced expression of FLG and other proteins involved in the differentiation and maturation of skin cells, the skin barrier integrity in AD is impaired [94]. Moreover, skin barrier dysfunction causes an increased colonization of microorganisms, such as *Staphylococcus aureus*, which further promotes skin inflammation [95].

Previous studies suggested that Th2-mediated immune response is associated with reduced production of the tryptophan-derived AhR ligand indole-3-aldehyde (IAld) by the skin microbiome. Yu et al. reported that IAld-induced AhR activation attenuated AD-like dermatitis [96]. Decreased inflammation was associated with the inhibition of thymic stromal lymphopoietin (TSLP) production in keratinocytes. TSLP is an inflammatory cytokine overexpressed in keratinocytes of AD patients. Upon IAld stimulation, AhR may interact with the *TSLP* promoter region and promote immune homeostasis in the skin of healthy subjects. TSLP expression is also observed in MC903-induced AD-like dermatitis mouse model, as it plays a crucial role in Th2-mediated inflammation. Although the inhibitory effect of IAld on TSLP expression reduces the inflammatory response in MC903-induced AD-like dermatitis in mice, this effect has not been observed in different models of AD-like skin inflammation, such as imiquimod (IMQ)-induced psoriatic dermatitis and oxazolone (OXA)-induced contact hypersensitivity. Due to aberrant skin microbiota, a reduced level of IAld may indicate alterations in TSLP expression, leading to skin inflammation in patients diagnosed with AD. Therefore, a deficiency of physiological AhR ligands in the Th2-deviated environment may underlie the skin lesions in AD [96].

Expression of FLG in keratinocytes is dependent on AhR activity as AhR ligation leads to OVOL1 nuclear translocation and subsequent FLG transcription [62]. The AhR/ARNT/-FLG signaling pathway may be activated by both rapidly metabolized AhR ligands, such as IAld or FICZ, and by dioxins (Figure 2) [97,98].

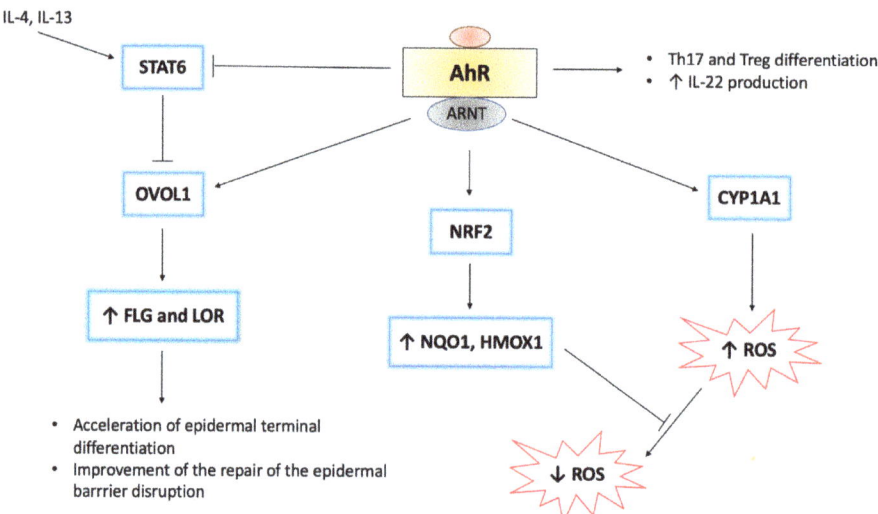

Figure 2. Molecular interactions within the AhR signaling pathway [99]. In the nucleus, AhR/ARNT complex binds to the XRE sequence, thus initiating the transcription of AhR-responsive genes, such as *CYP1A1*. CYP1A1 activity is associated with ROS production. Furthermore, AhR ligands are degraded by CYP1A1. Regarding chemically stable ligands, such as dioxins, sustained CYP1A1 activation leads to enhanced ROS generation. On the other hand, several AhR ligands activate nuclear factor-erythroid 2-related factor-2 (Nrf2), a transcription factor, which induces expression of antioxidative enzymes (e.g., heme oxygenase 1 (HMOX1) and NQO1). Moreover, AhR signaling is associated with the differentiation of immune cells, such as Th17 and Treg. Regarding inflammatory skin diseases, such as psoriasis and atopic dermatitis, AhR-mediated IL-22 production plays a crucial role in alleviating skin lesions. AhR/ARNT interaction upregulates filaggrin (FLG) and loricrin (LOR) expression via activation of the OVO-like 1 (OVOL1) transcription factor. Both FLG and LOR play a key role in epidermal differentiation. However, IL-4/IL-13-mediated signal transducer and activator of transcription 6 (STAT6) activation inhibits the OVOL1/FLG/LOR pathway. AhR stimulation may inhibit STAT6 and upregulate FLG and LOR expression. The pathogenic implication of AhR signaling in inflammatory skin diseases is not fully understood as the activation of the AhR/OVOL1/FLG/LOR pathway may become harmful. As the use of rapid metabolizing AhR ligands, such as FICZ, may alleviate skin inflammation, sustainable activation of this pathway by dioxins exacerbates epidermal barrier dysfunction. The scheme is based on previously reported data [9,10,84,99]. ↑-activation, upregulation, ↓-downregulation, T-arrow-inhibition.

Therefore, dioxin-mediated or persistent AhR activation may promote skin barrier dysfunction and exacerbate the course of AD [97]. However, topically applied FICZ reduced inflammation in skin lesions in a murine dermatitis model by AhR activation [98]. Moreover, a decrease in *Il 22* expression and an increase in *FLG* transcription were observed [98].

However, the role of AhR in AD pathogenesis is not fully understood. Kim et al. showed an increase in *ARNT* and *CYP1A1* messenger RNA (mRNA) expression in AD skin [100]. On the other hand, Hong et al. revealed an increased protein level of AhR and ARNT but not CYP1A1 in skin lesions of AD patients [97]. Hu et al. demonstrated higher expression of *AHR* in serum and increased protein level of AhR in skin lesions of AD patients compared to healthy controls. Moreover, mRNA levels of *AHR*, *AHRR*, and *CYP1A1* in peripheral blood mononuclear cells (PBMCs) of AD patients were higher in comparison to healthy controls. Thus, *AHR* expression level in PBMCs may be associated with eczema area and severity index score in AD patients [101].

The antioxidative transcription factor Nrf2 may be activated by some AhR ligands, and recent studies indicated a therapeutic effect of this group of AhR agonists. For instance, coal tar attenuates inflammatory response in AD and psoriasis patients by *NRF2* activation upon AhR interaction [92,102]. However, excessive activation of AhR leads to abnormally accelerated keratinization of cells and the formation of pruritic artemin [103,104].

One of genes encoding nerve elongation factors that may be related to epidermal hyperinnervation is *ARNT*. *ARNT*, encoding artemin, acts as pruritus-related AhR target gene. Edamitsu et al. suggest that besides *ARNT* overexpression, constitutive AhR activation may exacerbate alterations in the epidermis in patients with AD [103]. Moreover, artemin expression and alloknesis may be enhanced by air pollutants via AhR activation [104]. Artemin expression is higher in patients with AD compared to healthy controls [104]. Topical application of 7,12-dimethylbenz[a]anthracene (DMBA), an exogenous AhR agonist, induced an AD-like phenotype, but this effect was not achieved when using endogenous AhR ligand FICZ. As FICZ is rapidly metabolized by CYP1A1, it cannot efficiently activate AD-related target genes. Therefore, prolonged AhR activation is crucial for pruritic AD symptoms induction [104].

A few reports indicate that some AhR agonists, such as FICZ, 2-(1H-Indol-3-ylcarbonyl)-4-thiazolecarboxylic acid methyl ester (ITE), and soybean tar Glyteer, may activate both canonical and noncanonical AhR signaling pathways. For instance, in human keratinocytes, FICZ promotes wound healing via extracellular signal-regulated kinase (ERK) signaling in an AhR-independent manner [105]. AhR endogenous ligand ITE also reduces transforming growth factor-beta (TGF-β) signaling without AhR activation. However, the recruitment of Th2 cells in AD skin lesions is regulated by chemokine (CC motif) ligand 17 (CCL17) and CCL22 expression. Both chemokines are produced via signal transducer and activator of transcription 6 (STAT6) activation in DCs. Takemura et al. demonstrated that soybean tar Glytter inhibits STAT6 expression; thus, CCL17 and CCL22 production in DCs is reduced [106]. Moreover, STAT6 expression is blocked by coal tar via AhR-mediated activation of the Nrf2 signaling pathway [92]. Interestingly, coal tar induces a shift in skin microbiome composition due to the microbiome-modulating properties of some AhR agonists. As the skin microbiome plays an important role in the development of inflammatory skin diseases, this biological mechanism of coal tar may have an essential therapeutic value [107].

Clinical studies confirm the efficacy of the AhR agonist tapinarof in the treatment of AD [108]. The action of tapinarof is based on the activation of the Nrf2-antioxidative pathway. Improvement in skin condition after tapinarof application is also associated with reduced IL-17A production and increased *FLG* expression [108].

5.2. Psoriasis

Psoriasis is a chronic inflammatory skin disease characterized by the thickened epidermis and skin infiltration of polymorphonuclear cells. The tumor necrosis factor-alpha (TNF-α)/IL-23/IL-17A axis plays a key role in induction and progression of psoriasis; thus, biological drugs against TNFα/IL-23/IL-17A have good therapeutic efficacy [109].

The interaction between AhR and endogenous ligands changes the inflammatory profile of skin lesions in psoriasis [110]. AhR-mediated Th17 activity controls the expression of IL-22 [111,112]. Monitoring of IL-22 plasma concentration allows the assessment of the severity of the disease [113]. Furthermore, the activity of IL-22 in keratinocytes is associated with increased expression of the transcription factor STAT3, which contributes to increased proliferation of epidermal cells [114]. IL-22 also affects the final stage of epidermal cell differentiation, leading to psoriasis-like skin lesions [115,116].

Interestingly, AhR activity is required for IL-22 production specifically by Th17 cells. AhR induction is not necessary for other types of IL-22-producing cells, including $\gamma\delta$ T cells, CD4(−)CD8(−)TCRβ(+) T cells, and innate lymphoid cells. It is still unclear why Th17 specifically requires AhR stimulation to produce IL-22. Nevertheless, the reason for it may indicate the diversity of interactions of AhR downstream effectors with other transcription factors. For instance, TGF-β, which induces c-Maf activity, is involved in the differentiation of Th17 cells. C-Maf inhibits IL-22 expression by binding to its promoter. Hence, AhR activity appears to be necessary to overcome the suppressive activity of TGF-β [115].

On the other hand, the interaction of AhR with endogenous ligand FICZ reduces the inflammatory response in the IMQ-induced model of skin lesions [117]. Moreover, *AHR*-null mice presented significant exacerbation of the disease when compared to the *AHR*-sufficient control. In addition, an increase in mRNA expression of several proinflammatory cytokines involved in psoriasis, such as *Il 17a*, *Il 17c*, *Il 23*, *Il 22*, and *Il 1b*, was observed in the skin lesions of *AHR*-deficient mice [117].

Nevertheless, the role of AhR in psoriasis is controversial [99]. Kim et al. reported an increase in AhR and ARNT protein level in skin lesions in psoriasis, whereas CYP1A1 level was decreased when compared to healthy skin [100]. However, the fact that AhR may induce the expression of other genes not involved in the metabolism of xenobiotics cannot be ignored. It should be underlined that AhR controls activation of several signaling pathways, including phosphoinositide 3-kinase/protein kinase B (PI3K/Akt) and ERK signaling pathways, and the expression of various genes contributing to proliferation, adhesion, migration, or immune response [118,119]. On the other hand, serum levels of AhR and CYP1A1 in psoriasis patients were significantly higher when compared to the control group in the study conducted by Beranek et al. [120].

One of the genes that is found to be consistently upregulated in psoriatic skin lesions is *KYNU*, encoding an enzyme of the tryptophan metabolism. Kynureninase (KYNU) degrades kynurenine, an endogenous AhR ligand [121]. Gudjonsson et al. revealed other genes (e.g., *IDO1*, *CYP2E1*, *CYP4B1*, *SMOX*, and *ALDH3A2*) of the tryptophan catabolism pathway to be differentially regulated in psoriasis [122]. Deregulation of tryptophan metabolism in the skin may lead to a reduction of AhR ligands, such as kynurenine, kynurenic acid, and FICZ [123–125].

In both human psoriasis samples and an IMQ-induced model of skin inflammation, FICZ-induced AhR activation ameliorates inflammatory response. Moreover, Di Meglio et al. revealed that the expression of 29 out of 41 genes upregulated in psoriasis, including inflammatory-related genes such as *IFIT*, *IFIT3*, *RSAD2* and *MX2*, was reduced after FICZ-induced AhR activation. Thus, decreased AhR activity in psoriatic skin lesions may be associated with increased expression of proinflammatory cytokines in this tissue leading to hyperinflammation [117]. Moreover, AhR activity seems to be crucial modulator of the severity of psoriasis [117]. In summary, the limited availability of endogenous AhR ligands could affect skin homeostasis regulated by this receptor.

It is not explicitly confirmed that a specific cytokine profile is responsible for the severity of skin lesions. This crosstalk between immune cells and nonhematopoietic cells involved in the inflammatory response is crucial for determining the pathogenesis of diseases such as psoriasis. However, the treatment of autoimmune inflammation is based on the modulation of the immune response [126]. An absence of AhR or blockade of its activity is associated with dysregulation of skin cell responses, mainly keratinocytes, to inflammatory stimuli. A number of inflammatory pathways are involved in the pathogenesis of psoriasis; thus, it is difficult to indicate the leading role of individual inflammatory mediators in the development of skin lesions. Recent studies indicate that the use of IL-17 blockers in an IMQ-induced psoriasis-like skin model is not sufficient to decrease the formation of skin lesions in *AHR*-deficient mice [127]. Moreover, AhR activity in the epidermal capillaries limits the recruitment of neutrophils, thus limiting the formation of skin lesions [128].

6. Skin Pigmentation Disorders

6.1. Hyperpigmentation

Hyperpigmentation of the skin is characteristic of tobacco smokers, and it may result from BaP-mediated AhR activation and enhanced melanogenesis [129]. The microphthalmia-associated transcription factor (MITF) is a major regulator of melanogenesis, which activates tyrosinase (TYR) and tyrosinase-related proteins (TYRPs). The expression of these melanogenic enzymes leads to melanin granules production [130]. The interaction of AhR with BaP or TCDD induces MITF activation, which in turn enhances TYR expression, resulting in increased melanogenesis [131]. Benzanthrone is another AhR ligand contributing to hyperpigmentation. Increased melanogenesis was observed in murine melanocytes treated with benzanthrone in vitro [132]. Skin pigmentation was also diagnosed in patients from Japan (Yusho) and Taiwan (Yucheng) after mass poisoning caused by cooking oil contaminated with PCBs and PCDFs [77,79]. Additionally, long-time exposure to high concentrations of PM2.5 may also be associated with hyperpigmentation [76].

6.2. Vitiligo

Vitiligo is an acquired pigmentary disorder, characterized by the loss of functioning melanocytes in skin, hair, or both. The pathogenesis of vitiligo is based on melanocyte defects, an innate immune response, and T-cell-mediated melanocyte destruction [133]. Vitiligo patients reveal a reduced expression of *AHR* in skin lesions compared to healthy controls [134]. However, furanochromones psoralen and khellin, in combination with UVA phototherapy, activate AhR, thus increasing melanogenesis [135].

AhR-mediated Treg cell differentiation and IL-10 expression may be associated with vitiligo pathogenesis, as IL-10 plays a crucial role in the development of self-tolerance [136,137]. Importantly, vitiligo is an autoimmune disease in which macrophages, T cells, cytokines, and other proinflammatory mediators play a prominent role [138]. Recent studies demonstrated increased TNF-α concentration and decreased IL-10 production in the serum of vitiligo patients [136,139]. Moreover, Tregs from *AHR*-null mice produced decreased level of IL-10 [27]. On the other hand, Taher et al. revealed that tacrolimus-induced increase in IL-10 level inhibited the degradation of melanocytes and might reduce disease symptoms [138].

The pathogenesis of this disease may be related to *AHR* − 129C > T polymorphism [139]. The T allele of this polymorphism increases the binding affinity of the SP1 transcription factor to *AHR*, thereby increasing the activity of the *AHR* promoter. Multiple TATA-less genes responsible for cell growth and immune response are controlled by SP1. *AHR* lacks TATA boxes, although its core promoter region possesses GC-rich fragments with several putative SP1 binding sites [140]. The abnormal binding affinity of the *AHR* promoter to SP1 (due to *AHR* hypermethylation or under the influence of an SP1 inhibitor) may decrease *AHR* expression. Interestingly, increased *AHR* expression was observed in carriers of the −129 T allele; thus, it could potentially be a genetic marker for vitiligo. On the other hand, −129 T allele possession is associated with higher expression of IL-10. Therefore, *AHR* − 129C > T polymorphism may be related to vitiligo by altering IL-10 production [139].

IL-22-producing cells, whose activity is dependent on AhR ligation, may also contribute to abnormal immune response underlying vitiligo [141]. Furthermore, IL-17 expression is correlated with vitiligo and may play a role in its pathogenesis [142]. However, the relationship between AhR-mediated IL-17 expression and vitiligo has not yet been stated. Similarly, the involvement of ROS in vitiligo pathogenesis remains controversial [143].

7. Skin Appendage Disorder: Chloracne

Chloracne is characterized by acne-like eruptions, blackheads, cysts, and pimples on the skin and may appear in response to permanent exposure to AhR ligands from polluted air, such as TCDD and PCDFs [77–79]. Chloracne skin lesions are located mainly on retroauricular and malar areas of the face as well as on the ear lobes and groin [80,144]. Increased *AHR* expression is observed in skin lesions of people exposed to dioxins present in polluted air [101]. Moreover, constitutive activation of AhR and excessive production of ROS may be crucial for the development of this disease [145]. Pathogenesis of chloracne is based on the accelerated process of final differentiation of keratinocytes induced by AhR agonists, although the molecular aspect of this mechanism is not fully understood [146,147].

Caputo et al. reported that exposure to high doses of TCDD contained in polluted air caused chloracne in children after the explosion in Seveso [78]. Similarly, massive poisoning of PCDFs and their derivatives induced chloracne in Japan (Yusho) [80] and

Taiwan (Yucheng) [81]. Skin lesions covering over 30% of body surface area and sebaceous gland involution are observed in people exposed to very high doses of TCDD [146].

In physiological conditions, AhR activation also leads to accelerated keratinocyte differentiation [148]. However, structural stability of dioxins may be crucial for chloracne development, as endogenous AhR ligands are rapidly degraded. The key role of AhR in keratinization was confirmed in studies in on *AHR*-deficient and *AHR*-transgenic mice [73,149].

Furthermore, lipophilic dioxins can accumulate in the sebaceous glands with high *AHR* expression and might be secreted with sebum [144,150,151]. Moreover, chloracne indicates hyperkeratinization of interfollicular epidermis hair follicle cells [152]. In addition, a change in the physiology of sebocytes is observed in the form of a gradual loss of sebaceous cells and involution of sebaceous glands, which leads to cyst formation [150,153]. AhR-dioxins interaction results in hyperkeratinization of keratinocytes and transformation of sebocytes into keratinocytes [151,152,154].

Pathogenesis of chloracne is related to upregulation of the expression of particular genes and proteins. The reduced number of sebaceous glands and sebocytes may be associated with the altered metabolism of the mature sterol-binding protein (mSREBP-1), resulting from AMP-activated protein kinase (AMPK) activation [155]. On the other hand, the MAPK signaling pathway is also crucial for skin lesion formation in chloracne patients, as AhR activation in chloracne induces the activation of EGFR and MAPK [156]. EGFR and AhR compete for common coactivator p300 for their transcriptional activity. Thus, the activation of the EGFR pathway results in inhibition of AhR-mediated *CYP1A1* expression [157].

A number of compounds coordinate the course of each stage of keratinization, which includes transglutaminase-1 and -3 of ceramides and various epidermal differentiation complex (EDC) proteins [158]. TCDD indirectly accelerates keratinization by interacting with EDC molecules such as LOR and FLG [159]. The expression of LOR and FLG increases due to the interaction with TCDD and induces earlier maturation of the epidermal barrier in the skin of mouse fetuses [160]. Application of TCDD directly on hairless mouse skin resulted in hyperkeratosis, epidermal hyperplasia, and sebaceous gland metaplasia [161].

Moreover, TCDD-induced activation of AhR increases the expression of genes involved in the keratinization process. This applies especially to EDC genes and genes responsible for ceramide synthesis [162]. Inflammation in chloracne results from increased expression of cytokines (IL-6, IL-8, and IL-1a) produced by keratinocytes and sebocytes [8,84,153].

The previous studies suggested that the severity of chloracne depends on the level of dioxins in the blood [80]. AhR stimulation is associated with impaired sebocyte proliferation and impaired lipid synthesis in these cells. As a result of dioxin-induced AhR activity, sebocytes lose their characteristic phenotype; thus, inhibition of lipogenesis and a decrease in the expression of keratin 7 and the epithelial antigen membrane occur. Moreover, the transformation of sebocytes into keratinocytes is associated with increased expression of the keratinocyte-specific molecules: keratin 10 and peroxisome proliferator-activated receptor-δ (PPAR-δ) [151–153]. Dedifferentiation of sebocytes may depend on the activity of the AhR/Blimp1 signaling pathway. Inhibition of lipogenesis and sebaceous gland atrophy is associated with inhibition of sebocyte proliferation and reduction of c-Myc expression mediated by Blimp1 activity. Furthermore, AhR–TCDD interaction induces the AhR/Blimp1 signaling activity [152].

8. Skin Cancer

AhR is also associated with carcinogenesis and tissue homeostasis [163,164]. However, its role in carcinogenesis is not clearly defined, and opposite effects of AhR on tumor progression have been reported. It is hypothesized that this seemingly contradictory function of AhR in tumor progression may be partially dependent on its cell-type-specific roles in cell migration [reviewed in [165]].

There is no clear confirmation that AhR activation leads to the development of skin cancers. However, the observed procarcinogenic effects of some AhR ligands and the biological role of down-effector genes activated by this receptor suggest its involvement in carcinogenesis. Long-term observations revealed that overexposure to some synthetic AhR ligands (e.g., polycyclic aromatic hydrocarbons) or UVB may lead to premalignant lesions or skin cancer [166–168].

Carcinogenicity might be associated with the activity of cytochrome P-450 enzymes, as it leads to either detoxification or potential carcinogens formation. Importantly, UVB is also involved in the induction of cytochrome P-450 subfamilies, including CYP1A1 and CYP1B1, and metabolic activation and transformation of organic procarcinogens to carcinogens. Moreover, several UV-induced mechanisms may be associated with skin carcinogenesis, such as direct UVB damage to skin cell DNA, reduced apoptosis, intensified keratinocyte proliferation, and chronic skin inflammation [169]. Finally, the stimulation of AhR leads to activation of MAPK signaling which may be involved in cancer cell proliferation.

8.1. Squamous Cell Carcinoma

Squamous cell carcinoma (SCC) is the most frequent skin malignancy in humans [170]. Importantly, AhR was identified as one of the genetic determinants of susceptibility to SCC in humans [171]. Several procarcinogenic and proinflammatory AhR-related genes potentially involved in carcinogenesis and cancer progression are upregulated in keratinocytes exposed to UVB, including *CYP1A1*, *CYP1B1*, *COX-2*, *CXCL5*, and matrix metalloproteinases (MMPs) (reviewed in [172]). It was suggested that the AhR signaling pathway is involved in the initiation of keratinocyte-derived skin cancers induced by UVB radiation [23]. Moreover, AhR signaling may contribute to the degradation of the cyclin-dependent kinase inhibitor p27Kip1 involved in cell cycle regulation, proliferation, and apoptosis in keratinocytes [172–174].

8.2. Melanoma

Surprisingly, there have been very few studies reported on the role of AhR in melanoma promotion and progression, although *AHR* is highly expressed in melanoma cell lines [175]. Furthermore, the interactions between tumor and stroma are mediated by AhR. It was reported that although *AHR* expression in the tumor inhibits melanoma growth and metastasis, the expression of this receptor in the stroma promotes melanomagenesis. AhR might act as tumor suppressor regarding melanoma cells, as its activity was associated with decreased migration and invasion, a reduced numbers of cancer stem-like cells, and aberrant β1-integrin and caveolin 1 concentrations. Human melanoma cell lines with the highest protein level of AhR have also inhibited migration and invasion activity. Moreover, AhR protein level is reduced in human melanomas with respect to nevi lesions. It is supposed that tumor progression and metastasis depend on stromal AhR in the case of *AHR* knockdown in melanoma cells [165]. Activation of AhR signaling in the tumor microenvironment may stimulate cancer cell proliferation, and migration by enhanced expression of proangiogenic mediators and factors increased cancer cell motility, including the vascular endothelial growth factor (VEGF) and TGF-β [48,176].

On the other hand, it was reported that environmental chemicals considered as AhR agonists contribute to melanoma progression and invasion through the stimulation and activity of MMPs [177]. Another study revealed that exposure to TCDD leads to upregulation of the melanogenic pathway not only in melanocytes but also in melanoma cells. However, no stimulation of melanoma cell proliferation was observed [131].

9. The Role of Tryptophan-Derived AhR Ligands in Skin Homeostasis

Previous studies revealed several ligands of AhR that can be grouped as follows:
- Exogenous/synthetic ligands (i.e., TCDD, biphenyls, DMBA, methylcholanthrene, and BaP);

- Exogenous/natural compounds, found in or metabolized from dietary plants (i.e., resveratrol and other glucosinolates, flavonoids, indolcarbinols, and kynurenic acid);
- Endogenous ligands formed in the body (i.e., kynurenine, kynurenic acid, ITE, a tryptopha cysteine dimer, and FICZ).

Several AhR agonists are derived from tryptophan, which is an essential amino acid that is considered as the strongest near-UV absorbing chromophore [65]. Thus, the role of these AhR ligands may be crucial for various processes in the skin. UV absorption by tryptophan leads to the production of several stable photoproducts that may have various biological activities. Some of these are considered as AhR ligands since conformational changes of tryptophan under exposure to UV radiation in the skin result in FICZ production [75,178]. Importantly, some other non-UV-induced tryptophan metabolites produced enzymatically in cells are also considered as AhR ligands (i.e., kynurenine and kynurenic acid) [123,179,180].

Three main ways by which tryptophan-derived AhR ligands reach the skin can be distinguished: topical application on the skin, as these ligands may be the compounds of skin care products; endogenous synthesis in cells of the skin [4]; and intragastric administration [181,182]. Furthermore, tryptophan-derived ligand activity affects various physiological and pathological processes.

9.1. FICZ

FICZ, a tryptophan oxidation product formed by exposure to UV or visible irradiation, binds with high affinity to AhR in mammalian cells, inducing expression of CYP1A1 [183]. UVB is the most efficient in FICZ formation from aqueous tryptophan, whereas visible light and UVA induce FICZ production with lower yields [184]. FICZ has a very high affinity for AhR but is quickly and efficiently degraded in cells by AhR-induced CYP1A1, CYP1A2, and CYP1B1, giving it low intracellular levels [185,186]. Importantly, FICZ has been found to be physiologically relevant in human skin [187]. However, the biological role of this tryptophan metabolite in physiological and pathological processes in the skin has not been fully studied. It was revealed that direct FICZ-mediated AhR activation alleviates inflammation in both human psoriasis samples and a mouse model of psoriasis-like skin lesions [117]. The FICZ–AhR interaction activates the AhR/ROS signaling pathway and increases the expression of inflammatory mediators (IL-1A, IL-1B, and IL-6) and, thus, may be associated with the dangerous effects of exposure to UVB radiation [8].

FICZ reveals a photosensitizing effect on keratinocytes. The simultaneous exposure to FICZ and UVA radiation induces apoptosis of keratinocytes due to caspase 3 activation and heat shock protein 70 (HSP70) production [183,184]. Moreover, FICZ reduces TGF-β-mediated collagen formation in human dermal fibroblasts [187,188]. These data indicate that FICZ may be responsible for the effect of photoaging after UVB exposure.

On the other hand, FICZ limits the production of IL-17 and IL-22 in skin lesions and reduces inflammation in dermatitis model [98,117]. Moreover, FICZ-mediated AhR activation is associated with increased expression of EDC, such as FLG and LOR [71]. FICZ promotes wound healing by increasing keratinocyte migration due to the activation of the MEK/ERK pathway in an AhR-independent manner [105]. Cell migration is supported by FICZ even in the conditions of AHR knockdown by small interfering RNAs (siRNAs) or an AhR inhibitor [105]. Therefore, inflammatory cell migration may result directly from interactions between FICZ and the TGF-β/ERK signaling pathway. However, the effect of FICZ may be associated with other molecular mechanisms stimulated by injury. These results shed a new light on the role of FICZ in skin homeostasis. Nevertheless, the mechanism of FICZ-mediated keratinocyte migration may be relevant to managing the treatment of skin wounds.

Mengoni et al. reported that *AHR* expression strictly correlates with the degree of dedifferentiation in both human melanoma samples and human and mouse melanoma cell lines [189]. Moreover, in the inflammatory environment, FICZ-mediated AhR activation induces the phenotypic switch of melanoma cells into the dedifferentiated state [189].

In addition, AhR-induced suppression of E-cadherin expression and induction of MMP activity resulted in reduced cell adhesion and increased cell motility [177,190]. Taken together, these data indicate that AhR activity may promote invasive features of tumor cells.

9.2. Kynurenine

Kynurenine, a key metabolite of the main route of tryptophan catabolism, is an endogenous agonist of AhR [191]. Although kynurenine activates the AhR using classical response genes such as *CYP1A* [191], it was previously revealed that kynurenine plays a more important role in AhR-dependent immunological responses rather than in the metabolism of xenobiotics. In a dose-dependent manner, kynurenine upregulates the expression of immunosuppressive genes, such as *TGFB1* and *IDO1* [192,193]. Kynurenine regulates T-cell differentiation and induces immunosuppressive strategies in cancer cells [124,191]. Moreover, kynurenine may display an immunosuppressive activity; thus, it takes part in disease tolerance pathways and represents a link between tryptophan catabolism and the AhR signaling pathway [192].

Although the impact of kynurenine on cancer cell proliferation is not fully understood, recent studies indicate that kynurenine activity is related to anticancer immune response. Kynurenine is produced by the tryptophan catabolizing enzymes, indoleamine 2,3-dioxygenase (IDO) and tryptophan 2,3-dioxygenase (TDO), in several types of cancer, including melanoma, to promote immune evasion [124,194]. Moreover, TCDD, one of the synthetic AhR ligand, determines tumor immunity as it promotes IDO activation, leading to kynurenine formation. *IDO* is constitutively expressed by many tumors and promotes immunosuppressive mechanisms due to depletion of tryptophan. Moreover, IDO promotes the formation of several tryptophan metabolites such as kynurenine with immunosuppressive activity. It was reported that *IDO* expression is associated with unfavorable prognosis in patients with various malignancies (reviewed in [195]). Importantly, expression and activity of IDO 1 and 2 are controlled by inflammatory mediators [196].

Similarly to tryptophan derivatives, AhR activity is associated with immune response regulation, as it was previously demonstrated in fibroblasts, endothelial cells, and macrophages [197–199]. Bessede et al. reported that tryptophan metabolites—AhR interaction contributes to the activation of Scr kinase, thus promoting IDO1 phosphorylation [192]. Furthermore, TGF-β expression is blocked, as kynurenic acid cannot induce its activation without IDO1. TGF-β is a major immune tolerance indicator; thus, AhR-mediated IDO1 phosphorylation affects immune response [192].

9.3. Kynurenic Acid

Kynurenic acid, a product of tryptophan metabolism enzymatically formed from kynurenine, is a natural ligand for AhR. Kynurenic acid is produced by kynurenine aminotransferases (KATs), which promotes L-kynurenine transamination. Moreover, the presence of ROS allows the direct transformation of tryptophan or kynurenine into kynurenic acid (reviewed in [179,196]). Kynurenic acid in nanomolar concentrations is an efficient agonist for the human AhR inducing IL-6 production and xenobiotic metabolism in cells [123]. Nevertheless, the role of kynurenine pathway metabolites in AhR-mediated skin homeostasis remains unclear. Recent studies indicate that AhR-kynurenic acid interaction may be relevant for maintaining the immunosuppressive microenvironment in several cancer types [179,200].

It has been revealed that kynurenic acid has various biological activities, including neuroprotective, anticonvulsant, anti-inflammatory, antioxidant, and antiulcer activity (reviewed in [179,196]). Importantly, kynurenic acid also has antiproliferative and antimigratory properties against various types of cancer cells (reviewed in [179]) by inhibition of signaling pathways (MAPK, PI3K/Akt) and overexpression of cell cycle regulatory proteins (p21Waf1/Cip1) [119,201]. Moreover, a recent study confirmed the biological activity of kynurenic acid towards melanoma A375 and RPMI-7951 cells [202].

Kynurenic acid is formed endogenously and is present in almost all human body fluids and tissues (reviewed in [179]) Importantly, kynurenic acid is also present in several products of human diet [181,182]. The intragastrically administered KYNA is absorbed and transported to peripheral organs via the bloodstream [203]. The role of kynurenic acid in the skin is not fully studied. It was reported that kynurenic acid is phototoxic for erythrocytes and glia cells, but no specific studies regarding skin cells have been performed [204,205].

Although *AHR* expression levels do not differ significantly in various types of skin cancer (Figure 3A), we observed a significant downregulation of *AHRR* expression in skin cutaneous melanoma (SKCN) (Figure 3B). Vogel et al. report that the upregulated AhRR expression inhibits the AhR-mediated antiapoptotic response in mouse embryonic fibroblasts [206]. As AhRR tends to play a significant role in suppressing inflammation, the downregulated *AHRR* expression may promote tumor growth.

Interestingly, the expression of genes encoding tryptophan catabolizing enzymes (e.g., *IDO1* and *KYNU*) is significantly upregulated in two types of skin cancer: head and neck squamous cell carcinoma (HNSC) and SKCN (Figure 3).

Although the reason for this phenomenon has not yet been revealed, a few hypotheses seems to be reliable and feasible. The kynurenine pathway is a major metabolic pathway involved in the formation of key a coenzyme, nicotinamide adenine dinucleotide (NAD+). As cancer cells display increased energy requirements, overexpression of *IDO1* and *KYNU* may arise from the need of an additional source of energy NAD+. On the other hand, it cannot be ignored that the increased activation of KYNU may be caused by the need to reduce the amount of kynurenine or kynurenic acid, which may have a negative effect on cancer cells.

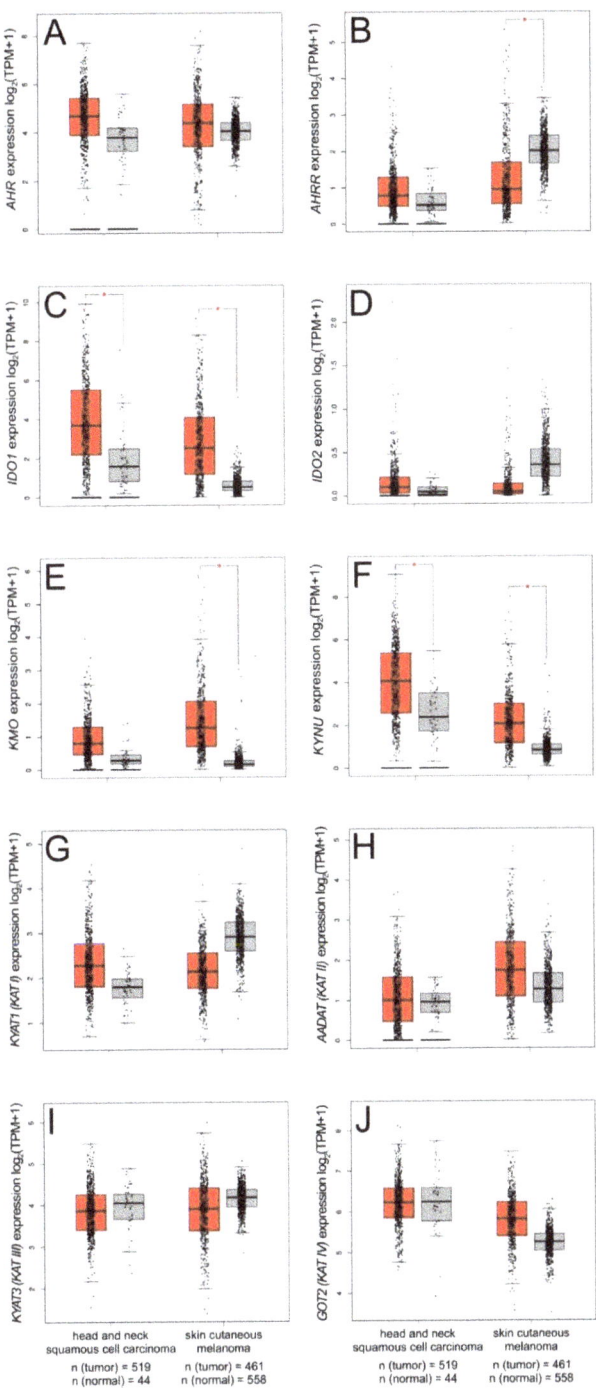

Figure 3. Expression pattern of genes encoding *AHR* (**A**), *AHRR* (**B**), tryptophan catabolizing enzymes (*IDO1* (**C**), *IDO2* (**D**), *KMO* (**E**), *KYNU*(**F**)) and kynurenine aminotransferases (*KAT I–IV* (**G–J**)) in human head and neck squamous cell carcinoma and skin cutaneous melanoma.

Expression of *KYAT1* (*KAT I*), *AADAT* (*KAT II*), *KYAT3* (*KAT III*), *GOT2* (*KAT IV*) was analyzed. Significant downregulation of *AHRR* expression was observed in skin cutaneous melanoma (**B**). Both head and neck squamous cell carcinoma and skin cutaneous melanoma showed significantly higher expression of *IDO1* (**C**). Moreover, significantly upregulated expression of *KMO* was found in skin cutaneous melanoma (SKCM) (**E**). Significantly upregulated *KYNU* expression was observed in both HNSC and SKCM (**F**). GEPIA2 was queried for skin cancers: human head and neck squamous cell carcinoma and skin cutaneous melanoma [207]. Differences in gene expression levels were studied using ANOVA. * $p < 0.01$ and fold-change threshold (| Log2FC | cutoff) of 1.

Theate et al. indicate that the expression of *IDO1* may act like a negative prognostic marker in various cancer types, including melanoma and carcinomas of the cervix, bladder, kidney, and lung [208]. Moreover, AhR regulates the expression of *IDO1* and *TDO*. Regarding the tumor microenvironment, a decreased level of tryptophan caused by IDO1 and TDO activity may result in loss of immune function through the suppression of antigen-specific T-cell response and stimulation of DC-mediated immune tolerance [124]. Thus, declined effectiveness of the anticancer immune response, resulting from deregulation of the kynurenine pathway, may be associated with cancer progression. Moreover, as the activation of the IDO/kynurenine/AhR pathway is associated with the resistance to immune checkpoint blockade, AhR may be involved in therapy resistance [209].

9.4. Skin Microbiome Metabolites

The epidermis may be colonized by various species of commensal microbes. For instance, lipophilic yeasts *Malassezia* are capable of converting tryptophan into indole compounds, some of which are AhR agonists. *Malassezia* furfur and *Malassezia globosa* colonize the skin of approximately 80% of the healthy population. However, their impact on skin physiology is controversial [12,210].

The activity of tryptophan-derived AhR agonists produced by *Malassezia* is associated with the hyperproliferation in seborrheic dermatitis and altered inflammatory in pityriasis versicolor [210]. Moreover, *Malassezia* metabolites affect cell cycle regulation and DNA repair, thus increasing the risk of skin cancer. Gaitanis et al. also reported that AhR ligands produced by Malassezia change ROS production and suppress the inflammatory response [211].

10. Conclusions

Previous studies confirmed at least a partial role of AhR in the pathogenesis of various skin diseases, including inflammatory diseases, skin pigmentation disorders, and cancer [84,89,154]. However, the function of AhR is complex as the outcome of AhR activation depends on the type of cell and ligand [13,15]. Furthermore, many different biological responses to AhR stimulation or inhibition in the skin are observed [56]. Most of the reported data are focused on the immunological and oncological effect of AhR stimulation. However, AhR ligation may induce excessive expression of proinflammatory cytokines and ROS production, leading to inflammatory disease development or carcinogenesis [55]. On the other hand, AhR-agonist-mediated activity may affect the differentiation of Treg cells, thus promoting immune tolerance [126,127]. Therefore, to determine the physiological mechanism of AhR and its role in skin disease development, more data are needed from both basic and clinical studies.

Importantly, tryptophan derivatives are a large group of AhR ligands that may potentially play a role in the pathogenesis or treatment of many skin diseases [7,169]. They are produced by enzymatic reactions or due to UV radiation in various skin cells; thus, skin is constantly exposed to tryptophan-derived AhR ligands. Additionally, some of them are present in herbs and plant extracts commonly used in skincare and treatment. However, their biological role requires further examination. In future studies, the involvement of tryptophan-derived AhR ligands in the initiation and progression of skin diseases should be clarified. The question of whether tryptophan-derived AhR ligands should be used in the prevention of skin diseases or whether we should avoid contact with them due to their negative impact on disease progression remains without a clear answer.

Author Contributions: Conceptualization, K.W.; writing—original draft preparation, M.S., K.W.; writing—review and editing, K.W., T.P.; visualization, M.S.; supervision, K.W., T.P.; project administration, K.W.; funding acquisition, K.W., T.P. All authors have read and agreed to the published version of the manuscript.

Funding: This study was supported by National Science Centre, Poland 2015/17/D/NZ7/02170 (DEC-2015/17/D/NZ7/02170) and Medical University of Lublin, Poland DS 544.

Acknowledgments: The results shown in this review are in part based upon data generated by the TCGA Research Network: https://www.cancer.gov/tcga.

Conflicts of Interest: The authors declare no conflict of interest.

Abbreviations

AD	atopic dermatitis
AhR	aryl hydrocarbon receptor
AhRR	aryl hydrocarbon receptor repressor
Akt	protein kinase B
AMPK	AMP-activated protein kinase
ARNT	aryl hydrocarbon receptor nuclear translocator
ATP	adenosine triphosphate
BaP	benzo[a]pyrene
cAMP	3'5'-cyclic adenosine monophosphate
CCL17	chemokine (C-C motif) ligand 17
CCL22	chemokine (C-C motif) ligand 22
CDK	cyclin-dependent kinase
DC	dendritic cell
DMBA	7,12-dimethylbenz[a]anthracene
EDC	epidermal differentiation complex
EGFR	epidermal growth factor receptor
ER	estrogen receptor
ERK	extracellular signal-regulated kinase
FICZ	6-formylindolo[3,2-b]carbazole
FLG	Filaggrin
HIF	hypoxia-induced factor
HNSC	head and neck squamous cell carcinoma
HMOX1	heme oxygenase 1
HSP	heat shock protein
IaId	indole-3-aldehyde
IDO	indoleamine 2,3-dioxygenase
IFN-γ	interferon gamma
ITE	2-(1H-Indol-3-ylcarbonyl)-4-thiazolecarboxylic acid methyl ester
KAT	kynurenine aminotransferases
KLF6	Kruppel-like factor 6
KYNA	kynurenic acid
KYNU	kynureninase
LC	Langerhans cell
LOR	Loricrin
LPR6	LDL receptor related protein 6
MAPK	mitogen-activated protein kinase
MITF	microphtalmia-associated transcription factor
Msrebp-1	mature sterol-binding protein

NAD	nicotinamide adenine dinucleotide
NF-κB	nuclear factor kappa-light-chain-enhancer of activated B cells
NQO1	NAD(P)H quinone dehydrogenase 1
Nrf2	nuclear factor-erythroid 2-related factor-2
PAH	polycyclic aromatic hydrocarbons
PBMC	peripheral blood mononuclear cell
PCB	polychlorinated biphenyls
PCDD	polychlorinated dibenzo-p-dioxins
PCDF	polychlorinated dibenzofurans
PI3K	phosphoinositide 3-kinase
PKA	protein kinase A
PKC	protein kinase C
PPAR-δ	peroxisome proliferator-activated receptor-δ
PTD	photodynamic therapy
Rb	retinoblastoma protein
ROS	reactive oxygen species
SCC	squamous cell carcinoma
siRNA	small interfering RNA
SKCN	skin cutaneous melanoma
SOS1	son of sevenless 1
STAT	signal transducer and activator of transcription
TCDD	2,3,7,8-tetrachlorodibenzo-p-dioxin
TDO	tryptophan 2,3-dioxygenase
TGF-β	transforming growth factor beta
TNF-α	tumor necrosis factor alpha
Treg	T regulatory cell
TRP	tryptophan
TSLP	thymic stromal lymphopoietin
TYR	tyrosinase
TYRP	tyrosinase-related protein
VEGF	vascular endothelial growth factor
XAP2	the hepatitis B virus X-associated protein 2
XRE	xenobiotic-responsive element

References

1. Abel, J.; Haarmann-Stemmann, T. An introduction to the molecular basics of aryl hydrocarbon receptor biology. *Biol. Chem.* **2010**, *391*, 1235–1248. [CrossRef] [PubMed]
2. Esser, C.; Rannug, A. The aryl hydrocarbon receptor in barrier organ physiology, immunology, and toxicology. *Pharmacol. Rev.* **2015**, *67*, 259–279. [CrossRef] [PubMed]
3. Denison, M.S.; Nagy, S.R. Activation of the aryl hydrocarbon receptor by structurally diverse exogenous and endogenous chemicals. *Annu. Rev. Pharmacol. Toxicol.* **2003**, *43*, 309–334. [CrossRef]
4. Sheipouri, D.; Braidy, N.; Guillemin, G.J. Kynurenine pathway in skin cells: Implications for UV-induced skin damage. *Int. J. Tryptophan. Res.* **2012**, *5*, 15–25. [CrossRef]
5. Mimura, J.; Fujii-Kuriyama, Y. Functional role of AhR in the expression of toxic effects by TCDD. *Biochim. Biophys. Acta Gen. Subj.* **2003**. [CrossRef]
6. Kawasaki, H.; Chang, H.W.; Tseng, H.C.; Hsu, S.C.; Yang, S.J.; Hung, C.H.; Zhou, Y.; Huang, S.K. A tryptophan metabolite, kynurenine, promotes mast cell activation through aryl hydrocarbon receptor. *Allergy* **2014**, *69*, 445–452. [CrossRef]
7. Wirthgen, E.; Hoeflich, A.; Rebl, A.; Günther, J. Kynurenic Acid: The janus-faced role of an immunomodulatory tryptophan metabolite and its link to pathological conditions. *Front. Immunol.* **2018**, *8*, 1957. [CrossRef]
8. Tanaka, Y.; Uchi, H.; Hashimoto-Hachiya, A.; Furue, M. Tryptophan Photoproduct FICZ Upregulates IL1A, IL1B, and IL6 Expression via Oxidative Stress in Keratinocytes. *Oxid. Med. Cell Longev.* **2018**, *2018*, 9298052. [CrossRef]
9. Nakahara, T.; Mitoma, C.; Hashimoto-Hachiya, A.; Takahara, M.; Tsuji, G.; Uchi, H.; Yan, X.; Hachisuka, J.; Chiba, T.; Esaki, H.; et al. Antioxidant opuntia ficus-indica extract activates AHR-NRF2 signaling and upregulates filaggrin and loricrin expression in human keratinocytes. *J. Med. Food* **2015**, *18*, 1143–1149. [CrossRef]
10. Takei, K.; Hashimoto-Hachiya, A.; Takahara, M.; Tsuji, G.; Nakahara, T.; Furue, M. Cynaropicrin attenuates UVB-induced oxidative stress via the AhR-Nrf2-Nqo1 pathway. *Toxicol. Lett.* **2015**, *234*, 74–80. [CrossRef]
11. Uchi, H.; Yasumatsu, M.; Morino-Koga, S.; Mitoma, C.; Furue, M. Inhibition of aryl hydrocarbon receptor signaling and induction of NRF2-mediated antioxidant activity by cinnamaldehyde in human keratinocytes. *J. Dermatol. Sci.* **2017**, *85*, 36–43. [CrossRef] [PubMed]

12. Krämer, H.J.; Podobinska, M.; Bartsch, A.; Battmann, A.; Thoma, W.; Bernd, A.; Kummer, W.; Irlinger, B.; Steglich, W.; Mayser, P. Malassezin, a novel agonist of the aryl hydrocarbon receptor from the yeast Malassezia furfur, induces apoptosis in primary human melanocytes. *Chembiochem* **2005**, *6*, 860–865. [CrossRef] [PubMed]
13. Wincent, E.; Bengtsson, J.; Bardbori, A.M.; Alsberg, T.; Luecke, S.; Rannug, U.; Rannug, A. Inhibition of cytochrome P4501-dependent clearance of the endogenous agonist FICZ as a mechanism for activation of the aryl hydrocarbon receptor. *Proc. Natl. Acad. Sci. USA* **2012**, *109*, 4479–4484. [CrossRef] [PubMed]
14. Mitchell, K.A.; Elferink, C.J. Timing is everything: Consequences of transient and sustained AhR activity. *Biochem. Pharmacol.* **2009**, *77*, 947–956. [CrossRef] [PubMed]
15. Sun, Y.V.; Boverhof, D.R.; Burgoon, L.D.; Fielden, M.R.; Zacharewski, T.R. Comparative analysis of dioxin response elements in human, mouse and rat genomic sequences. *Nucleic Acids Res.* **2004**, *32*, 4512–4523. [CrossRef] [PubMed]
16. Frericks, M.; Meissner, M.; Esser, C. Microarray analysis of the AHR system: Tissue-specific flexibility in signal and target genes. *Toxicol. Appl. Pharmacol.* **2007**, *220*, 320–332. [CrossRef] [PubMed]
17. Wakx, A.; Nedder, M.; Tomkiewicz-Raulet, C.; Dalmasso, J.; Chissey, A.; Boland, S.; Vibert, F.; Degrelle, S.A.; Fournier, T.; Coumoul, X.; et al. Expression, localization, and activity of the aryl hydrocarbon receptor in the human placenta. *Int. J. Mol. Sci.* **2018**, *19*, 3762. [CrossRef] [PubMed]
18. Moreno-Marín, N.; Barrasa, E.; Morales-Hernández, A.; Paniagua, B.; Blanco-Fernández, G.; Merino, J.M.; Fernández-Salguero, P.M. Dioxin receptor adjusts liver regeneration after acute toxic injury and protects against liver carcinogenesis. *Sci. Rep.* **2017**, *7*, 10420. [CrossRef] [PubMed]
19. Schmidt, J.V.; Su, G.H.; Reddy, J.K.; Simon, M.C.; Bradfield, C.A. Characterization of a murine Ahr null allele: Involvement of the Ah receptor in hepatic growth and development. *Proc. Natl. Acad. Sci. USA* **1996**, *93*, 6731–6736. [CrossRef] [PubMed]
20. Carreira, V.S.; Fan, Y.; Wang, Q.; Zhang, X.; Kurita, H.; Ko, C.I.; Naticchioni, M.; Jiang, M.; Koch, S.; Medvedovic, M.; et al. Ah receptor signaling controls the expression of cardiac development and homeostasis genes. *Toxicol. Sci.* **2015**, *147*, 425–435. [CrossRef] [PubMed]
21. Krock, B.L.; Eisinger-Mathason, T.S.; Giannoukos, D.N.; Shay, J.E.; Gohil, M.; Lee, D.S.; Nakazawa, M.S.; Sesen, J.; Skuli, N.; Simon, M.C. The aryl hydrocarbon receptor nuclear translocator is an essential regulator of murine hematopoietic stem cell viability. *Blood* **2015**, *125*, 3263–3272. [CrossRef] [PubMed]
22. Yu, M.; Wang, Q.; Ma, Y.; Li, L.; Yu, K.; Zhang, Z.; Chen, G.; Li, X.; Xiao, W.; Xu, P.; et al. Aryl hydrocarbon receptor activation modulates intestinal epithelial barrier function by maintaining tight junction integrity. *Int. J. Biol. Sci.* **2018**, *14*, 69–77. [CrossRef] [PubMed]
23. Agostinis, P.; Garmyn, M.; van Laethem, A. The Aryl hydrocarbon receptor: An illuminating effector of the UVB response. *Sci. STKE* **2007**, *2007*, pe49. [CrossRef] [PubMed]
24. Levine-Fridman, A.; Chen, L.; Elferink, C.J. Cytochrome P4501A1 promotes G1 phase cell cycle progression by controlling aryl hydrocarbon receptor activity. *Mol. Pharmacol.* **2004**, *65*, 461–469. [CrossRef]
25. Ishihara, Y.; Kado, S.Y.; Hoeper, C.; Harel, S.; Vogel, C.F.A. Role of NF-kB RelB in Aryl hydrocarbon receptor-mediated ligand specific effects. *Int. J. Mol. Sci.* **2019**, *20*, 2652. [CrossRef]
26. Kim, D.W.; Gazourian, L.; Quadri, S.A.; Romieu-Mourez, R.; Sherr, D.H.; Sonenshein, G.E. The RelA NF-kappaB subunit and the aryl hydrocarbon receptor (AhR) cooperate to transactivate the c-myc promoter in mammary cells. *Oncogene* **2000**, *19*, 5498–5506. [CrossRef]
27. Kimura, A.; Naka, T.; Nohara, K.; Fujii-Kuriyama, Y.; Kishimoto, T. Aryl hydrocarbon receptor regulates Stat1 activation and participates in the development of Th17 cells. *Proc. Natl. Acad. Sci. USA* **2008**, *105*, 9721–9726. [CrossRef]
28. Swedenborg, E.; Pongratz, I. AhR and ARNT modulate ER signaling. *Toxicology* **2010**, *268*, 132–138. [CrossRef]
29. Evans, B.R.; Karchner, S.I.; Allan, L.L.; Pollenz, R.S.; Tanguay, R.L.; Jenny, M.J.; Sherr, D.H.; Hahn, M.E. Repression of aryl hydrocarbon receptor (AHR) signaling by AHR repressor: Role of DNA binding and competition for AHR nuclear translocator. *Mol. Pharmacol.* **2008**, *73*, 387–398. [CrossRef]
30. Hahn, M.E.; Karchner, S.I.; Evans, B.R.; Franks, D.G.; Merson, R.R.; Lapseritis, J.M. Unexpected diversity of aryl hydrocarbon receptors in non-mammalian vertebrates: Insights from comparative genomics. *J. Exp. Zool. A Comp. Exp. Biol.* **2006**, *305*, 693–706. [CrossRef]
31. Wilson, S.R.; Joshi, A.D.; Elferink, C.J. The tumor suppressor Kruppel-like factor 6 is a novel aryl hydrocarbon receptor DNA binding partner. *J. Pharmacol. Exp. Ther.* **2013**, *345*, 419–429. [CrossRef] [PubMed]
32. Jackson, D.P.; Li, H.; Mitchell, K.A.; Joshi, A.D.; Elferink, C.J. Ah receptor-mediated suppression of liver regeneration through NC-XRE-driven p21Cip1 expression. *Mol. Pharmacol.* **2014**, *85*, 533–541. [CrossRef] [PubMed]
33. Joshi, A.D.; Carter, D.E.; Harper, T.A., Jr.; Elferink, C.J. Aryl hydrocarbon receptor-dependent stanniocalcin 2 induction by cinnabarinic acid provides cytoprotection against endoplasmic reticulum and oxidative stress. *J. Pharmacol. Exp. Ther.* **2015**, *353*, 201–212. [CrossRef] [PubMed]
34. Vogel, C.F.; Sciullo, E.; Matsumura, F. Involvement of RelB in aryl hydrocarbon receptor-mediated induction of chemokines. *Biochem. Biophys. Res. Commun.* **2007**, *363*, 722–726. [CrossRef]
35. Ge, N.L.; Elferink, C.J. A direct interaction between the aryl hydrocarbon receptor and retinoblastoma protein. Linking dioxin signaling to the cell cycle. *J. Biol. Chem.* **1998**, *273*, 22708–22713. [CrossRef]

36. Huang, G.; Elferink, C.J. A novel nonconsensus xenobiotic response element capable of mediating aryl hydrocarbon receptor-dependent gene expression. *Mol. Pharmacol.* **2012**, *81*, 338–347. [CrossRef]
37. Fritsche, E.; Schäfer, C.; Calles, C.; Bernsmann, T.; Bernshausen, T.; Wurm, M.; Hübenthal, U.; Cline, J.E.; Hajimiragha, H.; Schroeder, P.; et al. Lightening up the UV response by identification of the arylhydrocarbon receptor as a cytoplasmatic target for ultraviolet B radiation. *Proc. Natl. Acad. Sci. USA* **2007**, *104*, 8851–8856. [CrossRef]
38. Soshilov, A.; Denison, M.S. Role of the Per/Arnt/Sim domains in ligand-dependent transformation of the aryl hydrocarbon receptor. *J. Biol. Chem.* **2008**, *283*, 32995–33005. [CrossRef]
39. Fernandez-Salguero, P.M.; Hilbert, D.M.; Rudikoff, S.; Ward, J.M.; Gonzalez, F.J. Aryl-hydrocarbon receptor-deficient mice are resistant to 2,3,7,8-tetrachlorodibenzo-p-dioxin-induced toxicity. *Toxicol. Appl. Pharmacol.* **1996**, *140*, 173–179. [CrossRef]
40. Kolluri, S.K.; Weiss, C.; Koff, A.; Göttlicher, M. p27(Kip1) induction and inhibition of proliferation by the intracellular Ah receptor in developing thymus and hepatoma cells. *Genes. Dev.* **1999**, *13*, 1742–1753. [CrossRef]
41. Pang, P.H.; Lin, Y.H.; Lee, Y.H.; Hou, H.H.; Hsu, S.P.; Juan, S.H. Molecular mechanisms of p21 and p27 induction by 3-methylcholanthrene, an aryl-hydrocarbon receptor agonist, involved in antiproliferation of human umbilical vascular endothelial cells. *J. Cell Physiol.* **2008**, *215*, 161–171. [CrossRef]
42. Barnes-Ellerbe, S.; Knudsen, K.E.; Puga, A. 2,3,7,8-Tetrachlorodibenzo-p-dioxin blocks androgen-dependent cell proliferation of LNCaP cells through modulation of pRB phosphorylation. *Mol. Pharmacol.* **2004**, *66*, 502–511. [CrossRef]
43. Huang, G.; Elferink, C.J. Multiple mechanisms are involved in Ah receptor-mediated cell cycle arrest. *Mol. Pharmacol.* **2005**, *67*, 88–96. [CrossRef]
44. Marlowe, J.L.; Knudsen, E.S.; Schwemberger, S.; Puga, A. The aryl hydrocarbon receptor displaces p300 from E2F-dependent promoters and represses S phase-specific gene expression. *J. Biol. Chem.* **2004**, *279*, 29013–29022. [CrossRef]
45. Abdelrahim, M.; Smith, R., 3rd; Safe, S. Aryl hydrocarbon receptor gene silencing with small inhibitory RNA differentially modulates Ah-responsiveness in MCF-7 and HepG2 cancer cells. *Mol. Pharmacol.* **2003**, *63*, 1373–1381. [CrossRef]
46. Watabe, Y.; Nazuka, N.; Tezuka, M.; Shimba, S. Aryl hydrocarbon receptor functions as a potent coactivator of E2F1-dependent transcription activity. *Biol. Pharm. Bull.* **2010**, *33*, 389–397. [CrossRef]
47. Pierre, S.; Bats, A.S.; Chevallier, A.; Bui, L.C.; Ambolet-Camoit, A.; Garlatti, M.; Aggerbeck, M.; Barouki, R.; Coumoul, X. Induction of the Ras activator Son of Sevenless 1 by environmental pollutants mediates their effects on cellular proliferation. *Biochem. Pharmacol.* **2011**, *81*, 304–313. [CrossRef]
48. Mulero-Navarro, S.; Pozo-Guisado, E.; Pérez-Mancera, P.A.; Alvarez-Barrientos, A.; Catalina-Fernández, I.; Hernández-Nieto, E.; Sáenz-Santamaria, J.; Martínez, N.; Rojas, J.M.; Sánchez-García, I.; et al. Immortalized mouse mammary fibroblasts lacking dioxin receptor have impaired tumorigenicity in a subcutaneous mouse xenograft model. *J. Biol. Chem.* **2005**, *280*, 28731–28741. [CrossRef]
49. Diry, M.; Tomkiewicz, C.; Koehle, C.; Coumoul, X.; Bock, K.W.; Barouki, R.; Transy, C. Activation of the dioxin/aryl hydrocarbon receptor (AhR) modulates cell plasticity through a JNK-dependent mechanism. *Oncogene* **2006**, *25*, 5570–5574. [CrossRef]
50. Hanieh, H.; Mohafez, O.; Hairul-Islam, V.I.; Alzahrani, A.; Ismail, M.B.; Thirugnanasambantham, K. Novel aryl hydrocarbon receptor agonist suppresses migration and invasion of breast cancer cells. *PLoS ONE* **2016**, *11*, e0167650. [CrossRef] [PubMed]
51. Jin, U.H.; Lee, S.O.; Pfent, C.; Safe, S. The aryl hydrocarbon receptor ligand omeprazole inhibits breast cancer cell invasion and metastasis. *BMC Cancer* **2014**, *14*, 498. [CrossRef] [PubMed]
52. Yamashita, N.; Saito, N.; Zhao, S.; Hiruta, N.; Park, Y.; Bujo, H.; Nemoto, K.; Kanno, Y. Heregulin-induced cell migration is promoted by aryl hydrocarbon receptor in HER2-overexpressing breast cancer cells. *Exp. Cell Res.* **2018**, *366*, 34–40. [CrossRef] [PubMed]
53. Stanford, E.A.; Ramirez-Cardenas, A.; Wang, Z.; Novikov, O.; Alamoud, K.; Koutrakis, P.; Mizgerd, J.P.; Genco, C.A.; Kukuruzinska, M.; Monti, S.; et al. Role for the aryl hydrocarbon receptor and diverse ligands in oral squamous cell carcinoma migration and tumorigenesis. *Mol. Cancer Res.* **2016**, *14*, 696–706. [CrossRef]
54. Vogel, C.F.; Sciullo, E.; Li, W.; Wong, P.; Lazennec, G.; Matsumura, F. RelB, a new partner of aryl hydrocarbon receptor-mediated transcription. *Mol. Endocrinol.* **2007**, *21*, 2941–2955. [CrossRef]
55. Tsuji, G.; Takahara, M.; Uchi, H.; Matsuda, T.; Chiba, T.; Takeuchi, S.; Yasukawa, F.; Moroi, Y.; Furue, M. Identification of ketoconazole as an AhR-Nrf2 activator in cultured human keratinocytes: The basis of its anti-inflammatory effect. *J. Investig. Dermatol.* **2012**, *132*, 59–68. [CrossRef]
56. Haarmann-Stemmann, T.; Abel, J.; Fritsche, E.; Krutmann, J. The AhR-Nrf2 pathway in keratinocytes: On the road to chemoprevention? *J. Investig. Dermatol.* **2012**, *132*, 7–9. [CrossRef]
57. Puga, A.; Hoffer, A.; Zhou, S.; Bohm, J.M.; Leikauf, G.D.; Shertzer, H.G. Sustained increase in intracellular free calcium and activation of cyclooxygenase-2 expression in mouse hepatoma cells treated with dioxin. *Biochem. Pharmacol.* **1997**, *54*, 1287–1296. [CrossRef]
58. Li, W.; Matsumura, F. Significance of the nongenomic, inflammatory pathway in mediating the toxic action of TCDD to induce rapid and long-term cellular responses in 3T3-L1 adipocytes. *Biochemistry* **2008**, *47*, 13997–14008. [CrossRef]
59. Matsumura, F. The significance of the nongenomic pathway in mediating inflammatory signaling of the dioxin-activated Ah receptor to cause toxic effects. *Biochem. Pharmacol.* **2009**, *77*, 608–626. [CrossRef]

60. Vorrink, S.U.; Severson, P.L.; Kulak, M.V.; Futscher, B.W.; Domann, F.E. Hypoxia perturbs aryl hydrocarbon receptor signaling and CYP1A1 expression induced by PCB 126 in human skin and liver-derived cell lines. *Toxicol. Appl. Pharmacol.* **2014**, *274*, 408–416. [CrossRef]
61. Wong, W.J.; Richardson, T.; Seykora, J.T.; Cotsarelis, G.; Simon, M.C. Hypoxia-inducible factors regulate filaggrin expression and epidermal barrier function. *J. Investig. Dermatol.* **2015**, *135*, 454–461. [CrossRef] [PubMed]
62. Mathew, L.K.; Sengupta, S.S.; Ladu, J.; Andreasen, E.A.; Tanguay, R.L. Crosstalk between AHR and Wnt signaling through R-Spondin1 impairs tissue regeneration in zebrafish. *FASEB J.* **2008**, *22*, 3087–3096. [CrossRef] [PubMed]
63. Rogers, J.M.; Denison, M.S. Analysis of the antiestrogenic activity of 2,3,7,8-tetrachlorodibenzo-p-dioxin in human ovarian carcinoma BG-1 cells. *Mol. Pharmacol.* **2002**, *61*, 1393–1403. [CrossRef] [PubMed]
64. Safe, S.; Wang, F.; Porter, W.; Duan, R.; McDougal, A. Ah receptor agonists as endocrine disruptors: Antiestrogenic activity and mechanisms. *Toxicol. Lett.* **1998**, *102–103*, 343–347. [CrossRef]
65. Denison, M.S.; Soshilov, A.A.; He, G.; DeGroot, D.E.; Zhao, B. Exactly the same but different: Promiscuity and diversity in the molecular mechanisms of action of the aryl hydrocarbon (dioxin) receptor. *Toxicol. Sci.* **2011**, *124*, 1–22. [CrossRef]
66. Esser, C.; Bargen, I.; Weighardt, H.; Haarmann-Stemmann, T.; Krutmann, J. Functions of the aryl hydrocarbon receptor in the skin. *Semin. Immunopathol.* **2013**, *35*, 677–691. [CrossRef]
67. Schmidt, J.V.; Bradfield, C.A. Ah receptor signaling pathways. *Annu. Rev. Cell Dev. Biol.* **1996**, *12*, 55–89. [CrossRef]
68. Dere, E.; Forgacs, A.L.; Zacharewski, T.R.; Burgoon, L.D. Genome-wide computational analysis of dioxin response element location and distribution in the human, mouse, and rat genomes. *Chem. Res. Toxicol.* **2011**, *24*, 494–504. [CrossRef]
69. Mimura, J.; Ema, M.; Sogawa, K.; Fujii-Kuriyama, Y. Identification of a novel mechanism of regulation of Ah (dioxin) receptor function. *Genes. Dev.* **1999**, *13*, 20–25. [CrossRef]
70. Wong, R.; Geyer, S.; Weninger, W.; Guimberteau, J.C.; Wong, J.K. The dynamic anatomy and patterning of skin. *Exp. Dermatol.* **2016**, *25*, 92–98. [CrossRef]
71. Tsuji, G.; Hashimoto-Hachiya, A.; Kiyomatsu-Oda, M.; Takemura, M.; Ohno, F.; Ito, T.; Morino-Koga, S.; Mitoma, C.; Nakahara, T.; Uchi, H.; et al. Aryl hydrocarbon receptor activation restores filaggrin expression via OVOL1 in atopic dermatitis. *Cell Death Dis.* **2017**, *8*, e2931. [CrossRef] [PubMed]
72. Jux, B.; Kadow, S.; Esser, C. Langerhans cell maturation and contact hypersensitivity are impaired in aryl hydrocarbon receptor-null mice. *J. Immunol.* **2009**, *182*, 6709–6717. [CrossRef] [PubMed]
73. Tauchi, M.; Hida, A.; Negishi, T.; Katsuoka, F.; Noda, S.; Mimura, J.; Hosoya, T.; Yanaka, A.; Aburatani, H.; Fujii-Kuriyama, Y.; et al. Constitutive expression of aryl hydrocarbon receptor in keratinocytes causes inflammatory skin lesions. *Mol. Cell Biol.* **2005**, *25*, 9360–9368. [CrossRef] [PubMed]
74. Veldhoen, M.; Hirota, K.; Westendorf, A.M.; Buer, J.; Dumoutier, L.; Renauld, J.C.; Stockinger, B. The aryl hydrocarbon receptor links TH17-cell-mediated autoimmunity to environmental toxins. *Nature* **2008**, *453*, 106–109. [CrossRef] [PubMed]
75. Ma, Q. Influence of light on aryl hydrocarbon receptor signaling and consequences in drug metabolism, physiology and disease. *Expert Opin. Drug Metab. Toxicol.* **2011**, *7*, 1267–1293. [CrossRef]
76. Peng, F.; Xue, C.H.; Hwang, S.K.; Li, W.H.; Chen, Z.; Zhang, J.Z. Exposure to fine particulate matter associated with senile lentigo in Chinese women: A cross-sectional study. *J. Eur. Acad. Dermatol. Venereol.* **2017**, *31*, 355–360. [CrossRef]
77. Guo, Y.L.; Yu, M.L.; Hsu, C.C.; Rogan, W.J. Chloracne, goiter, arthritis, and anemia after polychlorinated biphenyl poisoning: 14-year follow-Up of the Taiwan Yucheng cohort. *Environ. Health Perspect.* **1999**, *107*, 715–719. [CrossRef]
78. Caputo, R.; Monti, M.; Ermacora, E.; Carminati, G.; Gelmetti, C.; Gianotti, R.; Gianni, E.; Puccinelli, V. Cutaneous manifestations of tetrachlorodibenzo-p-dioxin in children and adolescents. Follow-up 10 years after the Seveso, Italy, accident. *J. Am. Acad. Dermatol.* **1988**, *19*, 812–819. [CrossRef]
79. Furue, M.; Uenotsuchi, T.; Urabe, K.; Ishikawa, T.; Kuwabara, M. Overview of Yusho. *J. Dermatol. Sci. Suppl.* **2005**, *1*, 3–10. [CrossRef]
80. Mitoma, C.; Mine, Y.; Utani, A.; Imafuku, S.; Muto, M.; Akimoto, T.; Kanekura, T.; Furue, M.; Uchi, H. Current skin symptoms of Yusho patients exposed to high levels of 2,3,4,7,8-pentachlorinated dibenzofuran and polychlorinated biphenyls in 1968. *Chemosphere* **2015**, *137*, 45–51. [CrossRef]
81. Dandekar, A.; Mendez, R.; Zhang, K. Cross talk between ER stress, oxidative stress, and inflammation in health and disease. *Methods Mol. Biol.* **2015**, *1292*, 205–214. [CrossRef] [PubMed]
82. Haarmann-Stemmann, T.; Bothe, H.; Abel, J. Growth factors, cytokines and their receptors as downstream targets of arylhydrocarbon receptor (AhR) signaling pathways. *Biochem. Pharmacol.* **2009**, *77*, 508–520. [CrossRef] [PubMed]
83. Puga, A.; Ma, C.; Marlowe, J.L. The aryl hydrocarbon receptor cross-talks with multiple signal transduction pathways. *Biochem. Pharmacol.* **2009**, *77*, 713–722. [CrossRef] [PubMed]
84. Tsuji, G.; Takahara, M.; Uchi, H.; Takeuchi, S.; Mitoma, C.; Moroi, Y.; Furue, M. An environmental contaminant, benzo(a)pyrene, induces oxidative stress-mediated interleukin-8 production in human keratinocytes via the aryl hydrocarbon receptor signaling pathway. *J. Dermatol. Sci.* **2011**, *62*, 42–49. [CrossRef]
85. Fortes, C.; Mastroeni, S.; Leffondré, K.; Sampogna, F.; Melchi, F.; Mazzotti, E.; Pasquini, P.; Abeni, D. Relationship between smoking and the clinical severity of psoriasis. *Arch. Dermatol.* **2005**, *141*, 1580–1584. [CrossRef]

86. Kohda, F.; Takahara, M.; Hachiya, A.; Takei, K.; Tsuji, G.; Yamamura, K.; Furue, M. Decrease of reactive oxygen species and reciprocal increase of nitric oxide in human dermal endothelial cells by Bidens pilosa extract: A possible explanation of its beneficial effect on livedo vasculopathy. *J. Derm. Sci.* **2013**, *72*, 75–77. [CrossRef]
87. Han, S.G.; Han, S.S.; Toborek, M.; Hennig, B. EGCG protects endothelial cells against PCB 126-induced inflammation through inhibition of AhR and induction of Nrf2-regulated genes. *Toxicol. Appl. Pharmacol.* **2012**, *261*, 181–188. [CrossRef]
88. Yin, Y.; Li, W.; Son, Y.O.; Sun, L.; Lu, J.; Kim, D.; Wang, X.; Yao, H.; Wang, L.; Pratheeshkumar, P.; et al. Quercitrin protects skin from UVB-induced oxidative damage. *Toxicol. Appl. Pharmacol.* **2013**, *269*, 89–99. [CrossRef]
89. Niestroy, J.; Barbara, A.; Herbst, K.; Rode, S.; van Liempt, M.; Roos, P.H. Single and concerted effects of benzo[a]pyrene and flavonoids on the AhR and Nrf2-pathway in the human colon carcinoma cell line Caco-2. *Toxicol. In Vitro* **2011**, *25*, 671–683. [CrossRef]
90. Tan, Y.Q.; Chiu-Leung, L.C.; Lin, S.M.; Leung, L.K. The citrus flavonone hesperetin attenuates the nuclear translocation of aryl hydrocarbon receptor. *Comp. Biochem. Physiol. C Toxicol. Pharmacol.* **2018**, *210*, 57–64. [CrossRef]
91. Mohammadi-Bardbori, A.; Bengtsson, J.; Rannug, U.; Rannug, A.; Wincent, E. Quercetin, resveratrol, and curcumin are indirect activators of the aryl hydrocarbon receptor (AHR). *Chem. Res. Toxicol.* **2012**, *25*, 1878–1884. [CrossRef] [PubMed]
92. Van den Bogaard, E.H.; Bergboer, J.G.; Vonk-Bergers, M.; van Vlijmen-Willems, I.M.; Hato, S.V.; van der Valk, P.G.; Schröder, J.M.; Joosten, I.; Zeeuwen, P.L.; Schalkwijk, J. Coal tar induces AHR-dependent skin barrier repair in atopic dermatitis. *J. Clin. Investig.* **2013**, *123*, 917–927. [CrossRef] [PubMed]
93. Roelofzen, J.H.; Aben, K.K.; Oldenhof, U.T.; Coenraads, P.J.; Alkemade, H.A.; van de Kerkhof, P.C.; van der Valk, P.G.; Kiemeney, L.A. No increased risk of cancer after coal tar treatment in patients with psoriasis or eczema. *J. Investig. Dermatol.* **2010**, *130*, 953–961. [CrossRef] [PubMed]
94. Furue, M.; Ulzii, D.; Vu, Y.H.; Tsuji, G.; Kido-Nakahara, M.; Nakahara, T. Pathogenesis of atopic dermatitis: Current paradigm. *Iran. J. Immunol.* **2019**, *16*, 97–107. [CrossRef]
95. Iwamoto, K.; Moriwaki, M.; Miyake, R.; Hide, M. Staphylococcus aureus in atopic dermatitis: Strain-specific cell wall proteins and skin immunity. *Allergol. Int.* **2019**, *68*, 309–315. [CrossRef]
96. Yu, J.; Luo, Y.; Zhu, Z.; Zhou, Y.; Sun, L.; Gao, J.; Sun, J.; Wang, G.; Yao, X.; Li, W. A tryptophan metabolite of the skin microbiota attenuates inflammation in patients with atopic dermatitis through the aryl hydrocarbon receptor. *J. Allergy Clin. Immunol.* **2019**, *143*, 2108–2119.e12. [CrossRef]
97. Hong, C.H.; Lee, C.H.; Yu, H.S.; Huang, S.K. Benzopyrene, a major polyaromatic hydrocarbon in smoke fume, mobilizes Langerhans cells and polarizes Th2/17 responses in epicutaneous protein sensitization through the aryl hydrocarbon receptor. *Int. Immunopharmacol.* **2016**, *36*, 111–117. [CrossRef]
98. Kiyomatsu-Oda, M.; Uchi, H.; Morino-Koga, S.; Furue, M. Protective role of 6-formylindolo[3,2-b]carbazole (FICZ), an endogenous ligand for arylhydrocarbon receptor, in chronic mite-induced dermatitis. *J. Dermatol. Sci.* **2018**, *90*, 284–294. [CrossRef]
99. Furue, M.; Hashimoto-Hachiya, A.; Tsuji, G. Aryl hydrocarbon receptor in atopic dermatitis and psoriasis. *Int. J. Mol. Sci.* **2019**, *20*, 5424. [CrossRef]
100. Kim, H.O.; Kim, J.H.; Chung, B.Y.; Choi, M.G.; Park, C.W. Increased expression of the aryl hydrocarbon receptor in patients with chronic inflammatory skin diseases. *Exp. Dermatol.* **2014**, *23*, 278–281. [CrossRef]
101. Hu, Y.Q.; Liu, P.; Mu, Z.L.; Zhang, J.Z. Aryl hydrocarbon receptor expression in serum, peripheral blood mononuclear cells, and skin lesions of patients with atopic dermatitis and its correlation with disease severity. *Chin. Med. J. (Engl.)* **2020**, *133*, 148–153. [CrossRef] [PubMed]
102. Takei, K.; Mitoma, C.; Hashimoto-Hachiya, A.; Uchi, H.; Takahara, M.; Tsuji, G.; Kido-Nakahara, M.; Nakahara, T.; Furue, M. Antioxidant soybean tar Glyteer rescues T-helper-mediated downregulation of filaggrin expression via aryl hydrocarbon receptor. *J. Dermatol.* **2015**, *42*, 171–180. [CrossRef] [PubMed]
103. Edamitsu, T.; Taguchi, K.; Kobayashi, E.H.; Okuyama, R.; Yamamoto, M. Aryl Hydrocarbon receptor directly regulates artemin gene expression. *Mol. Cell Biol.* **2019**, *39*, e00190-19. [CrossRef] [PubMed]
104. Hidaka, T.; Ogawa, E.; Kobayashi, E.H.; Suzuki, T.; Funayama, R.; Nagashima, T.; Fujimura, T.; Aiba, S.; Nakayama, K.; Okuyama, R.; et al. The aryl hydrocarbon receptor AhR links atopic dermatitis and air pollution via induction of the neurotrophic factor artemin. *Nat. Immunol.* **2017**, *18*, 64–73. [CrossRef] [PubMed]
105. Morino-Koga, S.; Uchi, H.; Mitoma, C.; Wu, Z.; Kiyomatsu, M.; Fuyuno, Y.; Nagae, K.; Yasumatsu, M.; Suico, M.A.; Kai, H.; et al. 6-Formylindolo[3,2-b]Carbazole accelerates skin wound healing via activation of ERK, but not aryl hydrocarbon receptor. *J. Investig. Dermatol.* **2017**, *137*, 2217–2226. [CrossRef]
106. Takemura, M.; Nakahara, T.; Hashimoto-Hachiya, A.; Furue, M.; Tsuji, G. Glyteer, soybean tar, impairs IL-4/Stat6 signaling in murine bone marrow-derived dendritic cells: The basis of its therapeutic effect on atopic dermatitis. *Int. J. Mol. Sci.* **2018**, *19*, 1169. [CrossRef]
107. Smits, J.P.H.; Ederveen, T.H.A.; Rikken, G.; van den Brink, N.J.M.; van Vlijmen-Willems, I.M.J.J.; Boekhorst, J.; Kamsteeg, M.; Schalkwijk, J.; van Hijum, S.A.F.T.; Zeeuwen, P.L.J.M.; et al. Targeting the cutaneous microbiota in atopic dermatitis by coal tar via AHR-dependent induction of antimicrobial peptides. *J. Investig. Dermatol.* **2020**, *140*, 415–424.e10. [CrossRef]
108. Smith, S.H.; Jayawickreme, C.; Rickard, D.J.; Nicodeme, E.; Bui, T.; Simmons, C.; Coquery, C.M.; Neil, J.; Pryor, W.M.; Mayhew, D.; et al. Tapinarof is a natural AhR agonist that resolves skin inflammation in mice and humans. *J. Investig. Dermatol.* **2017**, *137*, 2110–2119. [CrossRef]

109. Kamata, M.; Tada, Y. Safety of biologics in psoriasis. *J. Dermatol.* **2018**, *45*, 279–286. [CrossRef]
110. Cibrian, D.; Saiz, M.L.; de la Fuente, H.; Sánchez-Díaz, R.; Moreno-Gonzalo, O.; Jorge, I.; Ferrarini, A.; Vázquez, J.; Punzón, C.; Fresno, M.; et al. CD69 controls the uptake of L-tryptophan through LAT1-CD98 and AhR-dependent secretion of IL-22 in psoriasis. *Nat. Immunol.* **2016**, *17*, 985–996. [CrossRef]
111. Qiu, J.; Heller, J.J.; Guo, X.; Chen, Z.M.; Fish, K.; Fu, Y.X.; Zhou, L. The aryl hydrocarbon receptor regulates gut immunity through modulation of innate lymphoid cells. *Immunity* **2012**, *36*, 92–104. [CrossRef] [PubMed]
112. Martin, B.; Hirota, K.; Cua, D.J.; Stockinger, B.; Veldhoen, M. Interleukin-17-producing gammadelta T cells selectively expand in response to pathogen products and environmental signals. *Immunity* **2009**, *31*, 321–330. [CrossRef] [PubMed]
113. Shimauchi, T.; Hirakawa, S.; Suzuki, T.; Yasuma, A.; Majima, Y.; Tatsuno, K.; Yagi, H.; Ito, T.; Tokura, Y. Serum interleukin-22 and vascular endothelial growth factor serve as sensitive biomarkers but not as predictors of therapeutic response to biologics in patients with psoriasis. *J. Dermatol.* **2013**, *40*, 805–812. [CrossRef] [PubMed]
114. Wolk, K.; Witte, E.; Wallace, E.; Döcke, W.D.; Kunz, S.; Asadullah, K.; Volk, H.D.; Sterry, W.; Sabat, R. IL-22 regulates the expression of genes responsible for antimicrobial defense, cellular differentiation, and mobility in keratinocytes: A potential role in psoriasis. *Eur. J. Immunol.* **2006**, *36*, 1309–1323. [CrossRef] [PubMed]
115. Cochez, P.M.; Michiels, C.; Hendrickx, E.; Van Belle, A.B.; Lemaire, M.M.; Dauguet, N.; Warnier, G.; de Heusch, M.; Togbe, D.; Ryffel, B.; et al. AhR modulates the IL-22-producing cell proliferation/recruitment in imiquimod-induced psoriasis mouse model. *Eur. J. Immunol.* **2016**, *46*, 1449–1459. [CrossRef]
116. Wolk, K.; Haugen, H.S.; Xu, W.; Witte, E.; Waggie, K.; Anderson, M.; Vom Baur, E.; Witte, K.; Warszawska, K.; Philipp, S.; et al. IL-22 and IL-20 are key mediators of the epidermal alterations in psoriasis while IL-17 and IFN-gamma are not. *J. Mol. Med. (Berl.)* **2009**, *87*, 523–536. [CrossRef]
117. di Meglio, P.; Duarte, J.H.; Ahlfors, H.; Owens, N.D.L.; Li, Y.; Villanova, F.; Tosi, I.; Hirota, K.; Nestle, F.O.; Mrowietz, U.; et al. Activation of the aryl hydrocarbon receptor dampens the severity of inflammatory skin conditions. *Immunity* **2014**, *40*, 989–1001. [CrossRef]
118. Goldsmith, Z.G.; Dhanasekaran, D.N. G protein regulation of MAPK networks. *Oncogene* **2007**, *26*, 3122–3142. [CrossRef]
119. Walczak, K.; Turski, W.A.; Rajtar, G. Kynurenic acid inhibits colon cancer proliferation in vitro: Effects on signaling pathways. *Amino Acids.* **2014**, *46*, 2393–2401. [CrossRef]
120. Beránek, M.; Fiala, Z.; Kremláček, J.; Andrýs, C.; Krejsek, J.; Hamáková, K.; Palička, V.; Borská, L. Serum levels of aryl hydrocarbon receptor, cytochromes P450 1A1 and 1B1 in patients with exacerbated psoriasis vulgaris. *Folia. Biol. (Praha.)* **2018**, *64*, 97–102.
121. Tian, S.; Krueger, J.G.; Li, K.; Jabbari, A.; Brodmerkel, C.; Lowes, M.A.; Suárez-Fariñas, M. Meta-analysis derived (MAD) transcriptome of psoriasis defines the "core" pathogenesis of disease. *PLoS ONE* **2012**, *7*, e44274. [CrossRef] [PubMed]
122. Gudjonsson, J.E.; Ding, J.; Johnston, A.; Tejasvi, T.; Guzman, A.M.; Nair, R.P.; Voorhees, J.J.; Abecasis, G.R.; Elder, J.T. Assessment of the psoriatic transcriptome in a large sample: Additional regulated genes and comparisons with in vitro models. *J. Investig. Dermatol.* **2010**, *130*, 1829–1840. [CrossRef] [PubMed]
123. DiNatale, B.C.; Murray, I.A.; Schroeder, J.C.; Flaveny, C.A.; Lahoti, T.S.; Laurenzana, E.M.; Omiecinski, C.J.; Perdew, G.H. Kynurenic acid is a potent endogenous aryl hydrocarbon receptor ligand that synergistically induces interleukin-6 in the presence of inflammatory signaling. *Toxicol. Sci.* **2010**, *115*, 89–97. [CrossRef] [PubMed]
124. Opitz, C.A.; Litzenburger, U.M.; Sahm, F.; Ott, M.; Tritschler, I.; Trump, S.; Schumacher, T.; Jestaedt, L.; Schrenk, D.; Weller, M.; et al. An endogenous tumour-promoting ligand of the human aryl hydrocarbon receptor. *Nature* **2011**, *478*, 197–203. [CrossRef]
125. Wei, Y.D.; Bergander, L.; Rannug, U.; Rannug, A. Regulation of CYP1A1 transcription via the metabolism of the tryptophan-derived 6-formylindolo[3,2-b]carbazole. *Arch. Biochem. Biophys.* **2000**, *383*, 99–107. [CrossRef]
126. di Meglio, P.; Perera, G.K.; Nestle, F.O. The multitasking organ: Recent insights into skin immune function. *Immunity* **2011**, *35*, 857–869. [CrossRef]
127. van der Fits, L.; Mourits, S.; Voerman, J.S.; Kant, M.; Boon, L.; Laman, J.D.; Cornelissen, F.; Mus, A.M.; Florencia, E.; Prens, E.P.; et al. Imiquimod-induced psoriasis-like skin inflammation in mice is mediated via the IL-23/IL-17 axis. *J. Immunol.* **2009**, *182*, 5836–5845. [CrossRef]
128. Zhu, Z.; Chen, J.; Lin, Y.; Zhang, C.; Li, W.; Qiao, H.; Fu, M.; Dang, E.; Wang, G. Aryl hydrocarbon receptor in cutaneous vascular endothelial cells restricts psoriasis development by negatively regulating neutrophil recruitment. *J. Investig. Dermatol.* **2020**, *140*, 1233–1243.e9. [CrossRef]
129. Nakamura, M.; Ueda, Y.; Hayashi, M.; Kato, H.; Furuhashi, T.; Morita, A. Tobacco smoke-induced skin pigmentation is mediated by the aryl hydrocarbon receptor. *Exp. Dermatol.* **2013**, *22*, 556–558. [CrossRef]
130. Nguyen, N.T.; Fisher, D.E. MITF and UV responses in skin: From pigmentation to addiction. *Pigment Cell Melanoma Res.* **2019**, *32*, 224–236. [CrossRef]
131. Luecke, S.; Backlund, M.; Jux, B.; Esser, C.; Krutmann, J.; Rannug, A. The aryl hydrocarbon receptor (AHR), a novel regulator of human melanogenesis. *Pigment Cell Melanoma Res.* **2010**, *23*, 828–833. [CrossRef] [PubMed]
132. Abbas, S.; Alam, S.; Singh, K.P.; Kumar, M.; Gupta, S.K.; Ansari, K.M. Aryl hydrocarbon receptor activation contributes to benzanthrone-induced hyperpigmentation via modulation of melanogenic signaling pathways. *Chem. Res. Toxicol.* **2017**, *30*, 625–634. [CrossRef] [PubMed]
133. Rashighi, M.; Harris, J.E. Vitiligo pathogenesis and emerging treatments. *Dermatol. Clin.* **2017**, *35*, 257–265. [CrossRef]

134. Rekik, R.; Ben Hmid, A.; Lajnef, C.; Zamali, I.; Zaraa, I.; Ben Ahmed, M. Aryl hydrocarbon receptor (AhR) transcription is decreased in skin of vitiligo patients. *Int. J. Dermatol.* **2017**, *56*, 1509–1512. [CrossRef] [PubMed]
135. Haarmann-Stemmann, T.; Esser, C.; Krutmann, J. The janus-faced role of aryl hydrocarbon receptor signaling in the skin: Consequences for prevention and treatment of skin disorders. *J. Investig. Dermatol.* **2015**, *135*, 2572–2576. [CrossRef]
136. Dwivedi, M.; Kemp, E.H.; Laddha, N.C.; Mansuri, M.S.; Weetman, A.P.; Begum, R. Regulatory T cells in vitiligo: Implications for pathogenesis and therapeutics. *Autoimmun. Rev.* **2015**, *14*, 49–56. [CrossRef]
137. Ahmed, M.B.; Zaraa, I.; Rekik, R.; Elbeldi-Ferchiou, A.; Kourda, N.; Belhadj Hmida, N.; Abdeladhim, M.; Karoui, O.; Ben Osman, A.; Mokni, M.; et al. Functional defects of peripheral regulatory T lymphocytes in patients with progressive vitiligo. *Pigment Cell Melanoma Res.* **2012**, *25*, 99–109. [CrossRef]
138. Taher, Z.A.; Lauzon, G.; Maguiness, S.; Dytoc, M.T. Analysis of interleukin-10 levels in lesions of vitiligo following treatment with topical tacrolimus. *Br. J. Dermatol.* **2009**, *161*, 654–659. [CrossRef]
139. Wang, X.; Li, K.; Liu, L.; Shi, Q.; Song, P.; Jian, Z.; Guo, S.; Wang, G.; Li, C.; Gao, T. AHR promoter variant modulates its transcription and downstream effectors by allele-specific AHR-SP1 interaction functioning as a genetic marker for vitiligo. *Sci. Rep.* **2015**, *5*, 13542. [CrossRef]
140. Eguchi, H.; Hayashi, S.I.; Watanabe, J.; Gotoh, O.; Kawajiri, K. Molecular cloning of the human ah receptor gene promoter. *Biochem. Biophys. Res. Commun.* **1994**, *203*, 615–622. [CrossRef]
141. Behfarjam, F.; Jadali, Z. Vitiligo patients show significant up-regulation of aryl hydrocarbon receptor transcription factor. *An. Bras. Dermatol.* **2018**, *93*, 302–303. [CrossRef] [PubMed]
142. Singh, R.K.; Lee, K.M.; Vujkovic-Cvijin, I.; Ucmak, D.; Farahnik, B.; Abrouk, M.; Nakamura, M.; Zhu, T.H.; Bhutani, T.; Wei, M.; et al. The role of IL-17 in vitiligo: A review. *Autoimmun. Rev.* **2016**, *15*, 397–404. [CrossRef]
143. Schallreuter, K.U.; Salem, M.A.; Gibbons, N.C.; Martinez, A.; Slominski, R.; Lüdemann, J.; Rokos, H. Blunted epidermal L-tryptophan metabolism in vitiligo affects immune response and ROS scavenging by Fenton chemistry, part 1: Epidermal H2O2/ONOO(-)-mediated stress abrogates tryptophan hydroxylase and dopa decarboxylase activities, leading to low serotonin and melatonin levels. *FASEB J.* **2012**, *26*, 2457–2470. [CrossRef]
144. Saurat, J.H.; Kaya, G.; Saxer-Sekulic, N.; Pardo, B.; Becker, M.; Fontao, L.; Mottu, F.; Carraux, P.; Pham, X.C.; Barde, C.; et al. The cutaneous lesions of dioxin exposure: Lessons from the poisoning of Victor Yushchenko. *Toxicol. Sci.* **2012**, *125*, 310–317. [CrossRef] [PubMed]
145. Furue, M.; Tsuji, G. Chloracne and hyperpigmentation caused by exposure to hazardous aryl hydrocarbon receptor ligands. *Int. J. Environ. Res. Public Health* **2019**, *16*, 4864. [CrossRef] [PubMed]
146. Panteleyev, A.A.; Bickers, D.R. Dioxin-induced chloracne—reconstructing the cellular and molecular mechanisms of a classic environmental disease. *Exp. Dermatol.* **2006**, *15*, 705–730. [CrossRef] [PubMed]
147. Suskind, R.R. Chloracne, "the hallmark of dioxin intoxication". *Scand. J. Work Environ. Health* **1985**, *11*, 165–171. [CrossRef]
148. Van den Bogaard, E.H.; Podolsky, M.A.; Smits, J.P.; Cui, X.; John, C.; Gowda, K.; Desai, D.; Amin, S.G.; Schalkwijk, J.; Perdew, G.H.; et al. Genetic and pharmacological analysis identifies a physiological role for the AHR in epidermal differentiation. *J. Investig. Dermatol.* **2015**, *135*, 1320–1328. [CrossRef]
149. Fernandez-Salguero, P.M.; Ward, J.M.; Sundberg, J.P.; Gonzalez, F.J. Lesions of aryl-hydrocarbon receptor-deficient mice. *Vet. Pathol.* **1997**, *34*, 605–614. [CrossRef]
150. Morokuma, S.; Tsukimori, K.; Hori, T.; Kato, K.; Furue, M. The vernix caseosa is the main site of dioxin excretion in the human foetus. *Sci. Rep.* **2017**, *7*, 739. [CrossRef]
151. Hu, T.; Wang, D.; Yu, Q.; Li, L.; Mo, X.; Pan, Z.; Zouboulis, C.C.; Peng, L.; Xia, L.; Ju, Q. Aryl hydrocarbon receptor negatively regulates lipid synthesis and involves in cell differentiation of SZ95 sebocytes in vitro. *Chem. Biol. Interact.* **2016**, *258*, 52–58. [CrossRef] [PubMed]
152. Ju, Q.; Fimmel, S.; Hinz, N.; Stahlmann, R.; Xia, L.; Zouboulis, C.C. 2,3,7,8-Tetrachlorodibenzo-p-dioxin alters sebaceous gland cell differentiation in vitro. *Exp. Dermatol.* **2011**, *20*, 320–325. [CrossRef] [PubMed]
153. Liu, Q.; Wu, J.; Song, J.; Liang, P.; Zheng, K.; Xiao, G.; Liu, L.; Zouboulis, C.C.; Lei, T. Particulate matter 2.5 regulates lipid synthesis and inflammatory cytokine production in human SZ95 sebocytes. *Int. J. Mol. Med.* **2017**, *40*, 1029–1036. [CrossRef] [PubMed]
154. Hu, T.; Pan, Z.; Yu, Q.; Mo, X.; Song, N.; Yan, M.; Zouboulis, C.C.; Xia, L.; Ju, Q. Benzo(a)pyrene induces interleukin (IL)-6 production and reduces lipid synthesis in human SZ95 sebocytes via the aryl hydrocarbon receptor signaling pathway. *Environ. Toxicol. Pharmacol.* **2016**, *43*, 54–60. [CrossRef]
155. Muku, G.E.; Blazanin, N.; Dong, F.; Smith, P.B.; Thiboutot, D.; Gowda, K.; Amin, S.; Murray, I.A.; Perdew, G.H. Selective Ah receptor ligands mediate enhanced SREBP1 proteolysis to restrict lipogenesis in sebocytes. *Toxicol. Sci.* **2019**, *171*, 146–158. [CrossRef]
156. Liu, J.; Zhang, C.M.; Coenraads, P.J.; Ji, Z.Y.; Chen, X.; Dong, L.; Ma, X.M.; Han, W.; Tang, N.J. Abnormal expression of MAPK, EGFR, CK17 and TGk in the skin lesions of chloracne patients exposed to dioxins. *Toxicol. Lett.* **2011**, *201*, 230–234. [CrossRef]
157. Sutter, C.H.; Yin, H.; Li, Y.; Mammen, J.S.; Bodreddigari, S.; Stevens, G.; Cole, J.A.; Sutter, T.R. EGF receptor signaling blocks aryl hydrocarbon receptor-mediated transcription and cell differentiation in human epidermal keratinocytes. *Proc. Natl. Acad. Sci. USA* **2009**, *106*, 4266–4271. [CrossRef]

158. Kypriotou, M.; Huber, M.; Hohl, D. The human epidermal differentiation complex: Cornified envelope precursors, S100 proteins and the 'fused genes' family. *Exp. Dermatol.* **2012**, *21*, 643–649. [CrossRef]
159. Loertscher, J.A.; Sattler, C.A.; Allen-Hoffmann, B.L. 2,3,7,8-Tetrachlorodibenzo-p-dioxin alters the differentiation pattern of human keratinocytes in organotypic culture. *Toxicol. Appl. Pharmacol.* **2001**, *175*, 121–129. [CrossRef]
160. Muenyi, C.S.; Carrion, S.L.; Jones, L.A.; Kennedy, L.H.; Slominski, A.T.; Sutter, C.H.; Sutter, T.R. Effects of in utero exposure of C57BL/6J mice to 2,3,7,8-tetrachlorodibenzo-p-dioxin on epidermal permeability barrier development and function. *Environ. Health Perspect.* **2014**, *122*, 1052–1058. [CrossRef]
161. Panteleyev, A.A.; Thiel, R.; Wanner, R.; Zhang, J.; Roumak, V.S.; Paus, R.; Neubert, D.; Henz, B.M.; Rosenbach, T. 2,3,7,8-tetrachlorodibenzo-p-dioxin (TCCD) affects keratin 1 and keratin 17 gene expression and differentially induces keratinization in hairless mouse skin. *J. Investig. Dermatol.* **1997**, *108*, 330–335. [CrossRef] [PubMed]
162. Kennedy, L.H.; Sutter, C.H.; Carrion, S.L.; Tran, Q.T.; Bodreddigari, S.; Kensicki, E.; Mohney, R.P.; Sutter, T.R. 2,3,7,8-Tetrachlorodibenzo-p-dioxin-mediated production of reactive oxygen species is an essential step in the mechanism of action to accelerate human keratinocyte differentiation. *Toxicol. Sci.* **2013**, *132*, 235–249. [CrossRef] [PubMed]
163. Pohjanvirta, R. *The AH Receptor in Biology and Toxicology*; John Wiley & Sons: New York, NY, USA, 2012; pp. 485–497. [CrossRef]
164. Furness, S.G.; Lees, M.J.; Whitelaw, M.L. The dioxin (aryl hydrocarbon) receptor as a model for adaptive responses of bHLH/PAS transcription factors. *FEBS Lett.* **2007**, *581*, 3616–3625. [CrossRef]
165. Contador-Troca, M.; Alvarez-Barrientos, A.; Barrasa, E.; Rico-Leo, E.M.; Catalina-Fernández, I.; Menacho-Márquez, M.; Bustelo, X.R.; García-Borrón, J.C.; Gómez-Durán, A.; Sáenz-Santamaría, J.; et al. The dioxin receptor has tumor suppressor activity in melanoma growth and metastasis. *Carcinogenesis* **2013**, *34*, 2683–2693. [CrossRef] [PubMed]
166. Aziz, M.H.; Reagan-Shaw, S.; Wu, J.; Longley, B.J.; Ahmad, N. Chemoprevention of skin cancer by grape constituent resveratrol: Relevance to human disease? *FASEB J.* **2005**, *19*, 1193–1195. [CrossRef]
167. Gawkrodger, D.J. Occupational skin cancers. *Occup. Med. (Lond.)* **2004**, *54*, 458–463. [CrossRef]
168. Richard, G.; Puisieux, A.; Caramel, J. Antagonistic functions of EMT-inducers in melanoma development: Implications for cancer cell plasticity. *Cancer Cell Microenviron.* **2014**, e61. [CrossRef]
169. Kostyuk, V.A.; Potapovich, A.I.; Lulli, D.; Stancato, A.; De Luca, C.; Pastore, S.; Korkina, L. Modulation of human keratinocyte responses to solar UV by plant polyphenols as a basis for chemoprevention of non-melanoma skin cancers. *Curr. Med. Chem.* **2013**, *20*, 869–879.
170. Rogers, H.W.; Weinstock, M.A.; Feldman, S.R.; Coldiron, B.M. Incidence estimate of nonmelanoma skin cancer (keratinocyte carcinomas) in the U.S. Population, 2012. *JAMA Dermatol.* **2015**, *151*, 1081–1086. [CrossRef]
171. Chahal, H.S.; Lin, Y.; Ransohoff, K.J.; Hinds, D.A.; Wu, W.; Dai, H.J.; Qureshi, A.A.; Li, W.Q.; Kraft, P.; Tang, J.Y.; et al. Genome-wide association study identifies novel susceptibility loci for cutaneous squamous cell carcinoma. *Nat. Commun.* **2016**, *7*, 12048. [CrossRef]
172. Vogeley, C.; Esser, C.; Tüting, T.; Krutmann, J.; Haarmann-Stemmann, T. Role of the aryl hydrocarbon receptor in environmentally induced skin aging and skin carcinogenesis. *Int. J. Mol. Sci.* **2019**, *20*, 6005. [CrossRef] [PubMed]
173. Kalmes, M.; Hennen, J.; Clemens, J.; Blömeke, B. Impact of aryl hydrocarbon receptor (AhR) knockdown on cell cycle progression in human HaCaT keratinocytes. *Biol. Chem.* **2011**, *392*, 643–651. [CrossRef] [PubMed]
174. Pollet, M.; Shaik, S.; Mescher, M.; Krutmann, J.; Haarmann-Stemmann, T. The AHR represses nucleotide excision repair and apoptosis and contributes to UV-induced skin carcinogenesis. *Cell Death Differ.* **2018**, *25*, 1823–1836. [CrossRef] [PubMed]
175. O'Donnell, E.F.; Kopparapu, P.R.; Koch, D.C.; Jang, H.S.; Phillips, J.L.; Tanguay, R.L.; Kerkvliet, N.I.; Kolluri, S.K. The aryl hydrocarbon receptor mediates leflunomide-induced growth inhibition of melanoma cells. *PLoS ONE* **2012**, *7*, e40926. [CrossRef]
176. Roman, A.C.; Carvajal-Gonzalez, J.M.; Rico-Leo, E.M.; Fernandez-Salguero, P.M. Dioxin receptor deficiency impairs angiogenesis by a mechanism involving VEGF-A depletion in the endothelium and transforming growth factor-beta overexpression in the stroma. *J. Biol. Chem.* **2009**, *284*, 25135–25148. [CrossRef]
177. Villano, C.M.; Murphy, K.A.; Akintobi, A.; White, L.A. 2,3,7,8-tetrachlorodibenzo-p-dioxin (TCDD) induces matrix metalloproteinase (MMP) expression and invasion in A2058 melanoma cells. *Toxicol. Appl. Pharmacol.* **2006**, *210*, 212–224. [CrossRef]
178. Helferich, W.G.; Denison, M.S. Ultraviolet photoproducts of tryptophan can act as dioxin agonists. *Mol. Pharmacol.* **1991**, *40*, 674–678.
179. Walczak, K.; Wnorowski, A.; Turski, W.A.; Plech, T. Kynurenic acid and cancer: Facts and controversies. *Cell Mol. Life Sci.* **2020**, *77*, 1531–1550. [CrossRef]
180. Cella, M.; Colonna, M. Aryl hydrocarbon receptor: Linking environment to immunity. *Semin. Immunol.* **2015**, *27*, 310–314. [CrossRef]
181. Turski, M.P.; Turska, M.; Zgrajka, W.; Bartnik, M.; Kocki, T.; Turski, W.A. Distribution, synthesis, and absorption of kynurenic acid in plants. *Planta Med.* **2011**, *77*, 858–864. [CrossRef]
182. Turski, M.P.; Turska, M.; Zgrajka, W.; Kuc, D.; Turski, W.A. Presence of kynurenic acid in food and honeybee products. *Amino Acids* **2009**, *36*, 75–80. [CrossRef] [PubMed]
183. Jönsson, M.E.; Franks, D.G.; Woodin, B.R.; Jenny, M.J.; Garrick, R.A.; Behrendt, L.; Hahn, M.E.; Stegeman, J.J. The tryptophan photoproduct 6-formylindolo[3,2-b]carbazole (FICZ) binds multiple AHRs and induces multiple CYP1 genes via AHR2 in zebrafish. *Chem. Biol. Interact.* **2009**, *181*, 447–454. [CrossRef] [PubMed]

184. Park, S.L.; Justiniano, R.; Williams, J.D.; Cabello, C.M.; Qiao, S.; Wondrak, G.T. The tryptophan-derived endogenous aryl hydrocarbon receptor ligand 6-Formylindolo[3,2-b]Carbazole is a nanomolar UVA photosensitizer in epidermal keratinocytes. *J. Investig. Dermatol.* **2015**, *135*, 1649–1658. [CrossRef] [PubMed]
185. Mukhtar, H.; DelTito, B.J., Jr.; Matgouranis, P.M.; Das, M.; Asokan, P.; Bickers, D.R. Additive effects of ultraviolet B and crude coal tar on cutaneous carcinogen metabolism: Possible relevance to the tumorigenicity of the Goeckerman regimen. *J. Investig. Dermatol.* **1986**, *87*, 348–353. [CrossRef] [PubMed]
186. Katiyar, S.K.; Matsui, M.S.; Mukhtar, H. Ultraviolet-B exposure of human skin induces cytochromes P450 1A1 and 1B1. *J. Investig. Dermatol.* **2000**, *114*, 328–333. [CrossRef] [PubMed]
187. Rannug, A.; Fritsche, E. The aryl hydrocarbon receptor and light. *Biol. Chem.* **2006**, *387*, 1149–1157. [CrossRef]
188. Murai, M.; Tsuji, G.; Hashimoto-Hachiya, A.; Kawakami, Y.; Furue, M.; Mitoma, C. An endogenous tryptophan photo-product, FICZ, is potentially involved in photo-aging by reducing TGF-β-regulated collagen homeostasis. *J. Dermatol. Sci.* **2018**, *89*, 19–26. [CrossRef]
189. Mengoni, M.; Braun, A.D.; Gaffal, E.; Tüting, T. The aryl hydrocarbon receptor promotes inflammation-induced dedifferentiation and systemic metastatic spread of melanoma cells. *Int. J. Cancer* **2020**, *147*, 2902–2913. [CrossRef]
190. Ikuta, T.; Kawajiri, K. Zinc finger transcription factor Slug is a novel target gene of aryl hydrocarbon receptor. *Exp. Cell Res.* **2006**, *312*, 3585–3594. [CrossRef]
191. Mezrich, J.D.; Fechner, J.H.; Zhang, X.; Johnson, B.P.; Burlingham, W.J.; Bradfield, C.A. An interaction between kynurenine and the aryl hydrocarbon receptor can generate regulatory T cells. *J. Immunol.* **2010**, *185*, 3190–3198. [CrossRef]
192. Bessede, A.; Gargaro, M.; Pallotta, M.T.; Matino, D.; Servillo, G.; Brunacci, C.; Bicciato, S.; Mazza, E.M.; Macchiarulo, A.; Vacca, C.; et al. Aryl hydrocarbon receptor control of a disease tolerance defence pathway. *Nature* **2014**, *511*, 184–190. [CrossRef] [PubMed]
193. Nuti, R.; Gargaro, M.; Matino, D.; Dolciami, D.; Grohmann, U.; Puccetti, P.; Fallarino, F.; Macchiarulo, A. Ligand binding and functional selectivity of L-tryptophan metabolites at the mouse aryl hydrocarbon receptor (mAhR). *J. Chem. Inf. Model.* **2014**, *54*, 3373–3383. [CrossRef] [PubMed]
194. Pilotte, L.; Larrieu, P.; Stroobant, V.; Colau, D.; Dolusic, E.; Frédérick, R.; De Plaen, E.; Uyttenhove, C.; Wouters, J.; Masereel, B.; et al. Reversal of tumoral immune resistance by inhibition of tryptophan 2,3-dioxygenase. *Proc. Natl. Acad. Sci. USA* **2012**, *109*, 2497–2502. [CrossRef] [PubMed]
195. Litzenburger, U.M.; Opitz, C.A.; Sahm, F.; Rauschenbach, K.J.; Trump, S.; Winter, M.; Ott, M.; Ochs, K.; Lutz, C.; Liu, X.; et al. Constitutive IDO expression in human cancer is sustained by an autocrine signaling loop involving IL-6, STAT3 and the AHR. *Oncotarget* **2014**, *5*, 1038–1051. [CrossRef] [PubMed]
196. Wirthgen, E.; Hoeflich, A. Endotoxin-induced tryptophan degradation along the kynurenine pathway: The role of indolamine 2,3-Dioxygenase and aryl hydrocarbon receptor-mediated immunosuppressive effects in endotoxin tolerance and cancer and its implications for immunoparalysis. *J. Amino Acids* **2015**, *2015*, 973548. [CrossRef] [PubMed]
197. Zhang, S.; Patel, A.; Chu, C.; Jiang, W.; Wang, L.; Welty, S.E.; Moorthy, B.; Shivanna, B. Aryl hydrocarbon receptor is necessary to protect fetal human pulmonary microvascular endothelial cells against hyperoxic injury: Mechanistic roles of antioxidant enzymes and RelB. *Toxicol. Appl. Pharmacol.* **2015**, *286*, 92–101. [CrossRef]
198. Baglole, C.J.; Maggirwar, S.B.; Gasiewicz, T.A.; Thatcher, T.H.; Phipps, R.P.; Sime, P.J. The aryl hydrocarbon receptor attenuates tobacco smoke-induced cyclooxygenase-2 and prostaglandin production in lung fibroblasts through regulation of the NF-kappaB family member RelB. *J. Biol. Chem.* **2008**, *283*, 28944–28957. [CrossRef]
199. Sekine, H.; Mimura, J.; Oshima, M.; Okawa, H.; Kanno, J.; Igarashi, K.; Gonzalez, F.J.; Ikuta, T.; Kawajiri, K.; Fujii-Kuriyama, Y. Hypersensitivity of aryl hydrocarbon receptor-deficient mice to lipopolysaccharide-induced septic shock. *Mol. Cell. Biol.* **2009**, *29*, 6391–6400. [CrossRef]
200. Murray, I.A.; Patterson, A.D.; Perdew, G.H. Aryl hydrocarbon receptor ligands in cancer: Friend and foe. *Nat. Rev. Cancer* **2014**, *14*, 801–814. [CrossRef]
201. Walczak, K.; Turski, W.A.; Rzeski, W. Kynurenic acid enhances expression of p21 Waf1/Cip1 in colon cancer HT-29 cells. *Pharmacol. Rep.* **2012**, *64*, 745–750. [CrossRef]
202. Walczak, K.; Langner, E.; Makuch-Kocka, A.; Szelest, M.; Szalast, K.; Marciniak, S.; Plech, T. Effect of tryptophan-derived Ahr ligands, kynurenine, kynurenic acid and FICZ, on proliferation, cell cycle regulation and cell death of melanoma cells-in vitro studies. *Int. J. Mol. Sci.* **2020**, *21*, 7946. [CrossRef] [PubMed]
203. Turski, M.P.; Turska, M.; Paluszkiewicz, P.; Parada-Turska, J.; Oxenkrug, G.F. Kynurenic Acid in the digestive system-new facts, new challenges. *Int. J. Tryptophan. Res.* **2013**, *6*, 47–55. [CrossRef] [PubMed]
204. Swanbeck, G.; Wennersten, G.; Nilsson, R. Participation of singlet state excited oxygen in photohemolysis induced by kynurenic acid. *Acta Derm. Venereol.* **1974**, *54*, 433–436. [PubMed]
205. Wennersten, G.; Brunk, U. Cellular aspects of phototoxic reactions induced by kynurenic acid. I. Establishment of an experimental model utilizing in vitro cultivated cells. *Acta Derm. Venereol.* **1977**, *57*, 201–209.
206. Vogel, C.F.A.; Ishihara, Y.; Campbell, C.E.; Kado, S.Y.; Nguyen-Chi, A.; Sweeney, C.; Pollet, M.; Haarmann-Stemmann, T.; Tuscano, J.M. A protective role of aryl hydrocarbon receptor repressor in inflammation and tumor growth. *Cancers (Basel)* **2019**, *11*, 589. [CrossRef] [PubMed]
207. Tang, Z.; Kang, B.; Li, C.; Chen, T.; Zhang, Z. GEPIA2: An enhanced web server for large-scale expression profiling and interactive analysis. *Nucleic Acids Res.* **2019**, *47(W1)*, W556–W560. [CrossRef]

208. Théate, I.; van Baren, N.; Pilotte, L.; Moulin, P.; Larrieu, P.; Renauld, J.C.; Hervé, C.; Gutierrez-Roelens, I.; Marbaix, E.; Sempoux, C.; et al. Extensive profiling of the expression of the indoleamine 2,3-dioxygenase 1 protein in normal and tumoral human tissues. *Cancer Immunol. Res.* **2015**, *3*, 161–172. [CrossRef]
209. Labadie, B.W.; Bao, R.; Luke, J.J. Reimagining IDO Pathway inhibition in cancer immunotherapy via downstream focus on the tryptophan-kynurenine-aryl hydrocarbon axis. *Clin. Cancer Res.* **2019**, *25*, 1462–1471. [CrossRef]
210. Vlachos, C.; Schulte, B.M.; Magiatis, P.; Adema, G.J.; Gaitanis, G. Malassezia-derived indoles activate the aryl hydrocarbon receptor and inhibit Toll-like receptor-induced maturation in monocyte-derived dendritic cells. *Br. J. Dermatol.* **2012**, *167*, 496–505. [CrossRef]
211. Gaitanis, G.; Velegraki, A.; Magiatis, P.; Pappas, P.; Bassukas, I.D. Could Malassezia yeasts be implicated in skin carcinogenesis through the production of aryl-hydrocarbon receptor ligands? *Med. Hypotheses* **2011**, *77*, 47–51. [CrossRef]

Article

Tryptophan Derivatives by *Saccharomyces cerevisiae* EC1118: Evaluation, Optimization, and Production in a Soybean-Based Medium

Michele Dei Cas [1,†], Ileana Vigentini [2,*,†], Sara Vitalini [3], Antonella Laganaro [2], Marcello Iriti [3], Rita Paroni [1] and Roberto Foschino [2]

[1] Department of Health Sciences, Università degli Studi di Milano, 20142 Milan, Italy; michele.deicas@unimi.it (M.D.C.); rita.paroni@unimi.it (R.P.)
[2] Department of Food, Environmental and Nutritional Sciences, Università degli Studi di Milano, Via G. Celoria 2, 20133 Milan, Italy; antonella.laganaro@gmail.com (A.L.); roberto.foschino@unimi.it (R.F.)
[3] Phytochem Lab, Department of Agricultural and Environmental Sciences, Center for Studies on Bioispired Agro-Environmental Technology (BAT Center), National Interuniversity Consortium of Materials Science and Technology, Università degli Studi di Milano, 20133 Milan, Italy; sara.vitalini@unimi.it (S.V.); marcello.iriti@unimi.it (M.I.)
* Correspondence: ileana.vigentini@unimi.it; Tel.: +39-02-5031-9165
† These authors contributed equally to this work.

Citation: Dei Cas, M.; Vigentini, I.; Vitalini, S.; Laganaro, A.; Iriti, M.; Paroni, R.; Foschino, R. Tryptophan Derivatives by *Saccharomyces cerevisiae* EC1118: Evaluation, Optimization, and Production in a Soybean-Based Medium. *Int. J. Mol. Sci.* **2021**, *22*, 472. https://doi.org/10.3390/ijms22010472

Received: 1 December 2020
Accepted: 30 December 2020
Published: 5 January 2021

Publisher's Note: MDPI stays neutral with regard to jurisdictional claims in published maps and institutional affiliations.

Copyright: © 2021 by the authors. Licensee MDPI, Basel, Switzerland. This article is an open access article distributed under the terms and conditions of the Creative Commons Attribution (CC BY) license (https://creativecommons.org/licenses/by/4.0/).

Abstract: Given the pharmacological properties and the potential role of kynurenic acid (KYNA) in human physiology and the pleiotropic activity of the neurohormone melatonin (MEL) involved in physiological and immunological functions and as regulator of antioxidant enzymes, this study aimed at evaluating the capability of *Saccharomyces cerevisiae* EC1118 to release tryptophan derivatives (dTRPs) from the kynurenine (KYN) and melatonin pathways. The setting up of the spectroscopic and chromatographic conditions for the quantification of the dTRPs in LC-MS/MS system, the optimization of dTRPs' production in fermentative and whole-cell biotransformation approaches and the production of dTRPs in a soybean-based cultural medium naturally enriched in tryptophan, as a case of study, were included in the experimental plan. Variable amounts of dTRPs, with a prevalence of metabolites of the KYN pathway, were detected. The LC-MS/MS analysis showed that the compound synthesized at highest concentration is KYNA that reached 9.146 ± 0.585 mg/L in fermentation trials in a chemically defined medium at 400 mg/L TRP. Further experiments in a soybean-based medium confirm KYNA as the main dTRPs, whereas the other dTRPs reached very lower concentrations. While detectable quantities of melatonin were never observed, two MEL isomers were successfully measured in laboratory media.

Keywords: kynurenine pathway; MEL biosynthesis; *Saccharomyces cerevisiae*; yeast; tryptophan extraction; LC-MS/MS; soybean

1. Introduction

L-tryptophan (L-TRP) is a non-polar amino acid and is the only amino acid containing an indole ring: in addition to being involved in the biosynthesis and turnover of proteins and peptides, tryptophan (TRP) after its absorption into the body is converted into a series of small bioactive, pleiotropic compounds, each capable of influencing certain cellular metabolic pathways and physiological responses. Concerning these bio-transformations, L-TRP is processed through different pathways: 90–97% lead to the breakdown of the indole ring with the formation of kynurenine (KYN) and derivatives; 3–10% keep the indole ring intact and produce chemical messengers of the indolamine family, among which is melatonin (MEL) [1].

The KYN pathway is a metabolic route in which TRP is converted into compounds with different biological functions [2]. In humans, the KYN pathway is involved in neurodegenerative and autoimmune disorders, including Alzheimer's, Huntington's disease, amyotrophic lateral sclerosis diseases, and in cancer and psychiatric syndromes, including schizophrenia [3–5]. KYN itself is a ligand of the aryl receptor and therefore regulates gene expression and immune function [6,7]. The metabolites deriving from the KYN pathway are considered cytoprotective (kynurenic acid, KYNA) or cytotoxic/pro-epilepsy (3OH kynurenine, 3OH KYN, 3OH anthranilic acid, 3OH AA, quinolinic acid, and nicotinic acid). KYNA is an endogenous neuroprotector that is usually present in the brain in nanomolar concentrations. It is an antagonist of the nicotinic cholinergic receptor α7 and a non-competitive antagonist of ionotropic N-methyl-D-aspartate (NMDA) receptors [8], the activation of which would facilitate the initiation of the processes leading to the death of neurons after a short period of anoxia. KYNA has been shown to be the endogenous agonist of the GPR35 receptor expressed mainly in immune cells and peripheral monocyte/macrophage cells where it is able to reduce the secretion of tumor necrosis factor α (TNFα) induced by lipopolysaccharides [9]. It is an antioxidant and neuroprotective compound [10]; in particular, altered KYNA levels may suggest an inflammatory response, as shown in Alzheimer's disease patients [11]. KYNA reduces the heart rate and glaucoma in mice and it has a role in carcinogenesis and cancer therapy [12–14].

Several studies have shown the existence of the KYN pathway in yeasts [15–18]. Panozzo and co-authors (2002) identified the genes that encode several KYN pathway enzymes in *Saccharomyces cerevisiae* [19]. Bna3p has been reported as a KYN aminotransferases (KAT) present in yeast, which converts KYN to KYNA [20]. According to Ohashi et al. (2017) [21], the KAT enzymes Aro8 and Aro9 reduce the toxic effects of some metabolites of TRP, allowing their conversion to less toxic metabolites for the cell. It has been shown that the biosynthesis of NAD$^+$ and niacin through the KYN pathway is present in *S. cerevisiae* and *S. uvarum* [15,17]. Actually, the synthesis of NAD$^+$ in yeasts occurs not only through de novo biosynthesis from TRP but also through the recovery pathway of the NAD$^+$ precursors which are well-known vitamins such as nicotinic acid and nicotinamide. Furthermore, nicotinamide-ribose, nicotinamide mononucleotide, and nicotinic acid-ribose have been identified as NAD$^+$ precursors [17,22–24]; they are transported into the yeast cells from the culture media and assimilated for the NAD$^+$ supply [25,26]. KYNA production has been optimized in *Y. lipolytica* [27], an unconventional yeast with a high biotechnological potential. The results showed that *Y. lipolytica* S12 strain is able to produce KYNA in high concentrations (up to 21.38 µg/mL in culture broth and 494.16 µg/g cell dry weight in biomass) in optimized conditions in a medium supplemented with tryptophan (200 mg/L). Recently. Wrobel-Kwiatkowska et al. 2020 reported that KYNA may be efficiently produced by the yeast *Y. lipolytica* S12 in media containing chestnut honey up to 68 mg/L in culture broth and 542 mg/kg in yeast biomass [28]. KYNA release has been also reported in brewing experiments performed by *S. cerevisiae* and *Saccharomyces pastorianus* [18].

A little information is available on the biosynthesis of MEL in fungi; it can be synthesized by *S. cerevisiae* via two routes. Serotonin (5OH TRY) can be acetylated to N-acetyl serotonin (NAC 5OH TRY) or it can be converted to MEL through 5-methoxytryptamine (5OME TRY) [29,30]. To date, only the *PAA1* gene in *S. cerevisiae*, which codes for a polyamine acetyltransferase, has been proposed as a homologue of the *AANAT* coding for the enzyme arylalkylamine *N*-acetyltransferase present in vertebrates [31]. Kinetic studies have shown that this enzyme has a higher catalytic efficiency for 5OME TRY than 5OH TRY, suggesting that this could be the final enzyme in the MEL biosynthesis pathway. In another study, a different strategy was used to investigate the MEL production by *S. cerevisiae*. Through the addition of intermediates of the pathway in the cells, at different growth phases, intracellular and extracellular samples were analyzed to evaluate the presence of indole compounds. The data suggest that the first step of the pathway is the decarboxylation of TRP to tryptamine (TRY). TRY is then hydroxylated to 5OH TRY which is converted into NAC 5OH TRY or 5OME TRY [32]. This implies that *S. cerevisiae* may use more than one pathway for MEL biosynthesis. MEL, however, is not the only existing N-acetyl-methoxindolamine. Its structural isomers have also been discovered in the past

10 years. It has been calculated that 42 combinations could be possible considering the position of the two side chains connected to the indole ring. From the point of view of biological functions, the MEL isomers (MIs) show antioxidant and cytoprotective activity depending on the modification of the position of the two side chains in the indole ring [33–35]. In recent years, investigations regarding the beneficial effects of fermented foods on human health have increased considerably. This effect is often associated with the metabolism of microorganisms that, proliferating in different matrices, lead to the synthesis of neuroactive compounds [36]. As introduced above, food supplementation with dTRPs could be considered a strategy in the prevention of several diseases. However, dTRPs in natural sources are often in a little amount [37,38]. Thus, the natural enrichment in TRP of food raw materials potentially fermentable by selected dTRPs high-producing microorganisms represent a novel and feasible approach in the setup of novel functional foods. The present study aimed at evaluating and optimizing the release of dTRPs by *S. cerevisiae* EC1118 in laboratory cultural media. Then, dTRPs production was assessed in a medium naturally enriched in TRP by soybean addition with potential functional properties on human health. This last approach can be considered as a "case of study" since it cannot be compared with the current literature; indeed, no studies have investigated so far about the use of raw materials of food origin as a natural source of TRP aimed at obtaining dTRPs.

2. Results and Discussion

This section presents the results obtained from laboratory scale cultures of *S. cerevisiae* EC1118 aimed at evaluating the potential production of dTRPs both from the KYN pathway and, partially, from the serotonin and indole degradation routes. This yeast strain is a well-studied wine yeast strain showing a relevant fermentative fitness [39] and it was found to release dTRPs in a previous investigation [40]. After the assessment of the main compounds released in YNBT100 medium, the increase of the production of dTRPs was successfully reached by optimization of the cultural medium and through experiments of whole-cell bioconversion (WCB). Then, fermentation experiments in a medium with soy flour, naturally rich in TRP, were performed as case of study in order to assess the effects of yeast fermentation in the synthesis of dTRPs.

2.1. Quantitation of dTRPs during the Fermentation of S. Cerevisiae EC1118 in YNBT100 Medium

TRP is one of the most unstable amino acids and the two parameters that can mainly influence its stability are temperature and pH. In order to evaluate the thermal stability of a chemically defined medium containing 100 mg/L TRP, a preliminary test was carried out in YNBT100 in absence of inoculation applying the experimental conditions adopted in study (25 °C, static, aerobic). The initial concentration of TRP was estimated at 98.49 ± 3.16 mg/L. The concentration was detected at 101.19 mg/L ± 4.32 mg/L after 144 h from the beginning of the experiment, demonstrating that there was no significant degradation of TRP; therefore, a decrease of TRP content in subsequent trials must be exclusively linked to the metabolism of *S. cerevisiae* EC1118, employed as selected yeast in this work.

The consumption of TRP and the cellular growth of *S. cerevisiae* EC1118 in YNBT100 medium during the fermentation is shown in Figure 1. The maximum level of yeast biomass was obtained after 24 h from the inoculum (7.00 ± 0.78 × 10^7 CFU/mL); this level remained stable up to 144 h. At the end of the experiment, viable cells were reduced at 2.25 ± 0.46 × 10^7 CFU/mL. The content of TRP at the starting time was 104.100 ± 10.800 mg/L. According to the yeast proliferation, its consumption mainly occurred in the first day of fermentation showing a 91.6% decrease. At 144 h from the inoculum, the TRP was almost depleted reaching a final amount of 0.549 ± 0.001 mg/L.

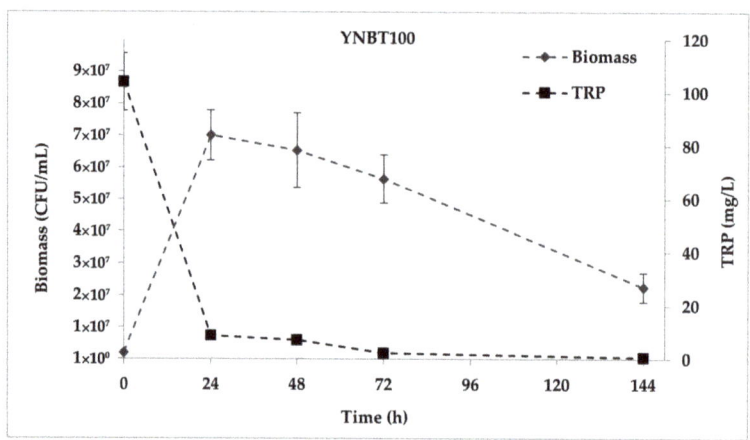

Figure 1. *S. cerevisiae* EC1118 growth curve in YNBT100 medium and TRP consumption.

The KYN pathway is an important metabolic pathway in which TRP is converted to compounds with different biological functions along three different branches (Figure 2a). The KYN route of the TRP degradation was followed throughout the kinetics production of TRP, KYN, 3OH KYN, KYNA, AA, and 3OH AA (Figure 2b).

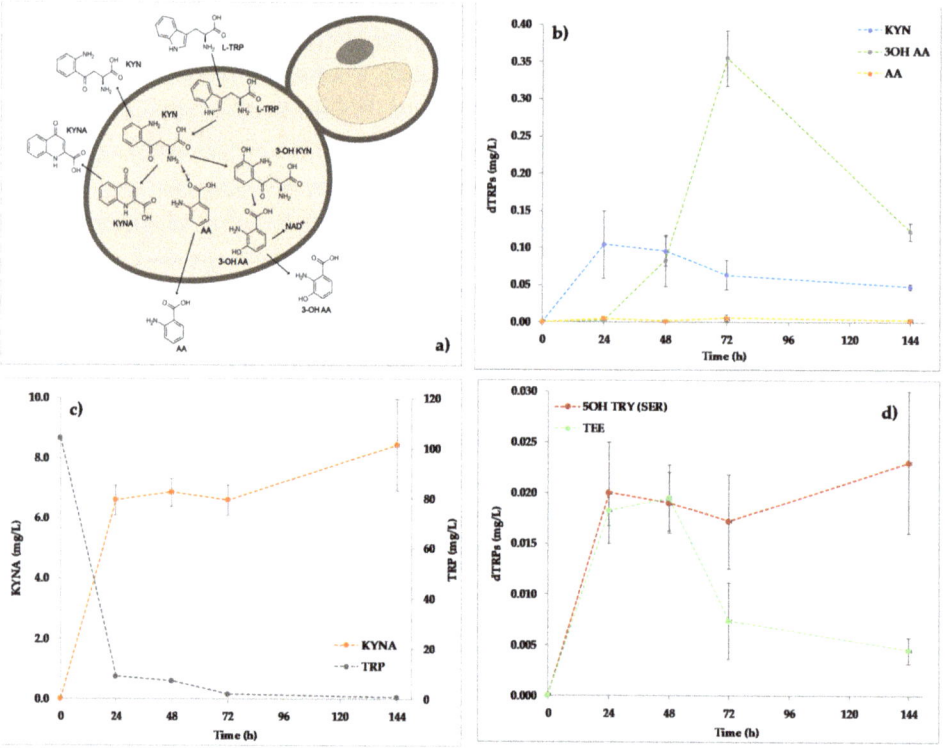

Figure 2. (**a**) Schematic view of the metabolites from the KYN pathway accumulated in YNBT100 medium and (**b**,**c**) kinetics production of the extracellular intermediates of the KYN pathway in *S. cerevisiae* EC1118. (**d**) Kinetics production of the extracellular intermediates of the MEL pathway in *S. cerevisiae* EC1118 in YNBT100.

Preliminary dosages demonstrated the absence of the analyzed molecules in the not inoculated cultural medium. Most of the assayed compounds increased accordingly with the yeast biomass level in the first 24 h after the inoculation. Indeed, KYN and KYNA accumulated in the supernatant up to 24 h (0.104 ± 0.046 and 6.603 ± 0.491 mg/L, respectively). Then, both compounds showed an accumulation plateau between 24–72 h. At 144 h, the trends of KYN and KYNA concentrations were opposite: while for KYN a significant decrease to 0.048 ± 0.040 mg/L ($p < 0.05$) was observed, KYNA conversely raised at 8.452 ± 1.516 mg/L. This finding confirms that in *S. cerevisiae* EC1118 the TRP consumption mainly leads to the breakdown of the indole ring with the formation of KYN and derivatives (KYN pathway) while a little proportion of reaction keep the indole ring intact and produce chemical messengers such as indolamines (serotonin and indole derivatives routes) [1]. Currently, there are no bibliographical references showing quantifications of KYN pathway metabolites by the yeast strain under study but Shin and collaborators (1991) [15] showed that *Saccharomyces uvarum* can potentially synthesize KYNA when TRP is present in the cultural medium. Recently, TRP and its derivatives were quantified in different beer samples and during the fermentation of wort by *S. cerevisiae* and *S. pastorianus* demonstrating that differences in KYN and KYNA levels could be related to the different fermentation conditions, process, and yeast strains [18]. In particular, *S. pastorianus* produced a smaller amount of the two compounds in comparison to *S. cerevisiae*. These results suggest that a significant portion of TRP is metabolized by the cells as a nitrogen source for the growth and the production of biomass; being an essential aromatic amino acid, TRP is used by yeast for the synthesis of proteins and other cellular compounds [41]. Also, the nitrogen requirement depends on the yeast species, explaining the observed difference in the consumption of TRP for *S. pastorianus* and *S. cerevisiae* [19,42]. Minor changes in AA production were detected; the evolution of this compound was quite stable during the experiment except for the sample point at 48 h where a significant but slight decrease of the molecule was found. On the contrary, a peak corresponding to the highest extracellular release of 3OH AA occurred at 72 h (0.355 ± 0.037 mg/L), when cells were in stationary phase. 3OH KYN failed to be accumulated in the medium.

Regarding the MEL biosynthesis pathway, 5OH TRP, serotonin (5OH TRY), NAC SER (NAC 5OH TRY), and MEL were measured, together with TEE (Figure 2c) and other potential MIs. Serotonin was the only intermediate detected. Its production trend was similar to the one of TEE up to 48 h. While the serotonin concentration was stable until the end of the trial, the TEE content decreased at the sampling time 72 h reaching its lowest value (between 0.004–0.007 mg/L). TEE was probably used by the yeast as a TRP source in the serotonin route since the amino acid in the medium was almost completely depleted already after 24 h from the beginning of the fermentation. 5OH TRP and NAC SER (NAC 5OH TRY) were never detected under the adopted experimental conditions. Moreover, unlike to what was observed for *S. cerevisiae* EC1118 [40], MEL levels were under the detection limit in the analyzed supernatants. This result could explain a poor technical caution in handling the samples that determined a rapid conversion of MEL to other indoles [43–45] or conversely, to confirm the already reported inconsistency in MEL release observed among biological replicates during yeast fermentations in standardized growth conditions [46]. However, the TRP metabolism of *S. cerevisiae* appears very sensitive to micro-environmental situation and physiological cell state. Indeed, Muniz-Calvo and collaborators (2019) were unable to detect melatonin formation in *S. cerevisiae* cultivated under different growth systems after tryptophan supplementation [32]; only by pulsing yeast cells with serotonin the melatonin production was observed. These results suggest that the TRP uptake and its further transformation to serotonin are heavily dependent by growth conditions and they are key elements in the MEL biosynthesis pathway.

The release of two MIs at 4.7 min (MI1) and 5.6 min (MI2) retention times was measured (Figure 3). Literature reports that *S. cerevisiae* EC1118 is capable to produce MIs [40]; however, also their accumulation can significantly vary among the biological replicates or under different experimental conditions. In order to verify whether the formation of MIs was consistent in the biological replicates, three independent six-well plate yeast cultures of *S. cerevisiae* EC1118, prepared inoculating different pre-cultures obtained from different glycerol stocks, were monitored over the time. The analysis of variance and the grouping test of the mean values and the relevant sum of the two isomers showed differences statistically significant ($p < 0.05$) at the sampling times 24–48, 72, and 144 h (Table 1). This outcome suggests that MIs production is potentially dependent by the yeast growth phase and that, in YNBT100 medium, the higher release of MIs is obtained at 72–144 h. Contrarily, the ANOVA of the means of the ratio (MI1/MI2) resulted not significant along the sampling times (Table 1) indicating that the N-acetylserotonin methyltransferase enzyme, involved in the synthesis of MEL and its possible isomers, maintained a stable accumulation of the derived products (MI1 and MI2) during the yeast growth.

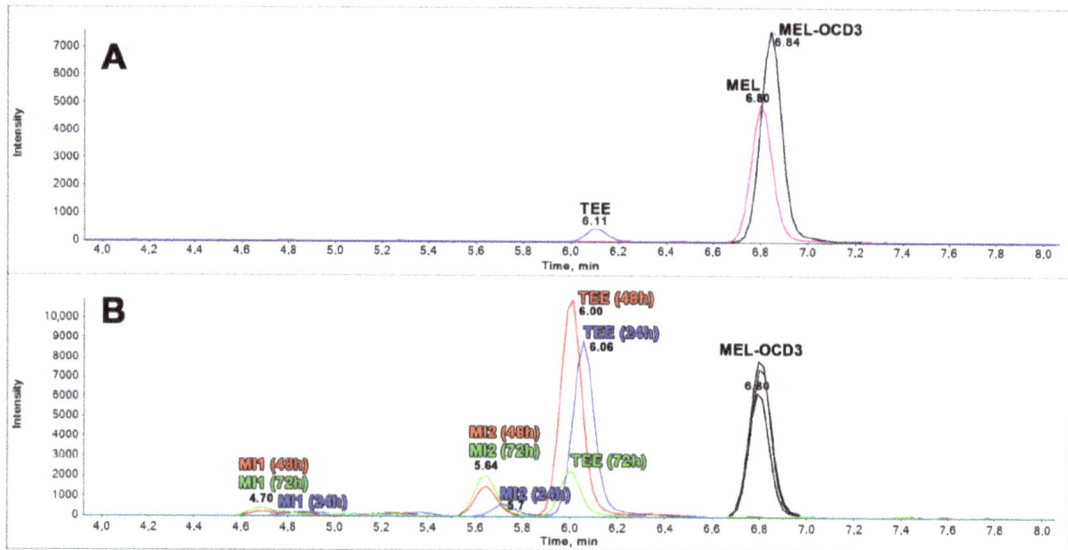

Figure 3. Melatonin isomers (MI) production by *S.cerevisiae* EC1118. Chromatogram in (**A**) shows the melatonin transition (m/z 233.2 > 174.2) of a pure standard containing melatonin (MEL, rt 6.80) tryptophan ethyl ester (TEE, rt 6.11) both at a concentration of 2 ng/mL and internal standard (MEL OCD3, rt 6.84, m/z 236.2 > 177.2, black). Chromatograms in (**B**) show the superimposition of melatonin trace (m/z 233.2 > 174.2) from three different *S. cerevisiae* EC1118 fermentation supernatants at 24 h (blue), 48 h (red), and 72 h (green). In (**B**) is evident the absence of the melatonin peak that should be expected at the same retention time of its labeled analogue (rt 6.80, black). Among melatonin isomers in fermentation medium it was evidenced the presence of tryptophan ethyl ester (TEE, rt 6.06) and two isomers, named MI1 (rt 4.70) and MI2 (rt 5.6), with a not-elucidated structure.

Table 1. MI1 and MI2 productions. Letters indicate different significant values ($p < 0.05$) at the yeast growth phases (24, 48, 72, and 144 h). Concentrations of MIs and their sums are here expressed in µg/L, considering the melatonin response. Letters indicate the grouping information using the Fisher LSD Method and 95% confidence.

Time (h)	YNBT100				YNBT400			
	MI1	MI2	Sum	MI1/MI2	MI1	MI2	Sum	MI1/MI2
24	14.6 ± 1.7 [a]	6.9 ± 1.0 [a]	21.8 ± 1.5 [a]	2.2 ± 0.4	54.8 ± 1.4 [a]	24.3 ± 6.2 [a]	79.1 ± 4.9 [a]	2.4 ± 0.7
48	14.2 ± 2.0 [a]	6.4 ± 1.5 [a]	20.6 ± 3.0 [a]	2.3 ± 0.4	99.5 ± 5.0 [b]	42.3 ± 6.0 [b]	141.8 ± 1.3 [b]	2.4 ± 0.4
72	20.4 ± 2.7 [b]	10.4 ± 1.6 [b]	30.7 ± 3.8 [b]	2.0 ± 0.3	109.5 ± 2.5 [c]	44.7 ± 3.7 [b]	154.2 ± 6.0 [c]	2.5 ± 0.2
144	27.8 ± 1.7 [c]	11.2 ± 1.2 [b]	38.9 ± 2.3 [c]	2.5 ± 0.3	106.2 ± 8.5 [bc]	37.5 ± 1.8 [b]	143.6 ± 9.9 [b,c]	2.8 ± 0.2

2.2. Production of dTRPs in YNBT400 Medium and in WCB Trials

In order to optimize the release of dTRPs by *S. cerevisiae* EC1118, fermentation and whole-cell biocatalysis processes were set up. In the first approach the growth test was performed in YNBT400 medium for 144 h, whereas in the second one the biotransformations were carried out in a buffer with TRP excess for 24 h. The comparison of these two approaches might be useful to assay the production of commercially valuable compounds that can be used in pharmaceutical and cosmetics fields or in the food and beverages sectors [47].

As far the fermentations in YNBT400 medium, the cultural broth was prepared with 400 mg/L TRP and increasing the carbon and nitrogen sources, vitamins, salts, and trace elements. In general, the fermentation trend YNBT400 medium in 144 h was comparable to the one obtained in YNBT100 medium for the entire duration of the experiment (data not shown), showing a considerable increase of cells in the first 24 h; however, at this time point a two-fold increase in yeast cell count was registered ($7.00 \pm 0.07 \times 10^7$ versus $1.40 \pm 0.11 \times 10^8$ CFU/mL in YNBT100 and YNBT400, respectively). At 48 h from the inoculation, glucose become a limiting factor for the yeast proliferation being almost depleted (0.0030 ± 0.0001 g/L). Accordingly, TRP consumption was rapid and it occurred almost completely within 48 h (Figure 4). However, the decrease of the cell viability did not correspond with a decrease in the optical density, which remained constant between 24–144 h (data not shown). This result suggests that cell autolysis did not occur under the investigated growth conditions.

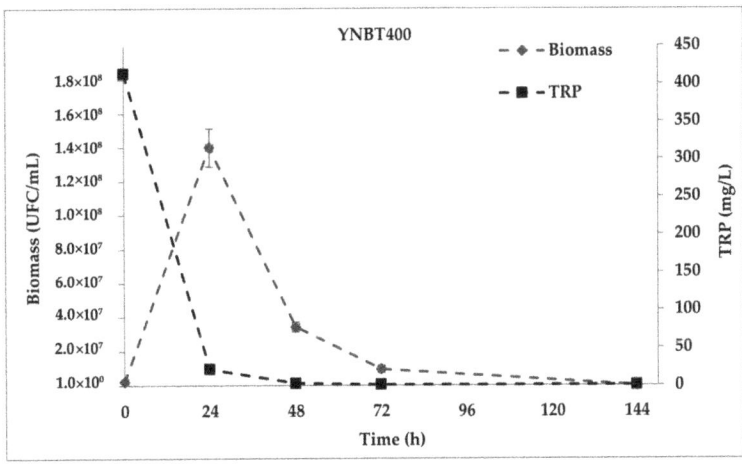

Figure 4. *S. cerevisiae* EC1118 growth curve in YNBT400 medium and TRP consumption.

As described for the YNBT100 medium, a TRP catabolism oriented towards the KYN pathway was confirmed (Figure 5). Despite the biomass improvement, a significant increase in all the dTRPs analyzed in this study was not always obtained. While KYNA and 3OH AA (9.146 ± 0.585 and 0.101 ± 0.013 mg/L, respectively) reached concentrations comparable to the levels found in YNBT100 medium, four-fold higher KYN and MIs amounts were detected in YNBT400 cultures (Table 1).

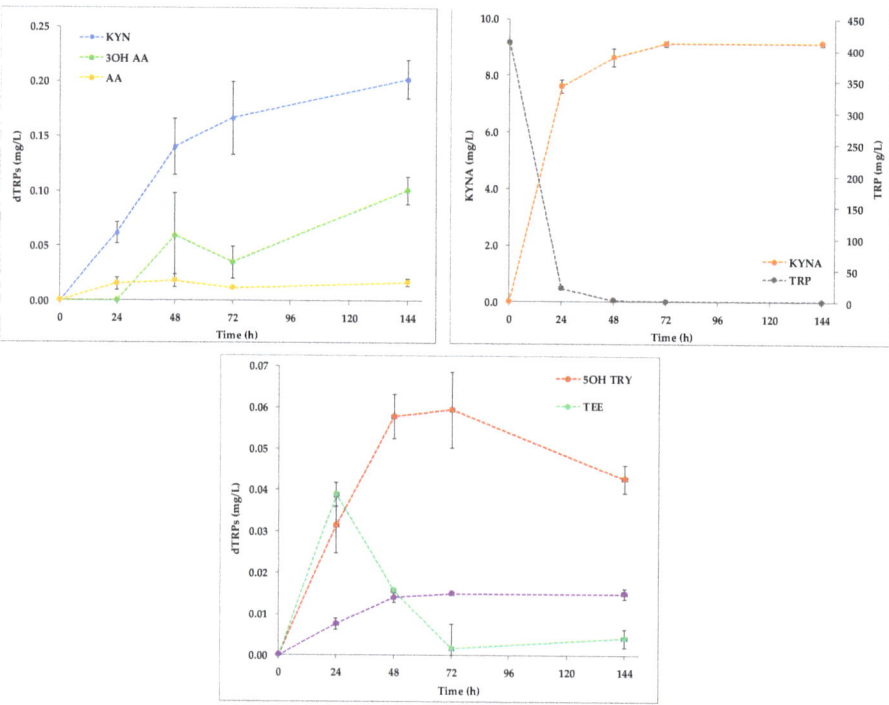

Figure 5. TRP derivatives produced by *S. cerevisiae* EC1118 in YNBT400 medium. Error bars indicate the standard deviation.

Concerning the whole-cell biotransformation experiments, after yeast biomass was harvested and washed from pre-cultures grown in a rich medium, cells were suspended in the biocatalysis buffer containing 1 g/L TRP and glucose. The high TRP concentration aimed at forcing the production of dTRPs. The presence of glucose in the biotransformation medium represents an important element for the metabolic activity of cells since it helps cells restoring the redox potential, given by the NAD^+/NADH ratio, and other cofactors such as ATP and those of the TCA cycle (i.e., 2-oxoglutarate, acetyl-CoA). After 24 h from the beginning of the WCB experiments glucose was depleted and a significant cell biomass increased was not detected (1.107 ± 0.012 $OD_{600/mL}$ at the inoculum time versus 1.123 ± 0.017 $OD_{600/mL}$ at 24 h). As observed in YNBT400 trials, cell autolysis did not occur (data not shown). TRP consumption was about 390 mg/L (from 923.41 ± 25.46 mg/L to 529.14 ± 46.57 mg/L); this amount was comparable with the one observed in fermentation tests in YNBT400 medium in 24 h, where about 410 mg/L TRP (from 413.32 ± 12.40 mg/L to 21.43 ± 0.047 mg/L) were used by cells. Only KYN, TRY, and 5OH TRY started to be accumulated immediately after 15 min from the beginning of the experiment, reaching a maximum concentration between 4 and 24 h of 0.653 ± 0.052 mg/L, 0.066 ± 0.001 mg/L, and 0.096 ± 0.010 mg/L, respectively (Figure 6).

The results confirm that *S. cerevisiae* EC1118 decarboxylates the TRP to TRY [32]. The extra and intracellular production of MEL and indoles in *S. cerevisiae* by adding intermediates to cells in different growth phases (exponential and stationary) was recently evaluated [32]; these authors revealed that serotonin mainly derived from the decarboxylation of TRP, followed by hydroxylation of TRY, as occurs in plants. Thus, MEL biosynthesis from serotonin can occur by N-acetylation followed by O-methylation or vice versa. KYNA, AA, NAC 5OH TRY, MI1, and MI2 were only detected at the end of the experiment (at 144 h) at 0.199 ± 0.014 mg/L, 0.052 ± 0.004 mg/L, 0.027 ± 0.006 mg/L, 0.201 ± 0.025 mg/L and 0.102 ± 0.016 mg/L, respectively. The synthesis of 5OH TRP, by TRP hydroxylase, was not detected. However, since 5OH TRP is an intermediate of the MEL biosynthesis pathway, further tests like a fluxomic analysis would be necessary to have more indications on its fate.

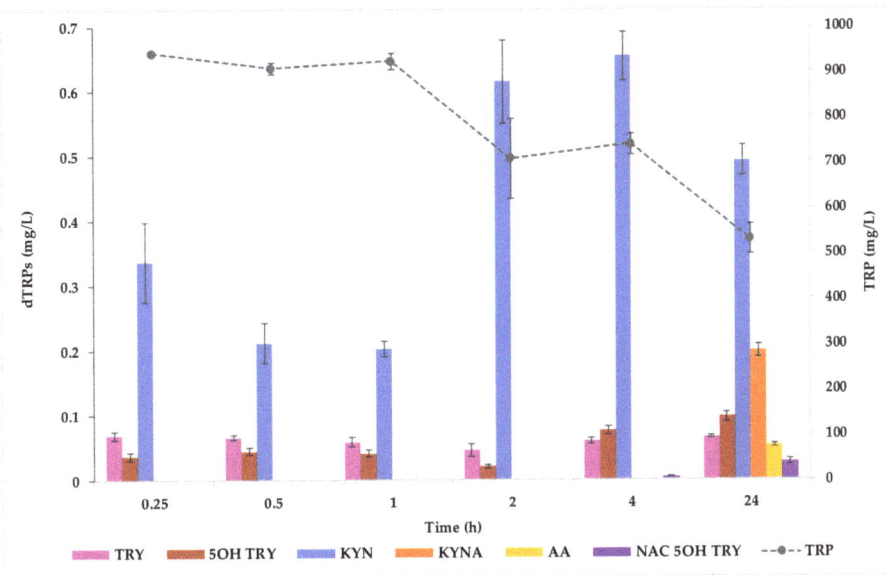

Figure 6. TRP derivatives produced by *S. cerevisiae* EC1118 in WCB experiments. Error bars indicate the standard deviation among the mean values of biological replicates.

In order to directly compare the concentrations of the dTRPs synthesized in fermentations in YNBT400 and in WCB experiments, normalized data against the biomass levels and the fold change between the considered approaches were calculated (Table 2). This estimation was possible because at 24 h from the beginning of both processes the obtained dTRPs may potentially derive from a comparable TRP consumption (about 400 mg/L TRP, as an average). KYNA was the only compound that showed a higher concentration in fermentations in YNBT400 compared to the WCB tests (about 8 times). The other dTRPs gained higher concentrations in WCB experiments. Thus, depending on the compound a fermentation or a bioconversion can be settled; briefly, KYN increased 35 times, TRY 15 times, 5OH TRY 14 times, MI1 and MI2 15 and 24 times, respectively in bioconversion in respect to the fermentation (Table 2).

Table 2. Comparison of the production of dTRPs in fermentations in YNBT400 medium and in WCB experiments at 24 h. The productions were normalized against the yeast biomass and expressed as mg L^{-1} 10^{-8} cells. The fold change of the main dTRPs was calculated as the ratio between the concentration of the corresponding metabolite in the two investigated approaches.

dTRPs	YNBT400 (mg L^{-1} 10^{-8} Cells)	WCB (mg L^{-1} 10^{-8} Cells)	Fold Change
KYNA	5.449 ± 0.276	0.622 ± 0.000	8.76
KYN	0.044 ± 0.003	1.538 ± 0.008	0.029
TRY	0.014 ± 0.004	0.206 ± 0.010	0.068
5OH TRY	0.022 ± 0.003	0.301 ± 0.008	0.073
MI1	0.040 ± 0.003	0.626 ± 0.035	0.064
MI2	0.015 ± 0.003	0.354 ± 0.024	0.042

2.3. dTRPs from YNB-Soybean-Based Culture Medium

In order to evaluate the capability of *S. cerevisiae* EC1118 to release dTRPs in a food-based matrix naturally rich in TRP, an experiment in a soybean-based cultural medium (YNBSOY) was set up as a case of study. Nowadays, the beneficial effects on human health of several raw materials fermented throughout spontaneous or controlled processes has been proposed [48,49]. Thus, showing that the supply of dTRPs with diet—including KYNA, 5OH TRY, or MEL and its isomers—could be implemented by the consumption of fermented foods represents a health and marketing plus value concept. Vitalini and co-authors (2020) [50] have recently demonstrated that toasted soybean flour treated by a water extraction at room temperature allows to obtain free TRP potentially bioavailable during yeast fermentation as precursor of dTRPs. The controlled fermentation of the YNBSOY medium with *S. cerevisiae* EC1118 showed a high fermentative power; indeed, a weight loss of about 10 times higher was obtained at 72 h in comparison to the control (not inoculated). Contrarily to what was observed in previous fermentations in YNBT100 and YNBT400 media, the yeast growth increased up to the end of the trials. Glucose was completely consumed in 72 h (0.015 ± 0.009 g/L); however, in the control sample it decreased (from 47.632 ± 0.188 to 37.880 ± 1.126 g/L) indicating the presence of an indigenous soy flour microbiota that could trigger uncontrolled spontaneous fermentations. An increase in acetic acid concentration (from 0.126 ± 0.004 to 0.192 ± 0.001 g/L) was detected in the inoculated YNBSOY medium. Generally, *S. cerevisiae* is characterized by low production of acetic acid, with an average value of 300 mg/L [51]; the yeast under study, being an oenological strain, produces smaller quantities in the conditions under study. On the other hands, a more consistent increase in acetic acid concentration (0.302 ± 0.001 g/L), possibly attributable to the development of microorganisms already present in the flour, was observed in the control.

Regarding the TRP available as substrate for indoles production, its initial concentration was detected at 61.301 ± 2.310 mg/L. Since YNBSOY is a YNB-based medium (already containing 20 mg/L TRP), it can be calculated that the soybean flour naturally enriched the growth culture of about 40 mg/L TRP. At the end of the experiment the TRP concentration was measured at 6.170 ± 0.003 mg/L. A preliminary investigation concerning the dTRPs content in YNBSOY medium revealed that some indoles were already present at the inoculum time; however, depending on the compound different fates were observed during the fermentation (Figure 7). While a decrease trend was found for KYN (from 0.039 ± 0.007 to 0.016 ± 0.004 mg/L), AA (from 0.027 ± 0.003 to 0.004 ± 0.001 mg/L), and 5OH TRP (from 0.090 ± 0.001 to 0.077 ± 0.005 mg/L), other compounds such as KYNA (from 0.009 ± 0.001 to 0.0137 ± 0.002 mg/L), TEE (from 0 to 0.006 ± 0.001 mg/L), and 5OH TRY (from 0.026 ± 0.001 to 0.036 ± 0.003 mg/L) increased. No significant variation of TRY and NAC 5OH TRY was observed. The two isomers, MI1 and MI2, detected in the fermentation trials in YNBT100 and YNBT400 media, failed to be released. Once again, the cultural conditions proved to be a key factor in the production of derivatives from the MEL biosynthesis pathway.

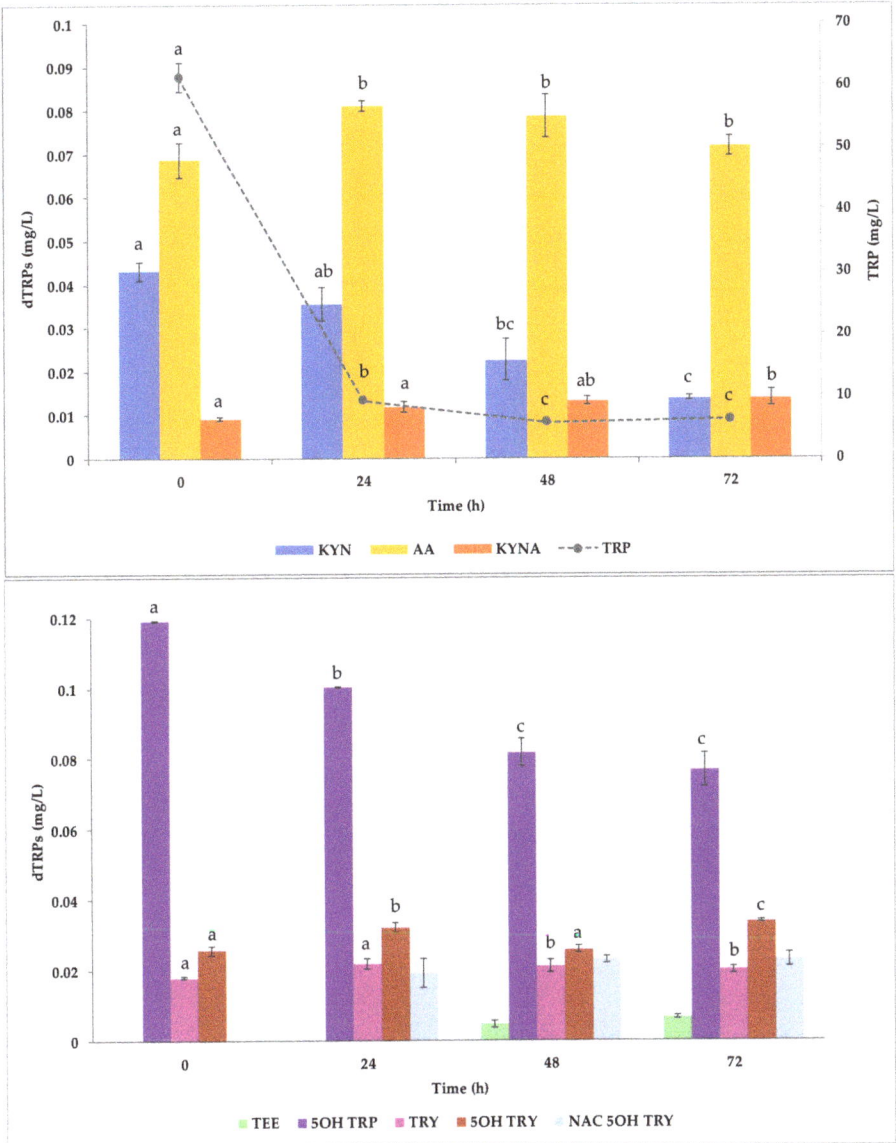

Figure 7. TRP derivatives produced by *S. cerevisiae* EC1118 in fermentation experiments in YNBSOY medium. Error bars indicate the standard deviation among the means of biological replicates. Letters indicate the grouping information using the Fisher LSD Method and 95% confidence.

3. Materials and Methods

3.1. Yeast Strain

The yeast strain used in this work is *Saccharomyces cerevisiae* EC1118, commercial wine strain. The pure culture is maintained in YPD medium (1% yeast extract, 2% peptone, 2% glucose (w/v)) with 20% (v/v) glycerol at $-80\ °C$.

3.2. Growth Conditions in YNB-Based Media

To evaluate the production of TRP derivatives (dTRPs) by *S. cerevisiae* EC1118, the strain was preliminary grown in a chemically defined cultural medium (20 g/L glucose; 6.7 g/L yeast nitrogen base (Difco, Heidelberg, Germany), pH 5.4) containing 20 mg/L TRP as a precursor (YNBT20). Then, the experiment was conducted in YNBT100 containing a final TRP concentration of 100 mg/L; to obtain this TRP amount, the amino acid was supplemented into YNBT20 medium as a stock solution (4 mg/mL) prepared in sterile demineralized water (sdH$_2$O) and sterilized by filtration with 0.22 µm filters, immediately before inoculation. Growth tests were performed in sterile six-well plates with 10 mL of YNBT100 medium at 25 °C, 110 rpm Briefly, from the glycerol stock the yeast strain was preliminarily inoculated in three separated tubes containing 5 mL of YPD medium at 25 °C for 24 h to obtain biological replicates. From the three pre-cultures, cells in the exponential growth phase [about 2–4 OD$_{600nm}$ measured in a UV–vis spectrophotometer, Jenway 7305, Stone, United Kingdom)] were inoculated at 1% (v/v) in three clean and sterile tubes containing 5 mL of YNB-based medium without amino acids [20 g/L glucose; 6.7 g/L YNB without amino acids, pH 5.4] at 25 °C for 72 h. Replicates were centrifuged at 4000 rpm (Hettich Zentrifugen, Rotina 380r, Tuttlingen, Germany) for 10 min, cells were washed with sdH$_2$O and further centrifuged under the above-mentioned conditions. Each cell pellet was resuspended in three clean and sterile tubes containing 5 mL of YNB medium without amino acids and glucose [6.7 g/L YNB without amino acids, pH 5.4] for 24 h to allow nutrients depletion. Then, starved yeasts were centrifuged at 4000 rpm (Hettich Zentrifugen, Rotina 380r, Tuttlingen, Germany) for 10 min and resuspended in 2 mL of sdH$_2$O. A 6-well plate was inoculated at 0.2 OD$_{600nm}$ per well from each pre-culture. The plates were sealed with sterile parafilm to avoid any possible contamination. In addition, they were covered with aluminum foil to preserve the dTRPs from oxidative photodegradation. The experiment lasted 144 h and the monitoring of the growth (by OD$_{600nm}$ measurement and by cell plate counting) and the production of dTRPs was carried out at 0, 24, 48, 72, and 144 h after inoculation. Regarding the cell count, 100 µL of appropriate decimal serial dilutions in sdH$_2$O water were spread on Petri dishes containing WL nutrient agar medium (Scharlab, Barcelona, Spain) and, after incubation at 25 °C for 3 days, the colonies were enumerated. For the quantification of dTRPs, 1 mL of culture was collected from two wells/plate, centrifuged at 10,000× g for 10 min and supernatants were 0.22 µm filtered and stored at −20 °C for further LC-MS/MS analysis. The analysis was carried out in triplicate on each biological replicate.

3.3. TRP Conversion Tests

A sterile six-well plate containing 10 mL of YNBT100 medium per well was placed at 25 °C, 110 rpm, the same experimental conditions further used in this study, for 6 days monitoring the content of TRP at 0, 24, 48, 72, and 144 h. The plate was sealed with sterile parafilm and covered with aluminum foil. One mL of medium was collected at each sampling time from three wells and the supernatant were 0.22 µm filtered and stored at −20 °C until to be subjected to the LC-MS/MS analysis.

3.4. Optimization of the Production of dTRPs: Approaches and Experimental Conditions

In order to increase the quantity of biomass produced and, consequently, the dTRPs formed by *S. cerevisiae* EC1118, two experiments were developed under different conditions: fermentation growth test in an optimized cultural medium (YNBT400) and whole-cell bioconversion (WCB) trials.

3.4.1. YNBT400 Trials

All tests were prepared in sterile 250 mL flasks, closed with sterile cotton caps, containing 100 mL of YNBT400 medium (100 g/L glucose; 26.8 g/L yeast nitrogen base (Difco, Heidelberg, Germany), pH 5.4) containing 400 mg/L TRP as precursor of dTRPs. The yeast strain was first inoculated into YPD medium. Flasks were inoculated at 0.2 OD$_{600nm}$ and incubated at 25 °C, 110 rpm for 6 days. Monitoring of cell growth was carried out at 0, 24, 48, 72, and 144 h by evaluating optical density at 600 nm (UV–vis spectrophotometer, Jenway 7305, Stone, United Kingdom) and plate count on WL nutrient agar medium. Residual glucose was determined with an enzymatic kit (K-FRUGL assay kit-Megazyme, Wicklow, Ireland). All tests were performed in triplicate.

3.4.2. WCB Trials

Experiments were carried out to evaluate the capability of S. cerevisiae EC1118 to produce dTRPs working as a cell factory using TRP as a precursor. WCB trials were carried out with a total yeast biomass of 350 OD_{600nm}. Three biological replicates were prepared in flasks at 25 °C, 110 rpm, in 300 mL bioconversion buffer containing 0.1 M phosphate buffer pH 6.7, 1 g/L TRP and 20 g/L glucose. Briefly, a pre-culture in 100 mL of YPD medium was obtained by inoculating the strain at 1% (v/v) from glycerol stocks and incubated for 24 h at 25 °C under stirring at 110 rpm. Then, cells were centrifuged at 4000 rpm (Hettich Zentrifugen, Rotina 380r, Tuttlingen, Germany) for 10 min, washed with sdH_2O and further centrifuged under the above-mentioned conditions. The cell pellet was resuspended in 100 mL of YNB without amino acids and glucose, overnight incubated at 25 °C at 110 rpm for nutrient depletion. Then, biomass was spectrophotometrically determined (600 OD/mL) and a volume of cell suspension corresponding to a total biomass of 350 OD_{600nm} was centrifuged at 4000 rpm (Hettich Zentrifugen, Rotina 380r, Tuttlingen, Germany) for 10 min, cells were washed with 20 mL sdH_2O, centrifuged again at the same above conditions and transferred to the bioconversion flask. The experiment lasted 24 h and the production of dTRPs was evaluated at 0, 15, and 30 min; and 1, 2, 4, and 24 h after the inoculation. One mL of culture was collected from the three biological replicates, centrifuged at $10,000 \times g$ for 10 min and the supernatants were 0.22 µm filtered and stored at −20 °C until to be subjected to the LC-MS/MS analysis. The final glucose concentration was measured via enzymatic kit (kit-Megazyme, Wicklow, Ireland).

3.5. TRP Extraction from Toasted Soybean

The organic soy flour was purchased from the "Fior di Loto" company (Turin, Italy). An aliquot of flour was placed in a fan oven for 24 h at 50 °C before extraction to eliminate as much water as possible. Different techniques were applied for the extraction of TRP from the food matrix, in order to compare their efficiency: Soxhlet extraction, methanol extraction at room temperature and extraction in water at room temperature [50]. All procedures were performed in low light conditions, to prevent degradation of metabolites. The extractions were set up in triplicates.

3.5.1. Soxhlet Extraction

The extraction with Soxhlet apparatus was performed following two different procedures: (i) 21 grams of toasted soybean flour were first defatted with petroleum ether and dichloromethane and subsequently extracted with methanol; (ii) The soy flour was also subjected to extraction in Soxhlet directly in methanol. Each obtained extract was dried with a rotary evaporator, frozen at −80 °C overnight and freeze-dried for about 5 h. The lyophilized sample was then stored at 4 °C until analysis.

3.5.2. Methanol Extraction at Room Temperature

A mixture of flour and methanol (1:10, w/v) was sonicated for 10 min, then stirred for 20 min (DLAB SK-O180-E). After decanting, the supernatant was centrifuged at $28,000 \times g$ for 10 min and filtered with 0.22 µm nylon filters. The extraction was repeated, and the combined supernatants were stored at −20 °C until analysis.

3.5.3. Extraction in Water at Room Temperature

A mixture of flour and distilled water (1:10, w/v) was sonicated, then stirred for 20 min (DLAB SK-O180-E). After decanting, the supernatant was centrifuged at $28,000 \times g$ for

10 min and filtered with 0.22 μm nylon filters. The obtained sample was stored at −20 °C until analysis.

3.6. Fermentation Tests in Soybean Flour

A triplicate destructive test was set up in 100 mL sterile flasks containing 15 g of toasted organic soybean flour and 45 mL of YNBT20 medium added with 50 g/L of glucose (YNBSOY). Experiments were run at 25 °C for 72 h, in anerobiosis with GasPack jar system (Merck Millipore, Burlington, MA, USA). The yeast was preliminarily inoculated in YPD at 25 °C for 48 h. Each flask was inoculated at 0.2 OD_{600nm}. Three non-inoculated flasks were also set up as a negative control. The microbial growth was monitored by flask weight loss due to CO_2 loss and cell plate count at 0, 24, 48, and 72 h after inoculation. The flasks were weighed daily with a technical balance until a constant weight. Cell count was determined on WL agar medium after incubation at 25 °C for 2–3 days. At each sampling time, supernatants for the determination of glucose, acetic acid, and dTRPS was collected as well: 1 mL of culture was collected and centrifuged at $11,000 \times g$ for 15 min. The supernatants were 0.22 μm filtered and frozen at −20 °C. Glucose and acetic acid were determined with enzymatic kits (kit-Megazyme, Wicklow, Ireland). All tests were performed in triplicate. The production of dTRPs was measured by LC-MS/MS analysis.

3.7. Chemical Analysis of TRP Derivatives

Methanol and formic acid (FA) were purchased from Sigma-Aldrich (Milan, Italy). All the aqueous solutions were prepared using Milli-Q purified water (Millipore, Milan, Italy). The pure chemical standards (analytical grade) of TRY, NAC TRY, 5OH TRY, 5OME TRY, NAC 5OH TRY, 5F TRY, 5OH TRP, TRP, TRP D5, TEE, AA, 3OH AA, KYNA, KYN, 3OH KYN, and MEL were purchased by Sigma-Aldrich (St. Louis, MO, USA). The isotopic isomer MEL OCD3 was synthesized by Prof. Andrea Penoni (Department of Science and High Technology, University of Insubria, Varese, Italy).

3.7.1. Purification of Supernatants Deriving from Yeast Fermentation

Supernatants deriving from fermentation (10 μL) were added to 50 μL of MIX IS (internal standards) (20 ng of TRP D5, 1 ng of 5F TRY and 1.1 ng of MEL OCD3) and 100 μL of precipitating solution (methanol + 0.5% FA). After stirring by vortexing and centrifugation 5 min at 12,000 rpm (Hettich Zentrifugen, MIKRO200, Tuttlingen, Germany), 100 μL were taken and dried under nitrogen. Subsequently, the sample was resuspended with 50 μL of dH_2O + 0.5% FA and 10 μL were injected directly into LC-MS/MS.

3.7.2. Selective Extraction of Melatonin from Yeast Fermentation

Supernatants deriving from the fermentations (500 μL) were added to 50 μL of MIX IS; subsequently, they were purified by solid-phase extraction (SPE) with Strata X-Polymeric Reversed Phase 30 mg/1 mL (Phenomenex, Torrance, CA, USA) [52]. Before use, the cartridges were conditioned with 1 mL of methanol and 1 mL of dH_2O. The samples, diluted with water to 1 mL as final volume, were loaded into the cartridges and then washed with 1 mL of water and 1 mL of 5% MeOH (v/v). After 5 min of vacuum drying to remove the excess of water, the residual compounds were eluted with 1 mL of methanol. The eluate was dried by nitrogen, the residue was resuspended with 100 μL of dH_2O + 0.5% FA and 10 μL was injected for LC-MS/MS analysis.

3.7.3. LC-MS/MS Analysis for the Detection of dTRPs

The analytical system consisted of an HPLC coupled to a tandem mass spectrometer, specifically a Dionex 3000 UltiMate instrument (Thermo Fisher Scientific, MA, USA) coupled to a tandem mass spectrometer (AB Sciex 3200 QTRAP, Concord, ON, Canada). The separation was obtained on reverse phase Zorbax Eclipse XDB-C18 (4.5 × 50 mm, 1.8 μm). The linear gradient was obtained by mixing the eluent A (dH_2O + 0.1% of formic acid) and eluent B (methanol + 0.1% of formic acid). The elution gradient was set as follows: 0–1 min

(20% B), 1–5 min (20–60% B), 5–7 min (60% B), 7.0–7.2 min (60–95% B), 7.2–8.2 min (95% B), 8.2–8.5 min (95–5% B), 10 (5% B). The flow rate was 0.4 mL/min, and the column temperature 40 °C. Multiple reaction monitoring (MRM) mode was used in electrospray positive mode (Table A1, Appendix A). A quantitative analysis was performed interpolating each area of the analyte peak/IS area with the calibration curve of each derivative of the TRP. Further analytical details can be found in [50].

3.7.4. Validation of the Analytical Method

The method performances (specificity, precision, accuracy, linearity, the limit of detection (LOD) and limit of quantification (LOQ)) were reported in full details elsewhere [50]. The LOQ (the lowest concentration that produces a signal-to-noise ratio greater than 10) of analytes extracted from medium following the described procedure ranged from 2–50 ng/mL. The performance of melatonin extracted by SPE was studied in a range of 2–0.25 ng/mL and resulted to be acceptable for our purposes (linearity equation $y = 0.7582x - 0.0062$, coefficient of determination R^2 0.994, LOQ 50 pg/mL and LOD 20 pg/mL).

3.8. Statistical Analysis

Where indicated the results were reported as mean ± standard deviation. Significant differences ($p < 0.05$) were evaluated by analysis of variance (one-way ANOVA) by using MINITAB Statistical Software 19 (Pennsylvania State University, PA, USA). Grouping information was determined using the Fisher LSD Method and 95% confidence.

4. Conclusions

Thanks to the advent of microbiological and chemical cutting-edge researches, many of the health benefits attributed to fermented foods and beverages are nowadays recognized. In particular, omics studies of traditional fermented foods have highlighted the microbiota of several spontaneous fermentations, revealing its role in the production of bioactive molecules [48]. The involvement of yeasts in the fermentation of several raw materials is well recognized in numerous productions, starting from the manufacture of alcoholic beverages (wine, beer, cider, and fruit wines) to fermented cereal-based doughs, and dairy, meat and fish products [53]. Yeasts share with animals several metabolic pathways which are relevant reservoirs of functional molecules, TRP derivatives above all [49]. The present study revealed the presence of variable amounts of dTRPs with a prevalence of metabolites of the KYN pathway, emphasizing that TRP metabolism of *S. cerevisiae* EC1118 pushed in this direction. The LC-MS/MS analysis showed that the compound synthesized in higher concentrations is KYNA, a molecule with neuroprotective and antioxidant properties. The fermentation in YNBT400 resulted the most promising approach for KYNA production which reached the highest concentrations (9.146 ± 0.585 mg/L). In comparison to literature data, the maximum concentration of KYNA measured in the post-culture medium was higher than KYNA levels in plasma of healthy subjects (5.8 ng/mL) [54] and milk of healthy breast-feeding women (3.9–56.6 microg/L) [55]. In addition, KYNA was detected in artificial baby formulas at lower concentrations (5.0–7.3 microg/L) [55]. Therefore, the supraphysiological levels of KYNA found in our experiments may exert therapeutic effects in human, even if the latter were mainly investigated in patients with neurological and psychiatric disorders in relation to brain and cerebrospinal fluid KYNA [56]. It is worth noting that the supplementation of rat maternal diet with KYNA in drinking water resulted in its increase in milk harvested, and the rat offspring fed with breast milk with artificially enhanced KYNA content demonstrated a lower mass gain during the first 21 days of life, which indicates that KYNA may act as an anti-obesity agent, as demonstrated from the body composition analysis [55]. Although the KYNA obtained in this way should be purified from the medium, probably with high costs compared to the organic synthesis, this work highlights the chance to obtain functional foods enriched with molecules with nutritional or therapeutic potential throughout the use of natural biological systems. Cells, for example, could be naturally oriented towards the production of specific dTRPs by enriching the cultures with more specific precursors other than TRP. However, the feasibility of the method proposed as an alternative approach of obtaining bioactive metabolites using biological systems needs to be evaluated, in terms of costs, instruments, time required, and yield.

As far MEL, detectable amounts were not measured in our experimental conditions (LOQ 50 pg/mL and LOD 20 pg/mL). However, two MEL isomers were successfully detected both in fermentation in YNBT400 and WCB experiments. In general, since the production of dTRPs could be a species- and strain-specific character [40] a full screening of dTRPs capable yeasts from wine, brewer and distillery should be performed in future investigations.

In conclusion, considering the high interest toward the dTRPs in the panorama of functional foods, the results of the present research provide the basis for in-depth studies on the effects of fermentation in the synthesis of compounds with important bioactive properties in novel food products.

Author Contributions: Conceptualization, R.F. and I.V.; Methodology, I.V. and R.P.; Software, M.D.C.; Validation, M.D.C.; Formal analysis, M.D.C., A.L., and S.V.; Investigation, M.D.C., I.V., A.L., and S.V.; Data curation, I.V., R.P., and M.I.; Writing—original draft preparation, I.V. and A.L.; Writing—review and editing, M.D.C., R.P., S.V., M.I., and R.F.; Visualization, M.D.C. and I.V.; Supervision, I.V., R.P., M.I., and R.F.; Project administration, R.F.; Funding acquisition, R.F. All authors have read and agreed to the published version of the manuscript.

Funding: This research was funded by the Bando Linea R&S per Aggregazioni, Regione Lombardia, Programma Operativo Regionale 2014–2020, Strategia "InnovaLombardia" (D.G.R. no. 2448/2014) Project number 145007.

Institutional Review Board Statement: Not applicable.

Informed Consent Statement: Not applicable.

Data Availability Statement: Not applicable.

Acknowledgments: MDC was supported by the PhD program in Molecular and Translational Medicine of the Università degli Studi di Milano, Milan, Italy. The Università degli Studi di Milano, partially covered the open access APC" (APC = Article processing charge).

Conflicts of Interest: The authors declare no conflict of interest.

Abbreviations

TRY	Tryptamine
NAC TRY	N-Ac Tryptamine
5OH TRY	5-OH Tryptamine or Serotonin
5OME TRY	5-OCH$_3$ Tryptamine
NAC 5OH TRY	N-Ac-5-OH Tryptamine
5F TRY; IS1	5-F Tryptamine
5OH TRP	5-OH Tryptophan
TRP	Tryptophan
TRP D5; IS2	Tryptophan D$_5$
TEE	Tryptophan Ethyl Ester
AA	Anthranilic acid
3OH AA	3-OH Anthranilic acid
KYNA	Kynurenic acid
KYN	Kynurenine
3OH KYN	3-OH Kynurenine
MEL	Melatonin or N-Ac-5- OCH3 Tryptamine
MEL OCD3, IS3	Melatonin OCD$_3$
SPE	Solid Phase Extraction

Appendix A

Table A1. MRM transitions for LC-MS/MS analysis of indolic species. Each subclass is preceded by the correspondent IS (in italics) used for quantification.

Name	MW	Q1	Q3	DP	CE	Rt
5F TRY(IS1)	*178*	*179.2*	*162.2*	*20*	*15.2*	*3.89*
TRY	160	161.2	144.2	20	14.8	2.89
5OH TRY (or Ser)	176	177.2	160.2	20	14.7	1.30
5OCH3 TRY	190	191.2	174.2	20	14.3	3.13
NAc TRY	202	203.2	144.2	20	19.1	6.87
NAc 5OHTRY (or N-Ac SER)	218	219.2	160.2	20	18.5	3.71
TRP D5 (IS2)	*209*	*210.2*	*192.2*	*20*	*14.8*	*3.03*
3OH AA	153	154.1	135.9	21	17.0	2.91
3OH KYN	224	225.1	208.1	16	13.0	1.30
AA	137	138.1	119.8	21	15.0	5.63
KYNA	189	190.1	89.1	36	51.0	4.56
KYN	208	209.2	192.2	21	15.0	1.72
TRP	204	205.2	188.2	20	13.4	3.12
TEE	232	233.2	216.2	20	13.1	6.06
5OH TRP	220	221.1	204.2	16	15.0	1.60
MEL OCD3 (IS3)	*235*	*236.2*	*177.2*	*20*	*19.1*	*6.80*
MEL	232	233.2	174.2	20	18.8	6.84

MW: molecular weight, Q1: molecular ion, Q3: daughter ion, DP: declustering potential, CE: collision energy, Rt: retention time.

References

1. Ruddick, J.P.; Evans, A.K.; Nutt, D.J.; Lightman, S.L.; Rook, G.A.; Lowry, C.A. Tryptophan metabolism in the central nervous system: Medical implications. *Expert Rev. Mol. Med.* **2006**, *8*, 1–27. [CrossRef] [PubMed]
2. Erhardt, S.; Olsson, S.K.; Engberg, G. Pharmacological manipulation of kynurenic acid: Potential in the treatment of psychiatric disorders. *CNS Drugs* **2009**, *23*, 91–101. [CrossRef] [PubMed]
3. Chen, Y.; Guillemin, G.J. Kynurenine pathway metabolites in humans: Disease and healthy states. *Int. J. Tryptophan Res.* **2009**, *2*, 1–19. [CrossRef] [PubMed]
4. Schwarcz, R.; Bruno, J.P.; Muchowski, P.J.; Wu, H.Q. Kynurenines in the mammalian brain: When physiology meets pathology. *Nat. Rev. Neurosci.* **2012**, *13*, 465–477. [CrossRef] [PubMed]
5. Sedlmayr, P.; Blaschitz, A.; Stocker, R. The role of placental tryptophan catabolism. *Front. Immunol.* **2014**, *5*, 230. [CrossRef] [PubMed]
6. Veldhoen, M.; Hirota, K.; Westendorf, A.M.; Buer, J.; Dumoutier, L.; Renauld, J.C.; Stockinger, B. The aryl hydrocarbon receptor links TH17-cell-mediated autoimmunity to environmental toxins. *Nature* **2008**, *453*, 106–109. [CrossRef] [PubMed]
7. Opitz, C.A.; Litzenburger, U.M.; Sahm, F.; Ott, M.; Tritschler, I.; Trump, S.; Schumacher, T.; Jestaedt, L.; Schrenk, D.; Weller, M.; et al. An endogenous tumour-promoting ligand of the human aryl hydrocarbon receptor. *Nature* **2011**, *478*, 197–203. [CrossRef]
8. Fuertig, R.; Ceci, A.; Camus, S.M.; Bezard, E.; Luippold, A.H.; Hengerer, B. LC–MS/MS-based quantification of kynurenine metabolites, tryptophan, monoamines and neopterin in plasma, cerebrospinal fluid and brain. *Bioanalysis* **2016**, *8*, 1903–1917. [CrossRef]
9. Wang, J.; Simonavicius, N.; Wu, X.; Swaminath, G.; Reagan, J.; Tian, H.; Ling, L. Kynurenic acid as a ligand for orphan G protein-coupled receptor GPR35. *J. Biol. Chem.* **2006**, *281*, 22021–22028. [CrossRef]
10. Lugo-Huitrón, R.; Blanco-Ayala, T.; Ugalde-Muniz, P.; Carrillo-Mora, P.; Pedraza-Chaverrí, J.; Silva-Adaya, D.; Maldonado, P.D.; Torres, I.; Pinzón, E.; Ortiz-Islas, E.; et al. On the antioxidant properties of kynurenic acid: Free radical scavenging activity and inhibition of oxidative stress. *Neurotoxicol. Teratol.* **2011**, *33*, 538–547. [CrossRef]
11. González-Sánchez, M.; Jiménez, J.; Narváez, A.; Antequera, D.; Llamas-Velasco, S.; Martín, A.H.S.; Arjona, J.A.M.; Munain, A.L.D.; Bisa, A.L.; Marco, M.; et al. Kynurenic Acid Levels are Increased in the CSF of Alzheimer's Disease Patients. *Biomolecules* **2020**, *10*, 571. [CrossRef] [PubMed]
12. Bądzyńska, B.; Zakrocka, I.; Turski, W.A.; Olszyński, K.H.; Sadowski, J.; Kompanowska-Jezierska, E. Kynurenic acid selectively reduces heart rate in spontaneously hypertensive rats. *Naunyn-Schmiedeberg's Arch. Pharmacol.* **2020**, *393*, 673–679. [CrossRef] [PubMed]
13. Nahomi, R.B.; Nam, M.H.; Rankenberg, J.; Rakete, S.; Houck, J.A.; Johnson, G.C.; Stankowska, D.L.; Pantcheva, M.B.; MacLean, P.S.; Nagaraj, R.H. Kynurenic Acid Protects Against Ischemia/Reperfusion-Induced Retinal Ganglion Cell Death in Mice. *Int. J. Mol. Sci.* **2020**, *21*, 1795. [CrossRef] [PubMed]

14. Walczak, K.; Wnorowski, A.; Turski, W.A.; Plech, T. Kynurenic acid and cancer: Facts and controversies. *Cell Mol. Life Sci.* **2020**, *77*, 1531–1550. [CrossRef] [PubMed]
15. Shin, M.; Sano, K.; Umezawa, C. Metabolism of tryptophan to niacin in *Saccharomyces uvarum*. *J. Nutr. Sci. Vitaminol.* **1991**, *37*, 269–283. [CrossRef] [PubMed]
16. Kucharczyk, R.; Zagulski, M.; Rytka, J.; Herbert, C.J. The yeast gene YJR025c encodes a 3-hydroxyanthranilic acid dioxygenase and is involved in nicotinic acid biosynthesis. *FEBS Lett.* **1998**, *424*, 127–130. [CrossRef]
17. Ohashi, K.; Kawai, S.; Murata, K. Secretion of quinolinic acid, an intermediate in the kynurenine pathway, for utilization in NAD^+ biosynthesis in the yeast *Saccharomyces cerevisiae*. *Eukaryot. Cell* **2013**, *12*, 648–653. [CrossRef]
18. Yılmaz, C.; Gökmen, V. Kinetic evaluation of the formation of tryptophan derivatives in the kynurenine pathway during wort fermentation using *Saccharomyces pastorianus* and *Saccharomyces cerevisiae*. *Food Chem.* **2019**, *297*, 124975. [CrossRef]
19. Panozzo, C.; Nawara, M.; Suski, C.; Kucharczyka, R.; Skoneczny, M.; Bécam, A.M.; Rytka, J.; Herbert, C.J. Aerobic and anaerobic NAD^+ metabolism in *Saccharomyces cerevisiae*. *FEBS Lett.* **2002**, *517*, 97–102. [CrossRef]
20. Wogulis, M.; Chew, E.R.; Donohoue, P.D.; Wilson, D.K. Identification of formyl kynurenine formamidase and kynurenine aminotransferase from *Saccharomyces cerevisiae* using crystallographic, bioinformatic and biochemical evidence. *Biochemistry* **2008**, *47*, 1608–1621. [CrossRef]
21. Ohashi, K.; Chaleckis, R.; Masak Takaine, M.; Wheelock, C.E.; Yoshida, S. Kynurenine aminotransferase activity of Aro8/Aro9 engage tryptophan degradation by producing kynurenic acid in *Saccharomyces cerevisiae*. *Sci. Rep.* **2017**, *7*, 12180. [CrossRef] [PubMed]
22. Belenky, P.; Christensen, K.C.; Gazzaniga, F.; Pletnev, A.A.; Brenner, C. Nicotinamide riboside and nicotinic acid riboside salvage in fungi and mammals. Quantitative basis for Urh1 and purine nucleoside phosphorylase function in NAD^+ metabolism. *J. Biol. Chem.* **2009**, *284*, 158–164. [CrossRef] [PubMed]
23. Bieganowski, P.; Brenner, C. Discoveries of nicotinamide riboside as a nutrient and conserved NRK genes establish a Preiss-Handler independent route to NAD^+ in fungi and humans. *Cell* **2004**, *117*, 495–502. [CrossRef]
24. Ghislain, M.; Talla, E.; François, J.M. Identification and functional analysis of the *Saccharomyces cerevisiae* nicotinamidase gene, PNC1. *Yeast* **2002**, *19*, 215–224. [CrossRef] [PubMed]
25. Belenky, P.A.; Moga, T.G.; Brenner, C. *Saccharomyces cerevisiae* YOR071C encodes the high affinity nicotinamide riboside transporter Nrt1. *J. Biol. Chem.* **2008**, *283*, 8075–8079. [CrossRef] [PubMed]
26. Llorente, B.; Dujon, B. Transcriptional regulation of the *Saccharomyces cerevisiae* DAL5 gene family and identification of the high affinity nicotinic acid permease TNA1 (YGR260w). *FEBS Lett.* **2000**, *475*, 237–241. [CrossRef]
27. Wróbel-Kwiatkowska, M.; Turski, W.; Kocki, T.; Rakicka-Pustułka, M.; Rymowicz, W. An efficient method for production of kynurenic acid by *Yarrowia lipolytica*. *Yeast* **2020**, *37*, 541–547. [CrossRef] [PubMed]
28. Wróbel-Kwiatkowska, M.; Turski, W.; Juszczyk, P.; Kita, A.; Rymowicz, W. Production of KYNA on media containing various honeys. *Sustainability* **2020**, *12*, 9424. [CrossRef]
29. Sprenger, J.; Hardeland, R.; Fuhrberg, B.; Han, S.Z. Melatonin and other 5-methoxylated indoles in yeast: Presence in high concentrations and dependence on tryptophan availability. *Cytologia* **1999**, *64*, 209–213. [CrossRef]
30. Tan, D.X.; Hardeland, R.; Back, K.; Manchester, L.C.; Alatorre-Jimenez, M.A.; Reiter, R.J. On the significance of an alternate pathway of melatonin synthesis via 5-methoxytryptamine: Comparisons across species. *J. Pineal Res.* **2016**, *61*, 27–40. [CrossRef]
31. Ganguly, S.; Mummaneni, P.; Steinbach, P.J.; Klein, D.C.; Coon, S.L. Characterization of the *Saccharomyces cerevisiae* homolog of the melatonin rhythm enzyme arylalkylamine N-acetyltransferase (EC 2.3.1.87). *J. Biol. Chem.* **2001**, *276*, 47239–47247. [CrossRef] [PubMed]
32. Muniz-Calvo, S.; Bisquert, R.; Fernandez-Cruz, E.; García-Parrilla, M.C.; Guillamón, J.M. Deciphering the melatonin metabolism in *Saccharomyces cerevisiae* by the bioconversion of related metabolites. *J. Pineal Res.* **2019**, *66*, e12554. [CrossRef] [PubMed]
33. Mor, M.; Spadoni, G.; Diamantini, G.; Bedini, A.; Tarzia, G.; Silva, C.; Vacondio, F.; Rivara, M.; Plazzi, P.V.; Franceschinit, D.; et al. Antioxidant and cytoprotective activity of indole derivatives related to melatonin. *Adv. Exp. Med. Biol.* **2003**, *527*, 567–575. [CrossRef] [PubMed]
34. Mor, M.; Silva, C.; Vacondio, F.; Plazzi, P.V.; Bertoni, S.; Spadoni, G.; Diamantini, G.; Bedini, A.; Tarzia, G.; Zusso, M.; et al. Indole-based analogs of melatonin: In vitro antioxidant and cytoprotective activities. *J. Pineal Res.* **2004**, *36*, 95–102. [CrossRef] [PubMed]
35. Spadoni, G.; Diamantini, G.; Bedini, A.; Tarzia, G.; Vacondio, F.; Silva, C.; Rivara, S.; Mor, M.; Plazzi, P.V.; Zusso, M.; et al. Synthesis, antioxidant activity and structure-activity relationships for a new series of 2-(N-acylaminoethyl)indoles with melatonin-like cytoprotective activity. *J. Pineal Res.* **2006**, *40*, 259–269. [CrossRef] [PubMed]
36. Lyte, J.M. Eating for 3.8 × 1013: Examining the impact of diet and nutrition on the microbiota-gut-brain axis through the lens of microbial endocrinology. *Front. Endocrinol.* **2019**, *9*, 796. [CrossRef] [PubMed]
37. Turski, M.P.; Turska, M.; Zgrajka, W.; Kuc, D.; Turski, W.A. Presence of kynurenic acid in food and honeybee products. *Amino Acids* **2009**, *36*, 75–80. [CrossRef]
38. Turski, M.P.; Kaminski, P.; Zgrajka, W.; Turska, M.; Turski, W.A. Potato-an important source of nutritional kynurenic acid. *Plant. Foods Hum. Nutr.* **2012**, *67*, 17–23. [CrossRef]

39. Novo, M.; Bigey, F.; Beyne, E.; Galeote, V.; Gavory, F.; Mallet, S.; Cambon, B.; Legras, J.L.; Wincker, P.; Casaregola, S.; et al. Eukaryote-to-eukaryote gene transfer events revealed by the genome sequence of the wine yeast *Saccharomyces cerevisiae* EC1118. *Proc. Natl. Acad. Sci. USA* **2009**, *106*, 16333–16338. [CrossRef]
40. Vigentini, I.; Gardana, C.; Fracassetti, D.; Gabrielli, M.; Foschino, R.; Simonetti, P.; Tirelli, A.; Iriti, M. Yeast contribution to melatonin, melatonin isomers and tryptophan ethyl ester during alcoholic fermentation of grape musts. *J. Pineal Res.* **2015**, *58*, 388–396. [CrossRef]
41. O'Connor-Cox, E.S.C.; Ingledew, W.M. Wort nitrogenous sources—Their use by brewing yeasts: A review. *J. Am. Soc. Brew. Chem.* **1989**, *47*, 102–108. [CrossRef]
42. Manginot, C.; Roustan, J.L.; Sablayrolles, J.M. Nitrogen demand of different yeast strains during alcoholic fermentation. Importance of the stationary phase. *Enzyme Microb. Technol.* **1998**, *23*, 511–517. [CrossRef]
43. Poeggeler, B.; Hardeland, R. Detection and quantification of melatonin in a dinoflagellate, Gonyuulax polyedru: Solutions to the problem of methoxyindole destruction in non-vertebrate material. *J. Pineal Res.* **1994**, *17*, 1–10. [CrossRef] [PubMed]
44. Tan, D.X.; Reiter, R.J.; Manchester, L.C.; Yan, M.T.; El-Sawi, M.; Sainz, R.M.; Mayo, J.C.; Kohen, R.; Allegra, M.C.; Hardeland, R. Chemical and physical properties and potential mechanisms: Melatonin as a broad spectrum antioxidant and free radical scavenger. *Curr. Top. Med. Chem.* **2002**, *2*, 181–197. [CrossRef] [PubMed]
45. Tan, D.X.; Manchester, L.C.; Esteban-Zubero, E.; Zhou, Z.; Reiter, R.J. Melatonin as a Potent and Inducible Endogenous Antioxidant: Synthesis and Metabolism. *Molecules* **2015**, *20*, 18886–18906. [CrossRef]
46. Motlhalamme, T. Characterization of Melatonin Production and Physiological Functions in Yeast. Ph.D. Thesis, Stellenbosch University, Stellenbosch, South Africa, March 2020.
47. Lin, B.; Tao, Y. Whole-cell biocatalysts by design. *Microb. Cell Fact.* **2017**, *16*, 106. [CrossRef] [PubMed]
48. Şanlier, N.; Gökcen, B.B.; Sezgin, A.C. Health benefits of fermented foods. *Crit. Rev. Food Sci. Nutr.* **2019**, *59*, 506–527. [CrossRef]
49. Shiferaw Terefe, N.; Augustin, M.A. Fermentation for tailoring the technological and health related functionality of food products. *Crit. Rev. Food Sci. Nutr.* **2020**, *60*, 2887–2913. [CrossRef]
50. Vitalini, S.; Dei Cas, M.; Rubino, F.M.; Vigentini, I.; Foschino, R.; Iriti, M.; Paroni, R. LC-MS/MS-Based Profiling of Tryptophan-Related Metabolites in Healthy Plant Foods. *Molecules* **2020**, *25*, 311. [CrossRef]
51. Vincenzini, M.; Romano, P.; Farris, G.A. *Microbiologia Del Vino*, 1st ed.; Casa Editrice Ambrosiana, Distribuzione Esclusiva Zanichelli: Bologna, Italy, 2005.
52. Paroni, R.; Dei Cas, M.; Rizzo, J.; Ghidoni, R.; Montagna, M.T.; Rubino, F.M.; Iriti, M. Bioactive phytochemicals of tree nuts. Determination of the melatonin and sphingolipid content in almonds and pistachios. *J. Food Compos. Anal.* **2019**, *82*, 103227. [CrossRef]
53. Tofalo, R.; Fusco, V.; Böhnlein, C.; Kabisch, J.; Logrieco, A.F.; Habermann, D.; Cho, G.S.; Benomar, N.; Abriouel, H.; Schmidt-Heydt, M.; et al. The life and times of yeasts in traditional food fermentations. *Crit. Rev. Food Sci. Nutr.* **2020**, *60*, 3103–3132. [CrossRef] [PubMed]
54. Huang, X.; Ding, W.; Wu, F.; Zhou, S.; Deng, S.; Ning, Y. Increased Plasma Kynurenic Acid Levels are Associated with Impaired Attention/Vigilance and Social Cognition in Patients with Schizophrenia. *Neuropsychiatr. Dis. Treat.* **2020**, *16*, 263–271. [CrossRef] [PubMed]
55. Milart, P.; Paluszkiewicz, P.; Dobrowolski, P.; Tomaszewska, E.; Smolinska, K.; Debinska, I.; Gawel, K.; Walczak, K.; Bednarski, J.; Turska, M.; et al. Kynurenic acid as the neglected ingredient of commercial baby formulas. *Sci. Rep.* **2019**, *9*, 6108. [CrossRef]
56. Linderholm, K.R.; Skogh, E.; Olsson, S.K.; Dahl, M.L.; Holtze, M.; Engberg, G.; Samuelsson, M.; Erhardt, S. Increased levels of kynurenine and kynurenic acid in the CSF of patients with schizophrenia. *Schizophr. Bull.* **2012**, *38*, 426–432. [CrossRef] [PubMed]

Review

5-Hydroxytryptophan (5-HTP): Natural Occurrence, Analysis, Biosynthesis, Biotechnology, Physiology and Toxicology

Massimo E. Maffei

Department of Life Sciences and Systems Biology, University of Turin, Via Quarello 15/a, 10135 Turin, Italy; massimo.maffei@unito.it; Tel.: +39-011-670-5967

Abstract: L-5-hydroxytryptophan (5-HTP) is both a drug and a natural component of some dietary supplements. 5-HTP is produced from tryptophan by tryptophan hydroxylase (TPH), which is present in two isoforms (TPH1 and TPH2). Decarboxylation of 5-HTP yields serotonin (5-hydroxytryptamine, 5-HT) that is further transformed to melatonin (*N*-acetyl-5-methoxytryptamine). 5-HTP plays a major role both in neurologic and metabolic diseases and its synthesis from tryptophan represents the limiting step in serotonin and melatonin biosynthesis. In this review, after an look at the main natural sources of 5-HTP, the chemical analysis and synthesis, biosynthesis and microbial production of 5-HTP by molecular engineering will be described. The physiological effects of 5-HTP are discussed in both animal studies and human clinical trials. The physiological role of 5-HTP in the treatment of depression, anxiety, panic, sleep disorders, obesity, myoclonus and serotonin syndrome are also discussed. 5-HTP toxicity and the occurrence of toxic impurities present in tryptophan and 5-HTP preparations are also discussed.

Keywords: 5-hydroxytryptophan; natural sources; microbial production; biosynthetic pathways; physiological effects; animal; human

Citation: Maffei, M.E. 5-Hydroxytryptophan (5-HTP): Natural Occurrence, Analysis, Biosynthesis, Biotechnology, Physiology and Toxicology. *Int. J. Mol. Sci.* **2021**, *22*, 181. https://dx.doi.org/10.3390/ijms22010181

Received: 4 December 2020
Accepted: 23 December 2020
Published: 26 December 2020

Publisher's Note: MDPI stays neutral with regard to jurisdictional claims in published maps and institutional affiliations.

Copyright: © 2020 by the author. Licensee MDPI, Basel, Switzerland. This article is an open access article distributed under the terms and conditions of the Creative Commons Attribution (CC BY) license (https://creativecommons.org/licenses/by/4.0/).

1. Introduction

L-5-hydroxytryptophan (5-HTP) is produced from tryptophan by tryptophan hydroxylase (TPH) and its decarboxylation yields serotonin (5-hydroxytryptamine, 5-HT), a monoamine neurotransmitter involved in the modulation of mood, cognition, reward, learning, memory, sleep and numerous other physiological processes [1]. 5-HT is further transformed to melatonin (*N*-acetyl-5-methoxytryptamine), the hormone primarily released by the pineal gland that regulates the sleep–wake cycle [2,3]. Therefore, the biosynthesis of 5-HTP is important and necessary for the production of key molecules such as 5-HT and melatonin. 5-HTP also plays a major role both in neurologic and metabolic diseases and its synthesis from tryptophan represents the limiting step in 5-HT and melatonin biosynthesis [4]. The occurrence of 5-HTP is not limited to animals or humans and the molecule is produced by lower and higher plants, mushrooms and microbes (see below).

5-HTP is both a drug and a natural component of some dietary supplements and occurrence in both synthetic tryptophan and 5-HTP of toxic impurities has caused eosinophilia myalgia syndrome cases [5–8]; therefore, accurate chemical analyses and characterization have been developed. Recently, microbial engineering allowed the production of 5-HTP from alternative biosynthetic routes, opening the interesting possibility of better controlling the presence of contaminants.

This review describes the natural sources of 5-HTP, which represent the raw material from which 5-HTP may be extracted and purified for its use in drugs and dietary supplements, as well as alternative microbial synthesis that is considered a stable and sustainable way to provide a constant supply of the molecule. Chemical synthesis and chemical analysis of 5-HTP are discussed, along with the major biochemical pathways involved in 5-HTP biosynthesis and transformation, with particular reference to tryptophan hydroxylase. The physiological role of 5-HTP is discussed by considering both animal studies and human clinical trials. The toxicology of 5-HTP and the potential occurrence of toxic impurities are also discussed.

2. Databases, Exclusion and Inclusion Criteria

The strategy that was implemented to carry out this review was based on a deep search in the databases Web of Science (1985–2020) and PubMed (1940–2020) by considering, as the main entry, the term 5-hydroxytryptophan. The total number of Web of Science Core Collection papers was 1857, whereas the total number in PubMed was 5431 papers. There were 816 selected papers in the first search (see Supplementary File S1) and the exclusion criteria were the impossibility to obtain a full text and the lack of specificity with respect to the selected areas of the review. Out of the 816 papers, 195 were used for this review because of the historical, breakthrough and innovative content.

3. Natural Sources of 5-HTP

Plants are a rich source of 5-HTP and *Griffonia simplicifolia* Baill. (Caesalpinaceae) (also known as the alternate incorrect name *Bandeiraea simplicifolia*) seeds are the most used for the extraction and the commercial production of 5-HTP. Preparations of *G. simplicifolia* containing high concentrations of 5-HTP are used to treat serotonin-related disorders, including motion sickness [9] and to increase the feeling of satiety associated with a significant reduction of food intake and body weight with decreasing body mass index (BMI) [10,11]. The extract also shows anxiolytic-like effects [12]. The authentication of *G. simplicifolia* seeds is based on methods that are often laborious, time-consuming and sensitive to interference from co-occurring materials. Rapid and simple estimation procedures have been developed for the identification and quantification of 5-HTP in *G. simplicifolia* extracts [13,14]. Recently, the combination of chemical analysis (with the characterization of 5-HTP along with the β-carboline alkaloid derivatives) and molecular DNA fingerprinting (DNA restriction fragment length polymorphism—PCR-RFLP—analysis which has been performed on the plant internal transcribed spacer—ITS) allowed the unequivocal identification of commercial *G. simplicifolia* seeds [15].

Both 5-HTP and 5-hydroxytryptophan hydrate have been identified in the root allelochemical exudates from the aggressive weed couch grass, *Elytrigia* (*Agropyron*) *repens* [16], whereas quack grass (*Agropyron repens* L. Beauv.) accumulates throughout the plant high levels of 5-HTP as glucosides attached in β-linkages to the 5-O-indolyl moiety [17]. St. John's wort (*Hypericum perforatum* cv. Anthos) stem explants were found to produce significant amounts of 5-HTP when plantlets were regenerated from thidiazuron-induced tissue grown on a basal culture medium for 2 months [18], whereas 5-HTP was detected during the alcoholic fermentation of some grape cultivars [19]. Food processing may alter the 5-HTP content. The cooking (boiling, steaming, and microwaving) of the cauliflower genotypes Forata (white inflorescence), Verde di Macerata (green inflorescence), Cheddar F1 (yellow inflorescence), and Graffiti (purple coloration) was found to increase tryptophan levels and to reduce the content of 5-HTP [20].

Interestingly, other organisms are potential sources of 5-HTP. The intertidal sponge *Hymeniacidon heliophila*, which survives under intense sunlight, contains 5-HTP as a major constituent [21], whereas methanolic extracts of the mushrooms *Boletus edulis*, *Suillus luteus*, and *Pleurotus ostreatus* contain, among other indole compounds, fairly good amounts of 5-HTP [22]. The fruiting bodies of the mushroom *Cantharellus cibarius* (the chanterelle) and the mycelium of this species cultured in vitro contain eight indole compounds, including 5-HTP [23], while the concentration of 5-HTP was higher in the stipes of five species of the fungal genus *Panaeolus* (*P. ater*, *P. rickenii*, *P. papilionaceous*, *P. sphinctrinus*, and *P. subbalteatus*) [24].

4. Qualitative and Quantitative Analysis of 5-HTP

The chemical analysis of 5-HTP is mainly performed by using high-performance liquid chromatography (HPLC) coupled to different detectors, including diode array

(DAD), fluorescence (FD), and mass spectrometry (MS) detectors. The separation is usually performed by reverse phase C-18 column chromatography [25]. The binary solvent system is H_2O acidified with 0.1% v/v formic acid (solvent A) and acetonitrile acidified with 0.1% v/v formic acid (solvent B). The chromatographic profiles are usually registered at 230 and 270 nm. Quantitative analysis is usually performed by positive ion mode, often using a selective ion monitoring (SIM) method with different ionization interfaces, the most used being electrospray ionization (EI) [15]. This method, along with the use of other detectors, has been used to detect and quantify 5-HTP in plant extracts [2,14,15,26], rat serum [25], mice whole brain tissue [27], human plasma [28], human urine [29], and in the central nervous system [30,31].

Besides HPLC, other chromatographic techniques have been used to detect and quantify 5-HPT. Capillary electrophoresis (CE) has been used with different detectors to determine 5-HTP in samples of commercial dietary supplements [32], in human platelet-rich plasma [33], and immortalized rat raphe nuclei neurons [34]. A rapid method for the separation of 5-HTP was developed by using micellar electrokinetic chromatography with diode array detection [35]. The identification and quantification of 5-HTP was also achieved by the combination of solid-phase extraction pretreatment and gas chromatography–mass spectrometry based on a modified method of derivatization by silanization [36].

An NMR-based approach showed that 5-HTP results in characteristic chemical shift correlations suited for its identification and quantification [37], whereas the quantitative analysis of 5-HTP of carcinoid tumors was assayed using gold nanoparticles as the assisted matrix in surface-assisted laser desorption/ionization time-of-flight mass spectrometry [38]. A green fluorescence transient 5-HTP is obtained by multiphoton near infrared excitation and this technique enables the detection of 5-HTP with extremely high sensitivity. The potential application of such a method is in the imaging of biological systems and the investigation of protein dynamics [39]. Both cyclic voltammetric and UV–visible spectroscopic methods have been demonstrated to show linear responses over a wide concentration range of 5-HTP, with low limits of detection also in dietary supplements [40].

Electrochemical sensors have also attracted much attention for the detection and quantification of 5-HTP. Their high sensitivity and miniaturization are expected to overcome the shortcomings of high-cost, time-consuming, and complicated operations linked to chromatographic methods. Various types of surface-modified electrodes have been developed to determine 5-HTP such as a carbon nanosheet-modified electrode [41], Ru^{II}terpyridine-doped composite electrode [42], gold-modified pencil graphite electrode [43], graphene-chitosan molecularly imprinted film modified on the surface of a glassy carbon electrode [44], sensors based on electro-polymerization to obtain a poly-(melamine)/poly-(o-aminophenol) co-polymeric film [45], nano-palladium decorated multi-walled carbon nanotubes [46], citrate-capped gold nanoparticles [47], a pyrolytic graphite electrode with the surface covered with a thin film of a nano-mixture of graphite/diamond [48], and by electrochemical microfluidic separation and sensing [49].

5. Chemical Synthesis of 5-HTP

The synthesis of 5-HTP by the condensation of 5-benzyloxygramine with diethyl formaminomalonate, followed by saponification, decarboxylation, and hydrogenolysis was described in 1951 [50] and 1954 [51] and was an application of gramine synthesis, developed by Snyder and Smith ten years before [52]. A few years later, another application of gramine synthesis was reported [53]. In the same year, Frangatos and Chubb [54] reported an application of the convenient tryptophan synthesis developed ten years before [55] by eliminating the difficult and tedious preparation of 5-benzyloxyindole. The *p*-benzyloxyphenylhydrazone of γ,γ-dicarbethoxy-γ-acetamido-butyraldehyde (Figure 1, **I**) was prepared and cyclized, without isolation, to form ethyl β-(5-benzyloxyindol-3-)-α-carbethoxy-α-acetamidopropioilate (Figure 1, **II**). Saponification and partial decarboxylation of **II**, followed by hydrolysis of the acetamido group, gave 5-benzyloxytryptophan (Figure 1, **III**). 5-HTP was obtained by hydrogenolysis of **III** (Figure 1). However, this synthetic method suffers from the difficulty involved in the regioselective hydroxylation of tryptophan.

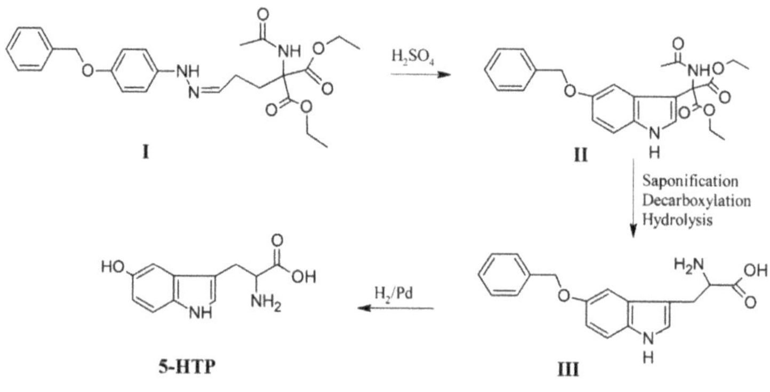

Figure 1. Chemical synthesis of 5-hydroxytryptophan (5-HTP). From [54], modified.

6. Biosynthesis of 5-HTP and Inhibition of Tryptophan Hydroxylase (TPH)

The biosynthesis of 5-HTP starts with the essential amino acid tryptophan, which is metabolized to 5-HTP by TPH in an initial, rate-limiting step in the biosynthesis of serotonin after the decarboxylation catalyzed by aromatic amino acid decarboxylase (AADC). TPH is a monooxygenase that belongs to the family of aromatic amino acid hydroxylases; it incorporates one atom of oxygen from molecular oxygen into the substrate and reduces the other atom to water. The two electrons required for the reduction of the second atom to water are supplied by tetrahydrobiopterin (BH_4), which acts as a substrate rather than a tightly bound cofactor, binding and dissociating each turnover [56]. The irreversible activation of O_2 is the initial step in this mechanism and utilizes two electrons from BH_4 to form a high-valent Fe(IV)O (ferryl) hydroxylating intermediate and 4a-hydroxypterin (4a-$HOPH_3$). The Fe(IV)O intermediate subsequently reacts with the side chain of the aromatic amino acid through electrophilic aromatic substitution [57,58]. The binding of both the amino acid and BH_4 results in a change in the coordination of the iron from six-coordinate to five-coordinate, presumably opening a coordination site for oxygen. The hydroxylating intermediate Fe(IV)O in TPH has been confirmed by rapid freeze-quench ^{57}Fe Mössbauer spectroscopy [57]. During L-tryptophan hydroxylation, BH_4 is oxidized to pterin-4α-carbinolamine (BH_3OH) and regenerated through the function of pterin-4α-carbinolamine dehydratase (PCD) and dihydropteridine reductase (DHPR) [59,60]. Figure 2 shows the involvement of BH_4 and the Fe intermediate in the reaction catalyzed by TPH.

In humans, BH_4 is synthesized from guanosine triphosphate (GTP) via a three-step pathway, containing GTP cyclohydrolase I (GCHI), 6-pyruvate-tetrahydropterin synthase (PTPS), and sepiapterin reductase (SPR) [61], as shown in Figure 3.

TPH activity is assayed by a continuous spectrophotometric method that exploits the different spectral properties of tryptophan and 5-HTP [62]. The sensitivity of the essay allows the use of relatively low enzyme concentrations and can be used to determine the steady-state kinetic parameters for each of the enzyme substrates [62].

TPH is composed of three functional domains, a regulatory N-terminal domain, a catalytic domain, and a C-terminal oligomerization domain [63]. TPH requires ferrous iron for activity. The activity of the enzyme is affected by phosphorylation and the requirement for Ca^{2+} suggests a need for a calcium-dependent kinase [56]. In the rat pineal gland, norepinephrine stimulates TPH synthesis and activity and cAMP through the activation of cAMP-dependent protein kinase A (PKA), phosphorylates the transcription factor cAMP response element binding protein (CREB), which starts the enzyme synthesis, and the incorporation of at least 1 mol of phosphate/mol of tetramer of native TPH is required for maximal activation [64].

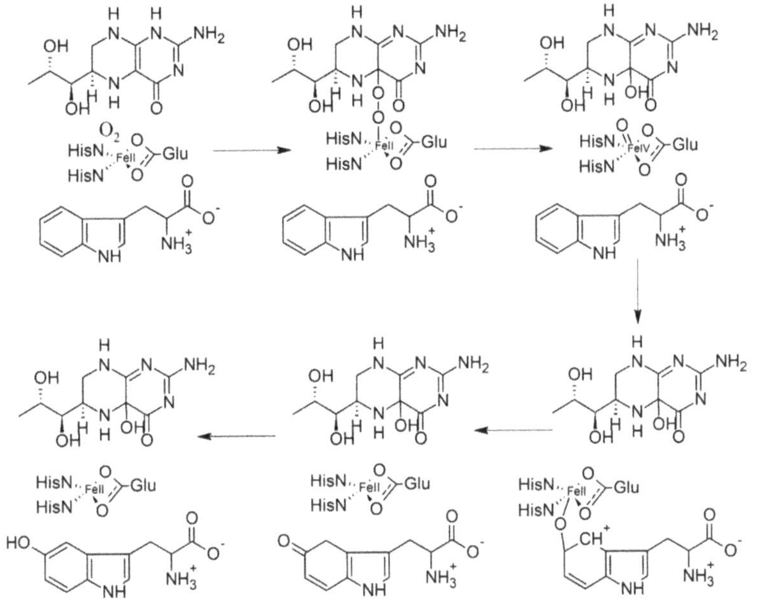

Figure 2. Hydroxylation of tryptophan by tryptophan hydroxylase (TPH). See text for explanation. From [58], modified.

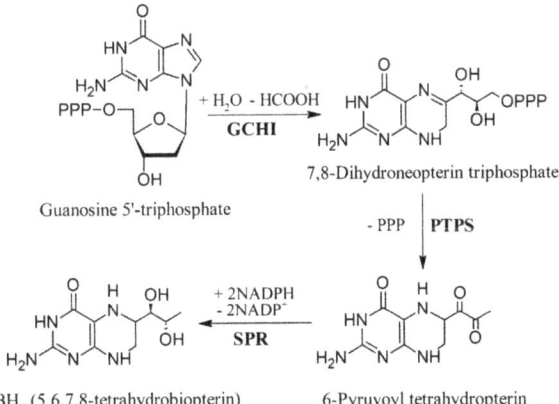

Figure 3. Biosynthesis of BH4 in mammals. Modified from [61].

In vertebrates, there are two molecular forms of TPH: TPH1 is responsible for serotonin synthesis in peripheral tissues and is mainly expressed in the enterochromaffin cells of the gut and in the pineal gland. TPH1-expressing cells of the gastrointestinal (GI) tract are responsible for blood 5-HT synthesis; 5-HTP then enters the circulation packed in dense granules of thrombocytes where it mediates its hormonal actions upon platelet release at the site of activation [65]. The second form, TPH2, is expressed in peripheral myenteric neurons in the gut and in the neurons of raphe nuclei in the brain stem but not in peripheral organs (lung, heart, kidney, or liver) [4]. Cells expressing TPH2 have rates of 5-HT synthesis which are affected by changes in tryptophan availability [66]. The presence of these two TPH forms justifies the duality of the 5-HT system, with two independently generated pools of 5-HT—one in the brain and another in the blood [65]. Antibodies that distinguish between the isoforms of TPH have been developed [67]. Figure 4 shows the central role of 5-HTP in the biochemical pathway of 5-HT.

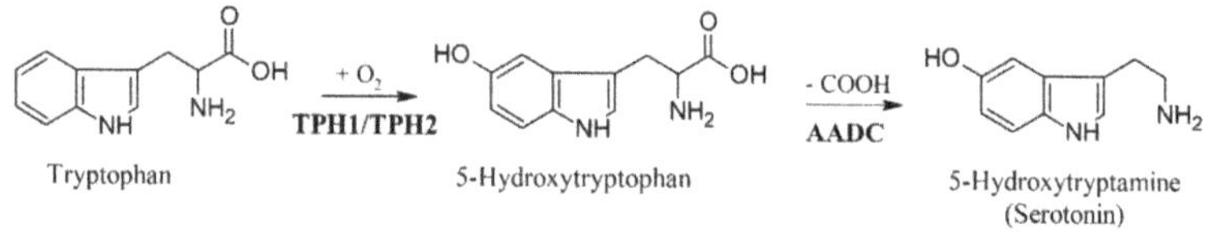

Figure 4. Biosynthetic pathway of serotonin (5-HT). Tryptophan is hydroxylated by the two forms of TPH to yield 5-HTP, which is then decarboxylated by the aromatic amino acid decarboxylase (AADC) to serotonin.

Because TPH inhibitors may provide novel treatments for various gastrointestinal disorders associated with dysregulation of the gastrointestinal serotonergic system, such as chemotherapy-induced emesis and irritable bowel syndrome, both academia and the pharmaceutical industry have worked on the search for specific TPH inhibitors. Naturally occurring unspecific inhibitors of TPH (and indoleamine metabolism) have been reported, including catecholamines, the food-derived carcinogenic heterocyclic amines 3-amino-1,4-dimethyl-5H-pyrido[4,3-b]indole (Trp-P-1), and 3-amino-1-methyl-5H-pyrido[4,3-b]indole (Trp-P-2), as well as the dopamine-derived tetrahydroisoquinolines, such as salsolinol and tetrahydropapaverine [68–70]. While the inhibition of TPH by salsolinols was found to be non-competitive with the substrate L-tryptophan [68], TPH was un-competitively inhibited by tetrahydropapaverine with the substrate L-tryptophan, and non-competitively inhibited with the cofactor DL-6-methyl-5,6,7,8-tetrahydropteridin in P-815 cells [69]. In the same cell system, the inhibition of TPH by Trp-P-2 was found to be competitive with the substrate L-tryptophan and non-competitive with the cofactor DL-6-methyl-5,6,7,8-tetrahydropteridin [70]. p-Ethynylphenylalanine (p-EPA) is a more potent TPH inhibitor; p-EPA injection induced a significant and gradual decrease in extracellular 5-HTP in the rat hippocampus, striatum, and frontal cortex. Moreover, p-EPA could also irreversibly interfere with the synthesis of TPH [71]. Selective inhibitors of TPH, such as LP-533401 [(2S)-2-amino-3-(4-(2-amino-6-(2,2,2-trifluoro-1-(3′-fluorobiphenyl-4-yl)ethoxy)pyrimidin-4-yl)phenyl)propanoic acid] and LP-615819 [(2S)-ethyl 2-amino-3-(4-(2-amino-6-(2,2,2-trifluoro-1-(3′-fluorobiphenyl-4-yl)ethoxy)pyrimidin-4-yl)phenyl)propanoic acid], were found to competitively bind to the tryptophan pocket of both TPH isoforms and to improve metabolic parameters, thus providing novel treatments for various gastrointestinal disorders associated with dysregulation of the gastrointestinal serotonergic system [72]. The kinetic analysis with these inhibitors showed that they are all competitive versus L-tryptophan but predominantly uncompetitive versus pterin [73]. Two other inhibitors of TPH are telotristat ethyl (the free base form of a hippurate salt called telotristat etiprate) and its active metabolite telotristat. In vitro, the inhibitory potency of telotristat was found to be 29-fold higher than that of its prodrug, with inhibition of TPH resulting in a reduced production of peripheral 5-HT [74].

A new series of acyl guanidines displaying potent TPH1 inhibition have been reported [75]; these molecules have been chemically and pharmacokinetically optimized and successfully tested in vivo [76]. 1-O-Galloylpedunculagin was screened as a drug-like compound from the traditional Chinese medicine (TCM) database for inhibitor activity on TPH. The molecule specifically inhibited TPH1 but was ineffective on TPH2, and the inhibitory action displayed characteristics of competitive inhibition [77]. The effects of nifedipine, an L-type calcium channel blocker, in noradrenergic-stimulated cultured rat pineal glands, showed that TPH activity was the main step inhibited by the molecule, demonstrating that the calcium influx through L-type high-voltage-activated calcium channels is essential for the full activation of the enzyme [78]. Figure 5 shows the chemical formulae of the TPH inhibitors cited above.

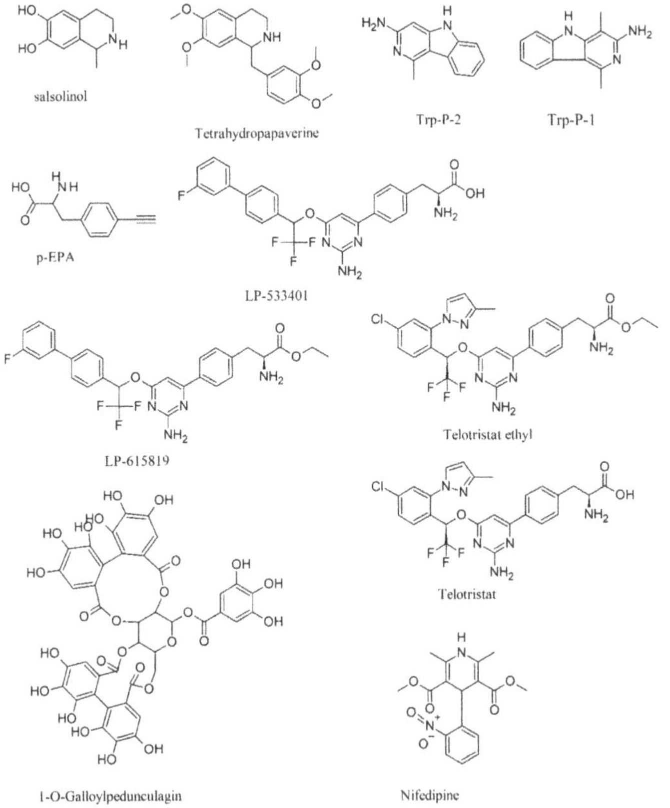

Figure 5. Structure formulae of some TPH inhibitors.

7. Metabolic Engineering and Heterologous Production of 5-HTP

As discussed above, extraction from the seeds of the African plant *Griffonia simplicifolia* is the typical approach for 5-HTP commercial production, because chemical synthesis is not economically feasible on a large scale. However, the material supply is seasonally and regionally dependent, which limits the output of 5-HTP. A promising alternative is the metabolic engineering of microorganisms. Among the many advantages, microorganisms grow quickly and their genetic engineering has been demonstrated to be a productive way to obtain important chemicals, such as aromatic compounds [79], fatty acids [80], carotenoids [81], flavors and fragrances [82], and many other natural products [83].

Although the microbial synthesis of 5-HTP has been achieved in different microorganisms, with particular reference to *Escherichia coli*, the first evidence of a microbial hydroxylation of tryptophan was found in *Chromobacterium violaceum*, about 70 years ago [84]. The function of tryptophan hydroxylation in this organism was considered important to provide the precursor for the characteristic blue pigment produced by this species, violacein, that is synthesized by a series of tryptophan metabolisms. Further characterization showed that the TPH from *C. violaceum* has, at least superficially, the characteristics of the TPH found in various mammalian tissues [85]. L-phenylalanine 4-hydroxylase from *C. violaceum* could convert L-tryptophan to 5-HTP and L-phenylalanine to L-tyrosine; however, the activity for L-tryptophan is extremely low compared to L-phenylalanine activity levels. The L-tryptophan hydroxylation activity of *C. violaceum* L-phenylalanine 4-hydroxylase (CviPAH) was enhanced using information on its crystal structures by introducing a saturation mutagenesis towards L101 and W180 in *C. vio-*

laceum phenylalanine 4-hydroxylase (PAH). Mutant libraries from the *L101* and *W180* produced several positive mutants, with *L101Y* and *W180F* showing the highest TPH activity, whereas the double mutant (*L101Y-W180F*) displayed higher TPH activity when compared with the wild type and the individual *W180F* and *L101Y* mutants [86]. Interestingly, pterin is still required as a cofactor for enzyme activity by the *CviPAH-L101Y-W180F* triple mutant, which is similar to the requirements of other types of aromatic amino acid hydroxylases [86]. A novel cofactor regeneration process to achieve enhanced synthesis of 5-HTP by using CviPAH was obtained by screening and investigating several key enzymes, including dihydropteridine reductase from *E. coli*, glucose dehydrogenase from *Bacillus subtilis*, and pterin-4α-carbinolamine dehydratase from *Pseudomonas syringae*. Genes encoding these three enzymes were overexpressed in an *E. coli* tryptophanase-deficient host, resulting in the synthesis of ten-fold (0.74 mM) 5-HTP in the presence of 0.1 mM pterin with respect to the absence of the regeneration of pterin [59].

The engineering of *E. coli* successfully improved the production of 5-HTP through bioprospecting and protein engineering approaches. An important achievement was the discovery that bacterial PAHs may utilize tetrahydromonapterin (MH_4, Figure 6) instead of BH_4 as the native pterin coenzyme; this resulted in a great advantage because BH_4 does not naturally occur in most bacteria [87].

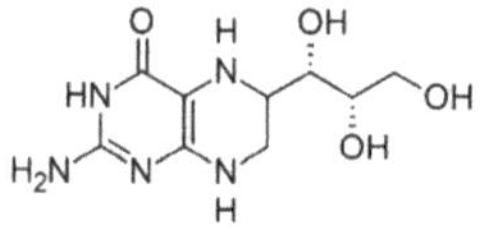

Tetrahydromonapterin

Figure 6. Chemical formula of tetrahydromonapterin.

Based on the potential ability of PAH to hydroxylate tryptophan, the development of PAH mutants highly active in converting tryptophan to 5-HTP allowed the establishment of an efficient 5-HTP production platform via further metabolic engineering efforts [88]. Whole-cell bioconversion allowed the high-level production of 5-HTP (1.1−1.2 g/L) from tryptophan in shake flasks, also allowing de novo 5-HTP biosynthesis from glucose [88] (Figure 7).

In a similar strategy, the tryptophan pathway was extended by using an engineered PAH from *Cupriavidus taiwanensis* (CtAAAH) and the production of 5-HTP was achieved by an endogenous cofactor with an artificial regeneration system [89].

In the search of a direct alternative precursor for the biosynthesis of 5-HTP, a novel salicylate 5-hydroxylase was used to convert the non-natural substrate anthranilate to 5-hydroxyanthranilate (5-HI) produced from glucose. To assess whether 5-HI may function as a precursor of 5-HTP, a medium copy number pCStrpDCBA plasmid was constructed in the *E. coli* strain BW2. Knockouts of *tnaA* (that prevents the products tryptophan and 5-HTP from degrading) and *trpE* (that blocks the native synthesis of anthranilic acid) harboring pCS-trpDCBA were used for the in vivo assay. *trpDCBA* was also cloned into a low copy number plasmid, yielding pSA-trpDCBA. The in vivo assay of the strain BW2 harboring pSA-trpDCBA accumulated the intermediate 5-hydroxyindole in the cultures, indicating that the reaction catalyzed by TrpB was a rate-limiting step. The de novo production of 5-HTP was then established by combining the full pathway and by adopting a two-stage strategy (see Figure 8) [90].

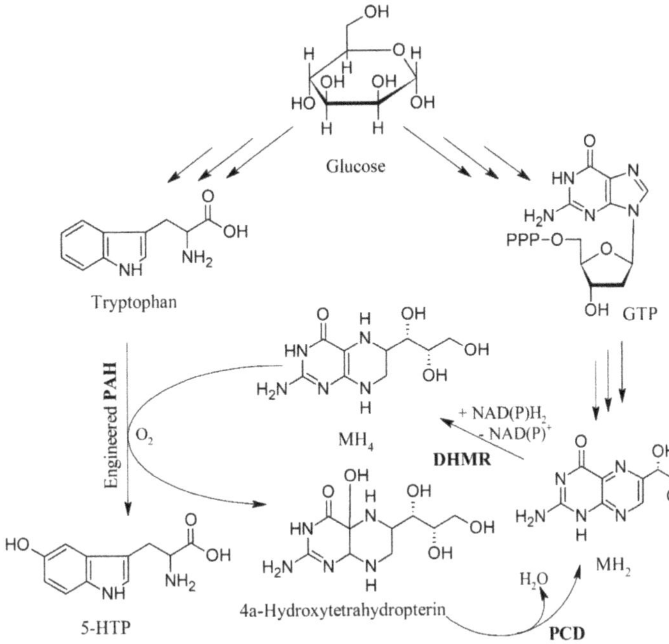

Figure 7. Production of 5-HTP from tryptophan in *E. coli* by engineered PAH and the utilization of tetrahydromonapterin (MH$_4$) instead of BH$_4$. DHMR, dihydromonapterin reductase; PAH, phenylalanine 4-hydroxylase; PCD, pterin-4α-carbinolamine dehydratase; MH$_2$, dihydromonapterin. Modified from [88].

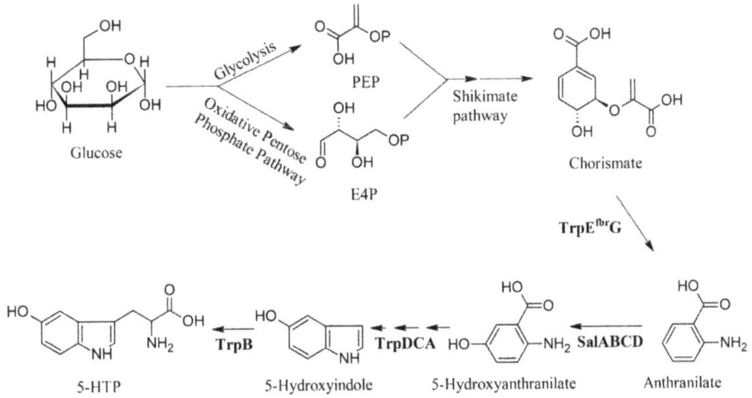

Figure 8. Production of 5-HTP from glucose in engineered *E. coli* by using a novel salicylate 5-hydroxylase. PEP, phosphoenol pyruvate; E4P, erythrose-4-phosphate; TrpE^{fbr}G, anthranilate synthase (from a feedback resistance mutant); SalABCD, salicylate 5-hydroxylase; TrpDCA and TrpB, *E. coli* native tryptophan biosynthetic enzymes. Adapted from [90].

By engineering *E. coli* with heterologous TPH by using a truncated form of human TrpH2 with an E2K mutation for improved protein abundance, 5-HTP conversion from tryptophan was improved by protein engineering TPH [91]. Recently, in order to increase 5-HTP yield and stability of TPH, the tryptophan biosynthetic pathway was integrated into the E. coli genome. A 24.8% improvement compared to the original strain was obtained by manipulating the replication origin of the hydroxylation plasmid and by substitution of the promoter of $aroH^{fbr}$ gene encoding 3-deoxy-7-phosphoheptulonate synthase (that catalyzes the first step of tryptophan biosynthesis). The resulted recombinant strain TRPmut/pSCHTP-LMT was able to produce 1.61 g/L 5-HTP in shake flasks, compared to 0.160 g/L of previous metabolic engineering productions [92].

A successful production of 5-HTP in microbial systems has been obtained in the yeast *Saccharomyces cerevisiae* BY4741 strain via LiAc-mediated yeast transformation. The strategy was based on the heterologous expression of either a prokaryotic PAH or eukaryotic tryptophan 3/5-hydroxylase, together with enhanced synthesis of the two cofactors MH_4 or BH_4 [93]. Interestingly, a native *S. cerevisiae* gene, *DFR1*, which encodes dihydrofolate reductase to catalyze tetrahydrofolate, played a pivotal role in 5-HTP synthesis by regenerating MH_4 [93]. Figure 9 summarizes the heterologous 5-HTP production.

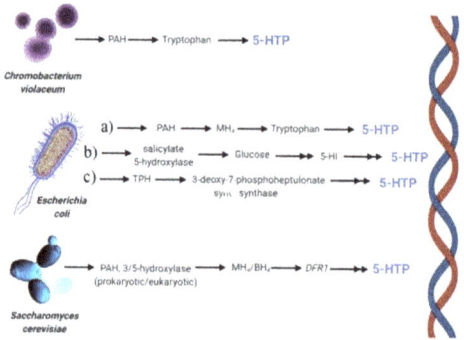

Figure 9. Summary of heterologous production of 5-HTP. *Chromobacterium violaceum*, with the first use of PAH for tryptophan hydroxylation to 5-HTP. *Escherichia coli*, with a) utilization of MH_4 instead of BH_4 as the native pterin coenzyme for 5-HTP synthesis, b) the biosynthesis of 5-HTP with the use of a novel salicylate 5-hydroxylase that uses glucose as the substrate for the production of the non-natural substrate anthranilate to 5-hydroxyanthranilate (5-HI), c) substitution of the promoter of $aroHfbr$ gene encoding 3-deoxy-7-phosphoheptulonate synthase to feed the biosynthetic pathway of tryptophan production. The use of yeasts with the heterologous expression of either a prokaryotic PAH or eukaryotic tryptophan 3/5-hydroxylase, together with enhanced synthesis of the two cofactors MH_4 or BH_4; the native *Saccharomyces cerevisiae* gene, *DFR1*, which encodes dihydrofolate reductase to catalyze tetrahydrofolate, plays a pivotal role in 5-HTP synthesis by regenerating MH_4. Created with BioRender.com.

8. Physiological Effects of 5-HTP

8.1. Animal Studies

Early experiments with 5-HTP in the 1950s have shown that this molecule is also capable of inhibiting gastric hydrochloric acid secretion [94] and that when administered to animals it increased peristaltic activity [95] and was rapidly taken up by most tissues [96], producing in dogs, cats, rabbits, rats, and mice somatic, autonomic, and behavioral effects which grossly resembled those of lysergic acid diethylamide [97].

Most of the animal experimentation on the effects of 5-HTP has been conducted on mice and rats.

In mice, injection of 5-HTP produces a characteristic head twitch due to a central action of 5-HT formed by decarboxylation of 5-HTP [98] and provokes characteristic behaviors, such as tremors, that become more frequent when doses are increased [99]. The accumulation of 5-HT in the mouse liver but not in the brain is causally related to the hypoglycemia induced by 5-HTP [100,101]. 5-HTP also suppresses inflammation and arthritis through decreasing the production of pro-inflammatory mediators [102] and the antihistaminic drugs chlorpyramine and, more strongly, chlorpheniramine potentiates the action of

5-HTP by inhibiting serotonin uptake [103]. Interestingly, dietary phenylalanine and 5-HTP were found to be mutually antagonistic in modulating mice audiogenic seizure susceptibility [104].

In rats, at least some of the behavioral effects of 5-HTP are due to increased levels or turnover of 5-HTP in peripheral serotonergic neuronal systems [105] and 5-HTP neurotoxicity caused by brain–blood barrier breakdown, edema formation, and NO production was found to be instrumental in causing adverse mental and behavioral abnormalities [106]. For instance, the wet dog shake behavior induced by 5-HTP was dose dependent and was mediated by the activation of 5-HT2 receptors [107], whereas in iproniazid-pretreated rats, a special form of stereotyped movements of the head and forelegs and a gnawing behavior were observed after intraperitoneal administration of 5-HTP [108]. 5-HTP produced a 6–11-fold increase in rat plasma prolactin [109,110]. Moreover, endogenous opioids also increase the serum level of prolactin induced by 5-HTP; therefore, different serotonergic neurotransmitter circuits might be capable of modulating the release of corticosterone and prolactin [111]. 5-HTP was also able to induce depression in rats working on an operant schedule for milk reinforcement, and this effect was mediated by serotonergic mechanisms involving 5-HT2 receptors [112]. Moreover, administration of 5-HTP prompted a synergistic increase in the synthesis and release of 5-HT by combining 5-HT uptake inhibition with the blockade of 5-HT1A autoreceptors [113]. Another interesting effect of 5-HTP in rats is the reduction in meal size and a slowing of eating, exerting an effect on the patterns of feeding [114].

Myoclonic twitches, jerks, or seizures are usually caused by brief lapses of contraction (negative myoclonus) or sudden muscle contractions (positive myoclonus). In guinea pigs, when 5-HTP was given 6 h after 0.5 mg progesterone in estradiol benzoate-primed males, myoclonus was enhanced, whereas progesterone reversed the facilitative effect of estradiol benzoate on 5-HTP-induced myoclonus in females [115]. It was also found that 5-HTP-induced myoclonus was influenced by the 5-HT1/2 receptor systems and that the absence of a significant change with a receptor antagonist implied that myoclonus was not related to diffuse activation of central serotonergic mechanisms [116]. In the guinea pig colon, 5-HTP facilitates the luminal 5-HT release from enterochromaffin cells, with no involvement of neuronal mechanisms and a non-neuronal cholinergic system [117].

In the terrestrial snail *Helix lucorum*, injection of 5-HTP alone did not restore the protein kinase M zeta (ZIP)- or the protein synthesis blocker anisomycin-impaired context memory, while the combination of 5-HTP and the reactivation of memory effectively reinstated the context memory [118].

In rabbits, 5-HTP injected intracisternally at a dose of 1.5–3 mg produced a fall in temperature often followed by a rise beyond the pre-injection level, and the anterior hypothalamus was supposed to be the site where 5-HTP acted [119]. To antagonize rabbit 5-HTP-induced hyperthermia, 5-HT receptor blockade is required and antagonism of *p*-methoxyamphetamine-induced hyperthermia is primarily a result of influence on the 5-HT system [120].

In cats, intrahypothalamic injection of 5-HTP allowed its detection in many fibers surrounding the injection site [121], whereas lysergic acid diethylamide was found to mimic the actions of 5-HTP by facilitating the stretch reflex and exciting extensor gamma motoneurons in the animal spine [122].

In dogs, the administration of 5-HTP increased both blood and tissue serotonin with a maximum effect evident in about one hour [123] and increased the propulsive activity, the contractile force, and the motility index [124]. In monoamine oxidase (MAO)-inhibited dogs, 5-HTP caused hypotension with variable effects on heart rate and the cerebral decarboxylation and formation of 5-HT was found to be responsible for this effect [125].

In sheep, the administration of 5-HTP substantially increases serum melatonin through a marked increase in pineal 5-HT and its metabolites, including *N*-acetylserotonin [126], whereas during the natural light period, a promotion of melatonin synthesis in the pineal gland and intestinal tract was found in rumen-protected 5-HTP [127]. In fetal lambs in late gestation, systemic infusion of 5-HTP during normoxia greatly increases the incidence of fetal breathing movements [128].

Finally, in the Holstein dairy cow liver, 5-HTP infusions stimulated an autocrine–paracrine adaptation to lactation [129].

Figure 10 summarizes some effects of 5-HTP on animals.

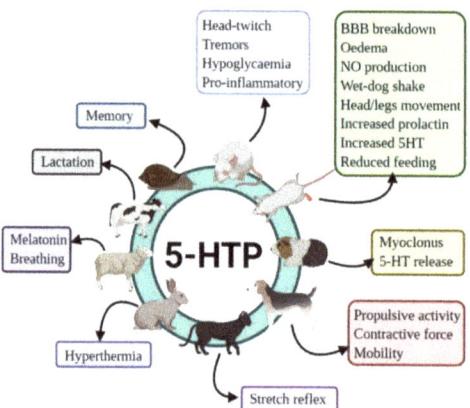

Figure 10. Summary of effects of 5-HTP on some animals. In terrestrial snails, 5-HTP reinstates the context memory, whereas in mice, 5-HTP produces a characteristic head twitch and tremors, induces hypoglycemia, and suppresses inflammation by inducing pro-inflammatory mediators. In rats, most of the listed effects of 5-HTP are dose dependent. In guinea pigs, 5-HTP enhances myoclonus and facilitates the luminal 5-HT release. In dogs, 5-HTP increases propulsive activity, contractive force, and mobility index. Injection of 5-HTP facilitates the stretch reflex in cats and produces a fall in temperature in rabbits. In sheep, 5-HTP increases serum melatonin and foetal breathing movements and stimulates an autocrine–paracrine adaptation to lactation in cows. Created with BioRender.com.

8.2. Effects of 5-HTP on Humans

8.2.1. Serotonin Syndrome

While on the one side, the lack of serotonin is responsible for several diseases, including depression, its excess may be problematic. Medication-induced serotonergic hyperactivity causes serotonin syndrome, which results from antidepressant medications and is characterized by the triad of altered mental status, autonomic dysfunction, and neuromuscular abnormalities. Serotonin syndrome may lead to misdiagnosis, in a similar way as for neuroleptic malignant syndrome. Serotonin syndrome may eventually result in death; however, supportive care alone is sufficient to recover completely for most of patients. Excessive 5-HTP stimulation is the main pathophysiologic mechanism involved and the use of a serotonin antagonist supports this finding [130]. Serotonin syndrome can trigger other clinical conditions; therefore, in order to detect the syndrome and prevent rapid clinical deterioration, it is important to better understand the molecular context of this condition [131].

8.2.2. Effect of 5-HTP on Depression, Anxiety, Dystonia, and Panic Disorders

The amount of endogenous 5-HTP available for serotonin synthesis depends on the availability of tryptophan and the activity of various enzymes, especially TPH, indoleamine 2,3-dioxygenase, and tryptophan 2,3-dioxygenase (TDO) [132]. In depressed patients, tryptophan, serotonin, kynurenine, and their metabolite levels remain unclear. Serotonin is involved in depressive pathophysiology and evidence indicates that the transport of 5-HTP across the blood–brain barrier is compromised in major depression [133], as found in childhood with major depression [134]. Early studies assessed the important role of tryptophan and 5-HTP as an antidepressant [135,136], and relatively few adverse effects are associated with its use in the treatment of depressed patients [137]. Among the side effects, the administration of 5-HTP may cause dose-dependent gastrointestinal problems, whereas the combination of 5-HTP with a peripheral decarboxylase inhibitor may cause psychopathological side effects, like acute anxiety [138]. Additionally, vomiting and nausea have been reported when 5-HTP was used at doses above 100 mg [139].

The antidepressant activity appears to be linked to the activation rather than suppression of monoaminergic activity; therefore, the decreased monoamine metabolism found in some types of depression is likely to depend on a primary metabolic deficit rather than receptor hypersensitivity [140]. Decreased serotonergic activity may be present in both depression and mania [141] and the therapeutic effect of 5-HTP has been correlated with an increase in serotonin at central serotonin receptors [142].

Evidence also supports the combined used of 5-HTP with other drugs. In 30 hospitalized patients affected by endogenous depression, the antidepressant action of the combination of nialamide and 5-HTP has been evaluated and compared with a control group which only received nialamide (along with placebo). The combination of nialamide and 5-HTP prompted a fuller recovery in treated patients with respect to those who were treated with nialamide alone [143]. L-deprenil (an irreversible selective MAO-B inhibitor) was used in an open trial study with patients with unipolar and bipolar depression receiving 5-HTP and benzerazide. The combination of L-deprenil and 5-HTP showed a significantly greater clinical improvement in treated patients with respect to placebo patients but not in patients treated with 5-HTP alone [144]. In women experiencing selective serotonin reuptake inhibitor (SSRI)- or serotonin–norepinephrine reuptake inhibitor (SNRIs)-resistant depression, the combination treatment with creatine and 5-HTP proved to be an effective augmentation strategy [145].

5-HT was found to mediate anxiety [146] and showed a moderate reduction of the symptomatology on the 90-item symptoms checklist (SCL-90) and the state scale of the Spielberger State–Trait Anxiety Inventory, suggesting that brain serotonergic pathways are involved in the pathogenesis of anxiety disorders, particularly in agoraphobia and panic disorders [147].

The serotonergic system of the central nervous system might play some role in the pathogenesis of dystonia in hereditary progressive dystonia [148], although results are at odds with the hypothesis that there is a supersensitivity of 5-HT2 receptors in panic disorder [149]. Nevertheless, in panic disorder patients, 5-HTP significantly reduced the reaction to the panic challenge, regarding number of panic attacks, panic symptom score, and subjective anxiety, when compared to placebo [150]. Furthermore, an increased availability of 5-HT may have a gender-dependent protective effect in cholecystokinin-tetrapeptide-induced panic [151].

8.2.3. Effect of 5-HTP on Sleep Disorders

In normal subjects treated with 5-HTP, rapid eye movement (REM) sleep increased from 5 to 53% of placebo baseline [152] and the effects on sleep were associated with different doses of 5-HTP and the diverse possibilities of metabolic transformation of the precursor [153]. In schizophrenic boys, the administration of 5-HTP was associated with an increase in REM sleep and eye movements [154]; however, in mongoloid infants who received oral 5-HTP for periods extending from 12 to 36 months, 5-HTP failed to induce any long-term differences in the eye movement frequencies. In fact, the drug has a short-term effect lasting up to 8 days and an increase in muscle tone and an improvement of motor behavior were the only long-lasting results [155]. In a group of children with sleep terrors, treatment with 5-HTP was able to modulate the arousal level and to induce a long-term improvement of sleep terrors [156]. Interestingly, it was found that a serotonergic abnormality is involved in affective disorders [157] and direct modulation of the serotonergic system with 5-HTP was useful for the treatment of psychological suffering associated with unreciprocated romantic love [158].

8.2.4. Effects of 5-HTP on Migraine, Ataxia, Fibromyalgia, Alzheimer's, and Parkinson's Disease

5-HTP was also found to be a treatment of choice in the prophylaxis of migraine [159,160]. In subjects who are predisposed to headache, 5-HTP can change the central nervous system (CNS) abnormalities underlying the mechanism of migraine [161]. In another study, two weeks after treatment with 5-HTP, a significant decrease in the number of days with headache was observed [162].

Ataxia is a clinical manifestation indicating dysfunction of the parts of the nervous system that coordinate movement, such as the cerebellum. Some features of cerebellar ataxia have been reported to regress partially with long-term administration of 5-HTP [163], with a significant decrease in the kinetic score, indicating an improvement in coordination, although the effect may sometimes be partial [164].

Lower levels of 5-HTP were found in women with fibromyalgia (FM) in comparison with controls, indicating that the dysregulation of the catecholamine and indolamine pathway in patients with FM may contribute to the physiopathology of this syndrome [165,166]. The treatment with 5-HTP significantly improved all the clinical parameters studied in 50 patients with primary FM syndrome, with only mild and transient side effects reported [167].

Concentrations of 5-HTP are lower in dementia of the Alzheimer's type (DAT) cerebrospinal fluid (CSF) than in a corresponding fraction of control CSF. Therefore, the serotoninergic system is involved in DAT and could be considered for a diagnostic test for DAT [168].

5-HTP has a long history of use as a therapy of Parkinson's disease (PD) [169,170]. In Parkinsonian patients, no effects on gastrointestinal absorption of 5-HTP were observed with co-administration of L-dopa with 5-HTP and decarboxylase inhibitors [171]. Moreover, in a single-center, randomized, double-blind, placebo-controlled, cross-over trial, patients receiving placebo and 50 mg of 5-HTP daily over a period of 4 weeks experienced a significant improvement of depressive symptoms during treatment compared with placebo, providing preliminary evidence of the clinical benefit of 5-HTP for treating depressive symptoms in PD [172].

8.2.5. Effect of 5-HTP on Myoclonus

5-HTP is useful in the treatment of patients with posthypoxic intention myoclonus [173], palatal myoclonus [174], and cherry red spot-myoclonus syndrome [175]. The administration of 5-HTP and carbidopa dramatically improved the action myoclonus and reduced the amplitude of giant somatosensory evoked potentials [176], whereas the combination of sodium valproate and 5-HTP was useful to control a spinal segmental myoclonus characterized by symmetric, rhythmic contractions of the abdomen [177].

8.2.6. Effect of 5-HTP on Obesity

The effect of 5-HTP on feeding behavior, mood state, and weight loss was studied. 5-HTP promoted decreased food intake and weight loss as well as typical anorexia-related symptoms without changes in mood state during the period of observation [178], with a consistent presence of early satiety and a consequent reduction in carbohydrate intake [179]. Moreover, treatment with 5-HTP prompted a decrease in BMI due to an increased feeling of satiety [10].

8.2.7. Effect of 5-HTP on Prolactin

Oral administration of 5-HTP significantly increases plasma human prolactin, suggesting that the serotonergic mechanism is involved in the regulation of prolactin secretion in humans [180]. The maximum downregulation of prolactin release occurs when 5-HTP is administered in 4 h intervals [181] and subacute serotonergic stimulation with oral 5-HTP with the peripheral decarboxylase inhibitor carbidopa resulted in prolactin but not aldosterone release [182].

8.2.8. Antioxidant, Anti-inflammatory, and Analgesic Effects of 5-HTP

5-HTP also exerts radical scavenging activities. 5-HTP showed higher hydroxyl radical scavenging effects when compared to vitamin C [183] and was also effective on hyperglycemia-induced oxidative stress [184]. Moreover, 5-HTP was found to preserve membrane fluidity in the presence of oxidative stress [185]. 5-HTP significantly reduced tert-butylhydroperoxide-induced oxidative damage in human fibroblast cells and protected these cells against oxidative DNA damage [186]. 5-HTP was also found to inhibit the lipopolysaccharide (LPS)-induced expression of NO and interleukine-6 (IL-6), playing a role in extracellular signal-regulated protein kinase (ERK) activation, cyclooxygenase-2 (COX-2,) and LPS-induced inducible nitric oxide synthase (iNOS). By acting as a reactive oxygen species (ROS) scavenger, 5-HTP has the potential for use in the treatment of inflammatory diseases and as an analgesic [187].

Figure 11 summarizes some effects of 5-HTP on humans.

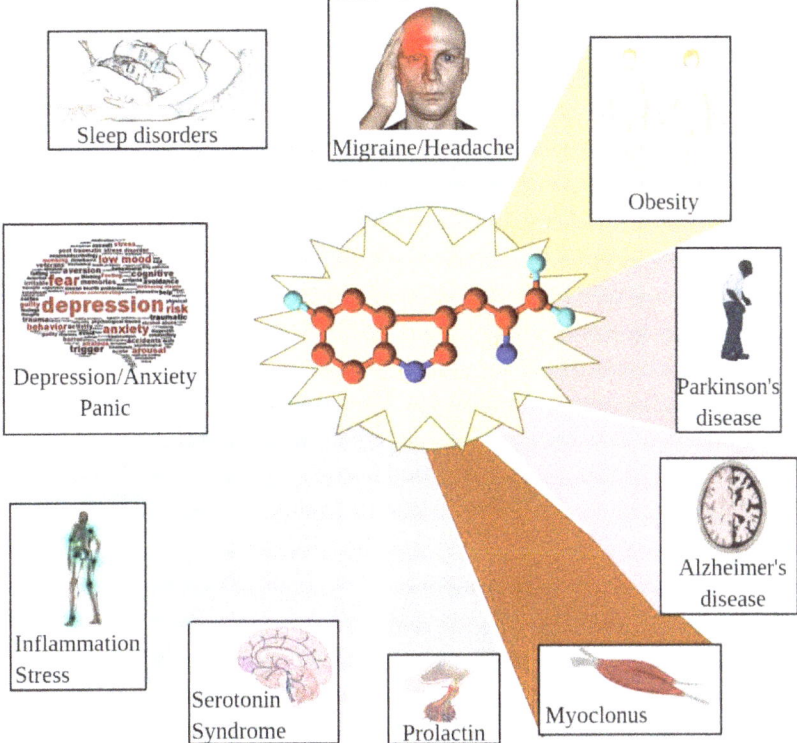

Figure 11. Summary of some effects of 5-HTP on humans. 5-HTP is a treatment of choice in the prophylaxis of migraine and headache and promotes decreased food intake and weight loss in obese patients. 5-HTP is used for treating depressive symptoms in Parkinson's disease and may be used as a diagnostic test for Alzheimer's disease. 5-HTP is useful to control some forms of myoclonus and significantly increases plasma human prolactin. Excessive 5-HTP generates serotonin syndrome. 5-HTP has the potential for use in the treatment of inflammatory diseases and oxidative stress. As a precursor of 5-HT, 5-HTP treatment is used to reduce depression, anxiety, and panic attacks. 5-HTP is associated with an increase in rapid eye movement (REM) sleep and reduces sleep disorder. Created with BioRender.com.

9. Toxicology of 5-HTP

An excess of 5-HTP may be responsible for serotonin syndrome (see Section 8.2.1) and an excessive treatment was found to be associated with severe side effects, including behavioral disturbances, abnormal mental functions, and intolerance. Clinically relevant ingestion of 5-HTP in dogs was found to result in a potentially life-threatening syndrome resembling serotonin syndrome in humans, which requires prompt and aggressive care [188]. After 5-HTP administration, the endogenous serotonin levels increased by fourfold in the rat plasma and brain, associated with profound hyperthermia, oxidative stress, and NO upregulation [106].

Toxicity issues have also been raised in 5-HTP. The use of L-tryptophan, the precursor of 5-HTP, as a dietary supplement was suspended in 1989 due to the occurrence of eosinophilia–myalgia syndrome (EMS), a rare, sometimes fatal neurological condition including debilitating myalgia and marked peripheral eosinophilia, that was traced to contaminated synthetic tryptophan from a single manufacturer (Showa Denko) [189,190]. Therefore, 5-HTP has been under vigilance by consumers, industry, academia, and government for its safety. With the possible exception of one unresolved case of a Canadian woman, no definitive cases of toxicity have emerged despite the worldwide usage of 5-HTP for the last 20 years. Neither toxic contaminants similar to those associated with L-tryptophan, nor the presence of any other significant impurities have been detected in several sources of 5-HTP. Speculations concerning the chemistry and toxicity of infinitesimal concentrations of a minor chromatographic peak (peak X) found in some 5-HTP samples lack credibility, due to possible chromatographic artifacts [5]. Further studies found no significant evidence of EMS in rats receiving high-dose 5-HTP for 1 year [191]. Based on accurate mass, tandem mass spectrometric analysis, and comparison with authentic standard compound analysis, peak X was determined to be 4,5-tryptophan-dione, a putative neurotoxin. Because 4,5-tryptophan-dione was found in case-implicated 5-HTP as well as six over the counter samples, some cause for concern in terms of the safety of such commercial preparations of 5-HTP was raised [7]. Indeed, it was demonstrated that samples of commercially available 5-HTP analyzed by HPLC-MS contained three or more contaminants of the peak X family [8]. Two other 5-HTP contaminants, peak E (1,1′-ethylidenebis(L-tryptophan)) and peak-UV5 (3-anilinoalanine), were found to contribute to the pathogenesis of EMS, or may be surrogates for other chemicals that induce EMS [192–194]. Another contaminant of tryptophan, peak AAA, was defined as a statistical contaminant and was identified as two distinct isomers: peak AA(1) as (S)-2-amino-3-(2((S,E)-7-methylnon-1-en-1-yl)-1H-indol-3-yl) propanoic acid and peak AAA(2) as (S)-2-amino-3-(2-((E)-dec-1-en-1-yl)-1H-indol-3-yl) propanoic acid [6]. Interestingly, after replacement with 5-HTP not containing impurities, eosinophilia was resolved [195].

10. Conclusions

The use of 5-HTP dates back to the first half of the 20th century, where it was recognized as an important precursor of the neurotransmitter serotonin. Its use increased after the discovery of some toxic impurities in commercial tryptophan, although further analyses also found similar impurities in 5-HTP preparation. The main application of 5-HTP is as a support for serotonin depletion and is considered an interesting alternative to the use of SSRIs. The main natural source of 5-HTP is the seeds of the African plant *Griffonia simplicifolia*, which also produces interesting β-carboline alkaloid derivatives. However, the seasonal and regional variations of this plant limit the output of 5-HTP. For this reason, recent biotechnological approaches have been developed by using recombinant genes and genetic engineering of both bacteria (mainly *E. coli*) and yeasts, for the economically and environmentally sustainable production of 5-HTP. These methods are based on biochemical reactions (such as TPH catalysis) and the use of alternative cofactors to improve the biosynthetic ability of bacterial and fungal cells to produce 5-HTP at industrial levels. While on the one side, the use of 5-HTP may be not recommended in humans (as in the case of serotonin syndrome), on the other, the molecule (drug or dietary supplement) has been proved to be effective to treat neurological and metabolic diseases. The majority of clinical trials are on depression, anxiety, panic attacks, and sleep disorders; however, the molecule has been demonstrated to be a promising support to reduce food/feed intake and to be potentially used for metabolic diseases like obesity and diabetes. A few studies also indicate its potential in neurodegenerative diseases like Alzheimer's and Parkinson's disease. Future developments are focused on safer 5-HTP production. The improvement of 5-HTP yield and the removal of toxic impurities by biotechnological transformation will allow a sustainable and safer 5-HTP production for both pharmaceutical and nutraceutical industries.

Supplementary Materials: The following are available online at https://www.mdpi.com/1422-0067/22/1/181/s1, Supplementary File S1, EndNote zipped file containing the 816 selected references.

Author Contributions: Conceptualization; writing; funding acquisition, M.E.M. All authors have read and agreed to the published version of the manuscript.

Funding: This research was funded by the University of Turin local research grants to M.E.M.

Conflicts of Interest: The author declares no conflict of interest.

Abbreviations

4a-HOPH3	4a-Hydroxypterin
5-HI	5-Hydroxyanthranilate
5-HT	5-Hydroxytryptamine, Serotonin
5-HTP	L-5-Hydroxytryptophan
AADC	Aromatic Amino Acid Decarboxylase
BH_3OH	Pterin-4α-carbinolamine
BH_4	Tetrahydrobiopterin
BMI	Body Mass Index
CE	Capillary Electrophoresis
CNS	Central Nervous System
COX-2	Cyclo oxygenase-2
CREB	cAMP Response Element-Binding Protein
CSF	Cerebrospinal Fluid
CviPAH	*C. violaceum* L-phenylalanine 4-hydroxylase
DAD	Diode Array Detector
DAT	Dementia of the Alzheimer's type
DHMR	Dihydromonapterin Reductase
DHPR	Dihydropteridine Reductase
E4P	Erythrose-4-phosphate
EI	Electrospray Ionization
EMS	Eosinophilia–Myalgia Syndrome
FD	Fluorescence Detector
FM	Fibromyalgia
GI	Gastrointestinal
HPLC	High-Performance Liquid Chromatography
IL-6	Interleukin-6
LP-533401	[(2S)-2-amino-3-(4-(2-amino-6-(2,2,2-trifluoro-1-(3′-fluorobiphenyl-4-yl)ethoxy)pyrimidin-4-yl)phenyl)propanoic acid]
LP-615819	[(2S)-ethyl 2-amino-3-(4-(2-amino-6-(2,2,2-trifluoro-1-(3′-fluorobiphenyl-4-yl)ethoxy)pyrimidin-4-yl)phenyl)propanoic acid]
LPS	Lipopolysaccharide
MAO	Monoamine oxidase
MH_2	Dihydromonapterin
MH_4	Tetrahydromonapterin
MS	Mass Spectrometry
PAH	L-Phenylalanine 4-hydroxylase

PCD	Pterin-4α-carbinolamine Dehydratase
PD	Parkinson's Disease
Peak AAA1	(S)-2-amino-3-(2((S,E)-7-methylnon-1-en-1-yl)-1H-indol-3-yl) propanoic acid
Peak AAA2	(S)-2-amino-3-(2-((E)-dec-1-en-1-yl)-1H-indol-3-yl) propanoic acid
Peak E	[1,1′-Ethylidenebis(L-tryptophan)]
Peak X	4,5-Tryptophan-dione
Peak-UV5	3-anilinoalanine
PEP	Phosphoenol Pyruvate
p-EPA	p-Ethynylphenylalanine
PKA	Protein Kinase A
PTPS	6-Pyruvate-tetrahydropterin Synthase
REM	Rapid Eye Movement
RFLP	Restriction Fragment Length Polymorphism
ROS	Reactive Oxygen Species
SalABCD	Salicylate 5-hydroxylase
SCL-90	90-Item Symptoms Checklist
SIM	Selective Ion Monitoring
SNRI	Serotonin–Norepinephrine Reuptake Inhibitor
SPR	Sepiapterin Reductase
SSRI	Selective Serotonin Reuptake Inhibitor
TCM	Traditional Chinese Medicine
TDO	Tryptophan 2,3-dioxygenase
TPH	Tryptophan Hydroxylase
TrpE^{fbr}G	Anthranilate Synthase
Trp-P-1	3-amino-1,4-dimethyl-5H-pyrido[4,3-b]indole
Trp-P-2	3-Amino-1-methyl-5H-pyrido[4,3-b]indole

References

1. Strac, D.S.; Pivac, N.; Muck-Seler, D. The serotonergic system and cognitive function. *Transl. Neurosci.* **2016**, *7*, 35–49.
2. Ye, T.T.; Yin, X.M.; Yu, L.; Zheng, S.J.; Cai, W.J.; Wu, Y.; Feng, Y.Q. Metabolic analysis of the melatonin biosynthesis pathway using chemical labeling coupled with liquid chromatography-mass spectrometry. *J. Pineal Res.* **2019**, *66*, 11. [CrossRef] [PubMed]
3. Samanta, S. Physiological and pharmacological perspectives of melatonin. *Arch. Physiol. Biochem.* **2020**, 1–22. [CrossRef] [PubMed]
4. Walther, D.J.; Bader, M. A unique central tryptophan hydroxylase isoform. *Biochem. Pharmacol.* **2003**, *66*, 1673–1680. [CrossRef]
5. Das, Y.T.; Bagchi, M.; Bagchi, D.; Preuss, H.G. Safety of 5-hydroxy-l-tryptophan. *Toxicol. Lett.* **2004**, *150*, 111–122. [CrossRef]
6. Klarskov, K.; Gagnon, H.; Boudreault, P.L.; Normandin, C.; Plancq, B.; Marsault, E.; Gleich, G.J.; Naylor, S. Structure determination of disease associated peak aaa from l-tryptophan implicated in the eosinophilia-myalgia syndrome. *Toxicol. Lett.* **2018**, *282*, 71–80. [CrossRef]
7. Klarskov, K.; Johnson, K.L.; Benson, L.M.; Cragun, J.D.; Gleich, G.J.; Wrona, M.; Jiang, X.R.; Dryhurst, G.; Naylor, S. Structural characterization of a case-implicated contaminant, "peak x," in commercial preparations of 5-hydroxytryptophan. *J. Rheumatol.* **2003**, *30*, 89–95.
8. Klarskov, K.; Johnson, K.L.; Benson, L.M.; Gleich, G.J.; Naylor, S. Eosinophilia-myalgia syndrome case-associated contaminants in commercially available 5-hydroxytryptophan. In *Tryptophan, Serotonin and Melatonin: Basic Aspects and Applications*; Huether, G., Kochen, W., Simat, T.J., Steinhart, H., Eds.; Springer: New York, NY, USA, 1999; pp. 461–468.
9. Esposito, M.; Precenzano, F.; Sorrentino, M.; Avolio, D.; Carotenuto, M. A medical food formulation of griffonia simplicifolia/magnesium for childhood periodic syndrome therapy: An open-label study on motion sickness. *J. Med. Food* **2015**, *18*, 916–920. [CrossRef]
10. Rondanelli, M.; Opizzi, A.; Faliva, M.; Bucci, M.; Perna, S. Relationship between the absorption of 5-hydroxytryptophan from an integrated diet, by means of griffonia simplicifolia extract, and the effect on satiety in overweight females after oral spray administration. *Eat. Weight Disord. Stud. Anorex. Bulim. Obes.* **2012**, *17*, E22–E28.
11. Carnevale, G.; Di Viesti, V.; Zavatti, M.; Benelli, A.; Zanoli, P. Influence of griffonia simplicifolia on male sexual behavior in rats: Behavioral and neurochemical study. *Phytomedicine* **2011**, *18*, 947–952. [CrossRef]
12. Carnevale, G.; Di Viesti, V.; Zavatti, M.; Zanoli, P. Anxiolytic-like effect of griffonia simplicifolia baill. Seed extract in rats. *Phytomedicine* **2011**, *18*, 848–851. [CrossRef] [PubMed]
13. Babu, S.K.; Ramakrishna, T.; Subbaraju, G.V. Hplc estimation of 5-hydroxytryptophan in griffonia simplicifolia extracts. *Asian J. Chem.* **2005**, *17*, 506–510.
14. Lemaire, P.A.; Adosraku, R.K. An hplc method for the direct assay of the serotonin precursor, 5-hydroxytrophan, in seeds of griffonia simplicifolia. *Phytochem. Anal.* **2002**, *13*, 333–337. [CrossRef] [PubMed]

15. Vigliante, I.; Mannino, G.; Maffei, M.E. Chemical characterization and DNA fingerprinting of griffonia simplicifolia baill. *Molecules* **2019**, *24*, 1032. [CrossRef] [PubMed]
16. Glinwood, R.; Pettersson, J.; Ahmed, E.; Ninkovic, V.; Birkett, M.; Pickett, J. Change in acceptability of barley plants to aphids after exposure to allelochemicals from couch-grass (elytrigia repens). *J. Chem. Ecol.* **2003**, *29*, 261–274. [CrossRef] [PubMed]
17. Hagin, R.D. Isolation and identification of 5-hydroxyindole-3-acetic acid and 5-hydroxytryptophan, major allelopathic aglycons in quackgrass (agropyron-repens l beauv). *J. Agric. Food Chem.* **1989**, *37*, 1143–1149. [CrossRef]
18. Murch, S.J.; KrishnaRaj, S.; Saxena, P.K. Tryptophan is a precursor for melatonin and serotonin biosynthesis in in vitro regenerated st. John's wort (hypericum perforatum l. Cv. Anthos) plants. *Plant Cell Rep.* **2000**, *19*, 698–704. [CrossRef]
19. Fernandez-Cruz, E.; Cerezo, A.B.; Cantos-Villar, E.; Troncoso, A.M.; Garcia-Parrilla, M.C. Time course of l-tryptophan metabolites when fermenting natural grape musts: Effect of inoculation treatments and cultivar on the occurrence of melatonin and related indolic compounds. *Aust. J. Grape Wine Res.* **2019**, *25*, 92–100. [CrossRef]
20. Diamante, M.S.; Borges, C.V.; da Silva, M.B.; Minatel, I.O.; Correa, C.R.; Gomez, H.A.G.; Lima, G.P.P. Bioactive amines screening in four genotypes of thermally processed cauliflower. *Antioxidants* **2019**, *8*, 311. [CrossRef]
21. Lysek, N.; Kinscherf, R.; Claus, R.; Lindel, T. L-5-hydroxytryptophan: Antioxidant and anti-apoptotic principle of the intertidal sponge hymeniacidon heliophila. *Z. Fur Nat. C J. Biosci.* **2003**, *58*, 568–572. [CrossRef]
22. Muszynska, B.; Sulkowska-Ziaja, K.; Ekiert, H. Indole compounds in fruiting bodies of some edible basidiomycota species. *Food Chem.* **2011**, *125*, 1306–1308. [CrossRef]
23. Muszynska, B.; Sulkowska-Ziaja, K.; Ekiert, H. Analysis of indole compounds in methanolic extracts from the fruiting bodies of cantharellus cibarius (the chanterelle) and from the mycelium of this species cultured in vitro. *J. Food Sci. Technol. Mysore* **2013**, *50*, 1233–1237. [CrossRef] [PubMed]
24. Gurevich, L.S. Indole-derivatives in certain panaeolus species from east europe and siberia. *Mycol. Res.* **1993**, *97*, 251–254. [CrossRef]
25. Koppisetti, G.; Siriki, A.; Sukala, K.; Subbaraju, G.V. Estimation of l-5-hydroxytryptophan in rat serum and griffonia seed extracts by liquid chromatography-mass spectrometry. *Anal. Chim. Acta* **2005**, *549*, 129–133. [CrossRef]
26. Guillen-Casla, V.; Rosales-Conrado, N.; Leon-Gonzalez, M.E.; Perez-Arribas, L.V.; Polo-Diez, L.M. Determination of serotonin and its precursors in chocolate samples by capillary liquid chromatography with mass spectrometry detection. *J. Chromatogr. A* **2012**, *1232*, 158–165. [CrossRef] [PubMed]
27. Magnussen, I. Effects of carbidopa on the cerebral accumulation of exogenous l-5-hydroxytryptophan in mice. *Acta Pharm. Toxicol. (Copenh)* **1984**, *55*, 199–202. [CrossRef]
28. Boulet, L.; Faure, P.; Flore, P.; Montereemal, J.; Ducros, V. Simultaneous determination of tryptophan and 8 metabolites in human plasma by liquid chromatography/tandem mass spectrometry. *J. Chromatogr. B Anal. Technol. Biomed. Life Sci.* **2017**, *1054*, 36–43. [CrossRef]
29. Konieczna, L.; Roszkowska, A.; Niedzwiecki, M.; Baczek, T. Hydrophilic interaction chromatography combined with dispersive liquid-liquid microextraction as a preconcentration tool for the simultaneous determination of the panel of underivatized neurotransmitters in human urine samples. *J. Chromatogr. A* **2016**, *1431*, 111–121. [CrossRef]
30. Morgan, L.D.; Baker, H.; Yeoman, M.S.; Patel, B.A. Chromatographic assay to study the activity of multiple enzymes involved in the synthesis and metabolism of dopamine and serotonin. *Analyst* **2012**, *137*, 1409–1415. [CrossRef]
31. Fickbohm, D.J.; Lynn-Bullock, C.P.; Spitzer, N.; Caldwell, H.K.; Katz, P.S. Localization and quantification of 5-hydroxytryptophan and serotonin in the central nervous systems of tritonia and aplysia. *J. Comp. Neurol.* **2001**, *437*, 91–105. [CrossRef]
32. Coelho, A.G.; Aguiar, F.P.C.; de Jesus, D.P. A rapid and simple method for determination of 5-hydroxytryptophan in dietary supplements by capillary electrophoresis. *J. Braz. Chem. Soc.* **2014**, *25*, 783–787. [CrossRef]
33. Peterson, Z.D.; Lee, M.L.; Graves, S.W. Determination of serotonin and its precursors in human plasma by capillary electrophoresis-electrospray ionization-time-of-flight mass spectrometry. *J. Chromatogr. B Anal. Technol. Biomed. Life Sci.* **2004**, *810*, 101–110. [CrossRef]
34. Wise, D.D.; Shear, J.B. Quantitation of nicotinamide and serotonin derivatives and detection of flavins in neuronal extracts using capillary electrophoresis with multiphoton-excited fluorescence. *J. Chromatogr. A* **2006**, *1111*, 153–158. [CrossRef] [PubMed]
35. Qiao, C.D.; Song, P.S.; Yan, X.; Jiang, S.X. Separation of biogenic amines by micellar electrokinetic chromatography. *Chin. J. Anal. Chem.* **2007**, *35*, 95–98.
36. Shi, H.M.; Wang, B.; Niu, L.M.; Cao, M.S.; Kang, W.J.; Lian, K.Q.; Zhang, P.P. Trace level determination of 5-hydroxytryptamine and its related indoles in amniotic fluid by gas chromatography-mass spectrometry. *J. Pharm. Biomed. Anal.* **2017**, *143*, 176–182. [CrossRef]
37. Hinterholzer, A.; Stanojlovic, V.; Regl, C.; Huber, C.G.; Cabrele, C.; Schubert, M. Identification and quantification of oxidation products in full-length biotherapeutic antibodies by nmr spectroscopy. *Anal. Chem.* **2020**, *92*, 9666–9673. [CrossRef]
38. Kuo, T.R.; Chen, J.S.; Chiu, Y.C.; Tsai, C.Y.; Hu, C.C.; Chen, C.C. Quantitative analysis of multiple urinary biomarkers of carcinoid tumors through gold-nanoparticle-assisted laser desorption/ionization time-of-flight mass spectrometry. *Anal. Chim. Acta* **2011**, *699*, 81–86. [CrossRef]
39. Bisby, R.H.; Arvanitidis, M.; Botchway, S.W.; Clark, I.P.; Parker, A.W.; Tobin, D. Investigation of multiphoton-induced fluorescence from solutions of 5-hydroxytryptophan. *Photochem. Photobiol. Sci.* **2003**, *2*, 157–162. [CrossRef]

40. Tunna, I.J.; Patel, B.A. Analysis of 5-hydroxytryptophan in the presence of excipients from dietary capsules: Comparison between cyclic voltammetry and uv visible spectroscopy. *Anal. Methods* **2013**, *5*, 2523–2528. [CrossRef]
41. Chen, Y.H.; Li, G.K.; Hu, Y.F. A sensitive electrochemical method for the determination of 5-hydroxytryptophan in rats' brain tissue based on a carbon nanosheets-modified electrode. *Anal. Methods* **2015**, *7*, 1971–1976. [CrossRef]
42. Ranganathan, D.; Zamponi, S.; Berrettoni, M.; Mehdi, B.L.; Cox, J.A. Oxidation and flow-injection amperometric determination of 5-hydroxytryptophan at an electrode modified by electrochemically assisted deposition of a sol-gel film with templated nanoscale pores. *Talanta* **2010**, *82*, 1149–1155. [CrossRef] [PubMed]
43. Kalachar, H.C.B.; Arthoba Naik, Y.; Basavanna, S.; Viswanatha, R.; Venkatesha, T.G.; Sheela, T. Amperometric and differential pulse voltammetric determination of 5-hydroxy-l-tryptophan in pharmaceutical samples using gold modified pencil graphite electrode. *J. Chem. Pharm. Res.* **2011**, *3*, 530–539.
44. Chen, L.; Lian, H.T.; Sun, X.Y.; Liu, B. Sensitive detection of l-5-hydroxytryptophan based on molecularly imprinted polymers with graphene amplification. *Anal. Biochem.* **2017**, *526*, 58–65. [CrossRef] [PubMed]
45. Kumar, N.; Goyal, R.N. Simultaneous determination of melatonin and 5-hydroxytrptophan at the disposable poly-(melamine)/poly-(o-aminophenol) composite modified screen printed sensor. *J. Electro. Anal. Chem.* **2020**, *874*, 114458. [CrossRef]
46. Kumar, N.; Sharma, R.; Goyal, R.N. Palladium nano particles decorated multi-walled carbon nanotubes modified sensor for the determination of 5-hydroxytryptophan in biological fluids. *Sens. Actuators B Chem.* **2017**, *239*, 1060–1068. [CrossRef]
47. Li, M.D.; Tseng, W.L.; Cheng, T.L. Ultrasensitive determination of indoleamines by combination of nanoparticle-based extraction with capillary electrophoresis/laser-induced native fluorescence. *J. Chromatogr. A* **2009**, *1216*, 6451–6458. [CrossRef]
48. Shahrokhian, S.; Bayat, M. Pyrolytic graphite electrode modified with a thin film of a graphite/diamond nano-mixture for highly sensitive voltammetric determination of tryptophan and 5-hydroxytryptophan. *Microchim. Acta* **2011**, *174*, 361–366. [CrossRef]
49. Seo, K.D.; Hossain, M.M.D.; Gurudatt, N.G.; Choi, C.S.; Shiddiky, M.J.A.; Park, D.S.; Shim, Y.B. Microfluidic neurotransmitters sensor in blood plasma with mediator-immobilized conducting polymer/n, s-doped porous carbon composite. *Sens. Actuators B Chem.* **2020**, *313*, 128017. [CrossRef]
50. Hamlin, K.E.; Fischer, F.E. The synthesis of 5-hydroxytryptamine. *J. Am. Chem. Soc.* **1951**, *73*, 5007–5008. [CrossRef]
51. Ek, A.; Witkop, B. The synthesis of labile hydroxytryptophan metabolites1. *J. Am. Chem. Soc.* **1954**, *76*, 5579–5588. [CrossRef]
52. Snyder, H.R.; Smith, C.W. A convenient synthesis of dl-tryptophan. *J. Am. Chem. Soc.* **1944**, *66*, 350–351. [CrossRef]
53. Koo, J.; Avakian, S.; Martin, G.J. Synthesis in the 5-hydroxyindole series. N-acetyl-5-hydroxytryptophan and related compounds. *J. Org. Chem.* **1959**, *24*, 179–183. [CrossRef]
54. Frangatos, G.; Chubb, F.L. A new synthesis of 5-hydroxytryptophan. *Can. J. Chem.* **1959**, *37*, 1374–1376. [CrossRef]
55. Moe, O.A.; Warner, D.T. 1,4-addition reactions. I. The addition of acylamidomalonates to acrolein1a. *J. Am. Chem. Soc.* **1948**, *70*, 2763–2765. [CrossRef]
56. Fitzpatrick, P.F. Tetrahydropterin-dependent amino acid hydroxylases. *Annu. Rev. Biochem.* **1999**, *68*, 355–381. [CrossRef] [PubMed]
57. Eser, B.E.; Barr, E.W.; Frantorn, P.A.; Saleh, L.; Bollinger, J.M.; Krebs, C.; Fitzpatrick, P.F. Direct spectroscopic evidence for a high-spin fe(iv) intermediate in tyrosine hydroxylase. *J. Am. Chem. Soc.* **2007**, *129*, 11334. [CrossRef] [PubMed]
58. Pavon, J.A.; Eser, B.; Huynh, M.T.; Fitzpatrick, P.F. Single turnover kinetics of tryptophan hydroxylase: Evidence for a new intermediate in the reaction of the aromatic amino acid hydroxylases. *Biochemistry* **2010**, *49*, 7563–7571. [CrossRef] [PubMed]
59. Hara, R.; Kino, K. Enhanced synthesis of 5-hydroxy-l-tryptophan through tetrahydropterin regeneration. *AMB Express* **2013**, *3*, 7. [CrossRef]
60. Wang, H.J.; Liu, W.Q.; Shi, F.; Huang, L.; Lian, J.Z.; Qu, L.; Cai, J.; Xu, Z.A. Metabolic pathway engineering for high-level production of 5-hydroxytryptophan in escherichia coli. *Metab. Eng.* **2018**, *48*, 279–287. [CrossRef]
61. Yamamoto, K.; Kataoka, E.; Miyamoto, N.; Furukawa, K.; Ohsuye, K.; Yabuta, M. Genetic engineering of escherichia coli for production of tetrahydrobiopterin. *Metab. Eng.* **2003**, *5*, 246–254. [CrossRef]
62. Moran, G.R.; Fitzpatrick, P.F. A continuous fluorescence assay for tryptophan hydroxylase. *Anal. Biochem.* **1999**, *266*, 148–152. [CrossRef] [PubMed]
63. Flatmark, T.; Stevens, R.C. Structural insight into the aromatic amino acid hydroxylases and their disease-related mutant forms. *Chem. Rev.* **1999**, *99*, 2137–2160. [CrossRef] [PubMed]
64. Ehret, M.; Cash, C.D.; Hamon, M.; Maitre, M. Formal demonstration of the phosphorylation of rat brain tryptophan hydroxylase by ca2+/calmodulin-dependent protein kinase. *J. Neurochem.* **1989**, *52*, 1886–1891. [CrossRef] [PubMed]
65. Matthes, S.; Bader, M. Peripheral serotonin synthesis as a new drug target. *Trends Pharmacol. Sci.* **2018**, *39*, 560–572. [CrossRef] [PubMed]
66. Hoglund, E.; Overli, O.; Winberg, S. Tryptophan metabolic pathways and brain serotonergic activity: A comparative review. *Front. Endocrinol.* **2019**, *10*, 158. [CrossRef]
67. Sakowski, S.A.; Geddes, T.J.; Thomas, D.M.; Levi, E.; Hatfield, J.S.; Kuhn, D.M. Differential tissue distribution of tryptophan hydroxylase isoforms 1 and 2 as revealed with monospecific antibodies. *Brain Res.* **2006**, *1085*, 11–18. [CrossRef]
68. Ota, M.; Dostert, P.; Hamanaka, T.; Nagatsu, T.; Naoi, M. Inhibition of tryptophan hydroxylase by (r)- and (s)-1-methyl-6,7-dihydroxy-1,2,3,4-tetrahydroisoquinolines (salsolinols). *Neuropharmacology* **1992**, *31*, 337–341. [CrossRef]

69. Kim, E.I.; Kang, M.H.; Lee, M.K. Inhibitory effects of tetrahydropapaverine on serotonin biosynthesis in murine mastocytoma p815 cells. *Life Sci.* **2004**, *75*, 1949–1957. [CrossRef]
70. Naoi, M.; Hosoda, S.; Ota, M.; Takahashi, T.; Nagatsu, T. Inhibition of tryptophan hydroxylase by food-derived carcinogenic heterocyclic amines, 3-amino-1-methyl-5h-pyrido[4,3-b]indole and 3-amino-1,4-dimethyl-5h-pyrido[4,3-b]indole. *Biochem. Pharmacol.* **1991**, *41*, 199–203. [CrossRef]
71. Zimmer, L.; Luxen, A.; Giacomelli, F.; Pujol, J.F. Short- and long-term effects of p-ethynylphenylalanine on brain serotonin levels. *Neurochem. Res.* **2002**, *27*, 269–275. [CrossRef]
72. Liu, Q.Y.; Yang, Q.; Sun, W.M.; Vogel, P.; Heydorn, W.; Yu, X.Q.; Hu, Z.X.; Yu, W.S.; Jonas, B.; Pineda, R.; et al. Discovery and characterization of novel tryptophan hydroxylase inhibitors that selectively inhibit serotonin synthesis in the gastrointestinal tract. *J. Pharmacol. Exp. Ther.* **2008**, *325*, 47–55. [CrossRef] [PubMed]
73. Cianchetta, G.; Stouch, T.; Yu, W.; Shi, Z.C.; Tari, L.W.; Swanson, R.V.; Hunter, M.J.; Hoffman, I.D.; Liu, Q. Mechanism of inhibition of novel tryptophan hydroxylase inhibitors revealed by co-crystal structures and kinetic analysis. *Curr. Chem. Genom.* **2010**, *4*, 19–26. [CrossRef] [PubMed]
74. Markham, A. Telotristat ethyl: First global approval. *Drugs* **2017**, *77*, 793–798. [CrossRef] [PubMed]
75. Goldberg, D.R.; De Lombaert, S.; Aiello, R.; Bourassa, P.; Barucci, N.; Zhang, Q.; Paralkar, V.; Stein, A.J.; Valentine, J.; Zavadoski, W. Discovery of acyl guanidine tryptophan hydroxylase-1 inhibitors. *Bioorganic Med. Chem. Lett.* **2016**, *26*, 2855–2860. [CrossRef]
76. Goldberg, D.R.; De Lombaert, S.; Aiello, R.; Bourassa, P.; Barucci, N.; Zhang, Q.; Paralkar, V.; Stein, A.J.; Holt, M.; Valentine, J.; et al. Optimization of spirocyclic proline tryptophan hydroxylase-1 inhibitors. *Bioorganic Med. Chem. Lett.* **2017**, *27*, 413–419. [CrossRef]
77. Shi, H.; Cui, Y.; Qin, Y. Discovery and characterization of a novel tryptophan hydroxylase 1 inhibitor as a prodrug. *Chem. Biol. Drug Des.* **2018**, *91*, 202–212. [CrossRef]
78. Barbosa, R.; Scialfa, J.H.; Terra, I.M.; Cipolla-Neto, J.; Simonneaux, V.; Afeche, S.C. Tryptophan hydroxylase is modulated by l-type calcium channels in the rat pineal gland. *Life Sci.* **2008**, *82*, 529–535. [CrossRef]
79. Braga, A.; Faria, N. Bioprocess optimization for the production of aromatic compounds with metabolically engineered hosts: Recent developments and future challenges. *Front. Bioeng. Biotechnol.* **2020**, *8*, 18. [CrossRef]
80. Xu, P.; Gu, Q.; Wang, W.Y.; Wong, L.; Bower, A.G.W.; Collins, C.H.; Koffas, M.A.G. Modular optimization of multi-gene pathways for fatty acids production in e. Coli. *Nat. Commun.* **2013**, *4*, 1409. [CrossRef]
81. Li, L.; Liu, Z.; Jiang, H.; Mao, X.Z. Biotechnological production of lycopene by microorganisms. *Appl. Microbiol. Biotechnol.* **2020**, *104*, 10307–10324. [CrossRef]
82. Chen, X.X.; Zhang, C.Q.; Lindley, N.D. Metabolic engineering strategies for sustainable terpenoid flavor and fragrance synthesis. *J. Agric. Food Chem.* **2020**, *68*, 10252–10264. [CrossRef] [PubMed]
83. Anarat-Cappillino, G.; Sattely, E.S. The chemical logic of plant natural product biosynthesis. *Curr. Opin. Plant Biol.* **2014**, *19*, 51–58. [CrossRef] [PubMed]
84. Mitoma, C.; Weissbach, H.; Udenfriend, S. 5-hydroxytryptophan formation and tryptophan metabolism in chromobacterium violaceum. *Arch. Biochem. Biophys.* **1956**, *63*, 122–130. [CrossRef]
85. Letendre, C.H.; Dickens, G.; Guroff, G. The tryptophan hydroxylase of chromobacterium violaceum. *J. Biol. Chem.* **1974**, *249*, 7186–7191. [PubMed]
86. Kino, K.; Hara, R.; Nozawa, A. Enhancement of l-tryptophan 5-hydroxylation activity by structure-based modification of l-phenylalanine 4-hydroxylase from chromobacterium violaceum. *J. Biosci. Bioeng.* **2009**, *108*, 184–189. [CrossRef] [PubMed]
87. Pribat, A.; Blaby, I.K.; Lara-Nunez, A.; Gregory, J.F.; de Crecy-Lagard, V.; Hanson, A.D. Folx and folm are essential for tetrahydromonapterin synthesis in escherichia coli and pseudomonas aeruginosa. *J. Bacteriol.* **2010**, *192*, 475–482. [CrossRef]
88. Lin, Y.H.; Sun, X.X.; Yuan, Q.P.; Yan, Y.J. Engineering bacterial phenylalanine 4-hydroxylase for microbial synthesis of human neurotransmitter precursor 5-hydroxytryptophan. *Acs Synth. Biol.* **2014**, *3*, 497–505. [CrossRef]
89. Mora-Villalobos, J.A.; Zeng, A.P. Synthetic pathways and processes for effective production of 5-hydroxytryptophan and serotonin from glucose in escherichia coli. *J. Biol. Eng.* **2018**, *12*, 3. [CrossRef]
90. Sun, X.X.; Lin, Y.H.; Yuan, Q.P.; Yan, Y.J. Precursor-directed biosynthesis of 5-hydroxytryptophan using metabolically engineered e. Coli. *Acs Synth. Biol.* **2015**, *4*, 554–558. [CrossRef]
91. Luo, H.; Schneider, K.; Christensen, U.; Lei, Y.; Herrgard, M.J.; Palsson, B.O. Microbial synthesis of human-hormone melatonin at gram scales. *Acs Synth. Biol.* **2020**, *9*, 1240–1245. [CrossRef]
92. Xu, D.; Fang, M.J.; Wang, H.J.; Huang, L.; Xu, Q.Y.; Xu, Z.N. Enhanced production of 5-hydroxytryptophan through the regulation of l-tryptophan biosynthetic pathway. *Appl. Microbiol. Biotechnol.* **2020**, *104*, 2481–2488. [CrossRef] [PubMed]
93. Zhang, J.T.; Wu, C.C.; Sheng, J.Y.; Feng, X.Y. Molecular basis of 5-hydroxytryptophan synthesis in saccharomyces cerevisiae. *Mol. Biosyst.* **2016**, *12*, 1432–1435. [CrossRef] [PubMed]
94. Smith, A.N.; Black, J.W.; Fisher, E.W. Inhibitory effect of 5-hydroxytryptophan on acid gastric secretion. *Nature* **1957**, *180*, 1127. [CrossRef] [PubMed]
95. Bulbring, E.; Lin, R.C. The effect of intraluminal application of 5-hydroxytryptamine and 5-hydroxytryptophan on peristalsis; the local production of 5-ht and its release in relation to intraluminal pressure and propulsive activity. *J. Physiol.* **1958**, *140*, 381–407. [PubMed]

96. Udenfriend, S.; Weissbach, H.; Bogdanski, D.F. Increase in tissue serotonin following administration of its precursor 5-hydroxytryptophan. *J. Biol. Chem.* **1957**, *224*, 803–810. [PubMed]
97. Bogdanski, D.F.; Weissbach, H.; Udenfriend, S. Pharmacological studies with the serotonin precursor, 5-hydroxytryptophan. *J. Pharmacol. Exp. Ther.* **1958**, *122*, 182–194.
98. Corne, S.J.; Pickering, R.W.; Warner, B.T. A method for assessing the effects of drugs on the central actions of 5-hydroxytryptamine. *Br. J. Pharmacol. Chemother.* **1963**, *20*, 106–120. [CrossRef]
99. Martin, P.; Frances, H.; Simon, P. Dissociation of head twitches and tremors during the study of interactions with 5-hydroxytryptophan in mice. *J. Pharmacol. Methods* **1985**, *13*, 193–200. [CrossRef]
100. Endo, Y. Evidence that the accumulation of 5-hydroxytryptamine in the liver but not in the brain may cause the hypoglycemia induced by 5-hydroxytryptophan. *Br. J. Pharmacol.* **1985**, *85*, 591–598. [CrossRef]
101. Endo, Y. Suppression and potentiation of 5-hydroxytryptophan-induced hypoglycemia by alpha-monofluoromethyldopa-correlation with the accumulation of 5-hydroxytryptamine in the liver. *Br. J. Pharmacol.* **1987**, *90*, 161–165.
102. Yang, T.H.; Hsu, P.Y.; Meng, M.H.; Su, C.C. Supplement of 5-hydroxytryptophan before induction suppresses inflammation and collagen-induced arthritis. *Arthritis Res. Ther.* **2015**, *17*, 364. [CrossRef] [PubMed]
103. Rogóz, Z.; Skuza, G.; Sowińska, H. The effect of the antihistaminic drugs on the central action of 5-hydroxytryptophan in mice. *Polish J. Pharm. Pharm.* **1981**, *33*, 459–465.
104. Truscott, T.C. Effects of phenylalanine and 5-hydroxytryptophan on seizure severity in mice. *Pharm. Biochem. Behav.* **1975**, *3*, 939–941. [CrossRef]
105. Carter, R.B.; Dykstra, L.A.; Leander, J.D.; Appel, J.B. Role of peripheral mechanisms in the behavioral effects of 5-hydroxytryptophan. *Pharmacol. Biochem. Behav.* **1978**, *9*, 249–253. [CrossRef]
106. Sharma, A.; Castellani, R.J.; Smith, M.A.; Muresanu, D.F.; Dey, P.K.; Sharma, H.S. 5-hydroxytryptophan: A precursor of serotonin influences regional blood-brain barrier breakdown, cerebral blood flow, brain edema formation, and neuropathology. In *New Therapeutic Strategies for Brain Edema and Cell Injury*; Sharma, H.S., Sharma, A., Eds.; Academic Press: Cambridge, MA, USA, 2019; pp. 1–44.
107. Yap, C.Y.; Taylor, D.A. Involvement of 5-ht2 receptors in the wet-dog shake behaviour induced by 5-hydroxytryptophan in the rat. *Neuropharmacology* **1983**, *22*, 801–804. [CrossRef]
108. Hadzović, S.; Ernst, A.M. The effect of 5-hydroxytryptamine and 5-hydroxytryptophan on extra-pyramidal function. *Eur. J. Pharmacol.* **1969**, *6*, 90–95. [CrossRef]
109. Meltzer, H.Y.; Fang, V.S. Effect of apomorphine plus 5-hydroxytryptophan on plasma prolactin levels in male rats. *Psychopharmacol. Commun.* **1976**, *2*, 189–198.
110. Ohgo, S.; Kato, Y.; Chihara, K.; Imura, H.; Maeda, K. Effect of hypothalamic surgery on prolactin release induced by 5-hydroxytryptophan (5-htp) in rats. *Endocrinol. Jpn.* **1976**, *23*, 485–491. [CrossRef]
111. Preziosi, P.; Cerrito, F.; Vacca, M. Effects of naloxone on the secretion of prolactin and corticosterone induced by 5-hydroxytryptophan and a serotonergic agonist, mcpp. *Life Sci.* **1983**, *32*, 2423–2430. [CrossRef]
112. Hingtgen, J.N.; Fuller, R.W.; Mason, N.R.; Aprison, M.H. Blockade of a 5-hydroxytryptophan-induced animal-model of depression with a potent and selective 5-ht2 receptor antagonist (ly53857). *Biol. Psychiatry* **1985**, *20*, 592–597. [CrossRef]
113. Dreshfield-Ahmad, L.J.; Thompson, D.C.; Schaus, J.M.; Wong, D.T. Enhancement in extracelllar serotonin levels by 5-hydroxytryptophan loading after administration of way 100635 and fluoxetine. *Life Sci.* **2000**, *66*, 2035–2041. [CrossRef]
114. Blundell, J.E.; Latham, C.J. Serotonergic influences on food intake: Effect of 5-hydroxytryptophan on parameters of feeding behaviour in deprived and free-feeding rats. *Pharmacol. Biochem. Behav.* **1979**, *11*, 431–437. [CrossRef]
115. Oconnor, L.H.; Feder, H.H. Estradiol and progesterone influence l-5-hydroxytryptophan-induced myoclonus in male guinea-pigs-sex-differences in serotonin steroid interactions. *Brain Res.* **1985**, *330*, 121–125. [CrossRef]
116. Pappert, E.J.; Goetz, C.G.; Stebbins, G.T.; Belden, M.; Carvey, P.M. 5-hydroxytryptophan-induced myoclonus in guinea pigs: Mediation through 5-ht1/2 receptor subtypes. *Eur. J. Pharmacol.* **1998**, *347*, 51–56. [CrossRef]
117. Kojima, M.; Ikeda, M.; Kamikawa, Y. Investigation into the 5-hydroxytryptophan-evoked luminal 5-hydroxytryptamine release from the guinea pig colon. *Jpn. J. Pharmacol.* **2000**, *84*, 174–178. [CrossRef] [PubMed]
118. Zuzina, A.B.; Vinarskaya, A.K.; Balaban, P.M. Increase in serotonin precursor levels reinstates the context memory during reconsolidation. *Invertebr. Neurosci.* **2019**, *19*, 8. [CrossRef] [PubMed]
119. Banerjee, U.; Burks, T.F.; Feldberg, W.; Goodrich, C.A. Temperature responses and other effects of 5-hydroxytryptophan and 5-hydroxytryptamine when acting from the liquor space in unanaesthetized rabbits. *Br. J. Pharmacol.* **1970**, *38*, 688–701. [CrossRef]
120. Fjalland, B. Neuroleptic influence on hyperthermia induced by 5-hydroxytryptophan and p-methoxy-amphetamine in maoi-pretreated rabbits. *Psychopharmacology* **1979**, *63*, 113–117. [CrossRef]
121. Denoyer, M.; Kitahama, K.; Sallanon, M.; Touret, M.; Jouvet, M. 5-hydroxytryptophan uptake and decarboxylating neurons in the cat hypothalamus. *Neuroscience* **1989**, *31*, 203–211. [CrossRef]
122. Ellaway, P.H.; Trott, J.R. The mode of action of 5-hydroxytryptophan in facilitating a stretch reflex in the spinal cat. *Exp. Brain Res.* **1975**, *22*, 145–162. [CrossRef]
123. Haverback, B.J.; Bogdanski, D.; Hogben, C.A. Inhibition of gastric acid secretion in the dog by the precursor of serotonin, 5-hydroxytryptophan. *Gastroenterology* **1958**, *34*, 188–195. [CrossRef]

124. Schemann, M.; Ehrlein, H.J. 5-hydroxytryptophan and cisapride stimulate propulsive jejunal motility and transit of chyme in dogs. *Digestion* **1986**, *34*, 229–235. [CrossRef] [PubMed]
125. Antonaccio, M.J.; Robson, R.D. Centrally-mediated cardiovascular effects of 5-hydroxytryptophan in mao-inhibited dogs: Modification by autonomic antagonists. *Arch. Int. Pharmacodyn. Ther.* **1975**, *213*, 200–210. [PubMed]
126. Sugden, D.; Namboodiri, M.A.A.; Klein, D.C.; Grady, R.K.; Mefford, I.N. Ovine pineal indoles-effects of l-tryptophan or l-5-hydroxytryptophan administration. *J. Neurochem.* **1985**, *44*, 769–772. [CrossRef] [PubMed]
127. Zhao, F.; Ma, C.; Zhao, G.D.; Wang, G.; Li, X.B.; Yang, K.L. Rumen-protected 5-hydroxytrytohan improves sheep melatonin synthesis in the pineal gland and intestinal tract. *Med. Sci. Monit.* **2019**, *25*, 3605–3616. [CrossRef] [PubMed]
128. Fletcher, D.J.; Hanson, M.A.; Moore, P.J.; Nijhuis, J.G.; Parkes, M.J. Stimulation of breathing movements by l-5-hydroxytryptophan in fetal sheep during normoxia and hypoxia. *J. Physiol. Lond.* **1988**, *404*, 575–589. [CrossRef]
129. Weaver, S.R.; Prichard, A.S.; Maerz, N.L.; Prichard, A.P.; Endres, E.L.; Hernandez-Castellano, L.E.; Akins, M.S.; Bruckmaier, R.M.; Hernandez, L.L. Elevating serotonin pre-partum alters the holstein dairy cow hepatic adaptation to lactation. *PLoS ONE* **2017**, *12*, e0184939. [CrossRef]
130. Martin, T.G. Serotonin syndrome. *Ann. Emerg. Med.* **1996**, *28*, 520–526. [CrossRef]
131. Francescangeli, J.; Karamchandani, K.; Powell, M.; Bonavia, A. The serotonin syndrome: From molecular mechanisms to clinical practice. *Int. J. Mol. Sci.* **2019**, *20*, 2288. [CrossRef]
132. Turner, E.H.; Loftis, J.M.; Blackwell, A.D. Serotonin a la carte: Supplementation with the serotonin precursor 5-hydroxytryptophan. *Pharmacol. Ther.* **2006**, *109*, 325–338. [CrossRef]
133. Agren, H.; Reibring, L.; Hartvig, P.; Tedroff, J.; Bjurling, P.; Hornfeldt, K.; Andersson, Y.; Lundqvist, H.; Langstrom, B. Low brain uptake of l- c-11 5-hydroxytryptophan in major depression-a positron emission tomography study on patients and healthy-volunteers. *Acta Psychiatr. Scand.* **1991**, *83*, 449–455. [CrossRef] [PubMed]
134. Ryan, N.D.; Birmaher, B.; Perel, J.M.; Dahl, R.E.; Meyer, V.; Alshabbout, M.; Iyengar, S.; Puigantich, J. Neuroendocrine response to l-5-hydroxytryptophan challenge in prepubertal major depression-depressed vs normal-children. *Arch. Gen. Psychiatry* **1992**, *49*, 843–851. [CrossRef] [PubMed]
135. Coppen, A.; Shaw, D.M.; Malleson, A. Changes in 5-hydroxytryptophan metabolism in depression. *Br. J. Psychiatry J. Ment. Sci.* **1965**, *111*, 105–107. [CrossRef] [PubMed]
136. Persson, T.; Roos, B.E. 5-hydroxytryptophan for depression. *Lancet* **1967**, *2*, 987–988. [CrossRef]
137. Byerley, W.F.; Judd, L.L.; Reimherr, F.W.; Grosser, B.I. 5-hydroxytryptophan-a review of its antidepressant efficacy and adverse-effects. *J. Clin. Psychopharmacol.* **1987**, *7*, 127–137.
138. Zmilacher, K.; Battegay, R.; Gastpar, M. L-5-hydroxytryptophan alone and in combination with a peripheral decarboxylase inhibitor in the treatment of depression. *Neuropsychobiology* **1988**, *20*, 28–35. [CrossRef]
139. Smarius, L.; Jacobs, G.E.; Hoeberechts-Lefrandt, D.H.M.; de Kam, M.L.; van der Post, J.P.; de Rijk, R.; van Pelt, J.; Schoemaker, R.C.; Zitman, F.G.; van Gerven, J.M.A.; et al. Pharmacology of rising oral doses of 5-hydroxytryptophan with carbidopa. *J. Psychopharmacol.* **2008**, *22*, 426–433. [CrossRef]
140. Van Praag, H.M. In search of the mode of action of antidepressants. 5-htp/tyrosine mixtures in depressions. *Neuropharmacology* **1983**, *22*, 433–440. [CrossRef]
141. Meltzer, H.Y.; Umberkoman-Wiita, B.; Robertson, A.; Tricou, B.J.; Lowy, M.; Perline, R. Effect of 5-hydroxytryptophan on serum cortisol levels in major affective disorders. I. Enhanced response in depression and mania. *Arch. Gen. Psychiatry* **1984**, *41*, 366–374. [CrossRef]
142. van Praag, H.M.; van den Burg, W.; Bos, E.R.; Dols, L.C. 5-hydroxytryptophan in combination with clomipramine in "therapy-resistant" depressions. *Psychopharmacologia* **1974**, *38*, 267–269. [CrossRef]
143. Aliño, J.J.; Gutierrez, J.L.; Iglesias, M.L. 5-hydroxytryptophan (5-htp) and a maoi (nialamide) in the treatment of depressions. A double-blind controlled study. *Int. Pharm.* **1976**, *11*, 8–15.
144. Mendlewicz, J.; Youdim, M.B. Antidepressant potentiation of 5-hydroxytryptophan by l-deprenil in affective illness. *J. Affect Disord.* **1980**, *2*, 137–146. [CrossRef]
145. Kious, B.M.; Sabic, H.; Sung, Y.H.; Kondo, D.G.; Renshaw, P. An open-label pilot study of combined augmentation with creatine monohydrate and 5-hydroxytryptophan for selective serotonin reuptake inhibitor- or serotonin-norepinephrine reuptake inhibitor-resistant depression in adult women. *J. Clin. Psychopharmacol.* **2017**, *37*, 578–583. [CrossRef] [PubMed]
146. Kahn, R.S.; Westenberg, H.G.M. L-5-hydroxytryptophan in the treatment of anxiety disorders. *J. Affect. Disord.* **1985**, *8*, 197–200. [CrossRef]
147. Kahn, R.S.; Westenberg, H.G.M.; Verhoeven, W.M.A.; Gispendewied, C.C.; Kamerbeek, W.D.J. Effect of a serotonin precursor and uptake inhibitor in anxiety disorders-a double-blind comparison of 5-hydroxytryptophan, clomipramine and placebo. *Int. Clin. Psychopharmacol.* **1987**, *2*, 33–45. [CrossRef]
148. Ishida, A.; Takada, G.; Kobayashi, Y.; Toyoshima, I.; Takai, K. Effect of tetrahydrobiopterin and 5-hydroxytryptophan on hereditary progressive dystonia with marked diurnal fluctuation-a suggestion of the serotonergic system involvement. *Tohoku J. Exp. Med.* **1988**, *154*, 233–239. [CrossRef]
149. Denboer, J.A.; Westenberg, H.G.M. Behavioral, neuroendocrine, and biochemical effects of 5-hydroxytryptophan administration in panic disorder. *Psychiatry Res.* **1990**, *31*, 267–278. [CrossRef]

150. Schruers, K.; van Diest, R.; Overbeek, T.; Griez, E. Acute l-5-hydroxytryptophan administration inhibits carbon dioxide-induced panic in panic disorder patients. *Psychiatry Res.* **2002**, *113*, 237–243. [CrossRef]
151. Maron, E.; Toru, I.; Vasar, V.; Shlik, J. The effect of 5-hydroxytryptophan on chotecystokinin-4-induced panic attacks in healthy volunteers. *J. Psychopharmacol.* **2004**, *18*, 194–199. [CrossRef]
152. Wyatt, R.J.; Zarcone, V.; Engelman, K.; Dement, W.C.; Snyder, F.; Sjoerdsma, A. Effects of 5-hydroxytryptophan on the sleep of normal human subjects. *Electroencephalogr. Clin. Neurophysiol.* **1971**, *30*, 505–509. [CrossRef]
153. Guilleminault, C.; Cathala, J.P.; Castaigne, P. Effects of 5-hydroxytryptophan on sleep of a patient with a brain-stem lesion. *Electroencephalogr. Clin. Neurophysiol.* **1973**, *34*, 177–184. [CrossRef]
154. Zarcone, V.; Kales, A.; Scharf, M.; Tan, T.L.; Simmons, J.Q.; Dement, W.C. Repeated oral ingestion of 5-hydroxytryptophan. The effect on behavior and sleep processes in two schizophrenic children. *Arch. Gen. Psychiatry* **1973**, *28*, 843–846. [CrossRef] [PubMed]
155. Petre-Quadens, O.; De Lee, C. 5-hydroxytryptophan and sleep in down's syndrome. *J. Neurol. Sci.* **1975**, *26*, 443–453. [CrossRef]
156. Bruni, O.; Ferri, R.; Miano, S.; Verrillo, E. L-5-hydroxytryptophan treatment of sleep terrors in children. *Eur. J. Pediatrics* **2004**, *163*, 402–407. [CrossRef]
157. Metz, J.T.; Holcomb, H.H.; Meltzer, H.Y. Effect of 5-hydroxytryptophan on h-reflex recovery curves in normal subjects and patients with affective-disorders. *Biol. Psychiatry* **1988**, *23*, 602–611. [CrossRef]
158. Emanuele, E.; Bertona, M.; Minoretti, P.; Geroldi, D. An open-label trial of l-5-hydroxytryptophan in subjects with romantic stress. *Neuroendocrinol. Lett.* **2010**, *31*, 663–666.
159. Titus, F.; Davalos, A.; Alom, J.; Codina, A. 5-hydroxytryptophan versus methysergide in the prophylaxis of migraine-randomized clinical-trial. *Eur. Neurol.* **1986**, *25*, 327–329. [CrossRef]
160. Maissen, C.P.; Ludin, H.P. Comparative efficacy of 5-hydroxytryptophan and propranolol in interval treatment of migraine. *Schweiz. Med. Wochenschr.* **1991**, *121*, 1585–1590.
161. Nicolodi, M.; Sicuteri, F. L-5-hydroxytryptophan can prevent nociceptive disorders in man. In *Tryptophan, Serotonin and Melatonin: Basic Aspects and Applications*; Huether, G., Kochen, W., Simat, T.J., Steinhart, H., Eds.; Springer: New York, NY, USA, 1999; pp. 177–182.
162. Ribeiro, C.A.F.; Portuguese Headache, S. L-5-hydroxytryptophan in the prophylaxis of chronic tension-type headache: A double-blind, randomized, placebo-controlled study. *Headache* **2000**, *40*, 451–456. [CrossRef]
163. Trouillas, P.; Brudon, F.; Adeleine, P. Improvement of cerebellar-ataxia with levorotatory form of 5-hydroxytryptophan-a double-blind-study with quantified data-processing. *Arch. Neurol.* **1988**, *45*, 1217–1222. [CrossRef]
164. Trouillas, P.; Serratrice, G.; Laplane, D.; Rascol, A.; Augustin, P.; Barroche, G.; Clanet, M.; Degos, C.F.; Desnuelle, C.; Dumas, R.; et al. Levorotatory form of 5-hydroxytryptophan in friedreichs ataxia-results of a double-blind drug-placebo cooperative study. *Arch. Neurol.* **1995**, *52*, 456–460. [CrossRef] [PubMed]
165. Rus, A.; Molina, F.; Del Moral, M.L.; Ramirez-Exposito, M.J.; Martinez-Martos, J.M. Catecholamine and indolamine pathway: A case-control study in fibromyalgia. *Biol. Res. Nurs.* **2018**, *20*, 577–586. [CrossRef] [PubMed]
166. Maffei, M.E. Fibromyalgia: Recent advances in diagnosis, classification, pharmacotherapy and alternative remedies. *Int. J. Mol. Sci.* **2020**, *21*, 7877. [CrossRef] [PubMed]
167. Caruso, I.; Puttini, P.S.; Cazzola, M.; Azzolini, V. Double-blind-study of 5-hydroxytryptophan versus placebo in the treatment of primary fibromyalgia syndrome. *J. Int. Med Res.* **1990**, *18*, 201–209. [CrossRef] [PubMed]
168. Volicer, L.; Langlais, P.J.; Matson, W.R.; Mark, K.A.; Gamache, P.H. Serotoninergic system in dementia of the alzheimer type-abnormal forms of 5-hydroxytryptophan and serotonin in cerebrospinal-fluid. *Arch. Neurol.* **1985**, *42*, 1158–1161. [CrossRef] [PubMed]
169. Sano, I.; Taniguchi, K. [l-5-hydroxytryptophan(l-5-htp) therapy of parkinson's disease. 2]. *Munch. Med. Wochenschr. (1950)* **1972**, *114*, 1717–1719.
170. Chase, T.N. 5-hydroxytryptophan in parkinsonism. *Lancet (London, England)* **1970**, *2*, 1029–1030. [CrossRef]
171. Magnussen, I.; Jensen, T.S.; Rand, J.H.; Van Woert, M.H. Plasma accumulation of metabolism of orally administered single dose l-5-hydroxytryptophan in man. *Acta Pharmacol. Toxicol. (Copenh)* **1981**, *49*, 184–189. [CrossRef]
172. Meloni, M.; Puligheddu, M.; Carta, M.; Cannas, A.; Figorilli, M.; Defazio, G. Efficacy and safety of 5-hydroxytryptophan on depression and apathy in parkinson's disease: A preliminary finding. *Eur. J. Neurol.* **2020**, *27*, 779–786. [CrossRef]
173. Growdon, J.H.; Young, R.R.; Shahani, B.T. L-5-hydroxytryptophan in treatment of several different syndromes in which myoclonus is prominent. *Neurology* **1976**, *26*, 1135–1140. [CrossRef]
174. Magnussen, I.; Dupont, E.; Prange-Hansen, A.; de Fine Olivarius, B. Palatal myoclonus treated with 5-hydroxytryptophan and a decarboxylase-inhibitor. *Acta Neurol. Scand.* **1977**, *55*, 251–253. [CrossRef] [PubMed]
175. Gascon, G.; Wallenberg, B.; Daif, A.K.; Ozand, P. Successful treatment of cherry red spot myoclonus syndrome with 5-hydroxytryptophan. *Ann. Neurol.* **1988**, *24*, 453–455. [CrossRef] [PubMed]
176. Nakano, K.; Hayakawa, T.; Shishikura, K.; Ohsawa, M.; Suzuki, H.; Fukuyama, Y. Improvement of action myoclonus by an administration of 5-hydroxytryptophan and carbidopa in a child with muscular subsarcolemmal hyperactivity. *Brain Dev.* **1990**, *12*, 516–520. [CrossRef]

177. Jimenezjimenez, F.J.; Roldan, A.; Zancada, F.; Molinaarjona, J.A.; Fernandezballesteros, A.; Santos, J. Spinal myoclonus-successful treatment with the combination of sodium valproate and l-5-hydroxytryptophan. *Clin. Neuropharmacol.* **1991**, *14*, 186–190. [CrossRef]
178. Ceci, F.; Cangiano, C.; Cairella, M.; Cascino, A.; Delben, M.; Muscaritoli, M.; Sibilia, L.; Fanelli, F.R. The effects of oral 5-hydroxytryptophan administration on feeding-behavior in obese adult female subjects. *J. Neural. Transm.* **1989**, *76*, 109–117. [CrossRef] [PubMed]
179. Cangiano, C.; Ceci, F.; Cascino, A.; Delben, M.; Laviano, A.; Muscaritoli, M.; Antonucci, F.; Rossifanelli, F. Eating behavior and adherence to dietary prescriptions in obese adult subjects treated with 5-hydroxytryptophan. *Am. J. Clin. Nutr.* **1992**, *56*, 863–867. [CrossRef] [PubMed]
180. Kato, Y.; Nakai, Y.; Imura, H.; Chihara, K.; Ogo, S. Effect of 5-hydroxytryptophan (5-htp) on plasma prolactin levels in man. *J. Clin. Endocrinol. Metab.* **1974**, *38*, 695–697. [CrossRef]
181. Sueldo, C.E.; Duda, M.; Kletzky, O.A. Influence of sequential doses of 5-hydroxytryptophan on prolactin-release. *Am. J. Obstet. Gynecol.* **1986**, *154*, 424–427. [CrossRef]
182. Vlasses, P.H.; Rotmensch, H.H.; Swanson, B.N.; Clementi, R.A.; Ferguson, R.K. Effect of repeated doses of l-5-hydroxytryptophan and carbidopa on prolactin and aldosterone secretion in man. *J. Endocrinol. Investig.* **1989**, *12*, 87–91. [CrossRef]
183. Keithahn, C.; Lerchl, A. 5-hydroxytryptophan is a more potent in vitro hydroxyl radical scavenger than melatonin or vitamin c. *J. Pineal Res.* **2005**, *38*, 62–66. [CrossRef]
184. Derlacz, R.A.; Sliwinska, M.; Piekutowska, A.; Winiarska, K.; Drozak, J.; Bryla, J. Melatonin is more effective than taurine and 5-hydroxytryptophan against hyperglycemia-induced kidney-cortex tubules injury. *J. Pineal Res.* **2007**, *42*, 203–209. [CrossRef] [PubMed]
185. Reyes-Gonzales, M.C.; Fuentes-Broto, L.; Martinez-Ballarin, E.; Miana-Mena, F.J.; Berzosa, C.; Garcia-Gil, F.A.; Aranda, M.; Garcia, J.J. Effects of tryptophan and 5-hydroxytryptophan on the hepatic cell membrane rigidity due to oxidative stress. *J. Membr. Biol.* **2009**, *231*, 93–99. [CrossRef] [PubMed]
186. Bae, S.J.; Lee, J.S.; Kim, J.M.; Lee, E.K.; Han, Y.K.; Kim, H.J.; Choi, J.; Ha, Y.M.; No, J.K.; Kim, Y.H.; et al. 5-hydroxytrytophan inhibits tert-butylhydroperoxide (t-bhp)-induced oxidative damage via the suppression of reactive species (rs) and nuclear factor-kappa b (nf-kappa b) activation on human fibroblast. *J. Agric. Food Chem.* **2010**, *58*, 6387–6394. [CrossRef] [PubMed]
187. Chae, H.S.; Kang, O.H.; Choi, J.G.; Oh, Y.C.; Lee, Y.S.; Jang, H.J.; Kim, J.H.; Park, H.; Jung, K.Y.; Sohn, D.H.; et al. 5-hydroxytryptophan acts on the mitogen-activated protein kinase extracellular-signal regulated protein kinase pathway to modulate cyclooxygenase-2 and inducible nitric oxide synthase expression in raw 264.7 cells. *Biol. Pharm. Bull.* **2009**, *32*, 553–557. [CrossRef] [PubMed]
188. Gwaltney-Brant, S.M.; Albretsen, J.C.; Khan, S.A. 5-hydroxytryptophan toxicosis in dogs: 21 cases (1989–1999). *J. Am. Vet. Med Assoc.* **2000**, *216*, 1937–1940. [CrossRef] [PubMed]
189. Shapiro, S. Tryptophan produced by showa denko and epidemic eosinophilia-myalgia syndrome-comment. *J. Rheumatol.* **1996**, *23*, 89–91.
190. Kilbourne, E.M.; Philen, R.M.; Kamb, M.L.; Falk, H. Tryptophan produced by showa denko and epidemic eosinophilia-myalgia syndrome. *J. Rheumatol.* **1996**, *23*, 81–88.
191. Preuss, H.G.; Echard, B.; Talpur, N.; Funk, K.A.; Bagchi, D. Does 5-hydroxytryptophan cause acute and chronic toxic perturbations in rats? *Toxicol. Mech. Methods* **2006**, *16*, 281–286. [CrossRef]
192. Belongia, E.A.; Hedberg, C.W.; Gleich, G.J.; White, K.E.; Mayeno, A.N.; Loegering, D.A.; Dunnette, S.L.; Pirie, P.L.; MacDonald, K.L.; Osterholm, M.T. An investigation of the cause of the eosinophilia-myalgia syndrome associated with tryptophan use. *N. Engl. J. Med.* **1990**, *323*, 357–365. [CrossRef]
193. Mayeno, A.N.; Lin, F.; Foote, C.S.; Loegering, D.A.; Ames, M.M.; Hedberg, C.W.; Gleich, G.J. Characterization of peak-e, a novel amino-acid associated with eosinophilia-myalgia-syndrome. *Science* **1990**, *250*, 1707–1708. [CrossRef]
194. Goda, Y.; Suzuki, J.; Maitani, T.; Yoshihira, K.; Takeda, M.; Uchiyama, M. 3-anilino-l-alanine, structural determination of uv-5, a contaminant in ems-associated l-tryptophan samples. *Chem. Pharm. Bull.* **1992**, *40*, 2236–2238. [CrossRef] [PubMed]
195. Michelson, D.; Page, S.W.; Casey, R.; Trucksess, M.W.; Love, L.A.; Milstien, S.; Wilson, C.; Massaquoi, S.G.; Crofford, L.J.; Hallett, M.; et al. An eosinophilia-myalgia-syndrome related disorder associated with exposure to l-5-hydroxytryptophan. *J. Rheumatol.* **1994**, *21*, 2261–2265. [PubMed]

Article

The Effect of Chronic Mild Stress and Escitalopram on the Expression and Methylation Levels of Genes Involved in the Oxidative and Nitrosative Stresses as Well as Tryptophan Catabolites Pathway in the Blood and Brain Structures

Paulina Wigner [1], Ewelina Synowiec [1], Paweł Jóźwiak [2], Piotr Czarny [3], Michał Bijak [4], Katarzyna Białek [1], Janusz Szemraj [3], Piotr Gruca [5], Mariusz Papp [5] and Tomasz Śliwiński [1,*]

[1] Laboratory of Medical Genetics, Faculty of Biology and Environmental Protection, University of Lodz, 90-136 Lodz, Poland; paulina.wigner@gmail.com (P.W.); ewelina.synowiec@biol.uni.lodz.pl (E.S.); biaalek.k@gmail.com (K.B.)
[2] Department of Cytobiochemistry, Faculty of Biology and Environmental Protection, University of Lodz, 90-136 Lodz, Poland; pawel.jozwiak@biol.uni.lodz.pl
[3] Department of Medical Biochemistry, Medical University of Lodz, 90-647 Lodz, Poland; piotr.czarny@umed.lodz.pl (P.C.); janusz.szemraj@umed.lodz.pl (J.S.)
[4] Biohazard Prevention Centre, Faculty of Biology and Environmental Protection, University of Lodz, 90-136 Lodz, Poland; michal.bijak@biol.uni.lodz.pl
[5] Institute of Pharmacology, Polish Academy of Sciences, 31-343 Krakow, Poland; gruca@if-pan.krakow.pl (P.G.); nfpapp@cyfronet.pl (M.P.)
* Correspondence: tomasz.sliwinski@biol.uni.lodz.pl; Tel.: +48-42-635-44-86; Fax: +48-42-635-44-84

Abstract: Previous studies suggest that depression may be associated with reactive oxygen species overproduction and disorders of the tryptophan catabolites pathway. Moreover, one-third of patients do not respond to conventional pharmacotherapy. Therefore, the study investigates the molecular effect of escitalopram on the expression of *Cat*, *Gpx1/4*, *Nos1/2*, *Tph1/2*, *Ido1*, *Kmo*, and *Kynu* and promoter methylation in the hippocampus, amygdala, cerebral cortex, and blood of rats exposed to CMS (chronic mild stress). The animals were exposed to CMS for two or seven weeks followed by escitalopram treatment for five weeks. The mRNA and protein expression of the genes were analysed using the TaqMan Gene Expression Assay and Western blotting, while the methylation was determined using methylation-sensitive high-resolution melting. The CMS caused an increase of *Gpx1* and *Nos1* mRNA expression in the hippocampus, which was normalised by escitalopram administration. Moreover, *Tph1* and *Tph2* mRNA expression in the cerebral cortex was increased in stressed rats after escitalopram therapy. The methylation status of the *Cat* promoter was decreased in the hippocampus and cerebral cortex of the rats after escitalopram therapy. The Gpx4 protein levels were decreased following escitalopram compared to the stressed/saline group. It appears that CMS and escitalopram influence the expression and methylation of the studied genes.

Keywords: depression; chronic mild stress; oxidative stress; tryptophan catabolites pathway; methylation; expression; escitalopram

1. Introduction

Depression is a serious mental illness which is believed to affect 350 million people worldwide, according to the WHO. It has also been recognized as the third leading cause of disability in 2015 [1]. Although the condition affects both sexes, women are approximately twice as likely to develop symptoms [2]. If untreated, depression can lead to suicide attempts, and approximately one million people commit suicide every year [3,4]. In addition, depression is associated with high economic costs, constituting about 60% of the total cost of treating all mental conditions [5]. Unfortunately, more than a third of patients suffer from treatment-resistant depression [6].

Despite being such a serious health problem, the pathogenesis of depression remains unclear. However, previous studies suggest that disorders of the tryptophan catabolites (TRYCATs) pathway and associated overproduction of reactive oxygen species (ROS) may contribute to depression development [7,8]. Patients with depression are characterised by a decreased level of tryptophan and increased activity of IDO1 and TDO2, which converts tryptophan into kynurenine. The following stages of the TRYCATs pathway generate quinolinic acid, 3-hydroxykynurenine and 3-hydroxyanthranilic acid; these may induce the production of ROS, such as hydrogen peroxide, and increase lipid oxidation [9–16]. Indeed, patients with depression are frequently characterised by an increased level of free radicals and a decreased level of nonenzymatic antioxidants, including zinc, glutathione, and vitamins E, C, and A [17]. However, more contradictory results have been obtained with enzymatic antioxidants such as glutathione peroxidase and superoxide dismutase [18–22].

Depression is also associated with changes in specific parts of the brain. Interestingly, the previous study used with magnetic resonance imaging showed that depressed patients were characterised by smaller volumes of the amygdala, hippocampus, inferior anterior cingulate, and the orbital prefrontal cortex (OPFC), components of the limbic-cortico-thalamic circuit [23]. Frodl et al. (2008) confirmed that the brain of patients with depression showed more decline in grey matter density of the hippocampus, anterior cingulum, left amygdala, and right dorsomedial prefrontal cortex [24]. In addition, animal studies suggest that increased level of glucocorticoid observed in patients with depression may negatively affect neurogenesis and lead to excitotoxic damage or be associated with reduced levels of key neurotrophins in the hippocampus. Antidepressants may neutralise these effects by the increase of the neurogenesis in the hippocampus and brain-derived neurotrophic factor levels [25]. Moreover, increased levels of free radicals may cause cell death and atrophy of the neurons in the hippocampus and cortex [26]. Therefore, antidepressant drugs may exert their effectiveness by acting as antioxidants. For example, the antidepressant escitalopram, a selective serotonin reuptake inhibitor (SSRI). Escitalopram inhibits the serotonin transporter protein and is widely accepted as first-line antidepressant therapy. Escitalopram is characterized by favorable safety profile and has been shown efficacious for both acute and long-term treatments. Interestingly, escitalopram may have antioxidant properties, indicated by increased GABA levels in the frontal cortices of rats exposed to chronic mild stress (CMS) [27]. Moreover, subchronic treatment with escitalopram caused the reduced plasma SOD, CAT, malondialdehyde (MDA) and NO levels in depressed patients. However, Sarandol et al. (2007) found no difference in the levels of polyunsaturated fatty acid peroxidation products in depressed patients before and after antidepressant therapy [28]. Interestingly, the parameters came close to the results of healthy controls [29].

Therefore, the aim of the present study was to investigate the effect of chronic mild stress and antidepressant treatment with escitalopram in peripheral blood mononuclear cells (PBMCs), hippocampus, amygdala and cerebral cortex of rats. The study evaluates mRNA and protein expression, and the methylation status of gene promoters involved in oxidative stress (*Gpx1*, *Gpx4*, *Cat*, *Nos1*, *Nos2*) and the tryptophan catabolites pathway (*Tph1*, *Tph2*, *Ido1*, *Kmo*, *Kynu*). All studied gene products are presented in Table 1.

Table 1. Characteristic of all studied genes in the presented paper.

Gene	Enzyme	Gene Location	Function of the Enzyme	Tissue mRNA Expression
			Oxidative and Nitrosative Stresses	
Gpx1	Glutathione peroxidase 1	8q32	Enzyme catalyses the reduction of organic hydroperoxides and hydrogen peroxide by glutathione, and thereby protect cells against oxidative damage.	Detected in all tissue
Gpx4	Glutathione peroxidase 4	7q11	Enzyme which catalyses the reduction of hydrogen peroxide, organic hydroperoxides and lipid hydroperoxides, and thereby protect cells against oxidative damage. Essential antioxidant peroxidase that directly reduces phospholipid hydroperoxide even if they are incorporated in membranes and lipoproteins (By similarity). Can also reduce fatty acid hydroperoxide, cholesterol hydroperoxide and thymine hydroperoxide.	Detected in all tissue
Cat	Catalase	3q32	The key antioxidant enzyme in the bodies defence against oxidative stress. Catalase is a heme enzyme that is present in the peroxisome of nearly all aerobic cells. Catalase converts the reactive oxygen species hydrogen peroxide to water and oxygen and thereby mitigates the toxic effects of hydrogen peroxide.	Detected in all tissue, however tissue enhanced–blood and liver
Nos1	Nitric oxide synthetase 1	12q16	Enzyme, which synthesize nitric oxide from L-arginine.	Brain, skeletal muscle
Nos2	Nitric oxide synthetase 2	10q25	Enzyme, which synthesize nitric oxide from L-arginine.	Detected in many tissue, however tissue enhanced–intestine, lymphoid tissue
			Tryptophan catabolites pathway	
Gene	Enzyme	Gene location	Function of the enzyme	Tissue mRNA expression
Tph1	Tryptophan hydroxylase 1	1q22	The enzyme catalyses the first and rate limiting step in the biosynthesis of serotonin, an important hormone and neurotransmitter.	Brain, intestine, pituitary gland, stomach
Tph2	Tryptophan hydroxylase 2	7q22	The encoded protein catalyses the first and rate limiting step in the biosynthesis of serotonin, an important hormone and neurotransmitter.	Brain
Ido1	Indolamine 2,3-dioxygenasse	16q12.5	Ido1 is heme enzyme that catalyses the first and rate-limiting step in tryptophan catabolism to N-formyl-kynurenine. This enzyme acts on multiple tryptophan substrates including D-tryptophan, L-tryptophan, 5-hydroxy-tryptophan, tryptamine, and serotonin.	Blood, placenta
Kmo	Kynurenine 3-monooxygenase	13q25	The enzyme catalyses the hydroxylation of L-kynurenine to form 3-hydroxy-L-kynurenine. It is Required for synthesis of quinolinic acid, a neurotoxic NMDA receptor antagonist and potential endogenous inhibitor of NMDA receptor signalling.	Blood, liver, placenta
Kynu	Kynureninase	3q12	It is enzyme that catalyses the cleavage of L-kynurenine and L-3-hydroxykynurenine into anthranilic and 3-hydroxyanthranilic acids, respectively. Kynureninase is involved in the biosynthesis of NAD cofactors from tryptophan through the kynurenine pathway	Detected in all tissue, however tissue enhanced–blood, liver, placenta

2. Results

2.1. The Effect of CMS Procedure and Escitalopram Treatment on Sucrose Intake

As shown in Figure 1B, after two weeks of initial stress, the stressed rats were characterised by a decrease in sucrose intake ($p < 0.05$), whereas the escitalopram-treated stressed animals showed an approximately 60% increase in sucrose intake (Week 7).

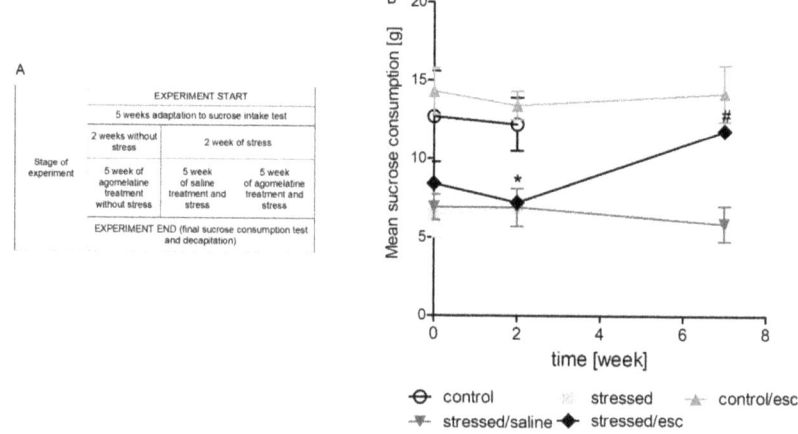

Figure 1. The course of the experiment of chronic mild stress and escitalopram therapy (**A**). Sucrose intake test in rats exposed to CMS for two weeks (week 2) and in animals exposed to CMS for seven weeks (week 7) and administered vehicle (1 mL/kg) or escitalopram (10 mg/kg) for five weeks (**B**). The consumption of 1.0% sucrose solution was measured in a 1-h test by weighing pre-weighed bottles. The data represents means ± SEM. $n = 6$. * $p < 0.01$ control group relative to the stressed group; # $p < 0.05$ stressed group relative to the stressed/esc group.

2.2. mRNA Expression

2.2.1. Gene Expression in PBMCs

The changes in mRNA expression are presented in Figure 2. The stressed rats receiving chronic administration of escitalopram demonstrated increased *Gpx1* (H = 12.130, df = 4, $p = 0.016$, Tukey test $p < 0.05$; H = 12.130, df = 4, $p = 0.016$, Tukey test $p < 0.001$, respectively) and *Gpx4* expression in PBMCs as compared to the stressed rats and the stressed rats after chronic administration of saline (F = 129.836, df = 4, $p < 0.001$, Tukey test $p < 0.001$; F = 129.836, df = 4, $p < 0.001$, Tukey test $p < 0.001$, respectively). In the case of the *Cat*, *Nos2*, and genes involved in TRYCATs pathway, no significant differences were observed between any studied groups.

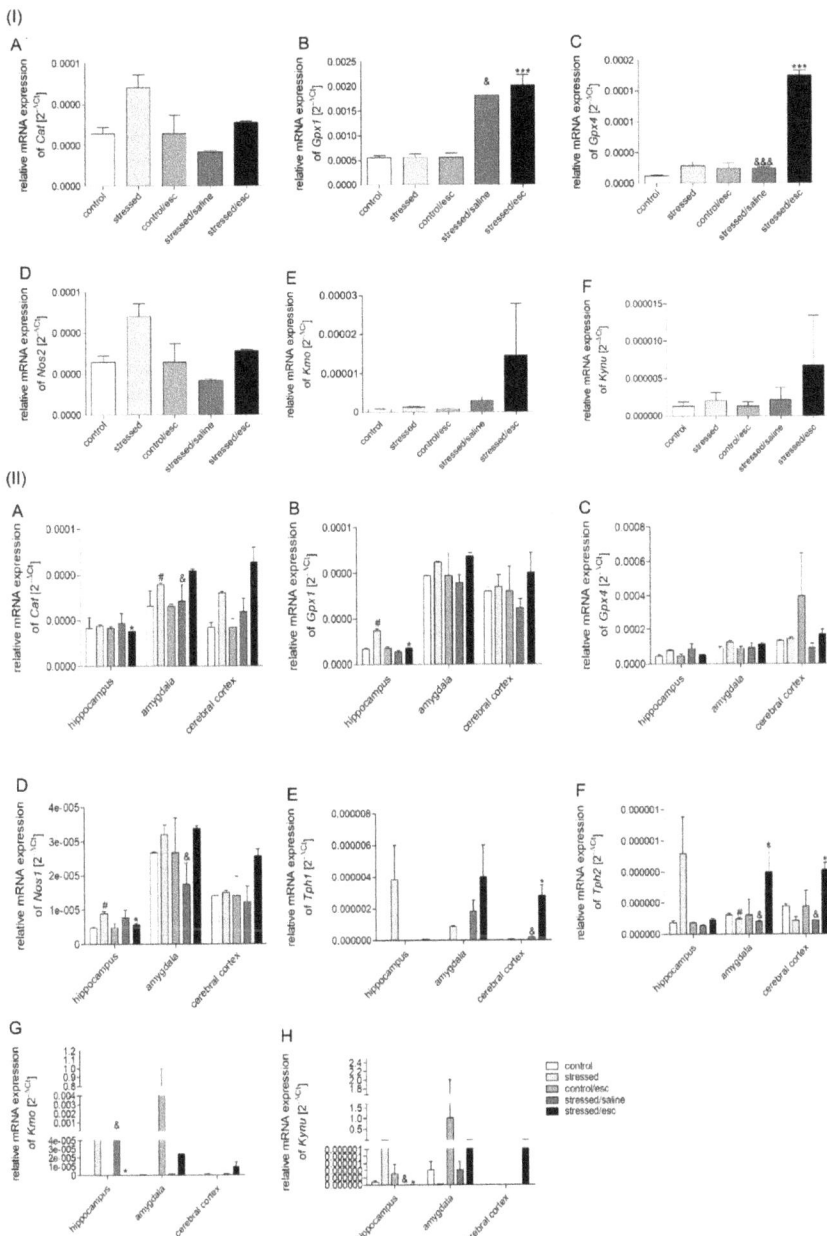

Figure 2. (**I**) mRNA expression of *Cat* (**A**), *Gpx1* (**B**), *Gpx4* (**C**), *Nos2* (**D**) *Kmo* (**E**), *Kynu* (**F**) in PBMCs and (**II**) mRNA expression of *CAT* (**A**), *Gpx1* (**B**), *Gpx4* (**C**), *Nos1* (**D**), *Tph1* (**E**), *Tph2* (**F**), *Kmo* (**G**), *Kynu* (**H**) in the brain structures (hippocampus, amygdala, cerebral cortex) of animals exposed to CMS for two weeks (control, stressed) and in animals exposed to CMS for seven weeks and administered vehicle (1 mL/kg) or escitalopram (10 mg/kg) for five weeks (control/esc, stressed/saline, stressed/esc). Relative gene expression levels were estimated using a $2^{-\Delta Ct\ (Ctgene-Ct18S)}$ method. Data represent means ± SEM. $n = 6$. * $p < 0.05$, *** $p < 0.001$ stressed group relative to stressed/esc group; # $p < 0.05$ stressed group relative to control group; & $p < 0.05$, &&& $p < 0.001$ stressed/esc group relative to stressed/saline group.

2.2.2. Gene Expression in Brain Structures

As shown in Figure 2, the effect of CMS procedure and chronic escitalopram administration on the expression of the studied genes varied according to brain structure. Reduced levels of *Cat* mRNA were observed in the hippocampus of the stressed rats after escitalopram therapy as compared to stressed animals (H = 12.233, df = 4, p = 0.016, Tukey test $p < 0.05$). In addition, increased *Cat* mRNA expression was observed in the amygdala after CMS (H = 12.100, df = 4, p = 0.017, Tukey test $p < 0.05$), as well as after chronic escitalopram administration, as compared to the stressed saline group (H = 12.100, df = 4, p = 0.017, Tukey test $p < 0.05$). Similarly, elevated *Gpx1* mRNA expression was recorded in the hippocampus of the stressed animals (H = 11.433, df = 4, p = 0.022, Tukey test $p < 0.05$), and this effect was normalised after chronic administration of escitalopram (H = 11.433, df = 4, p = 0.022, Tukey test $p < 0.05$). Additionally, *Nos1* mRNA expression was elevated in the hippocampus following CMS ($p < 0.05$) and this effect was normalized in stressed rats after escitalopram treatment (H = 9.462, df = 4, p = 0.024, Tukey test $p < 0.05$). In addition, *Nos1* was also elevated in the amygdala of stressed rats after antidepressant therapy compared to the stressed rats after saline therapy (H = 9.462, df = 4, p = 0.024, Tukey test $p < 0.05$).

In the case of genes involved in TRYCATs pathway, *Tph1* expression in the cerebral cortex was elevated following escitalopram treatment compared to the stressed group (H = 12.433, df = 4, p = 0.014, Tukey test $p < 0.05$) and the stressed group receiving saline treatment (H = 12.433, df = 4, p = 0.014, Tukey test $p < 0.05$). In contrast, *Tph2* mRNA expression in the amygdala fell following CMS. However, this effect was normalised by chronic administration of escitalopram (H = 8.692, df = 4, p = 0.034, Tukey test $p < 0.05$). In addition, higher *Tph2* mRNA expression was observed in the amygdala of rats after antidepressant therapy than in those after saline treatment (H = 8.692, df = 4, p = 0.034, Tukey test $p < 0.05$). Elevated *Tph2* expression was also observed in the cerebral cortex of animals after escitalopram treatment as compared to stressed group and animals after saline treatment (F = 7.123, df = 4, p = 0.006, Tukey test $p < 0.05$). The chronic administration of escitalopram caused a decrease of *Kmo* (H = 9.688, df = 4, p = 0.046, Tukey test $p < 0.05$) and *Kynu* mRNA (H = 10.937, df = 4, p = 0.027, Tukey test $p < 0.05$) expression in the hippocampus as compared to saline treatment.

2.2.3. The Effect of Escitalopram Treatment on Gene Expression in PBMCs and Brain Structures

Escitalopram caused an increase in *Tph2* expression in the amygdala and cerebral cortex compared to PBMCs ($p < 0.001$) (Supplementary Figure S1). No significant differences were found between any of the studied groups regarding the effect of escitalopram treatment on the expression of the other studied genes involved in oxidative stress and the TRYCATs pathway.

2.3. Methylation Status

2.3.1. Methylation Status of Promoter Regions in PBMCs

Interestingly, no significant change in the methylation status of the promoter regions of the studied genes was observed between the studied groups following CMS and escitalopram therapy (Supplementary Table S1).

2.3.2. Methylation Status of Promoter Regions in Brain Structures

The methylation status of the *Cat* promoter region was reduced in the hippocampus and the cerebral cortex of the stressed group after escitalopram treatment as compared to the stressed group after saline administration (H = 11.412, df = 4, p = 0.022, Tukey test $p < 0.05$) (Figure 3). Moreover, the methylation status of the *Ido1* promoter region in the hippocampus and the cerebral cortex was increased in the escitalopram-treated group compared to the stressed group and saline-treated group (F = 18.681, df = 4, $p < 0.001$, Tukey test $p < 0.001$; H = 12.247, df = 4, p = 0.016, Tukey test $p < 0.05$, respectively). In addition, the *Kmo* promoter region demonstrated increased methylated status in the cerebral cortex of stressed animals after escitalopram treatment compared to the stressed rats after saline therapy (H = 9.829, df = 4, p = 0.043, Tukey test $p < 0.05$). No significant changes in the methylation status of promoter regions were observed for the other studied genes (Supplementary Table S1).

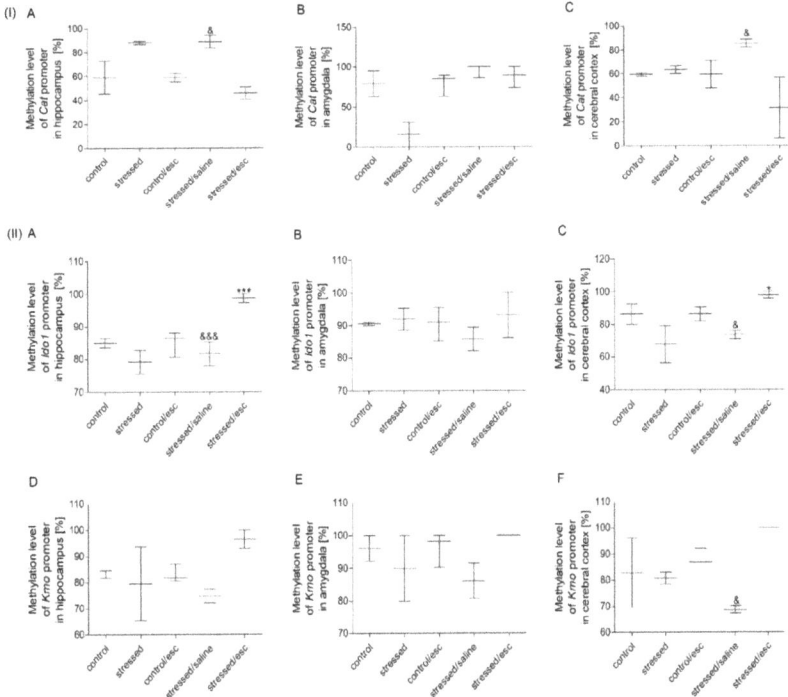

Figure 3. Methylation status of *Cat* promoter region (**I**) in the hippocampus (**A**), amygdala (**B**) and cerebral cortex (**C**) *Ido1* and *Kmo* promoter regions (**II**) in the hippocampus (**A,D**), amygdala (**B,E**) and cerebral cortex (**C,F**) of animals exposed to CMS for two weeks (control, stressed) and in animals exposed to CMS for seven weeks and administered vehicle (1 mL/kg) or escitalopram (10 mg/kg) for five weeks (control/esc, stressed/saline, stressed/esc). Data represents median and maxiumum-minimum values. $n = 6$. * $p < 0.05$, *** $p < 0.001$ stressed/esc group relative to stressed group; & $p < 0.05$, &&& $p < 0.001$ stressed/esc group relative to stressed/saline group.

2.3.3. The Effect of Escitalopram Treatment on the Methylation Status of Promoter Regions in PBMCs and Brain Structures

No significant differences were found between the studied groups with regard to the effect of escitalopram treatment on methylation status of the promoter regions of all studied genes involved in oxidative stress and the TRYCATs pathway (Supplementary Figure S2).

2.4. Protein Expression in Brain Structures

In the brain structures, the only change observed in protein expression associated with escitalopram treatment was observed in the cerebral cortex, where the level of Gpx4 protein was reduced compared to stressed animals after saline treatment (H = 11.445, df = 4, p = 0.022, Tukey test $p < 0.05$) (Figure 4; Supplementary Figure S3). No other significant differences were observed between the studied groups regarding the other genes involved in oxidative stress and the TRYCATs pathway.

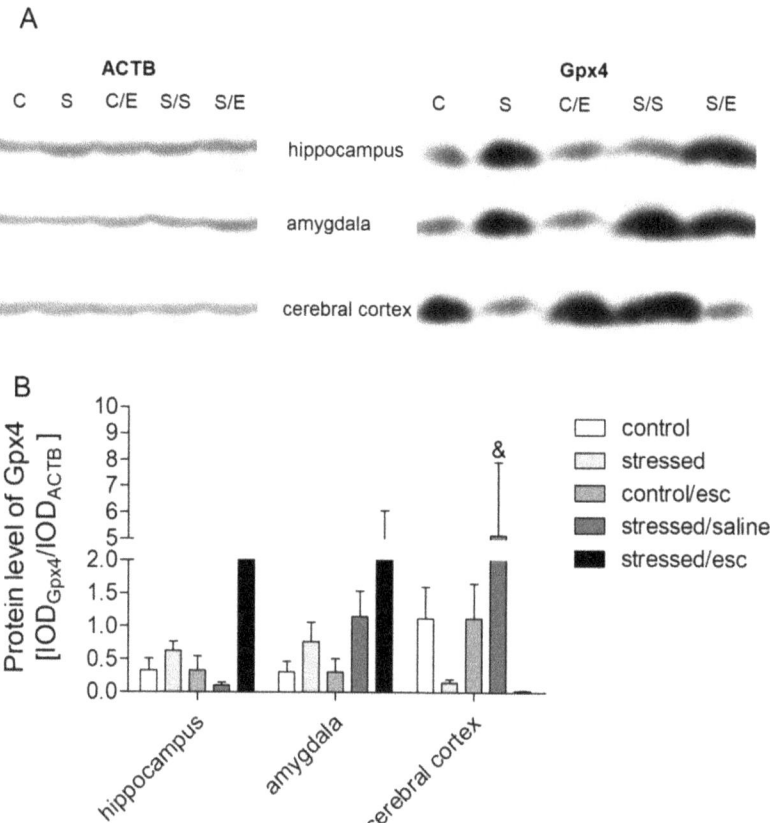

Figure 4. Protein expression of Gpx4 in brain structures of animals exposed to CMS for two weeks (control, stressed) and in animals exposed to CMS for seven weeks and administered vehicle (1 mL/kg) or escitalopram (10 mg/kg) for five weeks (control/esc, stressed/saline, stressed/esc). (**A**) Representative western blot analysis in midbrain and cerebral cortex. C–controls, S–stressed for two weeks, C/E–control/escitalopram, S/S–stressed/saline, S/E–stressed/escitalopram. (**B**) Levels of Gpx4 proteins measured in hippocampus, amygdala and cerebral cortex. Samples containing 25 µg of proteins were resolved by SDS-PAGE. The intensity of bands corresponding to Gpx4 was analysed by densitometry. Integrated optical density (IOD) was normalized by protein content and a reference sample (see the Methods Section for details). The graphs show the mean IODs of the bands from all analysed samples. The IOD_{gene}/IOD_{ACTB} method was used to estimate the relative protein expression levels in the analysed samples. Data represent means ± SEM. $n = 6$. & $p < 0.05$ for the difference between stressed/saline and stressed/escitalopram groups.

3. Discussion

Our findings demonstrate the impact of CMS and antidepressant treatment with escitalopram on the level of protein and mRNA expression in PBMCs and the hippocampus, amygdala and cerebral cortex. We also elucidate the methylation status of the promoter regions of the genes involved in oxidative and nitrosative stress, as well as the TRYCATs pathway. However, we didn't evaluate any oxidative and nitrosative stress markers, including MDA, 8-oxoguanine, and 8-iso-prostaglandin F2α. The increased levels of the above-mentioned markers have been repeatedly confirmed in studies involving humans and animals. Therefore, we focused our research on an attempt to explain the molecular causes of changes in the levels of oxidative stress markers, i.e., changes in methylation and expression of genes encoding antioxidant defence enzymes [30–33].

In the presented study, the chronic mild stress was used as an animal depression model. Previous studies have shown that the CMS model is associated with the development of depression-like behaviour; one such form of behaviour is anhedonia, which is manifested in a reduction of 1% saccharose solution consumption [34–38]. Similarly, the present results indicate that the stressed rats demonstrated decreased consumption of sucrose solution and were hence deficient in sensitivity to rewards. On the other hand, five-week antidepressant therapy with escitalopram caused this effect to be normalised in the stressed rats. Unfortunately, there is no ideal animal model for depression and also the one used in this study has its limitations and disadvantages. The CMS model used in the presented research is based on the reflection of only one symptom of depression, anhedonia. Thus, the model is rated low in terms of the face validity whereas the criteria of the construct validity and predictive validity are highly rated. Moreover, in the case of animal models of depression, it should be remembered that some symptoms cannot be obtained in rats, e.g., suicidal thoughts. Due to the multifactorial nature of depression and the variety of psychological symptoms, each animal model used has some imperfections. In the future, the presented study results may be used to develop new animal models that better reflect depression [39].

As well as changes in sensitivity to reward, the application of CMS and escitalopram therapy resulted in changes in the expression and methylation status of promoter regions of genes involved in oxidative and nitrosative stress, as well as the TRYCATs pathway. Similarly, previous animal studies suggest that such disorders may contribute to the development of depression [40–43]. In addition, patients with depression also demonstrate exacerbation of the oxidative process and insufficient antioxidant response, as well as overproduction of neurotoxic tryptophan metabolites such as quinolinic acid and 3-hydroxykynurenine [44,45]. Interestingly, the metabolites of the TRYCATs pathway may also induce the generation of reactive oxygen species [9–16]. In addition, CMS and venlafaxine treatment, an antidepressant belonging to the serotonin–norepinephrine reuptake inhibitor group, has also been found to influence the expression and methylation of genes involved in the TRYCATs pathway, and on oxidative stress and nitrosative stress [43]. The present study continues this line of research on the impact of antidepressant treatment on changes at the molecular level.

The first key finding of the current study is that escitalopram treatment increased *Gpx1*, *Gpx4*, and *Nos2* mRNA expression in PBMCs. Enzymes encoded by *Gpx1* and *Gpx4* genes catalyse the reduction of hydrogen peroxide to water whereas *Nos1* and *Nos2* takes part in NO synthesis [46,47]. Thus, the results suggest that antidepressant treatment with escitalopram may be associated with the reduction of intensifying of the oxidative stress process or elevated expression of genes encoding antioxidant enzymes in PBMCs. In contrast, *Cat*, *Gpx1*, and *Nos1* mRNA expression was increased in the hippocampus of rats after CMS, and this effect was normalised by chronic administration of escitalopram. Moreover, the stressed animals were found to demonstrate decreased protein expression of Gpx4 in the cerebral cortex after escitalopram treatment. Similarly, stressed rats were previously found to display decreased *Gpx1* mRNA expression in the hippocampus after antidepressant therapy with venlafaxine [37]. In addition, superoxide generation was found to be increased in the cerebral cortex and hippocampus of stressed rats [48]. Such intensification of oxidative stress has been found to cause increased lipid peroxidation in the cerebral cortex and hippocampus and protein peroxidation in the cortex of stressed animals [49]. Therefore, the increased mRNA expression of *Gpx1* observed in the hippocampus of rats after CMS observed in the present study may be associated with an intensification of antioxidant defence in response to the development of oxidative stress, as indicated previously [42]. Elsewhere, *Nos1* and *Gpx1* mRNA expression in the midbrain and basal ganglia were found to be increased in stressed rats, while this effect was normalised by antidepressant therapy with venlafaxine [42].

Another animal study found that an oxidative imbalance causes an increase of ROS levels, resulting in disturbed *Tph1/2* mRNA expression and the reduction of serotonin synthesis [50]. Similarly, our study confirmed that CMS caused a decrease of *Tph2* mRNA expression in the amygdala. Tph converts tryptophan into 5-hydroxytryptophan and determines the concentration of serotonin [51,52]. Thus, increased expression of *Tph1* and *Tph2* can provide an adequate level of serotonin synthesis [53]. Our data confirms that antidepressant therapy with escitalopram caused an increase of *Tph1* mRNA expression in the cerebral cortex of the stressed rats, as well as *Tph2* mRNA in the amygdala. This is an important finding in the light of the "serotonin hypothesis" of clinical depression, first proposed 50 years ago. Previous studies have found an impairment of serotonin function to play a crucial role in the pathophysiology of depression.

Depression may also be associated with elevated levels of the neurotoxic metabolites of the TRYCATs pathway, such as quinolinic acid, 3-hydroxykynurenine, and 3-hydroxyanthranilic acid. The former is a toxic compound that may induce the generation of ROS oxidation [9–16]. 3-hydroxykynurenine is a product of the reaction catalysed by Kmo while 3-hydroxyanthranilic acid is formed in a reaction catalysed by Kynu [54]. Thus, depression may be associated with increased levels of Kmo and Kynu, which may act as target for antidepressant therapy [55]. Moreover, our present study confirmed that the mRNA expression of Kmo and Kynu is decreased, while the methylation status of the *Kmo* promoter region is increased, in the cerebral cortex of stressed rats after escitalopram treatment.

Regarding *Ido1*, the gene encoding the rate-limiting enzyme of tryptophan metabolism and catalysed the oxidation of L-tryptophan to N-formylkynurenine [56], increased activity has been observed in the plasma of patients with depression and rats with anhedonia [57]. Our present findings indicate that the stressed rats demonstrated increased methylation status in the *Ido1* promoter region after escitalopram therapy, which can limit *Ido1* mRNA expression.

4. Conclusions

Our findings confirm that the depression and antidepressant therapy may be associated with the disorders of the interrelated biochemical pathways, including oxidative and nitrosative stress, as well as the TRYCATs pathway. Disorders in the TRYCATs pathway may induce oxidative stress processes and vice versa. In addition, it was found that the changes in mRNA and protein expression and methylation status of promoter regions appear to be dependent on the type of the tissue or brain structure. Our observations demonstrate that analyses of mRNA and protein expression, and promoter methylation status, can shed light on the mechanisms of depression development and the action of drugs for antidepressant therapy. Unfortunately, our study has some limitations. First, our research focused only on the expression and methylation of selected genes related to oxidative stress. Therefore, it cannot be conclusively assessed if escitalopram has any effects on oxidative stress and damage. Further analyses should be performed to support this hypothesis. Secondly, it has to be stressed that the depression model used is based only on anhedonia, expressed by reducing sucrose intake. Therefore, studies with other animal models of depression should also be conducted in the future.

5. Materials and Methods

5.1. Animals

Male Wistar Han rats (Charles River, Lindau/Bodensee, Germany), weighing approximately 200–220 g, were individually housed under standard conditions, i.e., room temperature (22 °C) with twelve-hour cycles of day and night, with unlimited access to food and water. Each studied group consisted of six animals. All experimental procedures were carried out in accordance with the rules of the 86/609/EEC Directive and were approved by the Local Bioethics Commission of the Institute of Pharmacology, Polish Academy of Sciences in Krakow.

5.2. CMS

In the first stage, rats were trained to consume a 1% solution of sucrose for lasted 1 h following 14 h water and food deprivation. After the adaptation and training period, the animals were divided into two matched groups and the sucrose intake test was performed weekly until the end of the experiment. One group of rats was exposed to a CMS procedure for two or seven weeks. The stress stimuli consisted of two periods of food or water deprivation, two periods of low-intensity stroboscopic illumination (150 flashes/min), two periods of 45° cage tilt, one period of paired housing, two periods of soiled cage (250 mL water in sawdust bedding), two periods of intermittent illumination (lights on and off every two hours), and three periods of no stress. The stressors were used individually and continuously, day and night, for 10–14 h periods. Non-stressed animals had unlimited access to food and water and were individually housed in cages without contact with the stressed animals. After two weeks of the CMS procedure, the animals were divided into two groups: one group was sacrificed and decapitated and the other was further divided into matched subgroups. The latter received daily administration of either vehicle (1 mL/kg, IP) or escitalopram (10 mg/kg, IP) for five weeks. After this five-week period and tha last sucrose intake test, all animals were sacrificed and decapitated. The scheme of the CMS procedure was presented in the Figure 1A.

5.3. Drug

Escitalopram (Carbosynth Ltd., Compton, Berkshire, United Kingdom), obtained commercially, was dissolved in 0.9% sterile saline and then injected at a dose of 10 mg/kg, IP (1 mL/kg of body weight).

5.4. Collection of PBMCs and Brain Structure Specimens

Blood samples were taken into vacutainer tubes with anticoagulants. PBMCs were isolated using Gradisol L (Aqua-Med, Lodz, Poland) and centrifugation and $400 \times g$ for 30 min at 4 °C. After PBMC isolation, the cell pellets were stored at −80 °C until required. In the case of the brain structures, the samples were frozen in liquid nitrogen and stored at −80 °C until required. The frozen brain samples were then suspended in PBS and manually homogenised using a FastGene® Tissue Grinder (Nippon Genetics Europe, Düren, Germany). The samples of PBMC pellets and brain homogenates were used for later experiments, including RNA and DNA isolation.

5.5. Determination of mRNA Expression Level in PBMCs and Brain Structures

The total RNA samples were isolated from PBMCs pellets and frozen brain structures using commercial kits (GenElute Mammalian Total RNA Miniprep Kit, Sigma-Aldrich, ISOLATE II RNA/DNA/Protein Kit, Bioline, respectively) according to the manufacturer's protocol. The quantity and purity of RNA samples were confirmed spectrophotometrically using a Synergy HTX Multi-Mode Microplate Reader, equipped with a Take3 Micro-Volume Plate (BioTek Instruments, Inc., Winooski, VT, USA).

The RNA samples were diluted to 5 ng/μl and transcribed into cDNA using a High-Capacity cDNA Reverse Transcription Kit (Applied Biosystems, Foster City, CA, USA) according to the manufacturer's recommendations. mRNA expression was determined by real-time PCR using a TaqMan Universal Master Mix, no UNG and species-specific TaqMan Gene Expression Assay–assay ID: Rn00560930_m1 (*Cat*), Rn00577994_g1 (*Gpx1*), Rn00820818_g1 (*Gpx4*), Rn00583793_m1 (*Nos1*) and Rn00561646_m1 (*Nos2*) (Thermo Fisher Scientific, Waltham, Massachusetts, USA) according to the manufacturer's instructions. Real-time PCR runs were performed using a CFX96TM Real-Time PCR Detection System Thermal Cycler (Bio Rad Laboratories Inc., Hercules, CA, USA).

The relative levels of mRNA expression of all studied genes were estimated as fold = $2^{-\Delta Ct}$, where ΔCt sample = $Ct_{\text{target gene}} - Ct_{\text{reference gene}}$, with the *18S* (18S ribosomal RNA) gene being used as the reference gene. Additionally, the $2^{-\Delta\Delta Ct}$ method was also used to estimate the fold change in expression caused by antidepressant administration [58].

5.6. Determination of Methylation Status in PBMCs and Brain Structures

The DNA samples were extracted from PBMCs using QIAamp DNA Mini Kit (Qiagen, Hilden, Germany) and from brain structures using ISOLATE II RNA/DNA/Protein Kit (Bioline Ltd., London, UK) following the manufacturer's instructions. The quantity and purity of DNA samples were estimated spectrophotometrically using a Synergy HTX Multi-Mode Microplate Reader, equipped with a Take3 Micro-Volume Plate (BioTek Instruments, Inc., Winooski, VT, USA). The DNA was then modified by bisulfite using a CiTi Converter DNA Methylation Kit (A&A Biotechnology, Gdynia, Poland) as indicated by the manufacturer. The primers for promoter regions that included CpG islands were designed using MethPrimer (http://www.urogene.org/methprimer2/) according Wojdacz et al. [59].

PCR amplification and MS-HRM assay were performed on the Bio-Rad CFX96 Real-Time PCR Detection System (BioRad Laboratories Inc., Hercules, CA, USA) equipped with Bio-Rad Precision Melt Analysis Software (BioRad Laboratories Inc., Hercules, CA, USA) to analyse the methylation status of all studied genes. The PCR reactions contained $5 \times$ HOT FIREPol® EvaGreen® HRM Mix (no ROX) (Solis BioDyne Tartu, Estonia), 500 nM of forward and reverse primers, DNA samples (10 ng/μL) after bisulphite modification and PCR-grade water. The sequences of the primers and conditions of reaction are presented in Table 2. Finally, the methylation status of the studied samples was estimated based on HRM profiles obtained from the amplification of methylated template DNA (CpGenomeTM Rat Methylated Genomic DNA Standard, Merck Millipore Burlington, MA, USA) and unmethylated DNA (CpGenomeTM Rat Unmethylated Genomic DNA Standard, Merck Millipore Burlington, MA, USA). Therefore, serial dilutions of template DNA were prepared: 0%, 10%, 25%, 50%, 75%, and 100% methylated DNA.

Table 2. Characteristic of primer and conditions of the MS-HRM protocol.

Gene	Starter Sequence (5'->3')	Product Size [bp]	The Initial Activation of the Polymerase Temperature [°C]	Time [min]	Denaturation Temperature [°C]	Time [s]	Annealing Temperature [°C]	Time [s]	Elongation Temperature [°C]	Time [s]	The HRM Analysis
					Oxidative and Nitrosative Stresses						
Cat	F: TTTGAGATTATTGTGTTTGAAA R: TACCTACACCCAAAAAAAATA	148	95	12	95	15	59	20	72	20	Denaturation at 95 °C for 15 s, reannealing at 60 °C for 1 min and melting from 60 to 95 °C at a ramp rate of 0.2 °C
Gpx1	F: GTTGTTTTAGGTTTTGTTGTTG R: AAAACTAAAATCCTCCAACTCT	102					65				
Gpx4 (promotor 2)	F: AGGTTGGAGGTTTAGAGGTTTA R: TCCCCTAAATACAAAAATCTCT	118					59				
Gpx4 (promotor 3)	F: AGGTTGGAGGTTTAGAGGTTTA R: AAAACATAACAAAAATCATCTCCC	147					65				
Nos1 (promotor 5)	F: GGGTTTTTAATTTTTTATTGTG R: CAACCCTCATTAAAAAACC	124					59				
Nos1 (promotor 7)	F: GTTTGAGATTGGAATTTTTGG R: CCAAAACATCCAAAAATACACA	124					59				

Gene	Starter sequence (5'->3')	Product size [bp]	The initial activation of the polymerase Temperature [°C]	Time [min]	Denaturation Temperature [°C]	Time [s]	Annealing Temperature [°C]	Time [s]	Elongation Temperature [°C]	Time [s]	The HRM analysis
					Tryptophan catabolites pathway						
Tph1	F: GGGACTTTTGTTTTGGTTTTA R: TCCTCAACCACAAAAAATCTAA	132	95	12	95	15	55	20	72	20	Denaturation at 95 °C for 15 s, reannealing at 60 °C for 1 min and melting from 60 to 95 °C at a ramp rate of 0.2 °C
Ido1	F: TTTGAGTTTTAGTGATTTTGGG R: TTAATATCTAATCCCAATCTCTAAAAC	100					59				
Tdo2 (promoter 1)	F: GATGATTTAGGTGTTTGACGT R: CAAAAAAACAAAATTCATCCA	123					59				
Tdo2 (promoter 2)	F: ATGAATTAGGTCGTTTGACGTT R: ACCCAATCTACCTAACTAACAAC	187					61.4				
Kmo	F: TTGTTTAGGGAAGGAAATR: ATAAAAAACTAAACCCAAAACAC	150					55.7				

5.7. Determination of Protein Expression in the Tested Brain Structures

The frozen brain tissue samples were lysed and sonicated in RIPA buffer containing serine protease inhibitor (1mM phenylmethylsulfonyl fluoride), and then centrifuged for 5 min at 5000 rpm in 4 °C. The concentration of protein samples was determined by the Lowry procedure using a Synergy HTX Multi-Mode Microplate Reader, equipped with a Take3 Micro-Volume Plate (BioTek Instruments, Inc., Winooski, VT, USA). Following this, 25 µg of the protein samples were separated by 10% SDS polyacrylamide gels electrophoresis and transferred onto Immobilon-P membrane (Millipore, Bedford, MA, USA). After transfer, the membranes were blocked in 5% non-fat dry milk solution and incubated with primary antibodies. Next, horseradish peroxidase-conjugated secondary antibodies were used to detect primary antibodies. A fuller description of the antibodies and incubation conditions is presented in Table 3. Finally, peroxidase substrate solution (Thermo Fisher Scientific, Waltham, MA, USA) was used for X-ray film visualization by enhanced chemiluminescence. The density of the bands was analysed using Gel-Pro® Analyzer Software (Media Cybernetics Inc., Rockville, MD, USA) and normalized to β-actin levels.

Table 3. Characteristics of antibody used in Western Blot.

Protein	Primary Antibody				Secondary Antibody			
	The Origin of Antibodies	Producent	Dilution	Condition of Incubation [h]	The Origin of Antibodies	Producent	Dilution	Condition of Incubation [h]
Oxidative Stress								
β-actin (a reference protein)	Mouse	Santa Cruz Biotechnology Inc, Dallas, Texas, USA	1:1000	1 h at room temperature	Anti-mouse	Cell Signalling Technologies Inc., Danvers, Massachusetts, USA	1:6000	1 h at room temperature
catalase	Mouse	Santa Cruz Biotechnology Inc, Dallas, Texas, USA	1:1000	overnight at 4 °C	Anti-mouse	Cell Signalling Technologies Inc., Danvers, Massachusetts, USA	1:6000	1 h at room temperature
glutathione peroxidase 4	Rabbit	Abcam, Cambridge, United Kingdom	1:6000	overnight at 4 °C	Anti-rabbit	Cell Signalling Technologies Inc., Danvers, Massachusetts, USA	1:6000	1 h at room temperature
superoxide dismutase 1	Mouse	Santa Cruz Biotechnology Inc, Dallas, Texas, USA	1:6000	2 h at room temperature	Anti-mouse	Cell Signalling Technologies Inc., Danvers, Massachusetts, USA	1:6000	1 h at room temperature
Tryptophan catabolites pathway								
Tryptophan hydroxylase 1	Rabbit	Cell Signalling Technologies Inc., Danvers, Massachusetts, USA	1:1000	overnight at 4 °C	Anti-rabbit	Cell Signalling Technologies Inc., Danvers, Massachusetts, USA	1:6000	1 h at room temperature
Tryptophan hydroxylase 2	Rabbit	Cell Signalling Technologies Inc., Danvers, Massachusetts, USA	1:6000	overnight at 4 °C	Anti-rabbit	Cell Signalling Technologies Inc., Danvers, Massachusetts, USA	1:6000	1 h at room temperature
Indoleamine 2,3-dioxygenase	Mouse	Santa Cruz Biotechnology Inc, Dallas, Texas, USA	1:1000	overnight at 4 °C	Anti-rabbit	Cell Signalling Technologies Inc., Danvers, Massachusetts, USA	1:6000	1 h at room temperature
Kynurenine aminotransferases	Mouse	Santa Cruz Biotechnology Inc, Dallas, Texas, USA	1:1000	overnight at 4 °C	Anti-rabbit	Cell Signalling Technologies Inc., Danvers, Massachusetts, USA	1:6000	1 h at room temperature
Kynureninase	Mouse	Santa Cruz Biotechnology Inc, Dallas, Texas, USA	1:1000	overnight at 4 °C	Anti-rabbit	Cell Signalling Technologies Inc., Danvers, Massachusetts, USA	1:6000	1 h at room temperature

5.8. Statistical Analysis

All statistical analyses were performed using Statistica 12 (Statsoft, Tulsa, OK, USA) and SigmaPlot 11.0 (Systat Software Inc., San Jose, CA, USA). The data were expressed as the mean ± standard error of the mean. The statistical analysis was started with the Shapiro–Wilk test which was used to evaluate data normality. Then, the one-way analysis of variance (ANOVA) was used to detect significant differences between samples with normal distribution whereas differences between probes with non-normal distribution were confirmed by the Kruskal–Wallis test. Finally, the Tukey test was used as post-hoc test. A value of $p < 0.05$ was considered to be significant.

Supplementary Materials: Included the addition results about methylation and expression level: The following are available online at https://www.mdpi.com/1422-0067/22/1/10/s1, Figure S1: mRNA expression of Cat (A), Gpx1 (B), Gpx4 (C), Tph2 (D), Kmo (E) and Kynu (F) in PBMCs and in the brain structures of animals exposed to CMS for two weeks (control, stressed) and in animals exposed to CMS for seven weeks and administered vehicle (1 mL/kg) or escitalopram (10 mg/kg) for five weeks (control/esc, stressed/saline, stressed/esc). The effects are presented as fold change ($2^{-\Delta\Delta Ct}$ method; Schmittgen and Livak, 2008). Data represent means ± SEM. $n = 6$. *** $p < 0.001$ for differences between blood and all studied brain structures. Figure S2. The methylation level of Tph1 (A), Ido1 (B), Kmo (C), Cat (D), Gpx1 (E), Gpx4 (F) promoter regions, Nos1 promoter 3 region (G), Nos1 promoter 7 region (H) between brain structures and PBMCs of animals exposed to CMS for two weeks control, stressed) and in animals exposed to CMS for seven weeks and treated vehicle (1 mL/kg) or escitalopram (10 mg/kg) for five weeks (control/esc, stressed/saline, stressed/esc). Data represent as means ± SEM. $n = 6$. Figure S3. Protein expression of Cat (A), Nos1 (B), Tph1 (C), Tph2 (D), Ido1 (E) and Kynu (F) in brain structures of animals exposed to CMS for two weeks (control, stressed) and in animals exposed to CMS for seven weeks and administered vehicle (1 mL/kg) or escitalopram (10 mg/kg) for five weeks (control/esc, stressed/saline, stressed/esc). Levels of Cat (A), Nos1 (B), Tph1 (C), Tph2 (D), Ido1 (E) and Kynu (F) proteins measured in hippocampus, amygdala and cerebral cortex. Samples containing 25 µg of proteins were resolved by SDS-PAGE. The intensity of bands corresponding to Gpx4 was analysed by densitometry. Integrated optical density (IOD) was normalized by protein content and a reference sample (see the Methods section for details). The graphs show the mean IODs of the bands from all analysed samples. The IODgene/IODACTB method was used to estimate the relative protein expression levels in the analysed samples. Data represent means ± SEM. $n = 6$. No significant changes were found between any groups. Table S1. Methylation level of, Gpx1 promoter (A), Gpx4 promoter 3 (B), Nos1 promoter 3 (C), Nos1 promoter 7 (D), Tph1 promoter (E), Ido1 promoter (F) Kmo promoter (G) in the hippocampus, amygdala, cerebral cortex and PBMCs of animals exposed to CMS for two weeks (control, stressed) and in animals exposed to CMS for seven weeks and administered vehicle (1 mL/kg) or escitalopram (10 mg/kg) for five weeks (control/esc, stressed/saline, stressed/esc). Data represents means ± SEM. $n = 6$. No significant changes were found between any groups.

Author Contributions: Conceptualization, T.S. and M.P.; Methodology, P.W., E.S., P.J., P.C., P.G. and M.P.; Validation, P.W., E.S., P.J., P.G. and M.P.; Formal Analysis, P.W., E.S.; Investigation, P.W., E.S., P.J., P.C., K.B., M.B., P.G. and M.P.; Data Curation, T.S.; Writing—Original Draft Preparation, P.W.; Writing—Review & Editing, J.S., M.P., T.S.; Supervision, T.S.; Project Administration, T.S.; Funding Acquisition, T.S. All authors have read and agreed to the published version of the manuscript.

Funding: Source of support: This study was funded by the National Science Centre of Poland grant (2015/19/BNZ7/00410) and grant (No. 1.01.01.000.XX) from the University of Lodz (funding for young scientists).

Conflicts of Interest: The authors declare no conflict of interest. All experimental procedures were carried out in accordance with the rules of the 86/609/EEC Directive and were approved by the Local Bioethics Commission of the Institute of Pharmacology, Polish Academy of Sciences in Krakow.

References

1. Vos, T.; Allen, C.; Arora, M.; Barber, R.M.; Bhutta, Z.A.; Brown, A.; Carter, A.; Casey, D.C.; Charlson, F.J.; Chen, A.Z.; et al. Global, regional, and national incidence, prevalence, and years lived with disability for 310 diseases and injuries, 1990–2015: A systematic analysis for the Global Burden of Disease Study 2015. *Lancet* **2016**, *388*, 1545–1602.
2. Seedat, S.; Scott, K.M.; Angermeyer, M.C.; Berglund, P.; Bromet, E.J.; Brugha, T.S.; Demyttenaere, K.; De Girolamo, G.; Haro, J.M.; Jin, R.; et al. Cross-National Associations Between Gender and Mental Disorders in the World Health Organization World Mental Health Surveys. *Arch. Gen. Psychiatry* **2009**, *66*, 785–795. [CrossRef]
3. Marcus, M.; Yasamy, M.T.; Van Van Ommeren, M.; Chisholm, D.; Saxena, S. Depression: A Global Public Health Concern. *Social Psychiatry Psychiatr. Epidemiol.* **2012**, *1*, 6–8.
4. Depression. Available online: http://www.who.int/news-room/fact-sheets/detail/depression (accessed on 1 March 2020).
5. Di Luca, M.; Olesen, J. The Cost of Brain Diseases: A Burden or a Challenge? *Neuron* **2014**, *82*, 1205–1208. [CrossRef]
6. Al-Harbi, K.S. Treatment-resistant depression: Therapeutic trends, challenges, and future directions. *Patient Prefer. Adherence* **2012**, *6*, 369–388. [CrossRef]
7. Wigner, P.; Czarny, P.; Galecki, P.; Su, K.-P.; Śliwiński, T. The molecular aspects of oxidative & nitrosative stress and the tryptophan catabolites pathway (TRYCATs) as potential causes of depression. *Psychiatry Res.* **2018**, *262*, 566–574. [CrossRef] [PubMed]
8. Wigner, P.; Czarny, P.; Galecki, P.; Sliwinski, T. Oxidative and nitrosative stress as well as the tryptophan catabolites pathway in depressive disorders. *Psychiatr. Danub.* **2017**, *29*, 394–400. [CrossRef]
9. Dykens, J.A.; Sullivan, S.G.; Stern, A. Oxidative reactivity of the tryptophan metabolites 3-hydroxyanthranilate, cinnabarinate, quinolinate and picolinate. *Biochem. Pharmacol.* **1987**, *36*, 211–217. [CrossRef]
10. Ríos, C.; Santamaría, A. Quinolinic acid is a potent lipid peroxidant in rat brain homogenates. *Neurochem. Res.* **1991**, *16*, 1139–1143. [CrossRef] [PubMed]
11. Okuda, S.; Nishiyama, N.; Saito, H.; Katsuki, H. 3-Hydroxykynurenine, an Endogenous Oxidative Stress Generator, Causes Neuronal Cell Death with Apoptotic Features and Region Selectivity. *J. Neurochem.* **2002**, *70*, 299–307. [CrossRef] [PubMed]
12. Guidetti, P.; Schwarcz, R. 3-Hydroxykynurenine potentiates quinolinate but not NMDA toxicity in the rat striatum. *Eur. J. Neurosci.* **1999**, *11*, 3857–3863. [CrossRef] [PubMed]
13. Goldstein, L.E.; Leopold, M.C.; Huang, X.; Atwood, C.S.; Saunders, A.J.; Hartshorn, M.; Lim, J.T.; Faget, K.Y.; Muffat, J.A.; Scarpa, R.C.; et al. 3-Hydroxykynurenine and 3-Hydroxyanthranilic Acid Generate Hydrogen Peroxide and Promote α-Crystallin Cross-Linking by Metal Ion Reduction. *Biochemistry* **2000**, *39*, 7266–7275. [CrossRef] [PubMed]
14. Murakami, K.; Haneda, M.; Yoshino, M. Prooxidant action of xanthurenic acid and quinoline compounds: Role of transition metals in the generation of reactive oxygen species and enhanced formation of 8-hydroxy-2′-deoxyguanosine in DNA†. *BioMetals* **2006**, *19*, 429–435. [CrossRef] [PubMed]
15. Santamaría, A.; Galván-Arzate, S.; Lisý, V.; Ali, S.F.; Duhart, H.M.; Osorio-Rico, L.; Rıos, C.; Sut'Astný, F. Quinolinic acid induces oxidative stress in rat brain synaptosomes. *NeuroReport* **2001**, *12*, 871–874. [CrossRef] [PubMed]
16. Smith, A.J.; Smith, R.A.; Stone, T.W. 5-Hydroxyanthranilic Acid, a Tryptophan Metabolite, Generates Oxidative Stress and Neuronal Death via p38 Activation in Cultured Cerebellar Granule Neurones. *Neurotox. Res.* **2009**, *15*, 303–310. [CrossRef] [PubMed]
17. Pandya, C.D.; Howell, K.R.; Pillai, A. Antioxidants as potential therapeutics for neuropsychiatric disorders. *Prog. Neuro Psychopharmacol. Biol. Psychiatry* **2013**, *46*, 214–223. [CrossRef]
18. Bilici, M.; Efe, H.; Köroğlu, M.; Uydu, H.A.; Bekaroğlu, M.; Değer, O. Antioxidative enzyme activities and lipid peroxidation in major depression: Alterations by antidepressant treatments. *J. Affect. Disord.* **2001**, *64*, 43–51. [CrossRef]
19. Herken, H.; Gurel, A.; Selek, S.; Armutcu, F.; Ozen, M.E.; Bulut, M.; Kap, O.; Yumru, M.; Savas, H.A.; Akyol, O. Adenosine Deaminase, Nitric Oxide, Superoxide Dismutase, and Xanthine Oxidase in Patients with Major Depression: Impact of Antidepressant Treatment. *Arch. Med Res.* **2007**, *38*, 247–252. [CrossRef]
20. Galecki, P.; Szemraj, J.; Bienkiewicz, M.; Florkowski, A.; Galecka, E. Lipid peroxidation andantioxidant protection in patients during acute depressive episodes and in remission after fluoxetine treatment. *Pharmacol. Rep.* **2009**, *61*, 436–447. [CrossRef]
21. Kodydková, J.; Vávrová, L.; Zeman, M.; Jirák, R.; Macášek, J.; Staňková, B.; Tvrzická, E.; Zák, A. Antioxidative enzymes and increased oxidative stress in depressive women. *Clin. Biochem.* **2009**, *42*, 1368–1374. [CrossRef]
22. Kotan, V.O.; Sarandol, E.; Kirhan, E.; Ozkaya, G.; Kirli, S. Effects of long-term antidepressant treatment on oxidative status in major depressive disorder: A 24-week follow-up study. *Prog. Neuro Psychopharmacol. Biol. Psychiatry* **2011**, *35*, 1284–1290. [CrossRef] [PubMed]
23. Hastings, R.S.; Parsey, R.V.A.; Oquendo, M.; Arango, V.; Mann, J.J. Volumetric Analysis of the Prefrontal Cortex, Amygdala, and Hippocampus in Major Depression. *Neuropsychopharmacology* **2003**, *29*, 952–959. [CrossRef] [PubMed]
24. Jain, S. Faculty Opinions recommendation of Depression-related variation in brain morphology over 3 years: Effects of stress? *Fac. Opin. Post Publ. Peer Rev. Biomed. Lit.* **2008**, *65*, 1156–1165. [CrossRef]
25. Campbell, S.; MacQueen, G. The role of the hippocampus in the pathophysiology of major depression. *J. Psychiatry Neurosci.* **2004**, *29*, 417–426.
26. Pulschen, D.; Thome, J. The Role of Oxidative Stress in Depressive Disorders. *Curr. Pharm. Des.* **2012**, *18*, 5890–5899. [CrossRef]
27. Shalaby, A.; Kamal, S. Effect of Escitalopram on GABA level and anti-oxidant markers in prefrontal cortex and nucleus accumbens of chronic mild stress-exposed albino rats. *Int. J. Physiol. Pathophysiol. Pharmacol.* **2009**, *1*, 154–161.

28. Sarandol, A.; Sarandol, E.; Eker, S.S.; Erdinc, S.; Vatansever, E.; Kirli, S. Major depressive disorder is accompanied with oxidative stress: Short-term antidepressant treatment does not alter oxidative–antioxidative systems. *Hum. Psychopharmacol. Clin. Exp.* **2007**, *22*, 67–73. [CrossRef]
29. Cimen, A.P.B.; Gumus, C.B.; Cetin, A.P.I.; Ozsoy, A.P.S.; Aydin, M.; Cimen, L. The Effects of Escitalopram Treatment on Oxidative/Antioxidative Parameters in Patients with Depression. *Klinik Psikofarmakol. Bülteni Bull. Clin. Psychopharmacol.* **2015**, *25*, 272–279. [CrossRef]
30. Duda, W.; Curzytek, K.; Kubera, M.; Iciek, M.; Kowalczyk-Pachel, D.; Bilska-Wilkosz, A.; Lorenc-Koci, E.; Leskiewicz, M.; Basta-Kaim, A.; Budziszewska, B.; et al. The Effect of Chronic Mild Stress and Imipramine on the Markers of Oxidative Stress and Antioxidant System in Rat Liver. *Neurotox. Res.* **2016**, *30*, 173–184. [CrossRef]
31. Liu, T.; Zhong, S.; Liao, X.; Chen, J.; He, T.; Lai, S.; Jia, Y. A Meta-Analysis of Oxidative Stress Markers in Depression. *PLoS ONE* **2015**, *10*, e0138904. [CrossRef]
32. Ahmadimanesh, M.; Abbaszadegan, M.R.; Rad, D.M.; Moallem, S.M.H.; Mohammadpour, A.H.; Ghahremani, M.H.; Hosseini, F.F.; Behdani, F.; Manteghi, A.A.; Jowsey, P.; et al. Effects of selective serotonin reuptake inhibitors on DNA damage in patients with depression. *J. Psychopharmacol.* **2019**, *33*, 1364–1376. [CrossRef] [PubMed]
33. Tohid, H.; Aleem, D.; Jackson, C. Major Depression and Psoriasis: A Psychodermatological Phenomenon. *Ski. Pharmacol. Physiol.* **2016**, *29*, 220–230. [CrossRef] [PubMed]
34. Gamaro, G.D.; Streck, E.L.; Matté, C.; Prediger, M.E.; Wyse, A.T.S.; Dalmaz, C. Reduction of hippocampal Na+, K+-ATPase activity in rats subjected to an experimental model of depression. *Neurochem. Res.* **2003**, *28*, 1339–1344. [CrossRef] [PubMed]
35. Bekris, S.; Antoniou, K.; Daskas, S.; Papadopoulou-Daifoti, Z. Behavioural and neurochemical effects induced by chronic mild stress applied to two different rat strains. *Behav. Brain Res.* **2005**, *161*, 45–59. [CrossRef]
36. Papp, M. Models of Affective Illness: Chronic Mild Stress in the Rat. *Curr. Protoc. Pharmacol.* **2012**, *57*, 5–9. [CrossRef]
37. Papp, M.; Gruca, P.; Lason-Tyburkiewicz, M.; Litwa, E.; Niemczyk, M.; Tota-Glowczyk, K.; Willner, P. Dopaminergic mechanisms in memory consolidation and antidepressant reversal of a chronic mild stress-induced cognitive impairment'. *Psychopharmacology* **2017**, *234*, 2571–2585. [CrossRef]
38. Papp, M.; Gruca, P.; Lason, M.; Niemczyk, M.; Willner, P. The role of prefrontal cortex dopamine D2 and D3 receptors in the mechanism of action of venlafaxine and deep brain stimulation in animal models of treatment-responsive and treatment-resistant depression. *J. Psychopharmacol.* **2019**, *33*, 748–756. [CrossRef]
39. Wang, Q.; Timberlake, M.A., 2nd; Prall, K.; Dwivedi, Y. The recent progress in animal models of depression. *Prog. Neuropsychopharmacol. Biol. Psychiatry* **2017**, *77*, 99–109. [CrossRef]
40. Tanke, M.A.C.; Alserda, E.; Doornbos, B.; Van Der Most, P.J.; Goeman, K.; Postema, F.; Korf, J. Low tryptophan diet increases stress-sensitivity, but does not affect habituation in rats. *Neurochem. Int.* **2008**, *52*, 272–281. [CrossRef]
41. Jacobsen, J.P.R.; Siesser, W.B.; Sachs, B.D.; Peterson, S.; Cools, M.J.; Setola, V.; Folgering, J.H.A.; Flik, G.; Caron, M.G. Deficient serotonin neurotransmission and depression-like serotonin biomarker alterations in tryptophan hydroxylase 2 (Tph2) loss-of-function mice. *Mol. Psychiatry* **2011**, *17*, 694–704. [CrossRef]
42. Wigner, P.; Synowiec, E.; Czarny, P.; Bijak, M.; Jóźwiak, P.; Szemraj, J.; Gruca, P.; Papp, M.; Śliwiński, T. Effects of venlafaxine on the expression level and methylation status of genes involved in oxidative stress in rats exposed to a chronic mild stress. *J. Cell. Mol. Med.* **2020**, *24*, 5675–5694. [CrossRef] [PubMed]
43. Wigner, P.; Synowiec, E.; Jóźwiak, P.; Czarny, P.; Bijak, M.; Białek, K.; Szemraj, J.; Gruca, P.; Papp, M.; Śliwiński, T. The Effect of Chronic Mild Stress and Venlafaxine on the Expression and Methylation Levels of Genes Involved in the Tryptophan Catabolites Pathway in the Blood and Brain Structures of Rats. *J. Mol. Neurosci.* **2020**, *70*, 1425–1436. [CrossRef] [PubMed]
44. Moylan, S.; Berk, M.; Dean, O.M.; Samuni, Y.; Williams, L.J.; O'Neil, A.; Hayley, A.C.; Pasco, J.A.; Anderson, G.M.; Jacka, F.N.; et al. Oxidative & nitrosative stress in depression: Why so much stress? *Neurosci. Biobehav. Rev.* **2014**, *45*, 46–62. [CrossRef] [PubMed]
45. Myint, A.-M. Kynurenines: From the perspective of major psychiatric disorders. *FEBS J.* **2012**, *279*, 1375–1385. [CrossRef] [PubMed]
46. Lubos, E.; Loscalzo, J.; Handy, D.E. Glutathione Peroxidase-1 in Health and Disease: From Molecular Mechanisms to Therapeutic Opportunities. *Antioxid. Redox Signal.* **2011**, *15*, 1957–1997. [CrossRef]
47. Knowles, R.G.; Moncada, S. Nitric oxide synthases in mammals. *Biochem. J.* **1994**, *298*, 249–258. [CrossRef]
48. Lucca, G.; Comim, C.M.; Valvassori, S.S.; Réus, G.Z.; Vuolo, F.; Petronilho, F.; Gavioli, E.C.; Dal-Pizzol, F.; Quevedo, J. Increased oxidative stress in submitochondrial particles into the brain of rats submitted to the chronic mild stress paradigm. *J. Psychiatr. Res.* **2009**, *43*, 864–869. [CrossRef]
49. Liu, J.; Wang, X.; Shigenaga, M.K.; Yeo, H.C.; Mori, A.; Ames, B.N. Immobilization stress causes oxidative damage to lipid, protein, and DNA in the brain of rats. *FASEB J.* **1996**, *10*, 1532–1538. [CrossRef]
50. Qiu, H.-M.; Yang, J.-X.; Jiang, X.-H.; Hu, X.-Y.; Liu, D.; Zhou, Q.-X. Enhancing tyrosine hydroxylase and tryptophan hydroxylase expression and improving oxidative stress involved in the antidepressant effect of sodium valproate on rats undergoing chronic unpredicted stress. *NeuroReport* **2015**, *26*, 1145–1150. [CrossRef]
51. Muller, C.P.; Jacobs, B.L. *Handbook of the Behavioral Neurobiology of Serotonin*; Academic Press: London, UK, 2010.
52. Nakamura, K.; Hasegawa, H. Developmental role of tryptophan hydroxylase in the nervous system. *Mol. Neurobiol.* **2007**, *35*, 45–53. [CrossRef]

53. Cowen, P.J.; Browning, M. What has serotonin to do with depression? *World Psychiatry* **2015**, *14*, 158–160. [CrossRef] [PubMed]
54. Phillips, R.S.; Iradukunda, E.C.; Hughes, T.; Bowen, J.P. Modulation of Enzyme Activity in the Kynurenine Pathway by Kynurenine Monooxygenase Inhibition. *Front. Mol. Biosci.* **2019**, *6*, 3. [CrossRef] [PubMed]
55. Laumet, G.; Zhou, W.; Dantzer, R.; Edralin, J.D.; Huo, X.; Budac, D.P.; O'Connor, J.C.; Lee, A.W.; Heijnen, C.J.; Kavelaars, A. Upregulation of neuronal kynurenine 3-monooxygenase mediates depression-like behavior in a mouse model of neuropathic pain. *Brain Behav. Immun.* **2017**, *66*, 94–102. [CrossRef] [PubMed]
56. Prendergast, G.C.; Emetz, R.; Muller, A.J.; Merlo, L.M.F.; Emandik-Nayak, L. IDO2 in Immunomodulation and Autoimmune Disease. *Front. Immunol.* **2014**, *5*, 585. [CrossRef]
57. Kim, H.; Chen, L.; Lim, G.; Sung, B.; Wang, S.; McCabe, M.F.; Rusanescu, G.; Yang, L.; Tian, Y.; Mao, J. Brain indoleamine 2,3-dioxygenase contributes to the comorbidity of pain and depression. *J. Clin. Investig.* **2012**, *122*, 2940–2954. [CrossRef]
58. Schmittgen, T.D.; Livak, K.J. Analyzing real-time PCR data by the comparative C(T) method. *Nat. Protoc.* **2008**, *3*, 1101–8110. [CrossRef] [PubMed]
59. Wojdacz, T.K.; Borgbo, T.; Hansen, L.L. Primer design versus PCR bias in methylation independent PCR amplifications. *Epigenetics* **2009**, *4*, 231–234. [CrossRef] [PubMed]

Review

The Uniqueness of Tryptophan in Biology: Properties, Metabolism, Interactions and Localization in Proteins

Sailen Barik

3780 Pelham Drive, Mobile, AL 36619, USA; barikfamily@gmail.com

Received: 2 November 2020; Accepted: 17 November 2020; Published: 20 November 2020

Abstract: Tryptophan (Trp) holds a unique place in biology for a multitude of reasons. It is the largest of all twenty amino acids in the translational toolbox. Its side chain is indole, which is aromatic with a binuclear ring structure, whereas those of Phe, Tyr, and His are single-ring aromatics. In part due to these elaborate structural features, the biosynthetic pathway of Trp is the most complex and the most energy-consuming among all amino acids. Essential in the animal diet, Trp is also the least abundant amino acid in the cell, and one of the rarest in the proteome. In most eukaryotes, Trp is the only amino acid besides Met, which is coded for by a single codon, namely UGG. Due to the large and hydrophobic π-electron surface area, its aromatic side chain interacts with multiple other side chains in the protein, befitting its strategic locations in the protein structure. Finally, several Trp derivatives, namely tryptophylquinone, oxitriptan, serotonin, melatonin, and tryptophol, have specialized functions. Overall, Trp is a scarce and precious amino acid in the cell, such that nature uses it parsimoniously, for multiple but selective functions. Here, the various aspects of the uniqueness of Trp are presented in molecular terms.

Keywords: tryptophan; indole; virus; immunity; serotonin; kynurenine; codon

---

1. Introduction

Tryptophan (Trp, W) is one of three aromatic amino acids that minimally contain a six-membered benzene ring in their side chains, the other two being phenylalanine (Phe, F) and tyrosine (Tyr, Y). Whereas Tyr is simply a p-hydroxy derivative of Phe, the side chain of Trp is indole, which is more complex, as it is a six-membered benzene ring fused to a five-membered pyrrole ring with an integrated NH group. Trp can also be viewed as a derivative of alanine (A), having an indole substituent on the β carbon (Table 1). The indole ring of Trp absorbs strongly in the near-ultraviolet wavelength of the spectrum, with an absorption maximum at 280 nm, which forms the basis of measuring A280 as a characteristic assay for proteins, distinguishable from nucleic acids that have an absorption maxima at 260 nm.

As presented later, the complex and nitrogenous aromatic side chain of Trp necessitates a large number of biosynthetic reactions, making Trp the most energetically expensive amino acid to synthesize. Discovered by Sir F.G. Hopkins in 1901 in milk casein hydrolysate, Trp was found to be an essential amino acid in experiments with mouse diet. Over the years, interest in Trp and its nutritional role in the mammalian diet has received enormous attention, much of which can be found in recent reviews [1–6]. Here, I have taken a complimentary approach, critically analyzing the major roles of Trp in cellular functions and in intermediary metabolism, such as translation, protein structure, adduct formation, the generation of important regulators, and most recently, RNA virus regulation.

Table 1. Codons that differ from the Trp codon in the third (wobble) base.

Codon	Amino Acid	Structure of Side Chain (-R)	Distinguishing Property
UGG	Trp	—CH$_2$—(indole)	Aromatic, hydrophobic
GGG	Gly	None	Smallest, flexible, no side chain
UCG	Ser	—CH$_2$—OH	Hydroxyl, hydrophilic
UGC, UGU	Cys	—CH$_2$—SH	Thiol, hydrophilic (Ser homolog)
UUG	Leu	—CH$_2$—CH(CH$_3$)$_2$	Large, hydrophobic
AGG, CGG	Arg	—CH$_2$—CH$_2$—CH$_2$—NH—C(NH)(NH$_2$)	Basic, charged, hydrophilic

2. Trp Codon, UGG

The standard genetic codon table assigns a total of 64 trinucleotide codons to the 20 amino acids used in translation. While a majority of amino acids are coded for by a set of four synonymous codons, some (e.g., Ser, Leu, Arg) are encoded by six codons, while several (e.g., His, Tyr, Gln, Asn, Lys, Asp, Glu, Cys) are encoded by two codons. The synonymous codons are thus functionally 'redundant', as they code for the same amino acid. Three nonsense codons, namely UAA, UAG, and UGA, do not code for any amino acid, but rather promote translational stop, and therefore, in this function, they are redundant as well. Methionine is an exclusive amino acid, since it is the nearly universal starting residue in translation, and is encoded by a single codon, AUG, both for translation initiation and for incorporation at internal sites during elongation. It is, therefore, intriguing that the only other amino acid, and the only internal amino acid, encoded by a single nonredundant codon, is Trp, encoded by UGG.

The mechanism and implications of the redundancy of codons have been a matter of intense speculation and research since the discovery of the complete genetic code [7–9]. The redundancy, which generally occurs at the third position of the codon, is possible due to 'wobble' base-pairing, as proposed by Crick in a set of rules, known as the 'wobble hypothesis' [10]. The central tenet of the hypothesis is that while the first and second positions of the codon obey the classic Watson–Crick base-pairing (A:U, G:C) in codon–tRNA (codon–anticodon) recognition during translation, the third position is relatively tolerant to nucleotide mismatch. This is largely due to the structural flexibility or 'wobble' of the tRNA molecule [10–12]. It is now believed that redundancy serves as a built-in genetic safety mechanism such that errors in the third position of the codon will be relatively silent, since it will still code for the same amino acid, thus maintaining the wild-type polypeptide sequence. Even outside of the synonymous codons, single nucleotide changes in many codons may lead to conservative replacements with similar amino acids; for example, changing UCG to ACG changes Ser to Thr, hydroxy amino acids that are often functionally similar (e.g., phosphorylated/dephosphorylated by the same Ser/Thr protein kinase/phosphatase). Similarly, GAU and GAA, differing in the third position, code for the acidic amino acids, Asp and Glu, respectively, and CUU and AUU code for Leu and Ile, which are structural isomers.

The nonredundant Trp codon does not enjoy this benefit, and in fact, single nucleotide substitutions at the third position—or at any other position of Trp codon—result in amino acids of very different physiochemical and functional properties, all of which, unlike Trp, are aliphatic (Table 1). Two changes (UGA, UAG) result in stop codons. Clearly, mutations in the Trp codon are highly likely to be deleterious, which reinforces the aforesaid suggestion that Trp is used judiciously in the protein, only where it is absolutely needed for its distinctive properties, which are detailed later.

It is tempting to speculate that the rarity of the Trp codon, combined with the low intracellular concentration of Trp, will promote ribosome pausing at the Trp codons during translation, allowing the proper folding of the nascent polypeptide [13–15]. While this is entirely possible, it would be a premature assumption, since many other factors influence translational dynamics, such as tRNA concentration, codon usage, and RNA secondary structure, which remain to be fully studied for Trp.

3. Trp Biosynthesis: Salient and Unique Features

With a molecular formula of $C_{11}H_{12}N_2O_2$ and molecular weight of 204.22 g/mol, Trp is the largest proteinogenic amino acid in the cell. As indicated earlier, Trp is not synthesized from simple molecules in animals, including humans, and is thus essential in their diet. Essential amino acids, in general, are synthesized by plants and microorganisms, and when present in the animal diet, are derived mainly from plants. The biosynthesis of all three amino acids, viz. Phe, Tyr and Trp, initially follows a common pathway that generates chorismate, known as the shikimate pathway [16,17]. From this point, the Trp biosynthetic pathway veers away from those of Phe and Tyr, in part due to the need to construct the indole ring of Trp. Chorismate is converted into prephonate, which then bifurcates into the Phe or the Tyr synthetic pathways; in contrast, chorismate is converted to anthranilate in the Trp synthesis pathway, which is not a shared precursor for Phe or Tyr synthesis. Subsequently, four additional steps are required to produce indole. The last step of Trp biosynthesis involves the condensation of indole with serine, the smallest hydroxy amino acid, which provides the features common to all amino acids, namely the alpha-carbon and amino and carboxylic groups. This step is catalyzed by tryptophan synthase (or synthetase), a complex multisubunit enzyme that occurs in microorganisms and plants but is absent in animals.

Trp synthase is a classic example of "substrate channeling" in which the indole is held in position within the α subunit until the Ser in the β subunit is converted to the highly reactive aminoacylate, which is followed by a fast reaction between the two [18]. Importantly, the coordinated channeling of the reactants prevents the release of indole from the enzyme core; without channeling, the indole could be released, and due to its hydrophobic nature, could traverse the cell membrane and exit out of the cell, thus abrogating Trp synthesis at the very last step.

The large number of reactions in the Trp biosynthetic pathway comes with a high energy bill. This was recognized early on in a pioneering study that calculated the energy cost of the twenty amino acids by adding the number of high-energy phosphate bonds (~P) required to generate the respective precursors of each pathway [19]. The bar graph, generated from the data (Figure 1), clearly reveals that the three aromatic amino acids are the most expensive to synthesize, and Trp, which requires an equivalent of 74~P bonds, tops the list by a sizeable margin over Phe (52~P) and Tyr (50~P). It has been postulated that the high energy cost is a major reason that the animals obtain the expensive amino acids from the diet instead of synthesizing them.

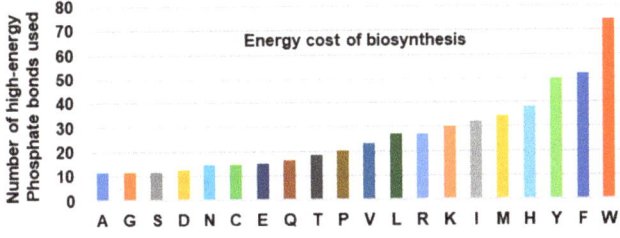

Figure 1. Total energy cost of amino acid biosynthesis. The total number of high-energy phosphates, equivalent to ATP and GTP, required for the biosynthesis of each amino acid, was plotted; the numbers were obtained from [19]. Note that the aromatic amino acids (Tyr, Phe, Trp) are energetically more expensive than the others, Trp being the highest of all.

4. Metabolism of Trp

4.1. Trp Degradation Pathways and the Gateway Enzymes

The major pathways of Trp catabolism and conversion to the key secondary metabolites are summarized here in a simplified diagram (Figure 2).

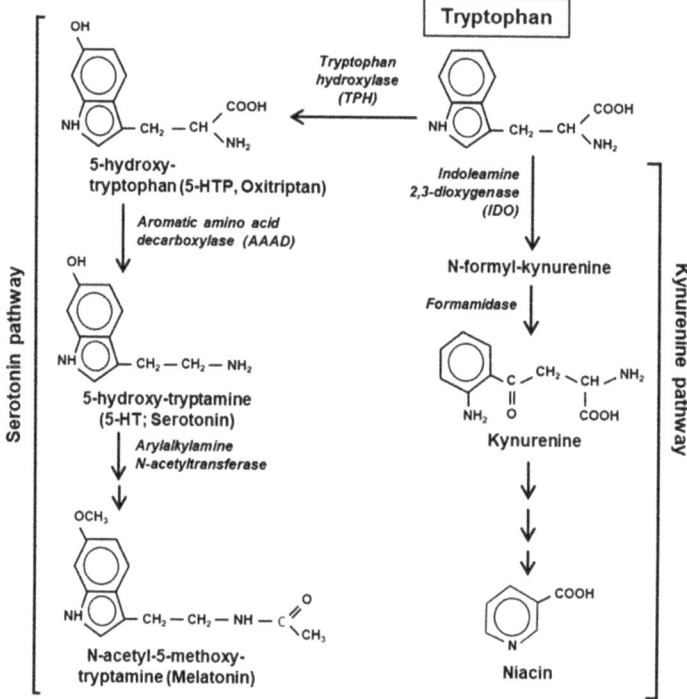

Figure 2. Two major pathways of Trp degradation. The serotonin and kynurenine pathways with their enzymes and products are shown. Note that the indole ring of Trp is destroyed at the very first step of the kynurenine but persists through the serotonin pathway.

Tryptophan is degraded via two parallel pathways, which can be named by their respective products or intermediates, viz. the serotonin pathway and the kynurenine pathway. These two pathways are non-overlapping and mutually exclusive, but differ in activity, the kynurenine pathway being substantially more active in mammalian cells. The two pathways compete with each other for the available pool of free Trp, not used in translation [20]. Fortunately, due to the low frequency of Trp residues in the proteins, nearly 99% of the total cellular Trp is available for non-translational use (Section 7). The serotonin pathway is initiated by tryptophan hydroxylase (TPH) that adds a hydroxy group to the 5 position of Trp to generate 5-HTP, also called oxitriptan. The kynurenine pathway is initiated by either of the two heme-containing oxidoreductases, indoleamine 2,3-dioxygenase (IDO) or the highly related tryptophan 2,3-dioxygenase (TDO). These two first steps are also the slowest—and hence rate-limiting—steps of the respective pathways, acting as gateways of Trp recruitment. An important distinction between the two pathways, which was noted earlier [21], is that the kynurenine pathway destroys the aromatic indole core of Trp, while the serotonin pathway retains the aromatic structure through all the compounds of the pathway (Figure 2).

Mammals possess two isoforms of TPH, viz. TPH1 and TPH2, which are ~70% identical in sequence, differing mainly in the regulatory domains, and are expressed in a tissue-specific manner [22,23].

While TPH1 predominates in peripheral tissues that express serotonin (a neurotransmitter; see later), such as the gastric system and skin, TPH2 is mostly expressed in neuronal cell types such as the central nervous system (CNS), specifically the brain. The difference in the regulatory domains likely allows them to have tissue-specific regulations. TPH is in fact a member of the amino acid hydroxylase superfamily that also comprises phenylalanine hydroxylase (PAH) and tyrosine hydroxylase (TH), all of which possess similar active sites and uses the same cofactors, and thus there is substantial overlap in their substrates along with preference. TPH, for example, hydroxylates both Trp and Phe with comparable kinetics; however, it hydroxylates Tyr at a ~5000-fold slower rate [23]. The full implication and molecular mechanism of the substrate overlap may shed important light on the distribution and evolution of these enzymes, which have remained unresolved.

Indoleamine 2,3-dioxygenase (IDO or IDO1), the first enzyme of the kynurenine pathway (Figure 2), serves as an important immunoregulatory checkpoint [24,25]. A closely related isoform (43% similar), referred as IDO2 [26], likely resulted from gene duplication, but is expressed in a very limited number of tissues, mainly liver, kidney and antigen-presenting cells in small amounts, and is also enzymatically much less active on Trp (~340-fold higher Km). It appears to have an accessory role in IDO1-mediated immune regulation and in inflammation [27,28]. In most literature, and in this review, the term IDO is to be considered synonymous to IDO1.

4.2. Secondary Metabolites of Trp

Because of the limiting concentration of intracellular Trp, it is reasonable to assume that the Trp degradation pathways indirectly regulate Trp levels by diverting some of the Trp to secondary metabolites. Thus, as in translation, Trp is frugally used in degradation, only to generate metabolites that are physiologically important, which will be briefly discussed here. Several metabolic products of Trp have received significant attention, the notable ones being 5-HTP (oxitriptan), serotonin (a neurotransmitter), melatonin, kynurenine, niacin (vitamin B3), and tryptophol. While melatonin and niacin are end-products of the two parallel pathways described above, 5-HTP, serotonin and kynurenine are intermediate metabolites (Figure 2). The dietary and pharmacological aspects of these chemicals have been extensively researched and reviewed, and therefore, only their molecular and regulatory roles and their relationship to Trp will be summarized, with emphasis on the underlying mechanisms, where available.

4.2.1. Metabolites of the Serotonin Pathway

Often branded as 'oxitriptan', 5-hydroxytryptophan (5-HTP) is a naturally occurring non-proteinogenic amino acid. As an immediate precursor of serotonin, a well-known monoamine neurotransmitter, 5-HTP is sold around the globe under many brand names as an over-the-counter (OTC) sleeping aid [29] as well as a suppressor of depression and appetite. Many clinical users prefer 5-HTP as it crosses the blood–brain barrier, whereas serotonin does not.

Serotonin (5-hydroxytrypatime) exhibits a multiplicity of complex physiological and clinical effects in diverse tissues, modulating mood, cognition, learning, and vasoconstriction [30,31]. Most of the regulatory roles of serotonin are triggered by binding to a large family of cell-surface receptors, known as serotonin receptors or 5-HT receptors [32,33]. However, serotonin can also cause the receptor-independent regulation of proteins through the post-translational addition of the serotonin moiety, in a process called 'serotonylation', in which the transglutaminase enzyme creates a glutamyl-amide bond between the primary amine group of serotonin and the carboxyl group of a glutamine residue in the acceptor protein [34–36]. The expanding list of serotonylated proteins regulate a variety of processes, such as thrombocyte production, vascular smooth muscle contraction, pulmonary hypertension, and release of insulin from pancreatic β cells [36].

In the human body, nearly 90% of serotonin is located in the enterochromaffin cells in the gastrointestinal tract, where it regulates intestinal movements, essential for the proper travel of food [30,32]. The 1–2% that is found in the CNS participates in the control of mood, sleep and hunger. Serotonin also regulates several cognitive functions, such as memory and learning. Drugs that alter serotonin levels of the body are used to treat a variety of psychiatric disorders, such as depression, anxiety, poor memory, migraine, nausea, and phobia, as well as sometimes obesity and Parkinson's disease [33]. Several serotonergic psychedelic drugs—naturally occurring as well as synthetic controlled substances—are agonists of 5-HT receptors [31].

Nonetheless, the pharmacological use and benefit of both 5-HTP and 5-HT (serotonin) have been matters of controversy [30,37,38], in part because many clinical studies of either metabolite lacked proper controls and were considered inconclusive in meta-analyses. Both exhibit a plethora of adverse effects, particularly in higher doses, including cardiovascular problems, upset stomach, headache, agitation, panic attacks, fatigue, sexual dysfunction, and suicidality. Not unexpectedly, similar side effects and a lack of scientific evidence also apply to the supplemental use of tryptophan, the parent compound.

Melatonin, a natural hormone, is the end-product of the serotonin pathway, and therefore, all the upstream compounds, from Trp to serotonin, can serve as its precursor. Melatonin helps to maintain the circadian clock of the body, particularly the wake–sleep cycle, comprising of ~16 h of daytime activity and ~8 h of nightly sleep [39]. It is generally safe to use and is used to treat insomnia, jet lag and various sleep disorders, thrombocytopenia (chemo-induced), 'winter blues' and seasonal affective disorder (SAD), and tardive dyskinesia [40–43]. Melatonin also has some immune-regulatory and anticancer effects, but these effects need further studies and validation [44,45]. Melatonin is also produced synthetically and is freely available as an OTC dietary supplement.

4.2.2. Metabolites of the Kynurenine Pathway

The kynurenine pathway (Figure 2) is responsible for processing >90% of the Trp in humans, producing kynurenine and niacin as the major metabolites. The build-up of these and other metabolites of this pathway can lead to multiple pathological conditions, such as AIDS-related dementia, multiple sclerosis, and ischemic brain injury [46–50], although the molecular mechanisms are not fully clear.

Niacin, synonymously called nicotinic acid, is the end-product of the pathway (Figure 2), and is a form of vitamin B3, essential in human health. Niacin occurs naturally in a variety of foods, such as meat, fish, and nuts, and is used to treat pellagra, a niacin-deficiency disease that shares some symptoms with Trp-deficiency. It is commonly seen among the poor and the malnourished in sub-Saharan Africa, Indonesia and China, as also in rural South America, where the staple food is maize, which is low in Trp and niacin. Interestingly, niacin is available in two forms, nicotinic acid (or niacin), and nicotinamide (or niacinamide), both have vitamin function, in which they act as precursors of the coenzymes nicotinamide adenine dinucleotide and nicotinamide adenine dinucleotide phosphate (NAD and NADP, respectively). However, niacin is also a medicine, prescribed for lowering cholesterol and triglyceride levels.

4.2.3. Tryptophol and Related Indole Derivatives

Several biologically important indole derivatives are also produced from separate branches of Trp metabolism [51]. A major one is tryptophol or indole-3-ethanol (indole–CH_2–CH_2OH), an aromatic alcohol. Its biosynthetic pathway (also known as the 'Ehrlich pathway' after its discoverer Felix Ehrlich) begins with the deamination of Trp to 3-indole pyruvate, followed by decarboxylation to indole acetaldehyde, and then reduction to the alcohol by alcohol dehydrogenase [52]. Tryptophol is produced mainly by plants and lower eukaryotes, such as yeast, fungus, marine sponge, and the unicellular protozoan parasite, Trypanosoma brucei, the agent of the deadly African 'sleeping sickness'. Specifically, tryptophol is found in wine and beer, as a secondary product of ethanol fermentation by yeast; it contributes to the distinctive taste of wine, and also acts as a quorum-sensing molecule for yeast population control [53]. As a strong 'soporific' (sleep promoting agent), it is the causative chemical of 'sleeping sickness', facilitated by its ability to cross the blood–brain barrier [54]; thus, it can be considered a functional analog of serotonin and melatonin, the two neuroactive products of the serotonin pathway of Trp degradation (Section 4.2.1). The 5-hydroxy and 5-methoxy derivatives of tryptophol are also sleep inducers, as tested in mice [55]. Note that the best studied plant growth hormone, auxin, is indole acetic acid (IAA), which is also produced from 3-indole pyruvate in plants, and via several other parallel pathways in *Neurospora crassa*, a filamentous fungus [52]. A small amount of Trp is converted into indole by the action of bacterial tryptophanase enzyme in the gut. A number of organisms convert tryptophol into other derivatives, such as indole acetaldehyde, and glucoside, galactoside and mannoside conjugates of unknown function [56–58]. Finally, tryptophol is a starting material for many natural and synthetic bioactive compounds that have been masterly described in a recent treatise [59]. A short list of the naturally occurring compounds include goniomitine, spermacoceine, and physovenine, whereas chemical synthesis generates a seemingly endless roster of products such as a novel inhibitor of hepatitis C virus NS5B polymerase, and etodolac, which in turn is a progenitor of a class of bioactive compounds with anti-inflammatory, analgesic and anticancer properties [59].

To sum up, cellular Trp metabolism generates several chemicals of paramount physiological importance that affect myriad aspects of health and behavior; however, their pharmaceutical applications and mechanism of action need further scrutiny and in-depth analysis.

4.2.4. Inhibition of Gluconeogenesis by Trp

Trp and some of its metabolites have a unique role in regulating gluconeogenesis, a pathway that generates glucose from non-carbohydrate carbon substrates, including several amino acids. Gluconeogenesis in vertebrates takes place mainly in the liver, and constitutes a major mechanism that maintains blood glucose levels in times of need, such as fasting, starvation and intense exercise, thus preventing hypoglycemia. Over half-a-century ago, it was noted that tryptophan acts as an inhibitor of gluconeogenesis [60–62]. Biochemical studies have since revealed that Trp inhibits phosphoenolpyruvate carboxykinase, a key enzyme of the gluconeogenesis pathway [61], but the molecular mechanism of this enzymatic inhibition has remained unknown. The metabolic rationale for the unique anti-gluconeogenesis effect of Trp has also remained a matter of speculation; however, it is logical to assume that a high concentration of a usually low cellular level of Trp would indicate adequate energy charge, making the synthesis of glucose unnecessary. Regardless of the mechanism, Trp appears to act as an important regulator of glucose and energy metabolism.

5. Trp Adducts in Proteins

Enzyme cofactors are typically preformed organic moieties or metals that are added to the enzyme polypeptide from external sources and are reversibly and noncovalently attached. However, a class of protein-derived cofactors are formed by the post-translational modification of one or more amino acid residues [63–67], a prototype example of which is tryptophan tryptophylquinone (TTQ). It is a unique adduct that functions as an essential cofactor of select amine dehydrogenases, such as methylamine dehydrogenase (MADH) and aromatic amine dehydrogenase (AADH) [64–66]. TTQ, generated by irreversible posttranslational conjugation of two Trp residues (hence, sometimes denoted as Trp–Trp) within the same polypeptide (Figure 3A,B), forms the catalytic or redox-active center of the dehydrogenase. This role of Trp is critical to prokaryotic energy metabolism, as the MADH holoenzyme catalyzes the oxidative deamination of methylamine, eventually allowing the organism to use methylamine as the sole source of carbon, nitrogen and energy [67].

The biosynthesis of TTQ involves the cross-linking of the indole rings of two Trp residues, brought to proximity by polypeptide folding, and the insertion of two oxygen atoms onto adjacent carbons of one of the indole rings [63–66] (Figure 3A,B). The synthesis is orchestrated through a complex series of steps, within a precursor protein of MADH (called pre-MADH), and is completed by the diheme enzyme, MauG. The details of the electron transfer that occurs during these steps are beyond the scope of this review, but it is sufficient to mention that strategically located Trp residues of MauG are also active participants in this process.

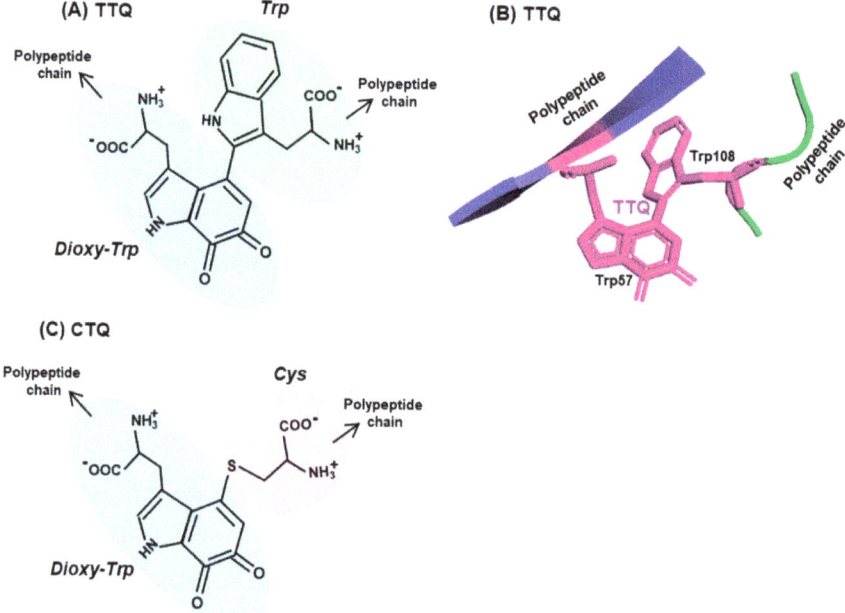

Figure 3. Two representative Trp adducts. Structures of two adducts, (**A**) tryptophan tryptophylquinone (TTQ) and (**C**) cysteine tryptophylquinone (CTQ) are shown, with each amino acid moieties circled in different colors (three Trp, one Cys), the dioxy-derivative of Trp being blue (compare with the Trp structure in Figure 2). Both TTQ and CTQ are part of the polypeptide chain, the continuity of which is indicated by arrowheads. PyMol representation of TTQ in the PDB entry 2BBK is shown in (**B**).

Lastly, another protein-derived cofactor, cysteine tryptophylquinone (CTQ), exemplifies the similarity as well as diversity in Trp-derived cofactors [68]. CTQ resembles TTQ in structure, but the oxygenated Trp residue is crosslinked to the sulfhydryl group of a Cys residue (Figure 3C). In contrast to TTQ, which are found exclusively in dehydrogenases, CTQ is found in both dehydrogenases and oxidases [64,68,69]. As observed earlier [64,70], these are the first amino acid oxidases that do not use flavin as cofactor. Collectively, these tryptophylquinone cofactors showcase the novelty and versatility of the use of Trp in biology.

6. Interacting Partners of Trp Residues in A Polypeptide

6.1. Nature of Trp Side Chain Interactions

As briefly stated before, the uniqueness of the molecular properties of Trp among the 20 amino acids is due to a combination of the following structural features [71]: the largest nonpolar (hydrophobic) area of the two-sided π-electron face that is also polarizable and highly accessible because of its planar topology; the strongest electrostatic potential for cation–π interactions; and an indole N–H moiety that can donate a H-bond. Indeed, redox-active Trp side chains play cardinal roles in electron transfer and protein function through regulated protonation state H-bonding. Lastly, Trp exhibits distinctive and sharp Raman spectral lines that have facilitated our understanding of the molecular mechanisms of protein structure and function [72,73].

Nevertheless, there has been no large-scale studies interrogating these properties of Trp for their interaction with amino acid residues in actual protein structures. To fill this vacuum, a number of X-ray crystallographic structures of diverse proteins were retrieved from the PDB protein databank (https://www.rcsb.org), and the side chain interactions of Trp were compared with those of two other aromatic residues, namely Phe and Tyr. These residues served as 'controls' of one another, since all three have π–electron surfaces, but Trp is the largest, and Tyr is the only one among the three that has a polar hydroxyl group. A set of example interactions is illustrated here (Figure 4), in which the nature of bonds was viewed in the secondary structure presentation in PyMol. For brevity, a single amino acid, namely Lys, was chosen as the interacting partner of all three, and the spacing was observed from multiple angles for an optimal view. The rational for choosing Lys was that it is amphipathic, as its side chain offers both a hydrophobic stretch and a terminal polar (amino) group. Thus, it is capable of interacting with the hydrophobic section of all three amino acids and also with the hydroxy group of Tyr. The results (Figure 4) show that this is indeed the case, i.e., the predicted hydrophobic and the ionic interactions could be discerned in appropriate spatial conformation.

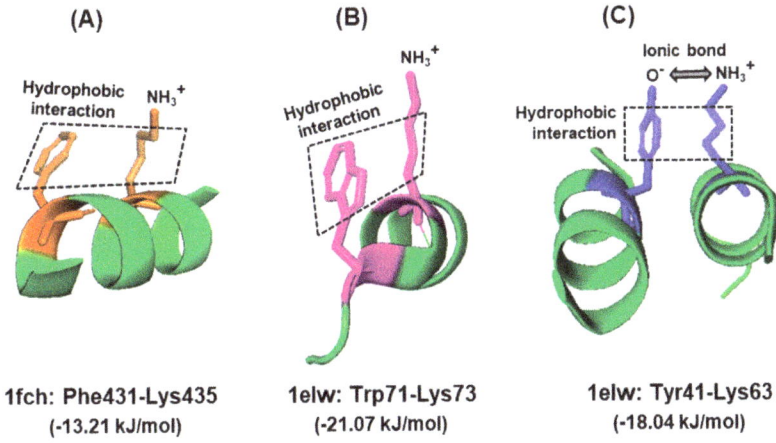

Figure 4. Interactions of aromatic side chains. Example of similarity and distinction among the three aromatic amino acids. In all three examples, the hydrophobic aromatic parts of the side chains of Phe (**A**), Trp (**B**), and Tyr (**C**) are engaged in hydrophobic interaction with the hydrocarbon part of Lys. In case of Tyr, the acidic (oxyanion, O^-) terminus additionally forms an ionic bond with the amino terminus of Lys side chain (NH_3^+), since they are negatively and positively charged, respectively, at physiological pH. The hydrophobic interaction is denoted by shaded rectangles. Other, interconnecting interactions are not shown for clarity; for example, the positively charged amino group of Lys435, which appears to be free in this diagram, is actually engaged in ionic interactions with the negatively charged carboxylic acid group of Glu232 (not shown), brought to proximity due to folding of the polypeptide. The amino acids in each interacting pair are indicated with the same color. The structures were obtained from the PDB entries written below, along with the corresponding energy of stabilizing interaction, as described in Section 6.1. Only the relevant portions of the proteins are shown as a ribbon diagram in the PyMol presentation.

Note that this is only a portion of the interactions, relevant for this query, and that each residue may interact with multiple other entities, such as water molecules, ions, and other neighboring residues. For example, the nitrogen (NH) of the indole ring can donate a hydrogen bond, and as a result, Trp can also facilitate the solvation of folded proteins, which cannot occur with Phe. Thus, the Trp side chain has a dipole moment of ~2D [74], whereas the Phe side chain does not have any. Lastly, cation–π interactions between the indole ring and Lys/Arg also stabilize a structure, provided they are in the appropriate spatial location. Nonetheless, the contribution of various environmental and structural parameters in a larger sample set is required to explore the significance of the energy difference patterns (also see Section 6.2).

6.2. The Energy Landscape of Trp Side Chain Interactions

The stabilizing free energy of interaction between amino acid side chains promotes and maintains the optimal higher order structure of a polypeptide backbone. These studies were, therefore, extended to analyses of the free energy operating between side chains involving Trp in several randomly selected proteins. The energy values were collected from the 'interaction energy matrix web server', described previously [75,76], manually entered in Excel, and those for Phe and Trp were separated out for comparison, since both are aromatic and hydrophobic. For proof-of-concept analysis, values for a total of 202 Phe and 92 Trp residues were tabulated. To illustrate, the 202 Phe residues interacted with a total of 105 Ala residues (N), and the total energy of the stabilizing interactions (E) was 244.16 kiloJoules/mole (kJ/mol). The energy per residue was then calculated as E/N, and this was similarly calculated for each of the 20 amino acids. In this manner, the strength of interaction between a pair of residues could be compared regardless of the total number of such pairs. The plotted results (Figure 5) revealed a similar interaction ensemble for Phe and Trp, i.e., their partner preferences were similar. Trp showed a higher E/N value for several interacting residues, but most notably for D and E. Interestingly, these residues, although acidic overall, possess hydrocarbon portions, which may promote stronger hydrophobic interactions with the larger hydrophobic area of Trp, as shown for Lys–Trp and Lys–Phe.

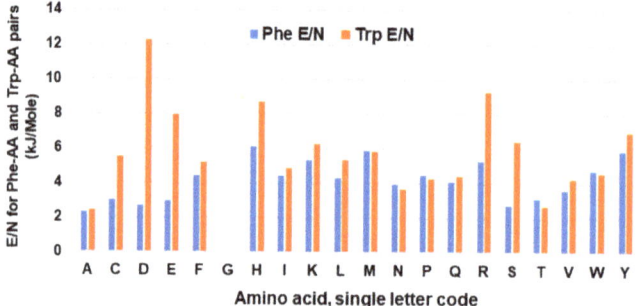

Figure 5. Average energy per interacting pair of amino acids. Energy of interaction of Trp or Phe with other amino acids were collected from multiple proteins, and the total energy (E) of a pair was divided by the total number of the pair (N) and plotted, as described in Section 6.2. Note the absence of values for Phe–Gly and Trp–Gly as Gly lacks a side chain.

Since the overall interaction energy survey did not show a clearly distinctive profile for Trp, it was inquired if perhaps the location of Trp in proteins exhibits a conspicuous signature. To this end, a previously collected large set of 22,999 pentatricopeptide (35 amino acid) repeats (PPRs) [77] was analyzed. The PPR is a class of nonidentical 35-amino acid-long repeats that contain signature amino acids in specific positions, and consist of repeat bihelical structures connected by flexible loops [77–79]. In previous studies, the Trp residues in the PPR were shown to be concentrated at residue numbers 3 and 16 [77]. Here, a similar study was performed for Phe and Tyr, and Leu as a dissimilar 'control', on the same set of PPRs, and the results were plotted for each (Figure 6), which revealed amino acid-specific patterns.

Trp was confirmed to occur at positions 3 and 16 (Figure 6A), which are near the beginning of the two signature PPR helices, designated as helix A and helix B [77]. In contrast, Phe was mainly concentrated at 23 (Figure 6B), and Tyr at 3 and 10 (Figure 6C). In other words, position 3 was popular with both Trp and Tyr, but not so much with Phe. Although the exact significance of the placement remains to be determined, in the more extensively studied and highly related tetratricopeptide repeats (TPRs) [80] the analogous Trp residue was shown to make extensive packing interactions with Leu in the preceding helix and H-bonding (by the ring NH) to the backbone carboxylate of the signature Pro32 in the following helix (not shown in Figure 2). A nonaromatic amino acid, Leu, showed no preference for any of those 'aromatic' positions, but was distributed over several other places (Figure 6D). It thus appears that Trp has its own preference for specific positions in these bihelical repeats, distinct from the others.

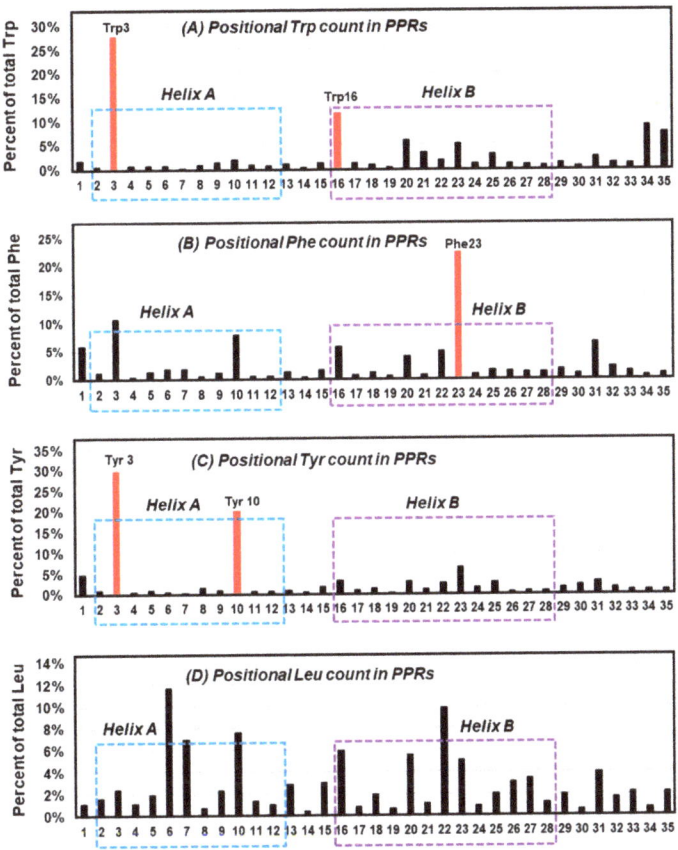

Figure 6. Locations of selected amino acids in all-helical repeats. The previously published collection of bihelical pentatricopeptide repeat (PPR) sequences [77] were visually analyzed for the location (1 to 35) of three aromatic residues, namely Trp (**A**), Phe (**B**), Tyr (**C**), and the control, nonaromatic but hydrophobic residue, Leu (**D**), and their percentage of occurrence in each position of the repeat was plotted. The most abundant location(s) of the aromatic amino acids are shown in red (Trp3/16, Phe23, Tyr3/10) and are indicated as red bars; note that Tyr3 and Trp3/16 tend to reside near the N-terminal end of the helices. Helix A and helix B [77,80] are indicated by dotted boxes, respectively, blue and purple.

6.3. Trp in Membrane Proteins and Antimicrobial Peptides

As indicated earlier, even though Trp is hydrophobic overall, the indole NH moiety can donate a hydrogen bond, and allow it to interact with the aqueous solvent. Moreover, cation–π interactions, such as when Lys or Arg is close to an aromatic ring, can provide the energy stabilization of several kilocalories/mole, which is often stronger than ionic bridges [81,82]. The combination of the physical properties of Trp makes it an ideal amphiphilic residue for the hydrophobic/hydrophilic interface of membrane proteins, where it likely plays functionally important roles [83]. In fact, while Trp is the least abundant residue in soluble proteins, accounting for only 1.1% of the amino acids expressed in cytoplasmic proteins, it is more prevalent in membrane proteins, with an abundance of 2.9% in transmembrane α-helical domains [84,85]. Several studies, including the pioneering use of UV Raman resonance spectroscopy, revealed that Trp residues act as anchors along the lipid bilayer interface, which stabilizes membrane-spanning proteins [86–88].

An impressive body of literature has documented that indole forms hydrogen bonds in membrane-associated proteins and peptides [88–93]. In the bacterial β-barrel outer membrane protein A (OmpA), mutating the Trp residues to Phe in fact destabilized the protein when folded into lipid bilayers [87,94]. The unique properties of Trp have been harnessed by a large family of short peptides with broad-spectrum antimicrobial activity, commonly referred to as 'antimicrobial peptides' or AMPs, essential for host defense and survival. Many AMPs also possess multiple other biological functions, and are therefore, also called host defense peptides (HDPs), such as the ability to regulate inflammation and immunity [93,95]. In several of them, such as members of the 'temporin' and 'aurein' family, the pairing of Trp and Arg has been shown to be essential for superior activity and bioavailability, as reviewed recently [95]. The Trp/Arg amino acid pair is a common theme in many AMPs, where it allows for cation–π interaction as well as the unique side chain properties of the Trp indole ring, mentioned earlier, thereby promoting the formation of the proper higher order structure and interaction with membrane lipid bilayer. A large number of diverse temporins and aureins, secreted from the skin of several frog species, protect these amphibians from infections in the wild. In another example, short Trp-rich AMPs provide antimicrobial defense to the water buffalo (*Bubalus bubalis*) [96] and have been subjected to mutational analysis. An optimally designed variant, WRK-12 (WRLRWKTRWRLK), was shown to efficiently target LPS and bacteria-specific phospholipids on the membrane [97], and as with many other AMPs, the placement of the Trp residues on one face of the helix (Figure 7A) was crucial for amphiphilicity and membrane interaction.

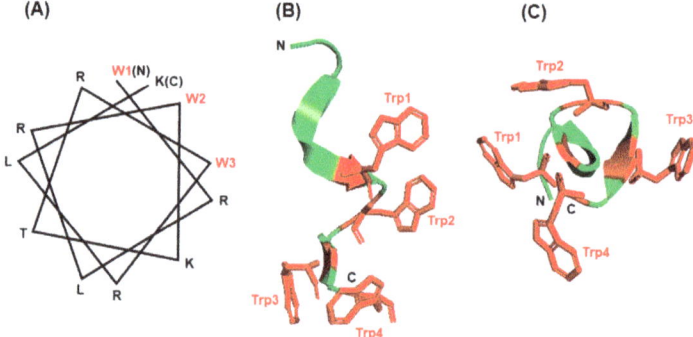

Figure 7. Structure of small Trp-containing peptides. (**A**) Helical wheel presentation of peptide WRK-12 (WRLRWKTRWRLK) and (**B**,**C**) three-dimensional structures of gramicidin A (PBD 1C4D) in PyMol ribbon presentation are shown. All Trp (W) residues are indicated in red. For gramicidin, two views of the Trp-containing portion of the peptide are presented: a side view (**B**) and a top view of the cylinder shape (**C**), showing the Trp side chains protruded around the channel. The amino- and carboxy-termini are indicated with N and C, respectively.

The channel peptide antibiotic, gramicidin A, has served as a prototype peptide for protein–lipid interactions [98]. Gramicidin A is an amphiphilic AMP consisting of 15 amino acids, of which four are Trp. The gramicidin family has several members of differing sequence, but common structural and functional features. All gramicidins exhibit a complex ensemble of structures, but in all of them the Trp residues are clustered on the surface of a helical conformation (Figure 7B,C), so that gramicidin locates at the membrane interface in the channel. As the Trp residues are substituted by Phe the channel conductance of the substituted gramicidin decreases along with a loss of antibacterial activity [99,100]. In other words, all four Trp residues are required for the full functionality of gramicidin A [101].

The formation of H-bond by indole NH, which cannot occur with Phe, is also essential for the native conformation and function of gramicidin and other ion channels and membrane proteins [102]. Ab initio calculations have also shown that the strength of the H-bonds formed between a proton donor and the π electron cloud of aromatic side chains, which represent a major class of stabilizing interactions, follow the order Trp > His > Tyr~Phe, i.e., Trp forms the strongest such bond [103]. Collectively, the examples presented in this section illustrate that the side chain features, location, and environment of Trp residues are all important for the folding and insertion of membrane proteins and membrane-associated peptides.

6.4. The Trp-Cage Family of Fast-Folding Peptides

Through studies of specific fragments of a naturally occurring 39-amino acid peptide, isolated from the oral secretions of a lizard species, Neidigh et al. designed a 20-residue peptide that efficiently achieved a compact and stable tertiary fold [104,105]. The folded structure was eventually named 'Trp cage' as its hydrophobic core shielded the side chain of the single Trp residue from solvent exposure. Kinetic measurements revealed that complete folding occurred within ~4 μs [106], establishing the Trp cage peptide as the fastest folding peptide known.

Although the detailed mechanism of the Trp-cage folding is still being researched, biophysical and molecular dynamics simulation studies implicated the side chain rotamer state of the Trp residue as a major contributor to the unusually fast folding rate. Subsequently, various residues of the Trp cage were mutated and several variations of the original Trp cage sequences were also synthesized, as summarized before [107]. Collectively, this led to the consensus (Figure 8) that the Trp indole ring forms the center hub of the cage, and the rest of the peptide, which consists of just one α-helix and one β-strand on either side of the Trp, forms a hydrophobic interior that cradles this side chain to shield it from the solvent. Several helix–strand interactions hold the cage together, but the notable ones are the hydrophobic interaction between Tyr3 and Pro19, the salt bridge between Asp9 and Arg16, and several H-bonded side chains, notably the one contributed by the -OH group of Ser14 (Figure 8). The hydrophobic large indole ring of Trp was crucial for the cage structure, as the peptides in which Trp6 was replaced by His or Phe, remain largely unfolded [108].

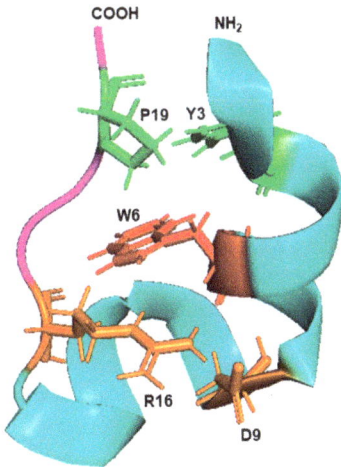

Figure 8. Trp cage fold. The 3D structure (PDB: 2JOF) of the 20 amino acid-long Trp cage peptide of the sequence DAYAQWLKDGGPSSGRPPPS was retrieved from the NCBI structure bank and shown in PyMol presentation. The amino- and carboxy-terminal ends are marked as such, and the single α-helix and the β-strand are colored in cyan and magenta, respectively. The central Trp (W6) is red, the Y3–P9 interacting pair is green, and the D9–R16 interacting pair is orange, as detailed in the text (Section 6.3).

7. Intersection between Trp Degradation and Immunity

Perhaps the most enigmatic aspect of Trp metabolism that has defied a clear molecular explanation for years is its intersection with various normal and pathological conditions, such as cancer, multiple sclerosis, transplantation, peripheral and CNS disorders, and recently in SARS-CoV-2 infection [47–50,109–111]. For the sake of brevity, the focus in this review is on selected branches of immune regulation, a key character in which is indole 2,3-dioxygenase (IDO), the rate-limiting first enzyme in the kynurenine pathway of Trp degradation (Figure 2).

It has been noted that only about 1% of dietary Trp is actually used in translation, in part due to its rarity in the polypeptides (~1.2% of total amino acids in the proteome) [9,84], while the rest is degraded or converted to multiple other compounds. As alluded to earlier, this makes Trp one of the largest contributors of non-amino acid metabolites in the cell. Approximately 90% of dietary Trp is metabolized through the kynurenine pathway alone (Figure 2), and the serotonin pathway utilizes another 1–2% [112].

IDO is strongly induced in inflammation; in fact, one of the best inducers of IDO is interferon-γ (IFN-γ) [21], which led to its designation as an "interferon-stimulated gene" (ISG). It was initially thought that IDO was a regulator of inflammatory response, and this was due to the depletion of Trp; however, the addition of Trp reversed the effect in some cell lines only, but not in others, suggesting that the mechanism is more complex [113,114]. Subsequent research demonstrated that IDO1-expressing immune cells, such as macrophages and dendritic cells, inhibit T-cell proliferation [115–117]. Mechanistic studies indicated that the T cell inhibition is caused by Trp depletion-induced GCN2 and/or mTOR signaling pathways. Studies in cell culture also demonstrated that exogenously added kynurenine acts as an immunosuppressive metabolite in combination with transforming growth factor. It thus appears that IDO, kynurenine accumulation, and Trp depletion, together with metabolic regulation, work together for the observed immunosuppressive effect of IDO [50,118], which is still an area of fervent research. Readers interested in further details may read the original papers, starting with the references cited here.

In its antimicrobial role, IDO was originally found to inhibit the replication of the protozoan parasite Toxoplasma gondii in immune cells [119–121]. Subsequently, an antiviral role of IDO was demonstrated against measles, herpes simplex, hepatitis B, influenza and respiratory syncytial viruses [122–127]. These studies were performed mostly in cells of immune origin, such as macrophages and dendritic cells, and thus an immune-regulatory role of IDO could be at play.

Very recently, a novel role antiviral of IDO was reported in cultured A549 cells, which are nonimmune, lung epithelial cells [21]. The induction of IDO by treatment with IFN-γ or by the expression of recombinant IDO in these cells was found to strongly inhibit the growth of human parainfluenza virus (PIV3), a negative-strand RNA virus and a major cause of sickness and death in children, the elderly, and the immunocompromised. The authors used several approaches to demonstrate that 5-hydoxytryptophan (5-HTP), the first product of the serotonin pathway of Trp degradation and the immediate precursor of serotonin, is essential to protect PIV3 growth against IDO in cell culture. The apparent antiviral effect of IDO on PIV3 growth, therefore, was not due to the generation of any of the kynurenine pathway metabolites, but rather due to the depletion of intracellular Trp by IDO, as a result of which Trp became unavailable for the alternative, serotonin pathway [21]. These studies established 5-HTP as a proviral Trp metabolite for PIV3 that is cell-intrinsic and not dependent on active immunity or humoral response. The mechanism of this proviral role of 5-HTP and its generality for other viruses remain to be determined.

8. Summary and Conclusions

This review offers a glimpse of the manifold uniqueness of tryptophan in biology, the major areas of which are summarized in Figure 9. The uniqueness of Trp in a polypeptide sequence derives from the side chain indole ring with its binuclear aromatic structure that allows it to find large hydrophobic pockets and support interactions that require a relatively large surface area. This is particularly evident in the preponderance of Trp in membrane proteins, membrane-active toxins and channels (Section 6.3), and in the formation of Trp cage (Section 6.4) and Trp adducts such as TTQ and CTQ (Section 5). It is not always clear if the indole ring also plays a pivotal role in the nearly two dozen physiologically functional small molecule metabolites that are derived from Trp, but it likely is, judging by the fact that many of them lose their bioactivity when the ring structure is opened up. For example, all three products of the serotonin pathway of Trp degradation (oxitriptan, serotonin, melatonin) (Figure 9) possess neuroregulatory and psychedelic activities, but all the products of the parallel kynurenine pathway retain only the six-membered benzene ring of indole and lack these activities.

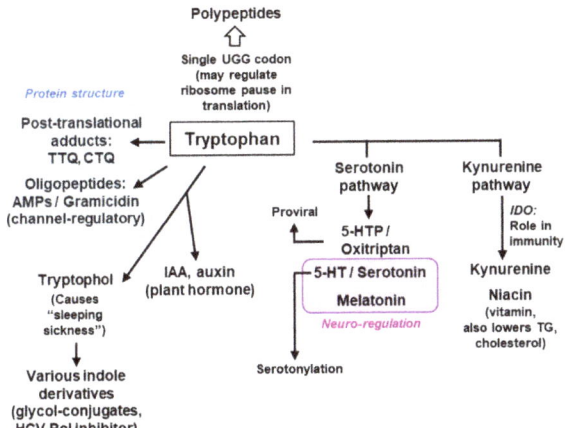

Figure 9. Brief schematic summary of the major roles of Trp. This is a concise diagram, showing only a few selected contributions of Trp, its metabolic products and their major roles and common usage. All compounds and their abbreviated names have been described at appropriate places in the review. The two-colored boxes indicate the main functions of the boxed Trp products, namely, the role in protein structure (Blue) and neuro-regulation (Brown). HCV = Hepatitis C virus. Note that this is by no means an exhaustive list and that many compounds have multiple roles, and many roles are played by multiple compounds, which are not shown here for the sake of brevity. AMP = Antimicrobial peptide.

The essential role of 5-HTP for the optimal replication of PIV3 was an unpredictable discovery that escaped attention for years, in part because the main focus was on the immunoregulatory role of kynurenine in the parallel pathway, operative in cells of myeloid origin [128]. The PIV3 studies are highly relevant for a potential antiviral regimen since PIV3 and many other negative-strand RNA viruses infect and grow in the nonimmune cells as their primary target, such as airway epithelial cells [129,130]. If 5-HTP is found to regulate the positive-strand RNA viruses as well, this could be pursued as a potential antiviral strategy against many newly emergent viruses that are lethal, such as hepatitis C, West Nile, dengue, and SARS-CoV. Lastly, the use of specific enzyme inhibitors of the two degradation pathways [21] strongly suggested that the steady-state level of 5-HTP is regulated by several factors; it is elevated by dietary Trp and TPH enzyme activity, and reduced by IDO and aromatic amino acid decarboxylase (AAAD) enzyme activity (Figure 2). The quantification of these balancing activities and their effects on virus replication under physiological and pathological conditions can be rewarding areas of future research of fundamental and clinical importance.

Since Trp is bioenergetically the most expensive amino acid, it makes sense that it is incorporated in exclusive sites in the protein where it is absolutely needed for protein structure and function, which is consistent with its rarity in proteins and its single codon. In parallel to its rare use in translation, Trp is used to generate a plethora of secondary metabolites. In other words, the translational and non-translational uses of Trp are mutually exclusive and likely compete with each other, since Trp is also the least abundant amino acid in the cell. It is tempting to wonder if Trp has other physiological roles that are still awaiting discovery.

Funding: This work received no external funding. The article publication cost and open access fee were paid by the personal funds of the author.

Conflicts of Interest: The author declares no conflict of interest.

Abbreviations

AAAD	Aromatic amino acid decarboxylase
AMP	Antimicrobial peptide
CNS	Central nervous system
CTQ	Cysteine tryptophylquinone
5-HT	5-hydroxy-tryptamine
5-HTP	5-hydroxy-tryptophan
IDO	Indole 2,3-dioxygenase
IFN	Interferon
LPS	Lipopolysaccharide
MADH	Methylamine dehydrogenase
OTC	Over the counter (non-prescription)
PIV3	Parainfluenza virus Type 3
PPR	Pentatricopeptide repeat
TDO	Tryptophan 2,3-dehydrogenase
TH	Tyrosine hydroxylase
TPH	Tryptophan hydroxylase
TPR	Tetratricopeptide
TTQ	Tryptophan tryptophylquinone

References

1. Wu, G.; Bazer, F.W.; Dai, Z.; Li, D.; Wang, J.; Wu, Z. Amino acid nutrition in animals: Protein synthesis and beyond. *Annu. Rev. Anim. Biosci.* **2014**, *2*, 387–417. [CrossRef] [PubMed]
2. Binks, H.; Vincent, G.E.; Gupta, C.; Irwin, C.; Khalesi, S. Effects of diet on sleep: A narrative. *Nutrients* **2020**, *12*, 936. [CrossRef] [PubMed]
3. Chojnacki, C.; Popławski, T.; Chojnacki, J.; Fila, M.; Konrad, P.; Blasiak, J. Tryptophan intake and metabolism in older adults with mood disorders. *Nutrients* **2020**, *12*, 3183. [CrossRef] [PubMed]
4. Kikuchi, A.M.; Tanabe, A.; Iwahori, Y. A systematic review of the effect of L-tryptophan supplementation on mood and emotional functioning. *J. Diet. Suppl.* **2020**, 1–18. [CrossRef]
5. Simonson, M.; Boirie, Y.; Guillet, C. Protein, amino acids and obesity treatment. *Rev. Endocr. Metab. Disord.* **2020**, *21*, 341–353. [CrossRef]
6. Taleb, S. Tryptophan dietary impacts gut barrier and metabolic diseases. *Front. Immunol.* **2019**, *10*, 2113. [CrossRef]
7. Khorana, H.G.; Buchi, H.; Ghosh, H.; Gupta, N.; Jacob, T.M.; Kossel, H.; Morgan, R.; Narang, S.A.; Ohtsuka, E.; Wells, R.D. Polynucleotide synthesis and the genetic code. *Cold Spring Harb. Symp. Quant. Biol.* **1966**, *31*, 39–49. [CrossRef]
8. Nirenberg, M.; Caskey, T.; Marshall, R.; Brimacombe, R.; Kellogg, D.; Doctor, B.; Hatfield, D.; Levin, J.; Rottman, F.; Pestka, S.; et al. The RNA code and protein synthesis. *Cold Spring Harb. Symp. Quant. Biol.* **1966**, *31*, 11–24. [CrossRef]
9. Gardini, S.; Cheli, S.; Baroni, S.; Di Lascio, G.; Mangiavacchi, G.; Micheletti, N.; Monaco, C.L.; Savini, L.; Alocci, D.; Mangani, S.; et al. On Nature's strategy for assigning genetic code multiplicity. *PLoS ONE* **2016**, *11*, e0148174. [CrossRef]
10. Crick, F.H.C. Codon-anticodon pairing: The wobble hypothesis. *J. Mol. Biol.* **1966**, *19*, 548–555. [CrossRef]
11. Limmer, S.; Reif, B.; Ott, G.; Arnold, L.; Sprinzl, M. NMR evidence for helix geometry modifications by a G-U wobble base pair in the acceptor arm of *E. coli* tRNA(Ala). *FEBS Lett.* **1996**, *385*, 15–20. [CrossRef]

12. Murphy, F.V.; Frank, V.; Ramakrishnan, V. Structure of a purine-purine wobble base pair in the decoding center of the ribosome. *Nat. Struct. Mol. Biol.* **2004**, *11*, 1251–1252. [CrossRef] [PubMed]
13. Ikeuchi, K.; Izawa, T.; Inada, T. Recent progress on the molecular mechanism of quality controls induced by ribosome stalling. *Front. Genet.* **2018**, *9*, 743. [CrossRef] [PubMed]
14. Bitran, A.; Jacobs, W.M.; Zhai, X.; Shakhnovich, E. Cotranslational folding allows misfolding-prone proteins to circumvent deep kinetic traps. *Proc. Natl. Acad. Sci. USA* **2020**, *117*, 1485–1495. [CrossRef]
15. Collart, M.A.; Weiss, B. Ribosome pausing, a dangerous necessity for co-translational events. *Nucleic Acids Res.* **2020**, *48*, 1043–1055. [CrossRef]
16. Braus, G.H. Aromatic amino acid biosynthesis in the yeast *Saccharomyces cerevisiae*: A model system for the regulation of a eukaryotic biosynthetic pathway. *Microbiol. Rev.* **1991**, *55*, 349–370. [CrossRef]
17. Radwanski, E.R.; Last, R.L. Tryptophan biosynthesis and metabolism: Biochemical and molecular genetics. *Plant Cell* **1995**, *7*, 921–934. [CrossRef]
18. Miles, E.W. The tryptophan synthase α2β2 complex: A model for substrate channeling, allosteric communication, and pyridoxal phosphate catalysis. *J. Biol. Chem.* **2013**, *288*, 10084–10091. [CrossRef]
19. Akashi, H.; Gojobori, T. Metabolic efficiency and amino acid composition in the proteomes of *Escherichia coli* and *Bacillus subtilis*. *Proc. Natl. Acad. Sci. USA* **2002**, *99*, 3695–3700. [CrossRef]
20. Le Floc'h, N.; Otten, W.; Merlot, E. Tryptophan metabolism, from nutrition to potential therapeutic applications. *Amino Acids* **2011**, *41*, 1195–1205. [CrossRef]
21. Rabbani, M.A.G.; Barik, S. 5-Hydroxytryptophan, a major product of tryptophan degradation, is essential for optimal replication of human parainfluenza virus. *Virology* **2017**, *503*, 46–51. [CrossRef] [PubMed]
22. Walther, D.J.; Peter, J.-U.; Winter, S.; Höltje, M.; Paulmann, N.; Grohmann, M.; Vowinckel, J.; Alamo-Bethencourt, V.; Wilhelm, C.S.; Ahnert-Hilger, G.; et al. Serotonylation of small GTPases is a signal transduction pathway that triggers platelet α-granule release. *Cell* **2003**, *115*, 851–862. [CrossRef]
23. Roberts, K.M.; Fitzpatrick, P.F. Mechanisms of tryptophan and tyrosine hydroxylase. *IUBMB Life* **2013**, *65*, 350–357. [CrossRef] [PubMed]
24. Yoshida, R.; Urade, U.; Tokuda, M.; Hayaishi, O. Induction of indoleamine 2,3-dioxygenase in mouse lung during virus infection. *Proc. Natl. Acad. Sci. USA* **1979**, *76*, 4084–4086. [CrossRef] [PubMed]
25. Munn, D.H.; Mellor, A.L. Indoleamine 2,3 dioxygenase and metabolic control of immune responses. *Trends Immunol.* **2013**, *34*, 137–143. [CrossRef]
26. Ball, H.J.; Sanchez-Perez, A.; Weiser, S.; Austin, C.J.D.; Astelbauer, F.; Miu, J.; McQuillan, J.A.; Stocker, R.; Jermiin, L.S.; Hunt, N.H. Characterization of an indoleamine 2,3-dioxygenase-like protein found in humans and mice. *Gene* **2007**, *396*, 203–213. [CrossRef]
27. Metz, R.; Smith, C.; DuHadaway, J.B.; Chandler, P.; Baban, B.; Merlo, L.M.F.; Pigott, E.; Keough, M.P.; Rust, S.; Mellor, A.L.; et al. IDO2 is critical for IDO1-mediated T-cell regulation and exerts a non-redundant function in inflammation. *Int. Immunol.* **2014**, *26*, 357–367. [CrossRef]
28. Merlo, L.M.F.; DuHadaway, J.B.; Montgomery, J.D.; Peng, W.-D.; Murray, P.J.; Prendergast, G.C.; Caton, A.J.; Muller, A.J.; Mandik-Nayak, M. Differential roles of IDO1 and IDO2 in T and B cell inflammatory immune responses. *Front. Immunol.* **2020**, *11*, 1861. [CrossRef]
29. Monti, J.M. Serotonin control of sleep-wake behavior. *Sleep. Med. Rev.* **2011**, *15*, 269–281. [CrossRef]
30. Berger, M.; Gray, J.A.; Roth, B.L. The expanded biology of serotonin. *Ann. Rev. Med.* **2009**, *60*, 355–366. [CrossRef]
31. Nichols, D.E. Psychedelics. *Pharmacol. Rev.* **2016**, *68*, 264–355. [CrossRef] [PubMed]
32. Gershon, M.D. Review article: Serotonin receptors and transporters–oles in normal and abnormal gastrointestinal motility. *Aliment. Pharmacol. Ther.* **2004**, *20* (Suppl. 7), 3–14. [CrossRef] [PubMed]
33. Hoyer, D. Targeting the 5-HT system: Potential side effects. *Neuropharmacology* **2020**, *179*, 108233. [CrossRef] [PubMed]
34. Sarkar, N.K.; Clarke, D.D.; Waelsch, H. An enzymically catalyzed incorporation of amines into proteins. *Biochim. Biophys. Acta* **1957**, *25*, 451–452. [CrossRef]
35. Walther, D.J.; Peter, J.U.; Bashammakh, S.; Hörtnagl, H.; Voits, M.; Fink, H.; Bader, M. Synthesis of serotonin by a second tryptophan hydroxylase isoform. *Science* **2003**, *299*, 76. [CrossRef]
36. Bader, M. Serotonylation: Serotonin signaling and epigenetics. *Front. Mol. Neurosci.* **2019**, *12*, 288. [CrossRef]
37. Javelle, F.; Lampit, A.; Bloch, W.; Häussermann, P.; Johnson, S.L.; Zimmer, P. Effects of 5-hydroxytryptophan on distinct types of depression: A systematic review and meta-analysis. *Nutr. Rev.* **2020**, *78*, 77–88. [CrossRef]

38. Shaw, K.A.; Turner, J.; Del Mar, C. Tryptophan and 5-Hydroxytryptophan for depression. *Cochrane Database Syst. Rev.* **2002**, *1*, CD003198. [CrossRef]
39. Claustrat, B.; Leston, J. Melatonin: Physiological effects in humans. *Neurochirurgie* **2015**, *61*, 77–84. [CrossRef]
40. Rios, E.R.V.; Venâncio, E.T.; Rocha, N.F.M.; Woods, D.J.; Vasconcelos, S.; Macedo, D.; de Sousa, F.C.F.; Fonteles, M.M.F. Melatonin: Pharmacological aspects and clinical trends. *Int. J. Neurosci.* **2010**, *120*, 583–590. [CrossRef]
41. Baglioni, C.; Bostanova, Z.; Bacaro, V.; Benz, F.; Hertenstein, E.; Spiegelhalder, K.; Rücker, G.; Frase, L.; Riemann, D.; Feige, B. A systematic review and network meta-analysis of randomized controlled trials evaluating the evidence base of melatonin, light exposure, exercise, and complementary and alternative medicine for patients with insomnia disorder. *J. Clin. Med.* **2020**, *9*, 1949. [CrossRef] [PubMed]
42. Srinivasan, V.; Spence, D.W.; Pandi-Perumal, S.R.; Trakht, I.; Cardinali, D.P. Jet lag: Therapeutic use of melatonin and possible application of melatonin analogs. *Travel Med. Infect. Dis.* **2008**, *6*, 17–28. [CrossRef] [PubMed]
43. Pandi-Perumal, S.R.; Srinivasan, V.; Maestroni, G.J.; Cardinali, D.P.; Poeggeler, B.; Hardeland, R. Melatonin: Nature's most versatile biological signal? *FEBS J.* **2006**, *273*, 2813–2838. [CrossRef] [PubMed]
44. Kong, X.; Gao, R.; Wang, Z.; Wang, X.; Fang, Y.; Gao, J.; Reiter, R.J.; Wang, J. Melatonin: A potential therapeutic option for breast cancer. *Trends Endocrinol. Metab.* **2020**, *31*, 859–871. [CrossRef]
45. Bondy, S.C.; Campbell, A. Melatonin and regulation of immune function: Impact on numerous diseases. *Curr. Aging Sci.* **2020**. Online ahead of print. [CrossRef]
46. Schwarcz, R.; Bruno, J.P.; Muchowski, P.J.; Wu, H.Q. Kynurenines in the mammalian brain: When physiology meets pathology. *Nat. Rev. Neurosci.* **2012**, *13*, 465–477. [CrossRef]
47. Proietti, E.; Rossini, S.; Grohmann, U.; Mondanelli, G. Polyamines and kynurenines at the intersection of immune modulation. *Trends Immunol.* **2020**, *41*, 1037–1050. [CrossRef]
48. Biernacki, T.; Sandi, D.; Bencsik, K.; Vécsei, L. Kynurenines in the pathogenesis of multiple sclerosis: Therapeutic perspectives. *Cells* **2020**, *9*, 1564. [CrossRef]
49. Huang, Y.S.; Ogbechi, J.; Clanchy, F.I.; Williams, R.O.; Stone, T.W. IDO and kynurenine metabolites in peripheral and CNS disorders. *Front. Immunol.* **2020**, *11*, 388. [CrossRef]
50. Zhai, L.; Bell, A.; Ladomersky, E.; Lauing, K.L.; Bollu, L.; Sosman, J.A.; Zhang, B.; Wu, J.D.; Miller, S.D.; Meeks, J.J.; et al. Immunosuppressive IDO in cancer: Mechanisms of action, animal models, and targeting strategies. *Front. Immunol.* **2020**, *11*, 1185. [CrossRef]
51. Lee, J.H.; Wood, T.K.; Lee, J. Roles of indole as an interspecies and interkingdom signaling molecule. *Trends Microbiol.* **2015**, *23*, 707–718. [CrossRef] [PubMed]
52. Sardar, P.; Kempken, F. Characterization of indole-3-pyruvic acid pathway-mediated biosynthesis of auxin in Neurospora crassa. *PLoS ONE* **2018**, *13*, e0192293. [CrossRef] [PubMed]
53. González, B.; Vázquez, J.; Morcillo-Parra, M.Á.; Mas, A.; Torija, M.J.; Beltran, G. The production of aromatic alcohols in non-Saccharomyces wine yeast is modulated by nutrient availability. *Food Microbiol.* **2018**, *74*, 64–74. [CrossRef] [PubMed]
54. Cornford, E.M.; Bocash, W.D.; Braun, L.D.; Crane, P.D.; Oldendorf, W.H.; MacInnis, A.J. Rapid distribution of tryptophol (3-indole ethanol) to the brain and other tissues. *J. Clin. Investig.* **1979**, *63*, 1241–1248. [CrossRef] [PubMed]
55. Feldstein, A.; Chang, F.H.; Kucharski, J.M. Tryptophol, 5-hydroxytryptophol and 5-methoxytryptophol induced sleep in mice. *Life Sci.* **1970**, *9*, 323–329. [CrossRef]
56. Laćan, G.; Magnus, V.; Jeričević, B.; Kunst, L.; Iskrić, S. Formation of tryptophol galactoside and an unknown tryptophol ester in Euglena gracilis. *Plant Physiol.* **1984**, *76*, 889–893. [CrossRef]
57. Hofsteenge, J.; Müller, D.R.; de Beer, T.; Löffler, A.; Richter, W.J.; Vliegenthart, J.F. New type of linkage between a carbohydrate and a protein: C-glycosylation of a specific tryptophan residue in human RNase Us. *Biochemistry* **1994**, *33*, 13524–13530. [CrossRef]
58. Gutsche, B.; Grun, C.; Scheutzow, D.; Herderich, M. Tryptophan glycoconjugates in food and human urine. *Biochem. J.* **1999**, *343*, 11–19. [CrossRef]
59. Palmieri, A.; Petrini, M. Tryptophol and derivatives: Natural occurrence and applications to the synthesis of bioactive compounds. *Nat. Prod. Rep.* **2019**, *36*, 490–530. [CrossRef]

60. Ray, P.D.; Foster, D.O.; Lardy, H.A. Paths of carbon in gluconeogenesis and lipogenesis. IV. Inhibition by L-tryptophan of hepatic gluconeogenesis at the level of phosphoenolpyruvate formation. *J. Biol. Chem.* **1966**, *241*, 3904–3908.
61. Smith, S.A.; Elliott, K.R.F.; Pogson, C.I. Differential effects of tryptophan on glucose synthesis in rats and guinea pigs. *Biochem. J.* **1978**, *176*, 817–825. [CrossRef] [PubMed]
62. Badawy, A.A. Kynurenine pathway of tryptophan metabolism: Regulatory and functional aspects. *Int. J. Tryptophan Res.* **2017**, *10*, 1178646917691938. [CrossRef] [PubMed]
63. Davidson, V.L.; Liu, A. Tryptophan tryptophylquinone biosynthesis: A radical approach to posttranslational modification. *Biochim. Biophys. Acta* **2012**, *1824*, 1299–1305. [CrossRef] [PubMed]
64. Davidson, V.L. Protein-derived cofactors revisited: Empowering amino acid residues with new functions. *Biochemistry* **2018**, *57*, 3115–3125. [CrossRef]
65. McIntire, W.S.; Wemmer, D.E.; Chistoserdov, A.; Lidstrom, M.E. A new cofactor in a prokaryotic enzyme: Tryptophan tryptophylquinone as the redox prosthetic group in methylamine dehydrogenase. *Science* **1991**, *252*, 817–824. [CrossRef]
66. Davidson, V.L.; Wilmot, C.M. Posttranslational biosynthesis of the protein-derived cofactor tryptophan tryptophylquinone. *Annu. Rev. Biochem.* **2013**, *82*, 531–550. [CrossRef]
67. Davidson, V.L. Pyrroloquinoline quinone (PQQ) from methanol dehydrogenase and tryptophan tryptophylquinone (TTQ) from methylamine dehydrogenase. *Adv. Protein. Chem.* **2001**, *58*, 95–140. [CrossRef]
68. Datta, S.; Mori, Y.; Takagi, K.; Kawaguchi, K.; Chen, Z.W.; Okajima, T.; Kuroda, S.; Ikeda, T.; Kano, K.; Tanizawa, K.; et al. Structure of a quinohemoprotein amine dehydrogenase with an uncommon redox cofactor and highly unusual crosslinking. *Proc. Natl. Acad. Sci. USA* **2001**, *98*, 14268–14273. [CrossRef]
69. Campillo-Brocal, J.C.; Chacon-Verdu, M.D.; Lucas-Elio, P.; Sanchez-Amat, A. Distribution in microbial genomes of genes similar to lodA and goxA which encode a novel family of quinoproteins with amino acid oxidase activity. *BMC Genom.* **2015**, *16*, 231. [CrossRef]
70. Campillo-Brocal, J.C.; Lucas-Elio, P.; Sanchez-Amat, A. Distribution in different amino acid oxidases with FAD or a quinone as cofactor and their role as antimicrobial proteins in marine bacteria. *Mar. Drugs* **2015**, *13*, 7403–7418. [CrossRef]
71. Schlamadinger, D.E.; Gable, J.E.; Kim, J.E. Hydrogen bonding and solvent polarity markers in the UV resonance Raman spectrum of tryptophan: Application to membrane proteins. *J. Phys. Chem. B* **2009**, *113*, 14769–14778. [CrossRef] [PubMed]
72. Thomas Jr, G.J. New structural insights from Raman spectroscopy of proteins and their assemblies. *Biopolymers* **2002**, *67*, 214–225. [CrossRef] [PubMed]
73. Takeuchi, H. UV Raman markers for structural analysis of aromatic side chains in proteins. *Anal. Sci.* **2011**, *27*, 1077–1086. [CrossRef] [PubMed]
74. Weiler-Feilchenfeld, H.; Pullman, A.; Berthod, H.; Giessner-Prettre, C. Experimental and quantum-chemical studies of the dipole moments of quinoline and indole. *J. Mol. Struct.* **1970**, *6*, 297–304. [CrossRef]
75. Galgonek, J.; Vymetal, J.; Jakubec, D.; Vondrášek, J. Amino Acid Interaction (INTAA) web server. *Nucleic Acids Res.* **2017**, *45*, W388–W392. [CrossRef] [PubMed]
76. Barik, S. Protein tetratricopeptide repeat and the companion non-tetratricopeptide repeat helices: Bioinformatic analysis of interhelical interactions. *Bioinform. Biol. Insights* **2019**, *13*, 1177932219863363. [CrossRef] [PubMed]
77. Barik, S. The nature and arrangement of pentatricopeptide domains and the linker sequences between them. *Bioinform. Biol. Insights* **2020**, *14*, 1177932220906434. [CrossRef]
78. Small, I.D.; Peeters, N. The PPR motif–a TPR-related motif prevalent in plant organellar proteins. *Trends Biochem. Sci.* **2000**, *25*, 46–47. [CrossRef]
79. Barkan, A.; Small, I. Pentatricopeptide repeat proteins in plants. *Annu. Rev. Plant. Biol.* **2014**, *65*, 415–442. [CrossRef]
80. Main, E.R.G.; Xiong, Y.; Cocco, M.J.; D'Andrea, L.; Regan, L. Design of stable alpha-helical arrays from an idealized TPR motif. *Structure* **2003**, *11*, 497–508. [CrossRef]
81. Gallivan, J.P.; Dougherty, D.A. Cation-pi interactions in structural biology. *Proc. Natl. Acad. Sci. USA* **1999**, *96*, 9459–9464. [CrossRef] [PubMed]
82. Olson, C.A.; Shi, Z.; Kallenbach, N.R. Polar interactions with aromatic side chains in alpha-helical peptides: Ch...O H-bonding and cation-pi interactions. *J. Am. Chem. Soc.* **2001**, *123*, 6451–6452. [CrossRef] [PubMed]

83. Hait, S.; Mallik, S.; Basu, S.; Kundu, S. Finding the generalized molecular principles of protein thermal stability. *Proteins* **2020**, *88*, 788–808. [CrossRef] [PubMed]
84. The UniProt Consortium. UniProt: A worldwide hub of protein knowledge. *Nucleic Acids Res.* **2019**, *47*, D506–D515. [CrossRef] [PubMed]
85. Jayasinghe, S.; Hristova, K.; White, S.H. MPtopo: A database of membrane protein topology. *Protein Sci.* **2001**, *10*, 455–458. [CrossRef]
86. Granseth, E.; von Heijne, G.; Elofsson, A. A study of the membrane-water interface region of membrane proteins. *J. Mol. Biol.* **2005**, *346*, 377–385. [CrossRef]
87. Sanchez, K.M.; Gable, J.E.; Schlamadinger, D.E.; Kim, J.E. Effects of tryptophan microenvironment, soluble domain, and vesicle size on the thermodynamics of membrane protein folding: Lessons from the transmembrane protein OmpA. *Biochemistry* **2008**, *47*, 12844–12852. [CrossRef]
88. Sanchez, K.M.; Kang, G.; Wu, B.; Kim, J.E. Tryptophan-lipid interactions in membrane protein folding probed by ultraviolet resonance Raman and fluorescence spectroscopy. *Biophys. J.* **2011**, *100*, 2121–2130. [CrossRef]
89. Persson, S.; Killian, J.A.; Lindblom, G. Molecular ordering of interfacially localized tryptophan analogs in ester- and ether-lipid bilayers studied by H-2-NMR. *Biophys. J.* **1998**, *75*, 1365–1371. [CrossRef]
90. van der Wel, P.C.A.; Reed, N.D.; Koeppe, R.E. Orientation and motion of tryptophan interfacial anchors in membrane-spanning peptides. *Biochemistry* **2007**, *46*, 7514–7524. [CrossRef]
91. Shafaat, H.S.; Sanchez, K.M.; Kim, J.E. Ultraviolet resonance Raman spectroscopy of a membrane-bound beta-sheet peptide as a model for membrane protein folding. *J. Raman Spectrosc.* **2008**, *40*, 1060–1064. [CrossRef]
92. Sun, H.; Greathouse, D.V.; Koeppe, R.E. The preference of tryptophan for membrane interfaces: Insights from N-methylation of tryptophans in gramicidin channels. *J. Biol. Chem.* **2008**, *283*, 22233–22243. [CrossRef] [PubMed]
93. Haney, E.F.; Straus, S.K.; Hancock, R.E.W. Reassessing the host defense peptide landscape. *Front. Chem.* **2019**, *7*, 43. [CrossRef] [PubMed]
94. Hong, H.; Park, S.; Jimenez, R.H.F.; Rinehard, D.; Tamm, L.K. Role of aromatic side chains in the folding and thermodynamic stability of integral membrane proteins. *J. Am. Chem. Soc.* **2007**, *129*, 8320–8327. [CrossRef]
95. Bhattacharjya, S.; Straus, S.K. Design, engineering and discovery of novel α-helical and β-boomerang antimicrobial peptides against drug resistant bacteria. *Int. J. Mol. Sci.* **2020**, *21*, 5773. [CrossRef]
96. Necelis, M.R.; Santiago-Ortiz, L.E.; Caputo, G.A. Investigation of the role of aromatic residues in the antimicrobial peptide BuCATHL4B. *Protein Pept. Lett.* **2020**. Online ahead of print. [CrossRef]
97. Liu, Y.; Shi, J.; Tong, Z.; Jia, Y.; Yang, K.; Wang, Z. Potent broad-spectrum antibacterial activity of amphiphilic peptides against multidrug-resistant bacteria. *Microorganisms* **2020**, *8*, 1398. [CrossRef] [PubMed]
98. Kelkar, D.A.; Chattopadhyay, A. The gramicidin ion channel: A model membrane protein. *Biochim. Biophys. Acta.* **2007**, *1768*, 2011–2025. [CrossRef]
99. Becker, M.D.; Greathouse, D.V.; Koeppe, R.E.; Andersen, O.S. Amino acid sequence modulation of gramicidin channel function: Effects of tryptophan-to-phenylalanine substitutions on the single-channel conductance and duration. *Biochemistry* **1991**, *30*, 8830–8839. [CrossRef]
100. Salom, D.; Pérez-Payá, E.; Pascal, J.; Abad, C. Environment- and sequence-dependent modulation of the double-stranded to single-stranded conformational transition of gramicidin A in membranes. *Biochemistry* **1998**, *37*, 14279–14291. [CrossRef]
101. Cotten, M.; Xu, F.; Cross, T.A. Protein stability and conformational rearrangements in lipid bilayers: Linear gramicidin, a model system. *Biophys. J.* **1997**, *73*, 614–623. [CrossRef]
102. Chaudhuri, A.; Haldar, S.; Sun, H.; Koeppe II, R.R.; Chattopadhyay, A. Importance of indole N-single bond-H hydrogen bonding in the organization and dynamics of gramicidin channels. *Biochim. Biophys. Acta.* **2014**, *1838*, 419–428. [CrossRef] [PubMed]
103. Scheiner, S.; Kar, T.; Pattanayak, J. Comparison of various types of hydrogen bonds involving aromatic amino acids. *J. Am. Chem. Soc.* **2002**, *124*, 13257–13264. [CrossRef] [PubMed]
104. Neidigh, J.W.; Fesinmeyer, R.M.; Prickett, K.S.; Andersen, N.H. Exendin-4 and glucagon-like-peptide-1: NMR structural comparisons in the solution and micelle-associated states. *Biochemistry* **2001**, *40*, 13188–13200. [CrossRef]
105. Neidigh, J.W.; Fesinmeyer, R.M.; Andersen, N.H. Designing a 20-residue protein. *Nat. Struct. Biol.* **2002**, *9*, 425–430. [CrossRef]

106. Qiu, L.; Pabit, S.A.; Roitberg, A.E.; Hagen, S.J. Smaller and faster: The 20-residue Trp-cage protein folds in 4 micros. *J. Am. Chem. Soc.* **2002**, *124*, 12952–12953. [CrossRef]
107. Byrne, A.; Williams, D.V.; Barua, B.; Hagen, S.J.; Kier, B.L.; Andersen, N.H. Folding dynamics and pathways of the trp-cage miniproteins. *Biochemistry* **2014**, *53*, 6011–6021. [CrossRef]
108. Barua, B.; Lin, J.C.; Williams, V.D.; Kummler, P.; Neidigh, J.W.; Andersen, N.H. The Trp-cage: Optimizing the stability of a globular miniprotein. *Protein Eng. Des. Sel.* **2008**, *21*, 171–185. [CrossRef]
109. Sobash, P.T.; Kolhe, R.; Karim, N.A.; Guddati, A.K.; Jillella, A.; Kota, V. Role of indoleamine 2,3-dioxygenase in acute myeloid leukemia. *Future Oncol.* **2020**. Online ahead of print. [CrossRef]
110. Blasco, H.; Bessy, C.; Plantier, L.; Lefevre, A.; Piver, E.; Bernard, L.; Marlet, J.; Stefic, K.; Benz-de Bretagne, I.; Cannet, P.; et al. The specific metabolome profiling of patients infected by SARS-COV-2 supports the key role of tryptophan-nicotinamide pathway and cytosine metabolism. *Sci. Rep.* **2020**, *10*, 16824. [CrossRef]
111. Ma, N.; He, T.; Johnston, L.J.; Ma, X. Host-microbiome interactions: The aryl hydrocarbon receptor as a critical node in tryptophan metabolites to brain signaling. *Gut Microbes* **2020**, *11*, 1203–1219. [CrossRef] [PubMed]
112. Hényková, E.; Vránová, H.P.; Amakorová, P.; Pospíšil, T.; Žukauskaitė, A.; Vlčková, M.; Urbánek, L.; Novák, O.; Mareš, J.; Kaňovský, P.; et al. Stable isotope dilution ultra-high performance liquid chromatography-tandem mass spectrometry quantitative profiling of tryptophan-related neuroactive substances in human serum and cerebrospinal fluid. *J. Chromatogr. A* **2016**, *1437*, 145–157. [CrossRef] [PubMed]
113. Ozaki, Y.; Edelstein, M.P.; Duch, D.S. Induction of indoleamine 2,3-dioxygenase: A mechanism of the antitumor activity of interferon gamma. *Proc. Natl. Acad. Sci. USA* **1988**, *85*, 1242–1246. [CrossRef] [PubMed]
114. Feng, G.S.; Taylor, M.W. Interferon gamma-resistant mutants are defective in the induction of indoleamine 2,3-dioxygenase. *Proc. Natl. Acad. Sci. USA* **1989**, *86*, 7144–7148. [CrossRef] [PubMed]
115. Frumento, G.; Rotondo, R.; Tonetti, M.; Damonte, G.; Benatti, U.; Ferrara, G.B. Tryptophan-derived catabolites are responsible for inhibition of T and natural killer cell proliferation induced by indoleamine 2,3-dioxygenase. *J. Exp. Med.* **2002**, *196*, 459–468. [CrossRef] [PubMed]
116. Hwu, P.; Du, M.X.; Lapointe, R.; Do, M.; Taylor, M.W.; Young, H.A. Indoleamine 2,3-dioxygenase production by human dendritic cells results in the inhibition of T cell proliferation. *J. Immunol.* **2000**, *164*, 3596–3599. [CrossRef]
117. Munn, D.H.; Sharma, M.D.; Lee, J.R.; Jhaver, K.J.; Johnson, T.S.; Keskin, D.B.; Marshall, B.; Chandler, P.; Antonia, S.J.; Burgess, R.; et al. Potential regulatory function of human dendritic cells expressing indoleamine 2,3-dioxygenase. *Science* **2002**, *297*, 1867–1870. [CrossRef]
118. Orabona, C.; Grohmann, U. Indoleamine 2,3-dioxygenase and regulatory function: Tryptophan starvation and beyond. *Methods Mol. Biol.* **2011**, *677*, 269–280. [CrossRef]
119. Dai, W.; Pan, H.; Kwok, O.; Dubey, J.P. Human indoleamine 2,3-dioxygenase inhibits *Toxoplasma gondii* growth in fibroblast cells. *J. Interferon Res.* **1994**, *14*, 313–317. [CrossRef]
120. Mehraj, V.; Routy, J.P. Tryptophan catabolism in chronic viral infections: Handling uninvited guests. *Int. J. Tryptophan Res.* **2015**, *8*, 41–48. [CrossRef]
121. Mellor, A.L.; Munn, D.H. Tryptophan catabolism and regulation of adaptive immunity. *J. Immunol.* **2003**, *170*, 5809–5813. [CrossRef] [PubMed]
122. Adams, O.; Besken, K.; Oberdörfer, C.; MacKenzie, C.R.; Rüssing, D.; Däubener, W. Inhibition of human herpes simplex virus type 2 by interferon gamma and tumor necrosis factor alpha is mediated by indoleamine 2,3-dioxygenase. *Microbes Infect.* **2004**, *6*, 806–812. [CrossRef] [PubMed]
123. Huang, L.; Li, L.; Klonowski, K.D.; Tompkins, S.M.; Tripp, R.A.; Mellor, A.L. Induction and role of indoleamine 2,3 dioxygenase in mouse models of influenza A virus infection. *PLoS ONE* **2013**, *8*, e66546. [CrossRef] [PubMed]
124. Mao, R.; Zhang, J.; Jiang, D.; Cai, D.; Levy, J.M.; Cuconati, A.; Block, T.M.; Guo, J.T.; Guo, H. Indoleamine 2,3-dioxygenase mediates the antiviral effect of gamma interferon against hepatitis B virus in human hepatocyte-derived cells. *J. Virol.* **2011**, *85*, 1048–1057. [CrossRef]
125. Obojes, K.; Andres, O.; Kim, K.S.; Däubener, W.; Schneider-Schaulies, J. Indoleamine 2,3-dioxygenase mediates cell type-specific anti-measles virus activity of gamma interferon. *J. Virol.* **2005**, *79*, 7768–7776. [CrossRef]

126. Sage, L.K.; Fox, J.M.; Mellor, A.L.; Tompkins, S.M.; Tripp, R.A. Indoleamine 2,3-dioxygenase (IDO) activity during the primary immune response to influenza infection modifies the memory T cell response to influenza challenge. *Viral Immunol.* **2014**, *27*, 112–123. [CrossRef]
127. Rabbani, M.A.G.; Ribaudo, M.; Guo, J.-T.; Barik, S. Identification of interferon-stimulated gene proteins that inhibit human parainfluenza virus type 3. *J. Virol.* **2016**, *90*, 11145–11156. [CrossRef]
128. Schmidt, S.V.; Schultze, J.L. New insights into IDO biology in bacterial and viral infections. *Front. Immunol.* **2014**, *5*, 384. [CrossRef]
129. Bitko, V.; Musiyenko, A.; Shulyayeva, O.; Barik, S. Inhibition of respiratory viruses by nasally administered siRNA. *Nat. Med.* **2005**, *11*, 50–55. [CrossRef]
130. Villenave, R.; Shields, M.D.; Power, U.F. Respiratory syncytial virus interaction with human airway epithelium. *Trends Microbiol.* **2013**, *21*, 238–244. [CrossRef]

Publisher's Note: MDPI stays neutral with regard to jurisdictional claims in published maps and institutional affiliations.

© 2020 by the author. Licensee MDPI, Basel, Switzerland. This article is an open access article distributed under the terms and conditions of the Creative Commons Attribution (CC BY) license (http://creativecommons.org/licenses/by/4.0/).

Review

Developmental Programming and Reprogramming of Hypertension and Kidney Disease: Impact of Tryptophan Metabolism

Chien-Ning Hsu [1,2] and You-Lin Tain [3,4,*]

1. Department of Pharmacy, Kaohsiung Chang Gung Memorial Hospital, Kaohsiung 833, Taiwan; chien_ning_hsu@hotmail.com
2. School of Pharmacy, Kaohsiung Medical University, Kaohsiung 807, Taiwan
3. Department of Pediatrics, Kaohsiung Chang Gung Memorial Hospital and Chang Gung University College of Medicine, Kaohsiung 833, Taiwan
4. Institute for Translational Research in Biomedicine, Kaohsiung Chang Gung Memorial Hospital and Chang Gung University College of Medicine, Kaohsiung 833, Taiwan
* Correspondence: tainyl@hotmail.com; Tel.: +886-975-056-995; Fax: +886-7733-8009

Received: 15 September 2020; Accepted: 17 November 2020; Published: 18 November 2020

Abstract: The concept that hypertension and chronic kidney disease (CKD) originate in early life has emerged recently. During pregnancy, tryptophan is crucial for maternal protein synthesis and fetal development. On one hand, impaired tryptophan metabolic pathway in pregnancy impacts fetal programming, resulting in the developmental programming of hypertension and kidney disease in adult offspring. On the other hand, tryptophan-related interventions might serve as reprogramming strategies to prevent a disease from occurring. In the present review, we aim to summarize (1) the three major tryptophan metabolic pathways, (2) the impact of tryptophan metabolism in pregnancy, (3) the interplay occurring between tryptophan metabolites and gut microbiota on the production of uremic toxins, (4) the role of tryptophan-derived metabolites-induced hypertension and CKD of developmental origin, (5) the therapeutic options in pregnancy that could aid in reprogramming adverse effects to protect offspring against hypertension and CKD, and (6) possible mechanisms linking tryptophan metabolism to developmental programming of hypertension and kidney disease.

Keywords: aryl hydrocarbon receptor; chronic kidney disease; developmental origins of health and disease (DOHaD); hypertension; indole; kynurenine; melatonin; serotonin; tryptophan; uremic toxin

1. Introduction

Hypertension affects more than one fourth of the global population [1]. Despite pharmacotherapy advances over the past decades, the prevalence of hypertension is still rising globally [2]. Like hypertension, chronic kidney disease (CKD) is another major public health concern around the world [3]. Approximately 10% of the population worldwide is affected by CKD. Hypertension and CKD are bidirectionally interlinked, because aspects of the pathophysiology are shared by both diseases in the kidneys. Hypertension is a risk factor for CKD and most patients with CKD have hypertension. Although hypertension and CKD are more common in adults, both of which can be driven by environmental insults in early life [4,5]. This has given rise to the concept of "developmental origins of health and disease" (DOHaD) [6]. Maternal nutrition is an important factor which determines fetal development. Imbalanced maternal nutrition during pregnancy and lactation produces fetal programming that permanently alter the body's morphology and function and leads to many adult diseases, including hypertension and CKD [7]. Adverse renal programming alters structure and function of the kidneys permanently and increases the risk for developing hypertension and kidney disease later in life [8,9]. Conversely, early-life nutritional interventions have recently started to gain importance to reverse the programming processes to prevent hypertension and CKD by so-called reprogramming [10].

Tryptophan is a nutritionally essential amino acid that must be provided through dietary sources. Given the complexity of tryptophan metabolic pathways, the diverse properties of tryptophan-derived metabolites have been linked to various pathophysiological states [11–13]. Additionally, tryptophan metabolism has emerged as a central hub for host–microbe interactions [14]. During pregnancy, tryptophan is mandatory because of increased demand for maternal protein synthesis and fetal growth and development [15]. Although the estimated average requirement (EAR) in pregnancy for total protein is recommended [16], there remains a lack of recommendation of a specific amino acid like tryptophan for pregnant women. Endogenous tryptophan metabolites (e.g., serotonin and melatonin) and microbial tryptophan catabolites (e.g., indole and indole-3-aldehyde) have been linked to hypertension and kidney disease [17,18]; however, the impact of maternal tryptophan metabolism on the development of hypertension and kidney disease in adult offspring is still largely unknown.

According to the two aspects of the DOHaD concept, tryptophan metabolic pathways may act as a double-edged sword in developmental programming of hypertension and CKD. This review, therefore, highlights evidence on the impact of tryptophan metabolism during pregnancy on offspring hypertension and kidney disease, as well as the role of tryptophan-related interventions as a reprogramming strategy in the prevention of hypertension of CKD in adult offspring.

We searched the PubMed/MEDLINE databases for studies published in English between January 1990 and July 2020, using the following search terms: "blood pressure," "chronic kidney disease" "developmental programming," "DOHaD," "gestation" "hypertension," "indole," "indoxyl sulfate," "kynurenine," "melatonin," "mother," "maternal," "offspring," "progeny," "pregnancy," "perinatal," "serotonin," and "tryptophan." Relevant studies were assessed for inclusion by combining the title and abstract screening, followed by a review of full-text studies.

2. Tryptophan Metabolism

2.1. Tryptophan Metabolic Pathways

Tryptophan is present in protein-based foods, particularly meat, milk, peanuts and fish [11]. Approximately one-third of the whole-body flux of tryptophan comes from a dietary source, and the rest is from protein degradation. Although tryptophan is found in the smallest concentrations of the 20 amino acids in the human body [11], it has complex and multifaceted biological effects due to its wide range of biologically active metabolites along various metabolic pathways. Tryptophan metabolism follows three major pathways in the gut: (1) the kynurenine pathway in both epithelial and immune cells; (2) the serotonin pathway in enterochromaffin cells; and (3) the indole pathway by the gut microbiota [14]. A schematic summarizing the major tryptophan metabolic pathways is presented in Figure 1. Tryptophan absorption is primarily mediated by the solute carrier (SLC) 6A19, encoding system B (0) neutral amino acid transporter 1 (B^0AT1). Over 95% of the absorbed tryptophan is catabolized via the kynurenine pathway, while only 1–2% and 2–3% of dietary tryptophan are converted into serotonin and indole pathways, respectively [19,20].

First, the kynurenine pathway plays a critical role in generating cellular energy in the form of nicotinamide adenine dinucleotide (NAD^+) [20]. The initial and rate-limiting step in the kynurenine pathway involves one of three enzymes, namely, the indoleamine 2-3-dioxygenase 1 and 2 (IDO1 and IDO2) and tryptophan 2,3-dioxygenase (TDO). The product of the IDO/TDO-catalyzed reaction, N-formylkynurenine, is then hydrolyzed to kynurenine. Following its synthesis, kynurenine can be further metabolized by various enzymes. Kynureninase (KYNU) produces anthranilic acid (AA) from kynurenine. Kynurenine-3-monooxygenase (KMO) converts KYN into 3-hydroxykynurenine (3-HK), which can be taken by kynurenine aminotransferase (KAT) to produce xanthurenic acid or by the KYNU to form 3-hydroxyanthranilic acid (3-HAA). Catabolism of 3-HAA can lead to the generation of picolinic acid, quinolinic acid, and NAD⁺. In addition, KAT can metabolize kynurenine into kynurenic acid.

Second, the serotonin pathway is initiated by tryptophan being hydroxylated by tryptophan hydroxylase (TPH) to the intermediate 5-hydroxytryptophan (5-HTP), which is subsequently decarboxylated by aromatic amino acid decarboxylase (AAAD) to become serotonin (5-hydroxytryptamine, 5-HT). The enterochromaffin cells in the gut account for almost 90% of the human body's serotonin synthesis. Serotonin can be acetylated to form N-acetylserotonin by arylalkylamine N-acetyltransferase (AANAT), followed by N-acetylserotonin O-methyltransferase (ASMT) to generate melatonin. One the other hand, serotonin can be metabolized by monoamine oxidase (MAO) to 5-hydroxyindoleacetic acid (5-HIAA).

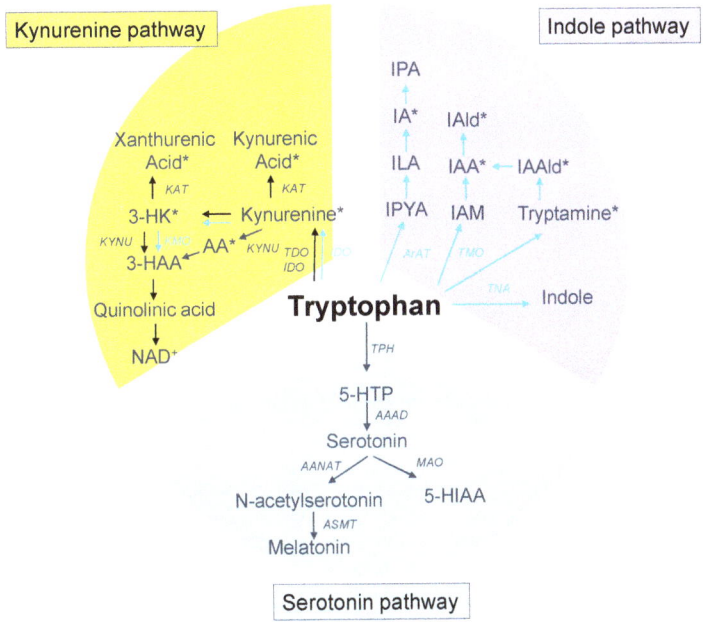

Figure 1. Overview of tryptophan metabolism through the kynurenine (yellow), serotonin (blue), and indole (purple) metabolic pathways. The black arrow lines indicate the host pathway, while the blue arrow lines indicate the microbial pathway. The asterisk indicates the aryl hydrocarbon receptor (AhR) ligand. IDO = indoleamine 2-3-dioxygenase; TDO = tryptophan 2,3-dioxygenase; AA = anthranilic acid; KMO = kynurenine-3-monooxygenase; KYNU = kynureninase; 3-HK = 3-hydroxykynurenine; KAT = kynurenine aminotransferase (KAT); 3-HAA = 3-hydroxyanthranilic acid; NAD⁺ = nicotinamide adenine dinucleotide; TPH = tryptophan hydroxylase; 5-HTP = 5-hydroxytryptophan; AAAD = aromatic amino acid decarboxylase; AANAT = arylalkylamine N-acetyltransferase; ASMT = N-acetylserotonin O-methyltransferase; MAO = monoamine oxidase; 5-HIAA = 5-hydroxyindoleacetic acid; TNA = tryptophanase; IAA = indoleacetic acid; IAld = indole-3-aldehyde; IAAld = indole-3-acetaldehyde (IAAld); IA = indoleacrylic acid; IPA = indole-3-propionic acid; ILA = indolelactic acid; IPYA = Indole-3-pyruvate; ArAT = acromatic amino acid aminotransferase; TMO = tryptophan 2-monooxygenase; IAM = indole-3-acetamide.

Last, a small amount of tryptophan is converted by gut microbiota into tryptamine, and through the action of the enzyme tryptophanase (TNA) into indole and its derivatives [13]. Tryptamine induces the release of serotonin by enterochromaffin cells. Indoles can be further metabolized to indoxyl sulfate (IS) and indoxyl-β-D glucuronide (IDG) in the liver [21]. Many indole derivatives, such as indoleacetic acid (IAA), indole-3-aldehyde (IAld), indole-3-acetaldehyde (IAAld), indoleacrylic acid

(IA), and indolelactic acid (ILA), are ligands for aryl hydrocarbon receptor (AhR) [22]. As shown in Figure 2, several microbial tryptophan catabolites like IS, IDG, and IAA are unable to be excreted in urine in patients with CKD. These tryptophan metabolites are accumulated as uremic toxins [23]. In addition to the indole pathway, tryptophan-derived uremic toxins can also come from the kynurenine pathway, such as kynurenine, kynurenic acid, 3-HK, 3-HAA, and quinolinic acid.

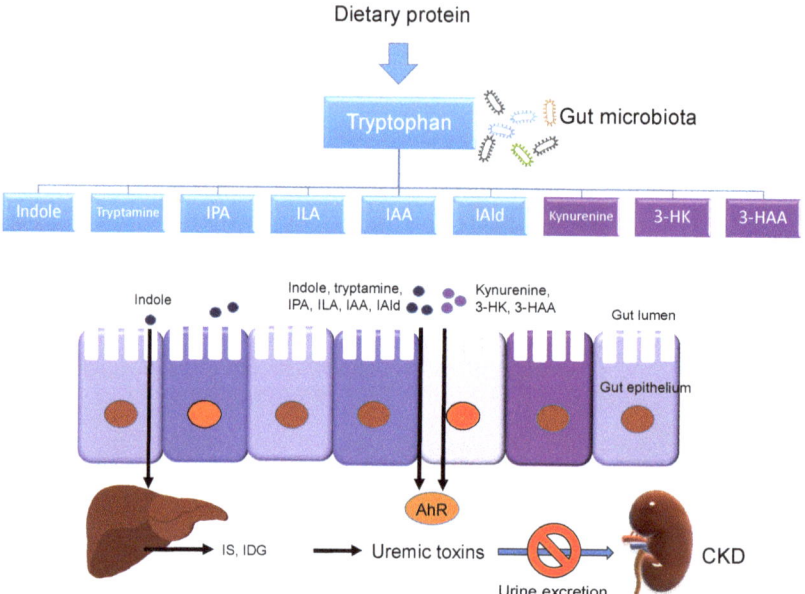

Figure 2. Schematic representation of the interplay occurring between tryptophan metabolites and gut microbiota on the production of uremic toxins in patients with chronic kidney disease (CKD). Microbial tryptophan catabolites are from the kynurenine (purple box) and indole (blue box) metabolic pathways. The black arrow lines indicate tryptophan metabolites are absorbed through the gut epithelium and enter the bloodstream. The indole can be further metabolized to indoxyl sulfate (IS) and indoxyl-β-D glucuronide (IDG) in the liver. Under chronic kidney disease (CKD), the kidneys are unable to excrete these tryptophan metabolites and cause the accumulation of uremic toxins. AhR = aryl hydrocarbon receptor; 3-HK = 3-hydroxykynurenine; 3-HAA = 3-hydroxyanthranilic acid; IPA = indole-3-propionic acid; ILA = indolelactic acid; IAA = indoleacetic acid; IAld = indole-3-aldehyde.

2.2. Tryptophan Metabolism in Pregnancy

The quantitative dietary tryptophan requirement varies broadly across species [24]. The recommended daily intake of tryptophan was 4 mg/kg body weight for adult humans by the WHO [25]. Tryptophan requirements in pregnancy are increased, because of the increased demand for maternal protein synthesis, fetal requirements for growth and development, serotonin release for signaling pathways, production of KA for neuronal protection, and NAD$^+$ synthesis [15]. Previous studies illustrated the increased requirement and transport of tryptophan to the fetus because tryptophan level in umbilical cord was higher than in maternal plasma [15]. Pregnancy-associated plasma hypoaminoacidemia develops early in gestation and persists throughout pregnancy [26]. Whole-body protein turnover studies demonstrated increased protein synthesis by 15% in the second trimester and 25% during the third trimester [25]. The current recommended pregnancy EAR for total protein is 0.88 g/kg/day, which is 1.33 times the EAR for non-pregnant adults; however, no recommendation has been developed for tryptophan requirements in human pregnancy [27]. In swine, the requirements for tryptophan have been revealed to increase by 35% during the late stages of pregnancy when compared to the early stages [28]. However, gestation-stage-specific dietary tryptophan recommendation awaits further evidence.

Excessive or insufficient consumption of a specific amino acid has been linked to adverse fetal outcomes [29,30]. Plasma tryptophan level is high at birth, but quickly declines within 24 h and reaches normal values by day 7 of life [31]. In terms of placentas at delivery, the KYN-to-tryptophan ratio measured in the chorionic plate is higher than that in the peripheral blood, suggesting placental IDO activity is highly active at delivery [32]. Additionally, the NAD pathway is enhanced in pregnant women and rats [33]. In the serotonin pathway, an adequate supply of serotonin and melatonin is crucial for fetal development [34–36]. Clinical and experimental evidence support the notion that perinatal selective serotonin reuptake inhibitors (SSRI) exposure can reduce body weight, impair brain development, and cause life-long adverse emotional health [37]. In a maternal melatonin deficient rat model, offspring had disrupted circadian rhythms and intrauterine growth retardation (IUGR), which were prevented by maternal melatonin treatment [38]. These findings indicate that tryptophan-derived metabolites in the serotonin pathway acts in different ways in the maternal–fetal system to bring on a successful pregnancy and fetal development.

3. Tryptophan Metabolism in Hypertension and Kidney Disease

Dietary tryptophan was shown to have a negative correlation with systolic BP in the TwinsUK study [39], while this finding was not supported by other evidence [40]. In hypertensive animal models, dietary tryptophan has been reported to attenuate the development of hypertension in the spontaneously hypertensive rats [41], Dahl salt-sensitive rats [42], and renovascular hypertensive rats [43]. Of note is that not only tryptophan but also its metabolites have vasodilatory property [44]. First, KYN was known to dilate coronary arteries or aorta in different animal species in a dose-dependent manner [44]. In a renovascular hypertensive rat model, higher plasma levels of several kynurenine metabolites such as KYN and AA were found in hypertensive rats than in sham rats [45]. In contrast, overexpression of IDO, a rate-limiting enzyme in the kynurenine pathway, in endothelium could protect against hypoxia-induced pulmonary hypertension in rodents [46]. These observations suggest hypertension is associated with activation of the kynurenine pathway (Figure 3). Second, the role played by serotonin in BP regulation is complex and still unclear [47]. Variations of serotonin levels in the plasma and platelet in different types of hypertensive patients can be increased [48], unaltered [49], or even decreased [50]. Although human and animal studies showed serotonin mainly results in acute and direct effect of arterial constriction [47], chronic administration of serotonin causes a long-term decrease in BP [51,52]. Melatonin is another important metabolite in the serotonin metabolic pathway. As we reviewed elsewhere, melatonin can prevent the development of hypertension via receptor-dependent and receptor-independent actions [53].

Uremic toxins derived from tryptophan fermentation by gut microbiota are associated with cardiovascular disease (CVD) in patients with CKD [17,23]. Tryptophan-derived uremic toxins, mainly coming from the indole and kynurenine pathways, have prooxidant, proinflammatory, procoagulant, and pro-apoptotic effects [23]; moreover, most of them are potent AhR ligands [23]. In patients with CKD, serum tryptophan level is decreased whereas metabolites of the kynurenine pathway are increased [54,55]. Despite the fact that hemodialysis can reduce kynurenine metabolites, plasma levels of KYN, 3-HK, AA, xanthurenic acid and quinolinic acid were still higher in uremic patients than those in healthy volunteers [55]. Uremic toxins from the kynurenine pathway like KYN and 3-HK are most frequent elevated in patients with CVD [56]. Moreover, tryptophan-derived uremic toxins from the indole pathway are also relevant to CVD in patients with CKD. IS was associated with the presence of CVD and cardiovascular mortality in uremic patients [57]. Moreover, bacterial tryptophan catabolites including IAA, IA, IAld, ILA are AhR ligands [13]. Previous studies demonstrated that exogenous AhR ligand induces a high BP [58–60] and activation of AhR modulating T helper 17 (TH17) axis is involved in the development of hypertension [61]. Indoxyl sulfate (IS) is one of the most extensively studied uremic toxin. IS has been shown to induce inflammation and fibrosis in proximal tubule cells, impair the proliferation of endothelial cells, promote calcification of vascular smooth muscle cells, induce oxidative stress in proximal tubular and endothelial cells, and increase AhR-regulated gene expression in endothelial cells [57]. Together, these mechanisms are thought to cause CVD and CKD progression [57].

A growing body of evidence has demonstrated three tryptophan metabolic pathways connect hypertension and kidney disease (Figure 3), but relatively little is known about the role of maternal tryptophan metabolism in the development of hypertension and CKD in offspring. Whether targeting on tryptophan metabolic pathways can be applied in pregnancy to improve offspring renal outcomes remains to be addressed.

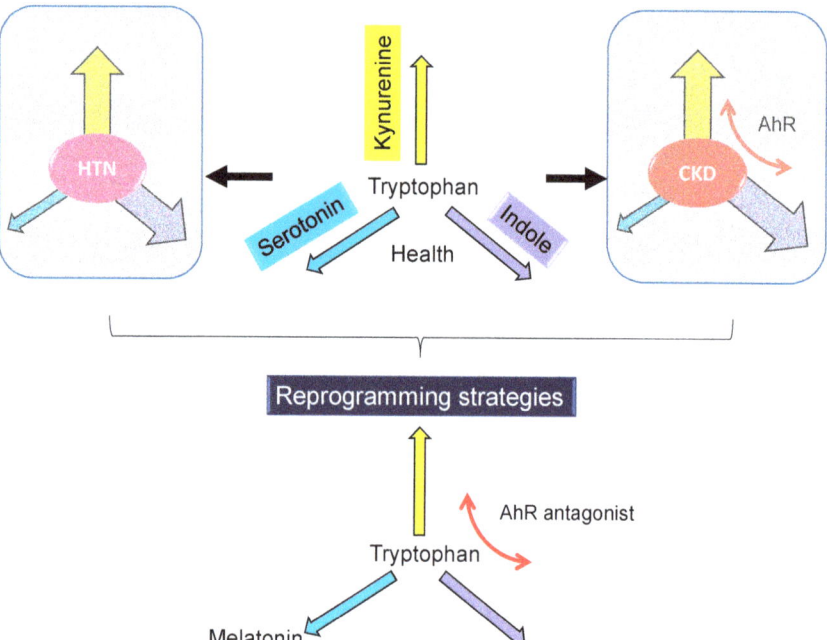

Figure 3. Schematic diagrams indicate repartitioning of tryptophan metabolic pathways in hypertension (HTN) and chronic kidney disease (CKD) based on the available literature data. The three major tryptophan metabolic pathways are kynurenine (yellow), serotonin (blue), and indole (purple) pathways. They are tightly interconnected to maintain good health and differentially affected in diseases. Weights of arrow lines indicate strength of pathway activation. The restoration of impaired tryptophan metabolic pathways using tryptophan supplementation, melatonin, or aryl hydrocarbon receptor (AhR) antagonist represents a promising reprogramming strategy.

4. Tryptophan Metabolic Pathways: Programming versus Reprogramming Effects

4.1. Tryptophan-Related Metabolites-Induced Hypertension and CKD of Developmental Origin

A maternal low protein diet has been reported to program hypertension-related disorders in adult offspring in rodents, pigs, sheep, and cows [62–64]. Some studies demonstrated that maternal dietary tryptophan deficiency caused adverse effects on the development of the brain, liver, and skeletal muscle in rats [65–67]. However, there were no studies showing the programming effect of specific amino acid deficiency like tryptophan in pregnancy on offspring's BP and renal outcome in humans and animals [28]. A plethora of tryptophan-derived metabolites have both detrimental and beneficial effects [11–13]. Therefore, excessive or deficit of a particular tryptophan-generating metabolite in pregnancy might be linked to hypertension and kidney disease in adult offspring. Maternal deficiency of melatonin, a tryptophan-derived metabolite produced in the serotonin pathway, has an increased risk for developing hypertension in adult offspring in a constant light exposure rat model [68]. Additionally, maternal CKD was reported to induce renal hypertrophy and hypertension in 12-week-old adult male rat offspring [69]. Since several uremic toxins from the kynurenine and indole pathways are endogenous ligands of AhR [58] and previous studies reporting maternal exposure to exogenous AhR ligand can induce hypertension and kidney disease in adult offspring [59,60], AhR activation might be an important target hub linking tryptophan metabolism and hypertension and kidney disease of developmental origin. Collectively, these observations suggest that dysregulated tryptophan metabolism in early-life is tightly linked to the risk for developing hypertension and kidney disease in adulthood.

4.2. Targeting on Tryptophan Metabolic Pathway as Reprogramming Strategies in Animal Models

Conversely, DOHaD theory offers a strategy to prevent the development of adult hypertension and kidney disease during early life, namely reprogramming [10]. Tryptophan supplementation has been used for the treatment of sleep disorders, pain, insomnia, depression, seasonal affective disorder, bulimia, attention deficit disorder, and chronic fatigue [11,70]. Nevertheless, less attention has been paid to study the potential beneficial effects of tryptophan supplementation during pregnancy and lactation on offspring health [11].

Since tryptophan-derived metabolites (e.g., serotonin and melatonin) and tryptophan-related signaling pathway (e.g., AhR) could be an alternative to obtaining the benefits provided by tryptophan, such tryptophan-related reprogramming interventions were recruited in the current review (Figure 3), with a focus on hypertension and kidney disease. We restricted this review to tryptophan-related interventions applied only during pregnancy or lactation periods, as there are critical periods for reprogramming strategies to prevent hypertension and kidney disease of developmental origin [71], which are listed in Table 1 [59,60,68,69,72–82].

Various adverse early-life environmental factors have been examined to induce hypertension and kidney disease in adult offspring, including a maternal CKD [69], SHR [72,80], a maternal caloric restriction [73], a maternal L-NAME exposure and/or postnatal high-fat diet [74,82], a maternal high-fructose diet [75], a maternal constant light exposure [68], a maternal high methyl-donor diet [76], a maternal high-fructose diet plus a post-weaning high-salt diet [77], a prenatal glucocorticoid (GC) exposure and/or post-weaning high-fat diet [78,79], a maternal plus post-weaning high-fructose diet [81], maternal 2,3,7,8-Tetrachlorodibenzo-p-dioxin (TCDD) and GC exposures [59], and a maternal bisphenol A (BPA) exposure and high-fat diet [60]. These diverse in utero insults cause adverse phenotypes in adult offspring including hypertension [68,69,72–82], altered transcriptome [76], and reduced nephron numbers [78]. All these adverse offspring outcomes can be prevented, or at least attenuated, by tryptophan-related interventions.

Table 1. Reprogramming effects protect adult offspring against hypertension and kidney disease by tryptophan-related interventions.

Interventions	Animal Models	Species/Gender	Age at Measure	Reprogramming Effects
Tryptophan				
Tryptophan 200 mg/kg BW/day via oral gavage during pregnancy	Maternal adenosine-induced CKD	SD rat/M	12 weeks	Prevented hypertension [69]
Melatonin				
10 mg/kg BW/day melatonin in drinking water during pregnancy	Genetic hypertension model	SHR/M	16 weeks	Prevented hypertension [72]
0.01% melatonin in drinking water during pregnancy and lactation	Maternal caloric restriction	SD rat/M	12 weeks	Prevented hypertension [73]
0.01% melatonin in drinking water during pregnancy and lactation	Maternal L-NAME exposure	SD rat/M	12 weeks	Prevented hypertension [74]
0.01% melatonin in drinking water during pregnancy and lactation	Maternal high-fructose diet	SD rat/M	12 weeks	Prevented hypertension [75]
0.01% melatonin in drinking water during pregnancy and lactation	Maternal constant light exposure	SD rat/M	12 weeks	Prevented hypertension [68]
0.01% melatonin in drinking water during pregnancy and lactation	Maternal methyl-donor diet	SD rat/M	12 weeks	Attenuated hypertension and altered renal transcriptome [76]
0.01% melatonin in drinking water during pregnancy and lactation	Maternal high-fructose diet plus post-weaning high-salt diet	SD rat/M	12 weeks	Attenuated hypertension [77]
0.01% melatonin in drinking water during pregnancy and lactation	Prenatal GC exposure	SD rat/M	16 weeks	Prevented hypertension and increased nephron number [78]
0.01% melatonin in drinking water during pregnancy and lactation	Prenatal GC exposure plus post-weaning high-fat diet	SD rat/M	16 weeks	Prevented hypertension [79]

Table 1. Cont.

Interventions	Animal Models	Species/Gender	Age at Measure	Reprogramming Effects
AhR antagonist				
4 g/kg diet resveratrol during pregnancy and lactation	Genetic hypertension model	SHR/M and F	20 weeks	Prevented hypertension [80]
50 mg/L resveratrol in drinking water during pregnancy and lactation	Maternal plus post-weaning high-fructose diet	SD rat/M	12 weeks	Prevented hypertension [81]
0.05% resveratrol in drinking water during pregnancy and lactation	Maternal TCDD and GC exposures	SD rat/M	16 weeks	Prevented hypertension [59]
50 mg/L resveratrol in drinking water during pregnancy and lactation	Maternal bisphenol A exposure and high-fat diet	SD rat/M	16 weeks	Prevented hypertension [60]
50 mg/L resveratrol in drinking water during pregnancy and lactation	Maternal L-NAME plus postnatal high-fat diet	SD rat/M	16 weeks	Prevented hypertension [82]

CKD = chronic kidney disease; SD = Sprague–Dawley; M = male; F = female; SHR = spontaneously hypertensive rat; L-NAM E = N^G-nitro-L-arginine methyl ester. GC = glucocorticoid; AhR = aryl hydrocarbon receptor; TCDD = 2,3,7,8-Tetrachlorodibenzo-p-dioxin.

As shown in Table 1, rats are the most used subjects among animal models of developmental programming. One study showed tryptophan supplementation in pregnancy was reported to protect adult offspring against hypertension programmed by maternal CKD [69]. Although many researchers have studied tryptophan requirement in pigs [24], little is known whether tryptophan supplementation is beneficial in preventing kidney disease and hypertension in large animal models. Although a wide-range of metabolites come from tryptophan metabolism, only melatonin has been studied as a reprogramming intervention in pregnancy and lactation to protect adult offspring against hypertension and kidney disease of developmental origin [53]. Melatonin has pleiotropic biofunctions, such as antioxidant, anti-inflammation, regulation of circadian rhythm, and epigenetic regulation [53,83]; it also plays a vital role in pregnancy and fetal growth [84,85]. Perinatal melatonin therapy not only prevents hypertension programmed by diverse early-life insults [68,72–79] but also affects nephron number and renal transcriptome [76,78]. Reviews elsewhere have highlighted that low nephron number increases later life risk of hypertension and kidney disease [71,86]. In rats, prenatal glucocorticoid (GC) exposure causes a reduced nephron number and hypertension in adult offspring [78,87]. Melatonin therapy during pregnancy and lactation can prevent the reduction in nephron number and the rise of BP [78]. These findings indicate there is a renal reprogramming effect of melatonin protecting against maternal GC exposure-induced adverse offspring outcomes. Another study reported that a maternal methyl-donor diet results in alterations of renal transcriptome and programmed hypertension in adult rat offspring [76]. Conversely, maternal melatonin therapy altered 677 genes in renal transcriptome by which the elevation of BP in adult offspring can be attenuated [76]. Although oral tryptophan supplementation was reported to increase nocturnal circulating melatonin levels in Wistar rats [88], whether the above-mentioned beneficial effects of melatonin can be reached by maternal tryptophan supplementation remains to be elucidated. Like melatonin, serotonin is another important tryptophan-generating metabolite from the serotonin metabolic pathway. Being a neurotransmitter, the impact of serotonin has been extensively studied in developmental programming of neuropsychiatric disorders [34,89,90]. Nevertheless, its reprogramming effects on hypertension and kidney disease of developmental origin have not been reported yet.

On the other hand, several bacterial tryptophan catabolites from the indole and kynurenine pathways are AhR ligands [13]. Since maternal AhR activation is related to programmed hypertension and kidney disease in adult offspring [59,60], AhR antagonists might be a potential reprogramming strategy to reverse programming processes and prevent adverse outcomes. Resveratrol, a natural AhR antagonist [91], has been proposed to reprogram hypertension-related disorders [92]. This review presents that resveratrol supplementation in pregnancy and lactation can aid in preventing the development of hypertension in various developmental hypertension models, including SHRs [80], a maternal high-fructose diet [81], maternal TCDD and GC exposures [59], and a maternal BPA and high-fat diet [60]. So far, a specific AhR antagonist still remains inaccessible in clinical practice. Because resveratrol has multiple biological functions not just an AhR antagonist, whereas not all tryptophan-derived metabolites are AhR ligands, additional studies are required to elucidate which metabolite(s)-induced hypertension and kidney disease is AhR-dependent and develop a specific AhR-targeting approach as a reprogramming intervention in the future.

5. Common Mechanisms Link Tryptophan Metabolism to Developmental Programming of Hypertension and Kidney Disease

Although several organ systems are involved in the regulation of BP, renal programming is considered crucial in the development of hypertension and kidney disease [8–10]. It is clear from the preceding sections that diverse early-life environmental insults lead to same offspring phenotype (i.e., hypertension and kidney disease) indicating that there may be common mechanisms underlying renal programming. To date, animal models have provided insight on certain pathways underlying renal programming [8–10]. Notably, some of these mechanisms that have been previously connected to tryptophan metabolism include oxidative stress, gut microbiota, activation of the renin–angiotensin system (RAS), and immunity/inflammation. Each will be discussed in turn.

5.1. Oxidative Stress

Pregnancy is characterized by a state of high oxidative stress owing to increased basal oxygen consumption [93]. Oxidative stress is an imbalance between oxidants and antioxidants in favor of oxidants. Previous works that haven been published support that oxidative stress is important for the developmental programming of hypertension and kidney disease [71,94]. Various animal models demonstrate oxidative stress involved in renal programming and hypertension, including prenatal GC exposure [43], maternal caloric restriction [73], maternal high fructose diet [75], maternal high methyl-donor diet [76], maternal NO depletion [82], and maternal diabetes [95]. Several tryptophan-derived metabolites from the kynurenine pathway like kynurenine, 3-HK, 3-HAA, and quinolinic acid have shown pro-oxidant effects [96–99]. However, evidence regarding the antioxidant effects of metabolites generated from the kynurenine pathway have also been reported [100]. Additionally, it is acknowledged that uremic toxins from the indole metabolic pathway like IS and IDG can induce oxidative stress, which in turn contributes to the progression of CKD [23].

Conversely, antioxidant therapy in pregnancy has been shown to protect offspring against hypertension and kidney disease programmed by different in utero environmental insults [94,101]. Melatonin, a potent antioxidant from the serotonin pathway, has shown beneficial effects on hypertension and kidney disease of developmental origin [68,72–79]. Though there is evidence on the correlation between oxidative stress and tryptophan metabolism, a single unifying theory that can show the beneficial or detrimental effects of various metabolites generating from different pathways is lacking.

In a maternal CKD model, tryptophan supplementation during pregnancy and lactation protects offspring hypertension and is associated with restoration of nitric oxide (NO) [69]. NO is a vasodilator and free radical and plays a role in oxidative stress. NO deficiency and increased oxidative stress in the kidney contribute to the pathogenesis of hypertension [102]. Accordingly, targeting NO has been reported as a reprogramming strategy to prevent hypertension and kidney disease of developmental programming [102]. There is a close interlink between NO and tryptophan metabolism. NO can inhibit IDO activity [103], inactivate TPH [104], mediate melatonin production [105], and counteract the inhibitory effect of indole-derived uremic toxin IS [106]. The potential for use of tryptophan for its antioxidant properties in pregnancy must be investigated for its metabolism interplay with oxidative stress in determining its impact on hypertension and kidney disease of developmental origin.

5.2. Gut Microbiota

Dsybiotic gut microbiome in early life may have adverse effects resulting in adulthood diseases, including hypertension [107]. The role of gut microbiota in the pathogenesis of hypertension and CKD is suspected [108]. A growing body of evidence proposes several possible mechanisms to link gut dysbiosis and hypertension, including alterations of gut microbiota composition and their metabolites, increased sympathetic activity, activation of the renin–angiotensin system (RAS), and inhibition of NO [109]. Of note is that the composition of the microbiota determines several tryptophan metabolites as they are gut microbial catabolites. These tryptophan-derived microbial catabolites are crucial signaling molecules in host–microbial crosstalk contributing to systemic homeostasis [13]. Several bacterial species have been reported to produce tryptophan catabolites, such as *Clostridium*, *Bifidobacterium*, *Lactobacillus*, *Ruminococcus*, *Ruminiclostridium*, *Bacteroides*, and *Peptostretococcus* [13,110,111]. We recently reported that maternal tryptophan supplementation protects offspring against hypertension programmed by maternal CKD is associated with alterations to several tryptophan-metabolizing microbes, including *Lactobacillus*, *Ruminococcus*, and *Clostridium* [69]. The involvement of tryptophan-metabolizing microbes is obvious in terms of the ability to produce tryptophan catabolites but might also account for the pathogenesis of maternal CKD-induced hypertension.

In experimental and clinical CKD, microbiota-derived uremic toxins from indole and kynurenine pathways are increased and contribute to the progression of CKD and CVD [17,23,112]. Recent studies support the notion that microbiota-targeted therapies can be applied to a variety of diseases [113], including CKD [114,115]. Manipulating the gut microbiota with prebiotics or probiotics has been reported to reduce gut microbiota-derived uremic toxins in CKD [114,115]. In our hands, targeting gut microbiota by prebiotics (i.e., a special form of dietary fiber), probiotics (i.e., beneficial bacteria in the gut), or postbiotics (i.e., microbial metabolites) is able to prevent hypertension programmed by various early-life insults [116–118]. However, the identification of microbes involved in the modulation of tryptophan metabolite signaling and developing microbiota-targeted therapy for hypertension and kidney disease of developmental origin demand further investigation.

5.3. Renin–Angiotensin System

RAS is a coordinated hormonal cascade in the control of BP and renal physiology [119]. The classical RAS, defined as the angiotensin converting enzyme (ACE)–Ang II-angiotensin type 1 receptor (AT1R) axis, promotes sodium retention and elevation of BP. Conversely, the non-classical RAS composed of the ACE2–Ang-(1-7)-Mas receptor axis leads to vasodilatation [119]. Pharmacological blockade of the classical RAS is currently used to treat hypertension and kidney disease [120]. A growing body of evidence indicates that dysregulated RAS is a common mechanism underlying renal programming and programmed hypertension [6,8–10]. Early blockade of the classical RAS can reprogram inappropriate activation of the RAS and reverse the adverse programmed processes [121,122].

Several lines of observation show that the interplay between tryptophan metabolism and the RAS has an impact on renal programming and hypertension. First, several tryptophan-containing peptides have abilities to inhibit ACE activity and may serve as a potential anti-hypertensive therapy [123]. Second, activation of the kynurenine pathway is connected in parallel with the RAS in a renovascular hypertension model [45]. Third, there are studies showing that tryptophan-derived uremic toxin IS upregulate Ang II signaling and downregulate Mas, contributing to CVD and CKD [124,125]. Last, the preceding sections show that the maternal melatonin therapy which protects offspring against hypertension is, at least in part, attributed to the RAS in a maternal constant light exposure model [68], a maternal caloric restriction model [73], a maternal L-NAME exposure model [74], and a maternal high-fructose diet model [75].

In a maternal CKD model, adult male offspring-developed hypertension is related to decreased renal mRNA expression of ACE2, MAS, and AT2R, which belong to the non-classical RAS pathway [69]. Nevertheless, maternal tryptophan treatment prevented the elevation of BP but had neglectable effects on the RAS. Detailed mechanisms that underlie the interactions between tryptophan metabolic pathways and the RAS and their impact on the programming process toward hypertension, however, remain to be clarified.

5.4. Immunity and Inflammation

Pregnancy is characterized as a physiologic systemic inflammatory response; compromised pregnancies and related complications are associated with inflammation [126]. The interrelationship between inflammation and tryptophan metabolism has been reported in pigs, mice, and humans [127,128]. Hypertension and kidney disease are associated with the accumulation of T cells, monocyte/macrophages, and T cell–derived cytokines in the kidney [129]. An imbalance of T regulatory cells (Treg) and T helper 17 (TH17) cells has been associated with hypertension [130], which can be protected by restoration the balance of Treg/TH17 by postbiotic therapy [131]. In CKD, the interplay between Treg/TH17 balance and inflammation has also been associated with hypertension and the progression of CKD [132].

Given that both Treg and TH17 cells are regulated by AhR [133], and that several microbial tryptophan catabolites are AhR ligands, AhR can serve as a mediator in inflammation and CVD in patients with CKD [132]. AhR signaling can trigger inflammation via several mechanisms, including participating in T cell differentiation, increasing monocyte adhesion, up-regulating pro-inflammatory gene expression, inducing the expression of endothelial adhesion molecules, reducing NO bioavailability, and increasing endothelial cyclooxygenase-2 expression [23]. In a maternal CKD-induced hypertension model, the BP-lowering effect of tryptophan therapy is associated with mediation of the AhR signaling pathway [69]. Additionally, AhR antagonist resveratrol has been reported to protect offspring against hypertension in several developmental hypertension models [59,60,80,81]. However, more research is needed to gain comprehensive insight into the role of immunity and inflammation in the modulation of hypertension and kidney disease of developmental origin. Specifically, future studies should focus on an investigation of reprogramming intervention targeting of the mechanism of inflammation.

5.5. Others

There are other potential mechanisms which link tryptophan metabolism to the development programming of hypertension and kidney disease. First, epigenetic regulation such as DNA methylation, histone modification, and miRNAs altering the expression of genes has been considered as an important mechanism underlying renal programming [71]. Tryptophan metabolism has been identified as a significantly regulated Kyoto Encyclopedia of Genes and Genomes (KEGG) pathway in two-week-old (right after the completion of nephrogenesis) offspring kidneys in models of maternal caloric restriction and diabetes [134]. Thus, the genes involved in tryptophan metabolism are likely epigenetically regulated by early-life insults leading to programmed hypertension and kidney disease. Next, nutrient-sensing signals also play a role in renal programming [94]. NAD^+ is a tryptophan metabolite generated from the kynurenine pathway. Increased $NAD^+/NADH$ ratio can activate silent information regulator transcript (SIRT) and cyclic adenosine monophosphate (AMP)-activated protein kinase (AMPK), consequently affecting PPARγ coactivator-1α (PGC-1α) activity to promote mitochondria biogenesis [135,136]. Since maternal resveratrol therapy protects hypertension programmed by maternal L-NAME plus postnatal high-fat exposure attributed to activation of the AMPK/PGC-1α pathway [82], whether tryptophan supplementation can increase NAD^+ synthesis and activate nutrient-sensing signals deserves further evaluation.

6. Conclusions

Although substantial progress has been made in understanding the role of tryptophan metabolism in pregnancy and offspring outcomes, there is always more to learn. Given tryptophan produces a plethora of biologically active metabolites, deciphering the complexity of different tryptophan metabolic pathways will aid in developing ideal reprogramming strategies targeting different tryptophan-related elements to open therapeutic opportunities for clinical translation. This review has provided an overview on reprogramming strategies against hypertension and kidney disease excepting tryptophan, which are related to the tryptophan metabolism, including melatonin and the AhR antagonist. Further research is needed to gain a clear understanding of the type of tryptophan-related molecules, the therapeutic dose and duration in pregnancy, and the microbial groups to metabolize tryptophan before the mother and child can benefit from reprogramming strategies targeting the tryptophan metabolism.

Author Contributions: C.-N.H.: contributed to concept generation, data interpretation, drafting of the manuscript, critical revision of the manuscript and approval of the article; Y.-L.T.: contributed to concept generation, data interpretation, drafting of the manuscript, critical revision of the manuscript and approval of the article. Both authors have read and agreed to the published version of the manuscript.

Funding: This work was supported by grants CMRPG8J0891, CMRPG8K0211, CMRPG8K1011, and CMRPG8J0252 from Chang Gung Memorial Hospital, Kaohsiung, Taiwan.

Conflicts of Interest: The authors declare no conflict of interest.

Abbreviations

AANAT	Arylalkylamine N-acetyltransferase
ACE	Angiotensin converting enzyme
AhR	Aryl hydrocarbon receptor
AMPK	Adenosine monophosphate activated protein kinase
ArAT	Acromatic amino acid aminotransferase
ASMT	N-acetylserotonin O-methyltransferase
AT1R	Angiotensin type 1 receptor
AT2R	Angiotensin type 2 receptor
CKD	Chronic kidney disease
CVD	Cardiovascular disease
NAD^+	Nicotinamide adenine dinucleotide
DOHaD	Developmental origins of health and disease
GC	Glucocorticoid
IAA	Indoleacetic acid
IAM	Indole-3-acetamide
IAlD	Indole-3-aldehyde
IDG	Indoxyl-β-D glucuronide
IDO	Indoleamine 2,3-dioxygenase
ILA	Indolelactic acid
IPA	Indole-3-propionic acid
KAT	Kynurenine aminotransferase
KMO	Kynurenine-3-monooxygenase
KYNU	Kynureninase
L-NAME	N^G-nitro-l-arginine-methyester
MAO	monoamine oxidase
Mas	Angiotensin-(1-7) receptor
NO	Nitric oxide
PGC-1α	PPARγ coactivator-1α
RAS	Renin-angiotensin system
ROS	Reactive oxygen species
SHR	Spontaneously hypertensive rat
SIRT	Silent information regulator transcript
SSRI	Selective serotonin reuptake inhibitor
TCDD	2,3,7,8-Tetrachlorodibenzo-p-dioxin
TDO	Tryptophan 2,3-dioxygenase
TPH	Tryptophan hydroxylase
3-HK	3-hydroxykynurenine
5-HIAA	5-hydroxyindoleacetic acid
5-HTP	5-hydroxytryptophan

References

1. Mills, K.T.; Bundy, J.D.; Kelly, T.N.; Reed, J.E.; Kearney, P.M.; Reynolds, K.; Chen, J.; He, J. Global Disparities of Hypertension Prevalence and Control: A Systematic Analysis of Population-Based Studies From 90 Countries. *Circulation* **2016**, *134*, 441–450. [CrossRef] [PubMed]
2. Mills, K.T.; Stefanescu, A.; He, J. The epidemiology of hypertension. *Nat. Rev. Nephrol.* **2020**, *16*, 223–237. [CrossRef] [PubMed]
3. Couser, W.G.; Remuzzi, G.; Mendis, S.; Tonelli, M. The contribution of chronic kidney disease to the global burden of major noncommunicable diseases. *Kidney Int.* **2011**, *80*, 1258–1270. [CrossRef]

4. Luyckx, V.A.; Bertram, J.F.; Brenner, B.M.; Fall, C.; Hoy, W.E.; Ozanne, S.E.; Vikse, B.E. Effect of fetal and child health on kidney development and long-term risk of hypertension and kidney disease. *Lancet* **2013**, *382*, 273–283. [CrossRef]
5. Chong, E.; Yosypiv, I.V. Developmental programming of hypertension and kidney disease. *Int. J. Nephrol.* **2012**, *2012*, 760580. [CrossRef]
6. Hanson, M.; Gluckman, P. Developmental origins of noncommunicable disease: Population and public health implications. *Am. J. Clin. Nutr.* **2011**, *94*, 1754S–1758S. [CrossRef]
7. Hsu, C.N.; Tain, Y.L. The Good, the Bad, and the Ugly of Pregnancy Nutrients and Developmental Programming of Adult Disease. *Nutrients* **2019**, *11*, 894. [CrossRef]
8. Kett, M.M.; Denton, K.M. Renal programming: Cause for concern? *Am. J. Physiol. Regul. Integr. Comp. Physiol.* **2011**, *300*, R791–R803. [CrossRef]
9. Nüsken, E.; Dötsch, J.; Weber, L.T.; Nüsken, K. Developmental Programming of Renal Function and Re-Programming Approaches. *Front. Pediatr.* **2018**, *6*, 36. [CrossRef]
10. Tain, Y.L.; Joles, J.A. Reprogramming: A preventive strategy in hypertension focusing on the kidney. *Int. J. Mol. Sci.* **2015**, *17*, 23. [CrossRef]
11. Richard, D.M.; Dawes, M.A.; Mathias, C.W.; Acheson, A.; Hill-Kapturczak, N.; Dougherty, D.M. L-tryptophan: Basic metabolic functions, behavioral research and therapeutic indications. *Int. J. Tryptophan Res.* **2009**, *2*, 45–60. [CrossRef]
12. Fernstrom, J.D. A perspective on the safety of supplemental tryptophan based on its metabolic fates. *J. Nutr.* **2016**, *146*, 2601S–2608S. [CrossRef]
13. Roager, H.M.; Licht, T.R. Microbial tryptophan catabolites in health and disease. *Nat. Commun.* **2018**, *9*, 3294. [CrossRef]
14. Agus, A.; Planchais, J.; Sokol, H. Gut Microbiota Regulation of Tryptophan Metabolism in Health and Disease. *Cell Host Microbe* **2018**, *23*, 716–724. [CrossRef]
15. Badawy, A. Tryptophan metabolism, disposition and utilization in pregnancy. *Biosci. Rep.* **2015**, *35*, e00261. [CrossRef]
16. Institute of Medicine, Food and Nutrition Board. *Dietary Reference Intakes: Energy, Carbohydrate, Fiber, Fat, Fatty Acids, Cholesterol, Protein and Amino Acids*; National Academies Press: Washington, DC, USA, 2005.
17. Addi, T.; Dou, L.; Burtey, S. Tryptophan-Derived Uremic Toxins and Thrombosis in Chronic Kidney Disease. *Toxins* **2018**, *10*, 412. [CrossRef]
18. Friedman, M. Analysis, Nutrition, and Health Benefits of Tryptophan. *Int. J. Tryptophan Res.* **2018**, *11*. [CrossRef]
19. Yao, K.; Fang, J.; Yin, Y.L.; Feng, Z.M.; Tang, Z.R.; Wu, G. Tryptophan metabolism in animals: Important roles in nutrition and health. *Front. Biosci. (Schol. Ed.)* **2011**, *3*, 286–297.
20. Stone, T.W.; Darlington, L.G. Endogenous kynurenines as targets for drug discovery and development. *Nat. Rev. Drug Discov.* **2002**, *1*, 609–620. [CrossRef]
21. Meyer, T.W.; Hostetter, T.H. Uremic solutes from colon microbes. *Kidney Int.* **2012**, *81*, 949–954. [CrossRef]
22. Hubbard, T.D.; Murray, I.A.; Perdew, G.H. Indole and Tryptophan Metabolism: Endogenous and Dietary Routes to Ah Receptor Activation. *Drug Metab. Dispos.* **2015**, *43*, 1522–1535. [CrossRef]
23. Sallée, M.; Dou, L.; Cerini, C.; Poitevin, S.; Brunet, P.; Burtey, S. The aryl hydrocarbon receptor-activating effect of uremic toxins from tryptophan metabolism: A new concept to understand cardiovascular complications of chronic kidney disease. *Toxins* **2014**, *6*, 934–949. [CrossRef]
24. Moehn, S.; Pencharz, P.B.; Ball, R.O. Lessons learned regarding symptoms of tryptophan deficiency and excess from animal requirement studies. *J. Nutr.* **2012**, *142*, 2231S–2235S. [CrossRef]
25. WHO. *Protein and Amino Acid Requirements in Human Nutrition. Report of a Joint WHO/FAO/UNU Expert Consultation*; WHO Technical Report Series 935; WHO Press: Geneva, Switzerland, 2007.
26. Schoengold, D.M.; DeFiore, R.H.; Parlett, R.C. Free amino acids in plasma throughout pregnancy. *Am. J. Obstet. Gynecol.* **1978**, *131*, 490–499. [CrossRef]
27. Duggleby, S.L.; Jackson, A.A. Protein, amino acid and nitrogen metabolism during pregnancy: How might the mother meet the needs of her fetus? *Curr. Opin. Clin. Nutr. Metab. Care* **2002**, *5*, 503–509. [CrossRef]
28. Elango, R.; Ball, R.O. Protein and amino acid requirements during pregnancy. *Adv. Nutr.* **2016**, *7*, 839S–844S. [CrossRef]
29. Gao, H. Amino Acids in Reproductive Nutrition and Health. *Adv. Exp. Med. Biol.* **2020**, *1265*, 111–131.

30. Hsu, C.N.; Tain, Y.L. Amino Acids and Developmental Origins of Hypertension. *Nutrients* **2020**, *12*, 1763. [CrossRef]
31. Tricklebank, M.D.; Pickard, F.J.; de Souza, S.W. Free and bound tryptophan in human plasma during the perinatal period. *Acta Paediatr. Scand.* **1979**, *68*, 199–204. [CrossRef]
32. Blaschitz, A.; Gauster, M.; Fuchs, D.; Lang, I.; Maschke, P.; Ulrich, D.; Karpf, E.; Takikawa, O.; Schimek, M.G.; Dohr, G.; et al. Vascular endothelial expression of indoleamine 2,3-dioxygenase 1 forms a positive gradient towards the feto-maternal interface. *PLoS ONE* **2011**, *6*, e21774. [CrossRef]
33. Fukuwatari, T.; Murakami, M.; Ohta, M.; Kimura, N.; Jin-No, Y.; Sasaki, R.; Shibata, K. Changes in the urinary excretion of the metabolites of the tryptopan-niacin pathway during pregnancy in japanese women and rats. *J. Nutr. Sci. Vitaminol.* **2004**, *50*, 392–398. [CrossRef]
34. Bonnin, A.; Levitt, P. Fetal, maternal, and placental sources of serotonin and new implications for developmental programming of the brain. *Neuroscience* **2011**, *197*, 1–7. [CrossRef]
35. Hsu, C.N.; Tain, Y.L. Light and Circadian Signaling Pathway in Pregnancy: Programming of Adult Health and Disease. *Int. J. Mol. Sci.* **2020**, *21*, 2232. [CrossRef]
36. Hsu, C.N.; Huang, L.T.; Tain, Y.L. Perinatal Use of Melatonin for Offspring Health: Focus on Cardiovascular and Neurological Diseases. *Int. J. Mol. Sci.* **2019**, *20*, 5681. [CrossRef]
37. Glover, M.E.; Clinton, S.M. Of rodents and humans: A comparative review of the neurobehavioral effects of early life SSRI exposure in preclinical and clinical research. *Int. J. Dev. Neurosci.* **2016**, *51*, 50–72. [CrossRef]
38. Mendez, N.; Abarzua-Catalan, L.; Vilches, N.; Galdames, H.A.; Spichiger, C.; Richter, H.G.; Valenzuela, G.J.; Seron-Ferre, M.; Torres-Farfan, C. Timed maternal melatonin treatment reverses circadian disruption of the fetal adrenal clock imposed by exposure to constant light. *PLoS ONE* **2012**, *7*, e42713. [CrossRef]
39. Louca, P.; Mompeo, O.; Leeming, E.R.; Berry, S.E.; Mangino, M.; Spector, T.D.; Padmanabhan, S.; Menni, C. Dietary Influence on Systolic and Diastolic Blood Pressure in the TwinsUK Cohort. *Nutrients* **2020**, *12*, 2130. [CrossRef]
40. Altorf-van der Kuil, W.; Engberink, M.F.; De Neve, M.; van Rooij, F.J.; Hofman, A.; van'tVeer, P.; Witteman, J.C.; Franco, O.H.; Geleijnse, J.M. Dietary amino acids and the risk of hypertension in a Dutch older population: The Rotterdam Study. *Am. J. Clin. Nutr.* **2013**, *97*, 403–410. [CrossRef]
41. Fregly, M.J.; Sumners, C.; Cade, J.R. Effect of chronic dietary treatment with L-tryptophan on the maintenance of hypertension in spontaneously hypertensive rats. *Can. J. Physiol. Pharmacol.* **1989**, *67*, 656–662. [CrossRef]
42. Lark, L.A.; Witt, P.A.; Becker, K.B.; Studzinski, W.M.; Weyhenmeyer, J.A. Effect of dietary tryptophan on the development of hypertension in the Dahl salt-sensitive rat. *Clin. Exp. Hypertens. A* **1990**, *12*, 1–13. [CrossRef]
43. Fregly, M.J.; Lockley, O.E.; Cade, J.R. Effect of chronic dietary treatment with L-tryptophan on the development of renal hypertension in rats. *Pharmacology* **1988**, *36*, 91–100. [CrossRef] [PubMed]
44. Wang, Y.; Liu, H.; McKenzie, G.; Witting, P.K.; Stasch, J.P.; Hahn, M.; Changsirivathanathamrong, D.; Wu, B.J.; Ball, H.J.; Thomas, S.R.; et al. Kynurenine is an endothelium-derived relaxing factor produced during inflammation. *Nat. Med.* **2010**, *16*, 279–285. [CrossRef] [PubMed]
45. Bartosiewicz, J.; Kaminski, T.; Pawlak, K.; Karbowska, M.; Tankiewicz-Kwedlo, A.; Pawlak, D. The activation of the kynurenine pathway in a rat model with renovascular hypertension. *Exp. Biol. Med.* **2017**, *242*, 750–761. [CrossRef] [PubMed]
46. Xiao, Y.; Christou, H.; Liu, L.; Visner, G.; Mitsialis, S.A.; Kourembanas, S.; Liu, H. Endothelial indoleamine 2,3-dioxygenase protects against development of pulmonary hypertension. *Am. J. Respir. Crit. Care Med.* **2013**, *188*, 482–491. [CrossRef]
47. Watts, S.W.; Morrison, S.F.; Davis, R.P.; Barman, S.M. Serotonin and blood pressure regulation. *Pharmacol. Rev.* **2012**, *64*, 59–88. [CrossRef]
48. Itskovitz, H.D.; Werber, J.L.; Sheridan, A.M.; Brewer, T.F.; Stier, C.T., Jr. 5-Hydroxytryptophan and carbidopa in spontaneously hypertensive rats. *J. Hypertens.* **1989**, *7*, 311–315. [CrossRef]
49. Brenner, B.; Harney, J.T.; Ahmed, B.A.; Jeffus, B.C.; Unal, R.; Mehta, J.L.; Kilic, F. Plasma serotonin levels and the platelet serotonin transporter. *J. Neurochem.* **2007**, *102*, 206–215. [CrossRef]
50. Topsakal, R.; Kalay, N.; Gunturk, E.E.; Dogan, A.; Inanc, M.T.; Kaya, M.G.; Ergin, A.; Yarlioglues, M. The relation between serotonin levels and insufficient blood pressure decrease during night-time in hypertensive patients. *Blood Press* **2009**, *18*, 367–371. [CrossRef]
51. Baron, A.; Riesselmann, A.; Fregly, M.J. Reduction in the elevated blood pressure of Dahl salt-sensitive rats treated chronically with L-5-hydroxytryptophan. *Pharmacology* **1991**, *42*, 15–22. [CrossRef]

52. Jelen, I.; Fananapazir, L.; Crawford, T.B. The possible relation between late pregnancy hypertension and 5-hydroxytryptamine levels in maternal blood. *Br. J. Obstet. Gynaecol.* **1979**, *86*, 468–471. [CrossRef]
53. Tain, Y.L.; Huang, L.T.; Chan, J.Y. Transcriptional regulation of programmed hypertension by melatonin: An epigenetic perspective. *Int. J. Mol. Sci.* **2014**, *15*, 18484–18495. [CrossRef]
54. Saito, K.; Fujigaki, S.; Heyes, M.P.; Shibata, K.; Takemura, M.; Fujii, H.; Wada, H.; Noma, A.; Seishima, M. Mechanism of increases in L-kynurenine and quinolinic acid in renal insufficiency. *Am. J. Physiol. Renal Physiol.* **2000**, *279*, F565–F572. [CrossRef] [PubMed]
55. Pawlak, D.; Pawlak, K.; Malyszko, J.; Mysliwiec, M.; Buczko, W. Accumulation of toxic products degradation of kynurenine in hemodialyzed patients. *Int. Urol. Nephrol.* **2001**, *33*, 399–404. [CrossRef] [PubMed]
56. Pawlak, K.; Domaniewski, T.; Mysliwiec, M.; Pawlak, D. The kynurenines are associated with oxidative stress, inflammation and the prevalence of cardiovascular disease in patients with end-stage renal disease. *Atherosclerosis* **2009**, *204*, 309–314. [CrossRef] [PubMed]
57. Leong, S.C.; Sirich, T.L. Indoxyl sulfate-review of toxicity and therapeutic strategies. *Toxins* **2016**, *8*, 358. [CrossRef] [PubMed]
58. Zhang, N. The role of endogenous aryl hydrocarbon receptor signaling in cardiovascular physiology. *J. Cardiovasc. Dis. Res.* **2011**, *2*, 91–95. [CrossRef] [PubMed]
59. Hsu, C.N.; Lin, Y.J.; Lu, P.C.; Tain, Y.L. Maternal Resveratrol Therapy Protects Male Rat Offspring against Programmed Hypertension Induced by TCDD and Dexamethasone Exposures: Is It Relevant to Aryl Hydrocarbon Receptor? *Int. J. Mol. Sci.* **2018**, *19*, 2459. [CrossRef]
60. Hsu, C.N.; Lin, Y.J.; Tain, Y.L. Maternal Exposure to Bisphenol A Combined with High-Fat Diet-Induced Programmed Hypertension in Adult Male Rat Offspring: Effects of Resveratrol. *Int. J. Mol. Sci.* **2019**, *20*, 4382. [CrossRef]
61. Wilck, N.; Matus, M.G.; Kearney, S.M.; Olesen, S.W.; Forslund, K.; Bartolomaeus, H.; Haase, S.; Mähler, A.; Balogh, A.; Markó, L.; et al. Salt-responsive gut commensal modulates T_H17 axis and disease. *Nature* **2017**, *551*, 585–589. [CrossRef]
62. Hsu, C.N.; Tain, Y.L. The Double-Edged Sword Effects of Maternal Nutrition in the Developmental Programming of Hypertension. *Nutrients* **2018**, *10*, 1917. [CrossRef]
63. McMillen, I.C.; Robinson, J.S. Developmental origins of the metabolic syndrome: Prediction, plasticity, and programming. *Physiol. Rev.* **2005**, *85*, 571–633. [CrossRef]
64. Herring, C.M.; Bazer, F.W.; Johnson, G.A.; Wu, G. Impacts of maternal dietary protein intake on fetal survival, growth, and development. *Exp. Biol. Med.* **2018**, *243*, 525–533. [CrossRef]
65. Penatti, E.M.; Barina, A.E.; Raju, S.; Li, A.; Kinney, H.C.; Commons, K.G.; Nattie, E.E. Maternal dietary tryptophan deficiency alters cardiorespiratory control in rat pups. *J. Appl. Physiol.* **2011**, *110*, 318–328. [CrossRef]
66. Omstedt, P.T.; von der Decken, A. Dietary amino acids: Effect of depletion and recovery on protein synthesis in vitro in rat skeletal muscle and liver. *Br. J. Nutr.* **1974**, *31*, 67–76. [CrossRef]
67. Lenis, N.P.; van Diepen, J.T.M.; Goedhart, P.W. Amino acid requirements of pigs. 1. Requirements for methionine + cystine, threonine and tryptophan for fast growing boars and gilts, fed ad libitum. *Neth. J. Agric. Sci.* **1990**, *38*, 577–595. [CrossRef]
68. Tain, Y.L.; Lin, Y.J.; Chan, J.Y.H.; Lee, C.T.; Hsu, C.N. Maternal melatonin or agomelatine therapy prevents programmed hypertension in male offspring of mother exposed to continuous light. *Biol. Reprod.* **2017**, *97*, 636–643. [CrossRef]
69. Hsu, C.N.; Lin, I.C.; Yu, H.R.; Huang, L.T.; Tiao, M.M.; Tain, Y.L. Maternal Tryptophan Supplementation Protects Adult Rat Offspring against Hypertension Programmed by Maternal Chronic Kidney Disease: Implication of Tryptophan-Metabolizing Microbiome and Aryl Hydrocarbon Receptor. *Int. J. Mol. Sci.* **2020**, *21*, 4552. [CrossRef] [PubMed]
70. Young, S.N. Behavioral effects of dietary neurotransmitter precursors: Basic and clinical aspects. *Neurosci. Biobehav. Rev.* **1996**, *20*, 313–323. [CrossRef]
71. Tain, Y.L.; Hsu, C.N. Developmental Origins of Chronic Kidney Disease: Should We Focus on Early Life? *Int. J. Mol. Sci.* **2017**, *18*, 381. [CrossRef]
72. Lee, S.K.; Sirajudeen, K.N.; Sundaram, A.; Zakaria, R.; Singh, H.J. Effects of antenatal, postpartum and post-weaning melatonin supplementation on blood pressure and renal antioxidant enzyme activities in spontaneously hypertensive rats. *J. Physiol. Biochem.* **2011**, *67*, 249–257. [CrossRef] [PubMed]

73. Tain, Y.L.; Huang, L.T.; Hsu, C.N.; Lee, C.T. Melatonin therapy prevents programmed hypertension and nitric oxide deficiency in offspring exposed to maternal caloric restriction. *Oxidative Med. Cell Longev.* **2014**, *2014*, 283180. [CrossRef] [PubMed]

74. Tain, Y.L.; Lee, C.T.; Chan, J.Y.; Hsu, C.N. Maternal melatonin or N-acetylcysteine therapy regulates hydrogen sulfide-generating pathway and renal transcriptome to prevent prenatal N(G)-Nitro-L-arginine-methyl ester (L-NAME)-induced fetal programming of hypertension in adult male offspring. *Am. J. Obstet. Gynecol.* **2016**, *215*, 636. [CrossRef] [PubMed]

75. Tain, Y.L.; Leu, S.; Wu, K.L.; Lee, W.C.; Chan, J.Y. Melatonin prevents maternal fructose intake-induced programmed hypertension in the offspring: Roles of nitric oxide and arachidonic acid metabolites. *J. Pineal Res.* **2014**, *57*, 80–89. [CrossRef]

76. Tain, Y.L.; Chan, J.Y.H.; Lee, C.T.; Hsu, C.N. Maternal Melatonin Therapy Attenuates Methyl-Donor Diet-Induced Programmed Hypertension in Male Adult Rat Offspring. *Nutrients* **2018**, *10*, 1407. [CrossRef] [PubMed]

77. Tain, Y.L.; Leu, S.; Lee, W.C.; Wu, K.L.H.; Chan, J.Y.H. Maternal Melatonin Therapy Attenuated Maternal High-Fructose Combined with Post-Weaning High-Salt Diets-Induced Hypertension in Adult Male Rat Offspring. *Molecules* **2018**, *23*, 886. [CrossRef] [PubMed]

78. Tain, Y.L.; Chen, C.C.; Sheen, J.M.; Yu, H.R.; Tiao, M.M.; Kuo, H.C.; Huang, L.T. Melatonin attenuates prenatal dexamethasone-induced blood pressure increase in a rat model. *J. Am. Soc. Hypertens.* **2014**, *8*, 216–226. [CrossRef] [PubMed]

79. Tain, Y.L.; Sheen, J.M.; Yu, H.R.; Chen, C.C.; Tiao, M.M.; Hsu, C.N.; Lin, Y.J.; Kuo, K.C.; Huang, L.T. Maternal Melatonin Therapy Rescues Prenatal Dexamethasone and Postnatal High-Fat Diet Induced Programmed Hypertension in Male Rat Offspring. *Front. Physiol.* **2015**, *6*, 377. [CrossRef]

80. Care, A.S.; Sung, M.M.; Panahi, S.; Gragasin, F.S.; Dyck, J.R.; Davidge, S.T.; Bourque, S.L. Perinatal Resveratrol Supplementation to Spontaneously Hypertensive Rat Dams Mitigates the Development of Hypertension in Adult Offspring. *Hypertension* **2016**, *67*, 1038–1044. [CrossRef]

81. Tain, Y.L.; Lee, W.C.; Wu, K.L.H.; Leu, S.; Chan, J.Y.H. Resveratrol Prevents the Development of Hypertension Programmed by Maternal Plus Post-Weaning High-Fructose Consumption through Modulation of Oxidative Stress, Nutrient-Sensing Signals, and Gut Microbiota. *Mol. Nutr. Food Res.* **2018**, *30*, e1800066. [CrossRef]

82. Chen, H.E.; Lin, Y.J.; Lin, I.C.; Yu, H.R.; Sheen, J.M.; Tsai, C.C.; Huang, L.T.; Tain, Y.L. Resveratrol prevents combined prenatal N^G-nitro-L-arginine-methyl ester (L-NAME) treatment plus postnatal high-fat diet induced programmed hypertension in adult rat offspring: Interplay between nutrient-sensing signals, oxidative stress and gut microbiota. *J. Nutr. Biochem.* **2019**, *70*, 28–37. [CrossRef]

83. Reiter, R.J.; Mayo, J.C.; Tan, D.X.; Sainz, R.M.; Alatorre-Jimenez, M.; Qin, L. Melatonin as an antioxidant: Under promises but over delivers. *J. Pineal Res.* **2016**, *61*, 253–278. [CrossRef] [PubMed]

84. Chen, Y.C.; Sheen, J.M.; Tiao, M.M.; Tain, Y.L.; Huang, L.T. Roles of melatonin in fetal programming in compromised pregnancies. *Int. J. Mol. Sci.* **2013**, *14*, 5380–5401. [CrossRef] [PubMed]

85. Tamura, H.; Nakamura, Y.; Terron, M.P.; Flores, L.J.; Manchester, L.C.; Tan, D.X.; Sugino, N.; Reiter, R.J. Melatonin and pregnancy in the human. *Reprod. Toxicol.* **2008**, *25*, 291–303. [CrossRef] [PubMed]

86. Luyckx, V.A.; Shukha, K.; Brenner, B.M. Low nephron number and its clinical consequences. *Rambam. Maimonides. Med. J.* **2011**, *2*, e0061. [CrossRef]

87. Ortiz, L.A.; Quan, A.; Weinberg, A.; Baum, M. Effect of prenatal dexamethasone on rat renal development. *Kidney Int.* **2001**, *59*, 1663–1669. [CrossRef]

88. Esteban, S.; Nicolaus, C.; Garmundi, A.; Rial, R.V.; Rodríguez, A.B.; Ortega, E.; Ibars, C.B. Effect of orally administered L-tryptophan on serotonin, melatonin, and the innate immune response in the rat. *Mol. Cell Biochem.* **2004**, *267*, 39–46. [CrossRef]

89. Oberlander, T.F. Fetal serotonin signaling: Setting pathways for early childhood development and behavior. *J. Adolesc. Health* **2012**, *51*, S9–S16. [CrossRef]

90. Siemann, J.K.; Green, N.H.; Reddy, N.; McMahon, D.G. Sequential photoperiodic programing of serotonin neurons, signaling and behaviors during prenatal and postnatal development. *Front. Neurosci.* **2019**, *13*, 459. [CrossRef]

91. Savouret, J.F.; Berdeaux, A.; Casper, R.F. The aryl hydrocarbon receptor and its xenobiotic ligands: A fundamental trigger for cardiovascular diseases. *Nutr. Metab. Cardiovasc. Dis.* **2003**, *13*, 104–113. [CrossRef]

92. Tain, Y.L.; Hsu, C.N. Developmental programming of the metabolic syndrome: Can we reprogram with resveratrol? *Int. J. Mol. Sci.* **2018**, *19*, 2584. [CrossRef]
93. Peter Stein, T.; Scholl, T.O.; Schluter, M.D.; Leskiw, M.J.; Chen, X.; Spur, B.W.; Rodriguez, A. Oxidative stress early in pregnancy and pregnancy outcome. *Free Radic. Res.* **2008**, *42*, 841–848. [CrossRef] [PubMed]
94. Tain, Y.L.; Hsu, C.N. Interplay between oxidative stress and nutrient sensing signaling in the developmental origins of cardiovascular disease. *Int. J. Mol. Sci.* **2017**, *18*, 841. [CrossRef] [PubMed]
95. Tain, Y.L.; Lee, W.C.; Hsu, C.N.; Lee, W.C.; Huang, L.T.; Lee, C.T.; Lin, C.Y. Asymmetric dimethylarginine is associated with developmental programming of adult kidney disease and hypertension in offspring of streptozotocin-treated mothers. *PLoS ONE* **2013**, *8*, e55420. [CrossRef] [PubMed]
96. Forrest, C.M.; Mackay, G.M.; Stoy, N.; Egerton, M.; Christofides, J.; Stone, T.W.; Darlington, L.G. Tryptophan loading induces oxidative stress. *Free Radic. Res.* **2004**, *38*, 1167–1171. [CrossRef] [PubMed]
97. Reyes Ocampo, J.; Huitr, L.R.; Gonzalez-Esquivel, D.; Ugalde-Muniz, P.; Jimenez-Anguiano, A.; Pineda, B.; Pedraza-Chaverri, J.; Rios, C.; Perez de la Cruz, V. Kynurenines with neuroactive and redox properties: Relevance to aging and brain diseases. *Oxid. Med. Cell. Longev.* **2014**, *2014*, 22. [CrossRef] [PubMed]
98. Okuda, S.; Nishiyama, N.; Saito, H.; Katsuki, H. 3-Hydroxykynurenine, an endogenous oxidative stress generator, causes neuronal cell death with apoptotic features and region selectivity. *J. Neurochem.* **1998**, *70*, 299–307. [CrossRef]
99. Stone, T.W. Kynurenines in the CNS: From endogenous obscurity to therapeutic importance. *Prog. Neurobiol.* **2001**, *64*, 185–218. [CrossRef]
100. Xu, K.; Liu, H.; Bai, M.; Gao, J.; Wu, X.; Yin, Y. Redox properties of tryptophan metabolism and the concept of tryptophan use in pregnancy. *Int. J. Mol. Sci.* **2017**, *18*, 1595. [CrossRef]
101. Bjørklund, G.; Chirumbolo, S. Role of oxidative stress and antioxidants in daily nutrition and human health. *Nutrition* **2017**, *33*, 311–321. [CrossRef]
102. Hsu, C.N.; Tain, Y.L. Regulation of Nitric Oxide Production in the Developmental Programming of Hypertension and Kidney Disease. *Int. J. Mol. Sci.* **2019**, *60*, 681. [CrossRef]
103. Alberati-Giani, D.; Malherbe, P.; Ricciardi-Castagnoli, P.; Köhler, C.; Denis-Donini, S.; Cesura, A.M. Differential regulation of indoleamine 2,3-dioxygenase expression by nitric oxide and inflammatory mediators in IFN-γ-activated murine macrophages and microglial cells. *J. Immunol.* **1997**, *159*, 419–426. [PubMed]
104. Kuhn, D.M.; Arthur, R., Jr. Molecular mechanism of the inactivation of tryptophan hydroxylase by nitric oxide: Attack on critical sulfhydryls that spare the enzyme iron center. *J. Neurosci.* **1997**, *17*, 7245–7251. [CrossRef] [PubMed]
105. Mukherjee, S. Novel perspectives on the molecular crosstalk mechanisms of serotonin and melatonin in plants. *Plant Physiol. Biochem.* **2018**, *132*, 33–45. [CrossRef] [PubMed]
106. Kharait, S.; Haddad, D.J.; Springer, M.L. Nitric oxide counters the inhibitory effects of uremic toxin indoxyl sulfate on endothelial cells by governing ERK MAP kinase and myosin light chain activation. *Biochem. Biophys. Res. Commun.* **2011**, *409*, 758–763. [CrossRef] [PubMed]
107. Chu, D.M.; Meyer, K.M.; Prince, A.L.; Aagaard, K.M. Impact of maternal nutrition in pregnancy and lactation on offspring gut microbial composition and function. *Gut Microbes* **2016**, *7*, 459–470. [CrossRef] [PubMed]
108. Meijers, B.; Jouret, F.; Evenepoel, P. Linking gut microbiota to cardiovascular disease and hypertension: Lessons from chronic kidney disease. *Pharmacol. Res.* **2018**, *133*, 101–107. [CrossRef] [PubMed]
109. Al Khodor, S.; Reichert, B.; Shatat, I.F. The microbiome and blood pressure: Can microbes regulate our blood pressure? *Front. Pediatr.* **2017**, *5*, 138. [CrossRef]
110. O'Mahony, S.M.; Clarke, G.; Borre, Y.E.; Dinan, T.G.; Cryan, J.F. Serotonin, tryptophan metabolism and the brain-gut-microbiome axis. *Behav. Brain Res.* **2015**, *277*, 32–48. [CrossRef]
111. Liang, H.; Dai, Z.; Kou, J.; Sun, K.; Chen, J.; Yang, Y.; Wu, G.; Wu, Z. Dietary l-Tryptophan supplementation enhances the intestinal mucosal barrier function in weaned piglets: Implication of Tryptophan-metabolizing microbiota. *Int. J. Mol. Sci.* **2019**, *20*, 20. [CrossRef]
112. Velasquez, M.T.; Centron, P.; Barrows, I.; Dwivedi, R.; Raj, D.S. Gut Microbiota and Cardiovascular Uremic Toxicities. *Toxins* **2018**, *10*, 287. [CrossRef]
113. Lankelma, J.M.; Nieuwdorp, M.; de Vos, W.M.; Wiersinga, W.J. The gut microbiota in internal medicine: Implications for health and disease. *Neth. J. Med.* **2015**, *73*, 61–68. [PubMed]

114. McFarlane, C.; Ramos, C.I.; Johnson, D.W.; Campbell, K.L. Prebiotic, probiotic, and synbiotic supplementation in chronic kidney disease: A systematic review and meta-analysis. *J. Ren. Nutr.* **2019**, *29*, 209–220. [CrossRef] [PubMed]
115. Yang, T.; Richards, E.M.; Pepine, C.J.; Raizada, M.K. The gut microbiota and the brain-gut-kidney axis in hypertension and chronic kidney disease. *Nat. Rev. Nephrol.* **2018**, *14*, 442–456. [CrossRef] [PubMed]
116. Hsu, C.N.; Lin, Y.J.; Hou, C.Y.; Tain, Y.L. Maternal administration of probiotic or prebiotic prevents male adult rat offspring against developmental programming of hypertension induced by high fructose consumption in pregnancy and lactation. *Nutrients* **2018**, *10*, 1229. [CrossRef] [PubMed]
117. Hsu, C.N.; Chang-Chien, G.P.; Lin, S.; Hou, C.Y.; Tain, Y.L. Targeting on gut microbial metabolite trimethylamine-n-oxide and short-chain fatty acid to prevent maternal high-fructose-diet-induced developmental programming of hypertension in adult male offspring. *Mol. Nutr. Food Res.* **2019**, *63*, e1900073. [CrossRef]
118. Hsu, C.N.; Hou, C.Y.; Chan, J.Y.H.; Lee, C.T.; Tain, Y.L. Hypertension programmed by perinatal high-fat diet: Effect of maternal gut microbiota-targeted therapy. *Nutrients* **2019**, *11*, 2908. [CrossRef]
119. Te Riet, L.; van Esch, J.H.; Roks, A.J.; van den Meiracker, A.H.; Danser, A.H. Hypertension: Renin-angiotensin aldosterone system alterations. *Circ. Res.* **2015**, *116*, 960–975. [CrossRef]
120. Santos, P.C.; Krieger, J.E.; Pereira, A.C. Renin-angiotensin system, hypertension, and chronic kidney disease: Pharmacogenetic implications. *J. Pharmacol. Sci.* **2012**, *120*, 77–88. [CrossRef]
121. Sherman, R.C.; Langley-Evans, S.C. Antihypertensive treatment in early postnatal life modulates prenatal dietary influences upon blood pressure in the rat. *Clin. Sci.* **2000**, *98*, 269–275. [CrossRef]
122. Hsu, C.N.; Lee, C.T.; Huang, L.T.; Tain, Y.L. Aliskiren in early postnatal life prevents hypertension and reduces asymmetric dimethylarginine in offspring exposed to maternal caloric restriction. *J. Renin Angiotensin Aldosterone Syst.* **2015**, *16*, 506–513. [CrossRef]
123. Khedr, S.; Deussen, A.; Kopaliani, I.; Zatschler, B.; Martin, M. Effects of tryptophan-containing peptides on angiotensin-converting enzyme activity and vessel tone ex vivo and in vivo. *Eur. J. Nutr.* **2018**, *57*, 907–915. [CrossRef] [PubMed]
124. Shimizu, H.; Hirose, Y.; Goto, S.; Nishijima, F.; Zrelli, H.; Zghonda, N.; Niwa, T.; Miyazaki, H. Indoxyl sulfate enhances angiotensin II signaling through upregulation of epidermal growth factor receptor expression in vascular smooth muscle cells. *Life Sci.* **2012**, *91*, 172–177. [CrossRef] [PubMed]
125. Ng, H.Y.; Yisireyili, M.; Saito, S.; Lee, C.T.; Adelibieke, Y.; Nishijima, F.; Niwa, T. Indoxyl sulfate downregulates expression of Mas receptor via OAT3/AhR/Stat3 pathway in proximal tubular cells. *PLoS ONE* **2014**, *9*, e91517. [CrossRef] [PubMed]
126. Challis, J.R.; Lockwood, C.J.; Myatt, L.; Norman, J.E.; Strauss, J.F.; Petraglia, F. Inflammation and pregnancy. *Reprod. Sci.* **2009**, *16*, 206–215. [CrossRef] [PubMed]
127. Le Floc'h, N.; Melchior, D.; Seve, B. Dietary tryptophan helps to preserve tryptophan homeostasis in pigs suffering from lung inflammation. *J. Anim. Sci.* **2008**, *86*, 3473–3479. [CrossRef]
128. Asp, L.; Johansson, A.S.; Mann, A.; Owe-Larsson, B.; Urbanska, E.M.; Kocki, T.; Kegel, M.; Engberg, G.; Lundkvist, G.B.; Karlsson, H. Effects of pro-inflammatory cytokines on expression of kynurenine pathway enzymes in human dermal fibroblasts. *J. Inflamm.* **2011**, *8*, 1476–9255. [CrossRef]
129. McMaster, W.G.; Kirabo, A.; Madhur, M.S.; Harrison, D.G. Inflammation, immunity, and hypertensive end-organ damage. *Circ. Res.* **2015**, *116*, 1022–1033. [CrossRef]
130. Ren, J.; Crowley, S.D. Role of T-cell activation in salt-sensitive hypertension. *Am. J. Physiol. Heart Circ. Physiol.* **2019**, *316*, H1345–H1353. [CrossRef]
131. Zhang, J.; Hua, G.; Zhang, X.; Tong, R.; Du, X.; Li, Z. Regulatory T cells/T-helper cell 17 functional imbalance in uraemic patients on maintenance haemodialysis: A pivotal link between microinflammation and adverse cardiovascular events. *Nephrology* **2010**, *15*, 33–41. [CrossRef]
132. Brito, J.S.; Borges, N.A.; Esgalhado, M.; Magliano, D.C.; Soulage, C.O.; Mafra, D. Aryl hydrocarbon receptor activation in chronic kidney disease: Role of uremic toxins. *Nephron* **2017**, *137*, 1–7. [CrossRef]
133. Stevens, E.A.; Mezrich, J.D.; Bradfield, C.A. The aryl hydrocarbon receptor: A perspective on potential roles in the immune system. *Immunology* **2009**, *127*, 299–311. [CrossRef] [PubMed]
134. Tain, Y.L.; Hsu, C.N.; Chan, J.Y.; Huang, L.T. Renal transcriptome analysis of programmed hypertension induced by maternal nutritional insults. *Int. J. Mol. Sci.* **2015**, *16*, 17826–17837. [CrossRef] [PubMed]

135. Sugden, M.C.; Caton, P.W.; Holness, M.J. PPAR control: It's SIRTainly as easy as PGC. *J. Endocrinol.* **2010**, *204*, 93–104. [CrossRef]
136. Efeyan, A.; Comb, W.C.; Sabatini, D.M. Nutrient-sensing mechanisms and pathways. *Nature* **2015**, *517*, 302–310. [CrossRef] [PubMed]

Publisher's Note: MDPI stays neutral with regard to jurisdictional claims in published maps and institutional affiliations.

 © 2020 by the authors. Licensee MDPI, Basel, Switzerland. This article is an open access article distributed under the terms and conditions of the Creative Commons Attribution (CC BY) license (http://creativecommons.org/licenses/by/4.0/).

Article

Effect of Tryptophan-Derived AhR Ligands, Kynurenine, Kynurenic Acid and FICZ, on Proliferation, Cell Cycle Regulation and Cell Death of Melanoma Cells—In Vitro Studies

Katarzyna Walczak [1,*], Ewa Langner [1,2], Anna Makuch-Kocka [1], Monika Szelest [1,†], Karolina Szalast [1], Sebastian Marciniak [1] and Tomasz Plech [1]

1. Department of Pharmacology, Medical University of Lublin, Chodźki 4a, 20-093 Lublin, Poland; ewa.langner@umlub.pl (E.L.); anna.makuch@umlub.pl (A.M.-K.); m.wlodarczyk214@gmail.com (M.S.); karolina.szalast@umlub.pl (K.S.); sebastian.marciniak@umlub.pl (S.M.); tomasz.plech@umlub.pl (T.P.)
2. Department of Medical Biology, Institute of Rural Health, Jaczewskiego 2, 20-090 Lublin, Poland
* Correspondence: katarzyna.walczak@umlub.pl; Tel.: +48-81-448-6774
† Monika Szelest is student, a volunteer in the Department of Pharmacology.

Received: 18 September 2020; Accepted: 24 October 2020; Published: 26 October 2020

Abstract: Tryptophan metabolites: kynurenine (KYN), kynurenic acid (KYNA) and 6-formylindolo [3,2-b]carbazole (FICZ) are considered aryl hydrocarbon receptor (AhR) ligands. AhR is mainly expressed in barrier tissues, including skin, and is involved in various physiological and pathological processes in skin. We studied the effect of KYN, KYNA and FICZ on melanocyte and melanoma A375 and RPMI7951 cell toxicity, proliferation and cell death. KYN and FICZ inhibited DNA synthesis in both melanoma cell lines, but RPMI7951 cells were more resistant to pharmacological treatment. Tested compounds were toxic to melanoma cells but not to normal human adult melanocytes. Changes in the protein level of cyclin D1, CDK4 and retinoblastoma tumor suppressor protein (Rb) phosphorylation revealed different mechanisms of action of individual AhR ligands. Importantly, all tryptophan metabolites induced necrosis, but only KYNA and FICZ promoted apoptosis in melanoma A375 cells. This effect was not observed in RPMI7951 cells. KYN, KYNA and FICZ in higher concentrations inhibited the protein level of AhR but did not affect the gene expression. To conclude, despite belonging to the group of AhR ligands, KYN, KYNA and FICZ exerted different effects on proliferation, toxicity and induction of cell death in melanoma cells in vitro.

Keywords: tryptophan; kynurenine; kynurenic acid; FICZ; AhR; melanoma; proliferation; cell death

1. Introduction

The skin is constantly exposed to various substances with pro- and anti-carcinogenic potential. Environmental pollution, skin care products and UV radiation can affect various processes in the skin. Environmental UV exposure, fair color of skin and hair, family history of melanoma and a high number of melanocytic nevi are well-characterized risk factors for developing melanoma [1]. Melanoma is rare among all types of skin cancers (less than 5%) but the most aggressive [2,3]. A median survival rate of patients with metastatic melanoma is approximately 6 months [2]. A poor prognosis for patients with advanced melanoma results from rapid metastasis, resistance to anti-cancer therapies and immunosuppressive abilities. Thus, there are two main strategies in melanoma prevention and treatment: inhibition of melanocyte-to-melanoma transition and inhibition of melanoma metastasis.

The aryl hydrocarbon receptor (AhR) is a major sensor of chemical signals and is mainly expressed in barrier tissues (including skin, liver, lungs and the gastrointestinal tract) which may be exposed to environmental factors [4]. AhR, a ligand-activated transcription factor, is a member of the evolutionarily conserved family of per-arnt-sim basic helix-loop-helix transcription factors (PAS-bHLH). AhR has been mainly associated with dioxin toxicity and detoxification of xenobiotic compounds, but it also plays a role in several physiological and pathological processes in skin such as detoxification, cellular homeostasis, skin pigmentation, skin immunity and carcinogenesis [5,6]. The role of AhR in melanomagenesis is not clear, however, previous studies revealed involvement of this receptor in melanoma cell dedifferentiation and metastases in response to inflammation [7].

Some tryptophan-derived metabolites including kynurenine (KYN), kynurenic acid (KYNA) and 6-formylindolo[3,2-b]carbazole (FICZ) are considered as AhR ligands [8,9]. Tryptophan, an essential amino acid, is well-known to be the most potent near-UV absorbing chromophore. Several tryptophan metabolites have various biological activities and may potentially affect processes in skin. However, the effect of KYN, KYNA and FICZ on melanomagenesis and melanoma progression has not been fully investigated.

KYN, an endogenous agonist of AhR, is produced enzymatically from tryptophan by indoleamine 2,3-dioxygenase (IDO) and tryptophan 2,3-dioxygenase (TDO) [10]. Previous studies revealed that both enzymes are expressed and upregulated in various cancer types, including melanoma [11,12]. The direct activity of KYN on cancer cell proliferation has not been fully studied, but it is supposed that this tryptophan metabolite plays a crucial role in the antitumor immune response. Opitz et al. revealed that KYN is produced during brain cancer progression and inflammation in the local microenvironment in amounts sufficient to activate the human AhR, suppressing antitumor immune responses and promoting cancer cell survival and motility [13]. Similarly, KYN induced an antiapoptotic response in breast cancer cells [14].

On the other hand, previous studies revealed antiproliferative and antimigratory properties of KYNA against various types of cancer cells [15]. However, its role in cellular processes in melanoma has not been revealed. KYNA, a natural ligand for AhR, is enzymatically formed from KYN and is present in almost all human body fluids and tissues [15]. Additionally, KYNA may be absorbed from the gastrointestinal tract due to its presence in many food products [16–18]. The role of KYNA in skin physiology and pathology has not been fully studied; however, phototoxic effects of KYNA on erythrocytes and glia cells have been reported [19,20]. Importantly, KYN and KYNA may be enzymatically synthesized by skin cells [21].

FICZ is a tryptophan photometabolite formed by exposure to UV or visible irradiation [22]. It plays a crucial role in several physiological processes including skin homeostasis, regulation of skin immunity and circadian rhythm, genomic stability and response to UV exposure [23]. FICZ is also considered to be the most potent UVA photosensitizer, leading to induction of oxidative stress and oxidative DNA lesions, which consequently leads to cell death of keratinocytes [23]. Some previous studies confirmed the potential role of FICZ in the differentiation-induction therapy of leukaemia [24]; however, its contribution to melanomagenesis has not been fully revealed. Mengoni et al., in in vitro studies, reported that FICZ is involved in inflammation-induced dedifferentiation [7]. On the other hand, FICZ is also considered as a sensitizer of photooxidative stress, leading to photodynamic elimination of skin cancer cells in vitro and in vivo [25].

There are three potential interactions between tryptophan-derived AhR ligands and skin cells: (1) direct contact or topical application on the skin surface, as they are present in several herbs, honey-bee products used in beauty and body treatments, and body care products [16,18,26], (2) endogenous production in skin cells, and (3) via the food chain (as KYNA is present in some food products [16,18]). Taking into consideration the continuous external and internal exposure of skin cells to the tryptophan-derived AhR ligands and the ambiguous role of AhR in melanomagenesis and melanoma progression, determining the negative or positive role of tested substances in carcinogenesis seems to be a priority. The aim of the study was to determine the biological activity of the selected tryptophan-derived AhR ligands on melanomagenesis by investigating the effect of KYN, KYNA and

FICZ on the proliferation, cytotoxicity and cell death of normal human melanocytes and melanoma cells in vitro.

2. Results

To determine the effect of L-KYN, KYNA and FICZ on proliferation of human melanocytes and melanoma cells, HEMa, A375 and RPMI7951 cells were exposed to serial dilutions of tested tryptophan-derived AhR ligands. Interestingly, L-KYN and FICZ, but not KYNA exerted antiproliferative activity towards human melanoma A375 and RPMI7951 cells, measured by means of BrdU Assay indicating the level of DNA synthesis (Figure 1). Importantly, the antiproliferative effect was enhanced in A375 cells in comparison to RPMI7951 cells representing metastatic melanoma. L-KYN at a concentration of only 1 pM statistically significantly inhibited cell proliferation of A375 cells, whereas only millimolar concentrations of L-KYN affected proliferation of RPMI7951 cells (Figure 1a). Similarly, A375 cells were more sensitive to the antiproliferative activity of FICZ in comparison to RPMI7951 cells (Figure 1c). However, this compound only moderately inhibited proliferation of melanoma cells by less than 20%. On the other hand, only L-KYN at the highest concentration (5 mM) among selected tryptophan-derived AhR ligands inhibited DNA synthesis in normal human melanocytes HEMa (Figure 1a).

Moreover, we tested the toxicity of L-KYN, KYNA and FICZ on human melanocytes and melanoma cells by means of LDH Assay (Figure 2). All tested tryptophan–derived AhR ligands did not induce LDH release and were not toxic to normal melanocytes HEMa. L-KYN and 5 mM KYNA increased LDH release in A375 cells (Figure 2a,b), whereas a toxic effect of FICZ was observed in RPMI7951 cells (Figure 2c).

To reveal the potential molecular mechanism of biological activity of selected tryptophan-derived AhR ligands in melanoma cells, the effect of L-KYN, KYNA and FICZ on activation and protein level of selected cell cycle regulators was determined by means of western blot (Figure 3). Similarly to results obtained from BrdU and LDH Assays, we reported the differences in the activation and level of selected proteins in melanoma A375 and RPMI7951 cells, representing successive stages of carcinogenesis.

L-KYN inhibited the protein level of cyclin D1 and cyclin-dependent kinase 4 (CDK4) in A375 cells, however, this effect was not observed in RPMI7951 cells (Figure 3a). Immunofluorescence staining confirmed inhibition of cyclin D1 and CDK4 in melanoma A375 cells exposed to L-KYN (Figure 4). No significant cellular relocalisation of cyclin D1 and CDK4 was observed (Figure 4). Moreover, L-KYN decreased phosphorylation of Rb in both A375 and RPMI7951 cells (Figure 3a). KYNA at a concentration of 5 mM significantly decreased the protein level of CDK4 in A375 cells, whereas it increased the protein level of this cell cycle regulator in RPMI7951 cells (Figure 3b). A similar effect was observed in Rb phosphorylation and the protein level of cyclin D1, but these moderate changes in cyclin D1 level were not significant (Figure 3b). Interestingly, we did not observe significant changes in the protein level of cyclin D1 and CDK4 in melanoma A375 and RPMI7951 cells exposed to FICZ (Figure 3c). However, this tryptophan-derived AhR ligand enhanced phosphorylation of Rb (Figure 3c).

To investigate the effect of selected tryptophan-derived AhR ligands on induction of cell death in melanoma cells, Cell Death Detection ELISA and fluorescent cell death analysis (Hoechst 33342 and propidium iodide staining) were applied. Taking into consideration the negative effect of 5 mM L-KYN on proliferation of HEMa cells (Figure 1a), this concentration was excluded from further analysis. KYNA 5 mM and FICZ 50 µM statistically significantly induced apoptosis, whereas all tested compounds in higher concentrations (L-KYN 1 mM, KYNA 5 mM, FICZ 50 µM) increased necrosis in melanoma A375 cells (Figure 5a,b). The effect was visualized by Hoechst 33342 and propidium iodide staining (Figure 6). On the other hand, we did not observed statistically significant changes in apoptosis and necrosis induction by selected tryptophan-derived AhR-ligands in melanoma RPMI7951 cells (Figure 5c,d; Figure 6).

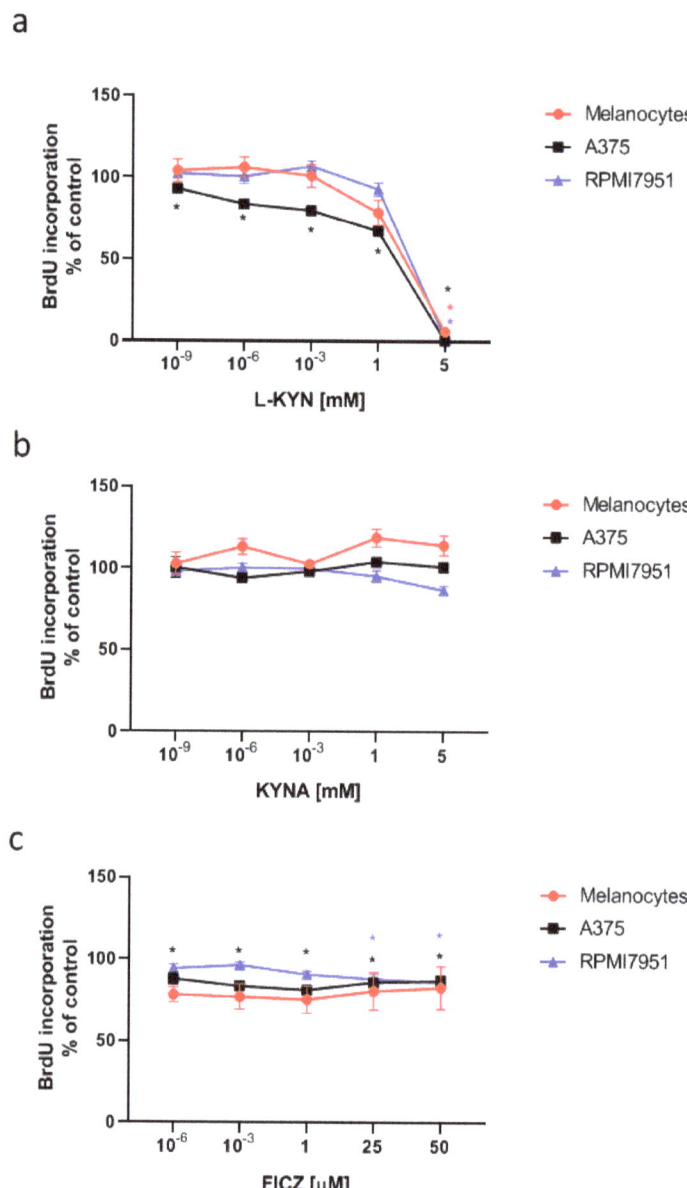

Figure 1. The effect of L-KYN (**a**), KYNA (**b**) and FICZ (**c**) on proliferation (DNA synthesis) of melanocytes and melanoma A375 and RPMI7951 cells. Normal human adult primary epidermal melanocytes (HEMa) and human melanoma A375 and RPMI7951 cells were exposed to fresh medium (control, C) or serial dilutions of L-KYN, KYNA and FICZ for 24 h. The effect of tested compounds on proliferation (DNA synthesis) was determined by means of BrdU Assay. Data represent a mean value (% of control; C = 100%) ± SEM of eight independent experiments. Values significant (*) in comparison to control (100%) with $p < 0.05$ (one-way ANOVA with Tukey post hoc test).

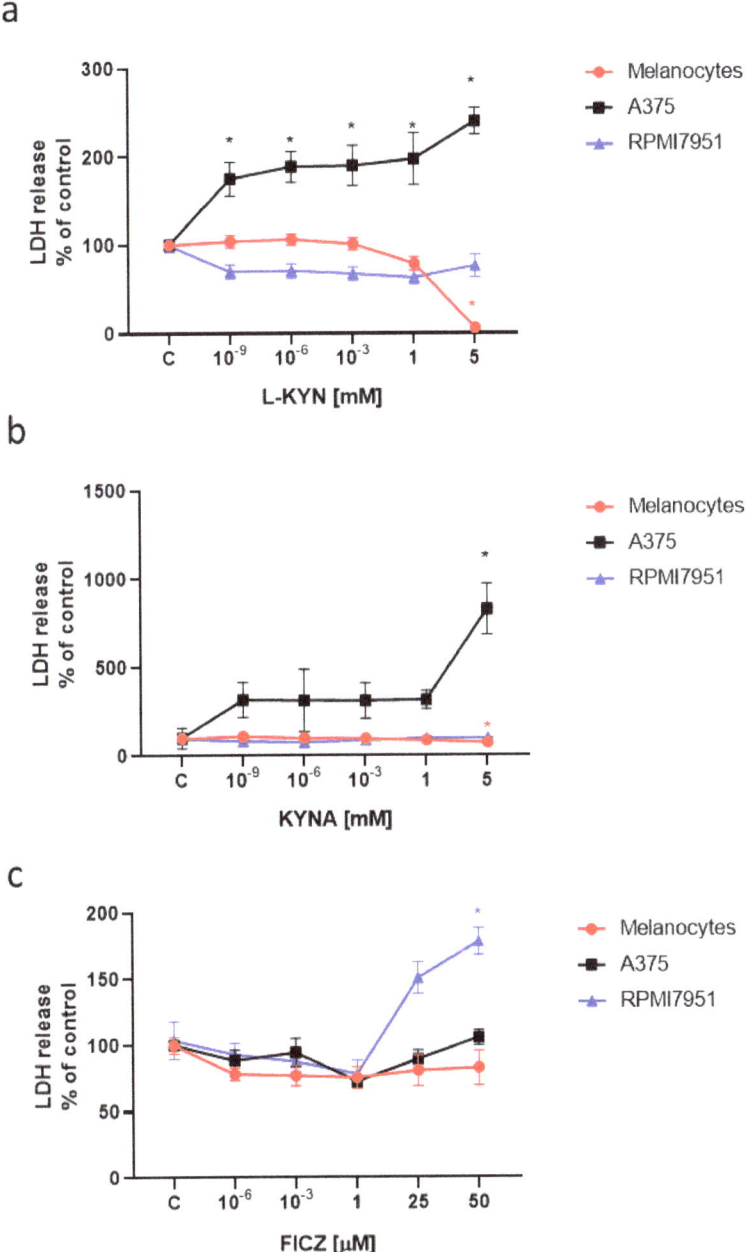

Figure 2. The toxicity of L-KYN (**a**), KYNA (**b**) and FICZ (**c**) towards melanocytes and melanoma A375 and RPMI7951 cells. Normal human adult primary epidermal melanocytes (HEMa) and human melanoma A375 and RPMI7951 cells were exposed to fresh medium (control, C) or serial dilutions of L-KYN, KYNA and FICZ for 24 h. The toxicity of tested compounds was assessed by means of LDH Assay measuring LDH release. Data represent a mean value (% of control) ± SEM of eight independent experiments. Values significant (*) in comparison to control (100%) with $p < 0.05$ (one-way ANOVA with Tukey post hoc test). Positive control for melanoma A375 cells (Total LDH) = 1720%.

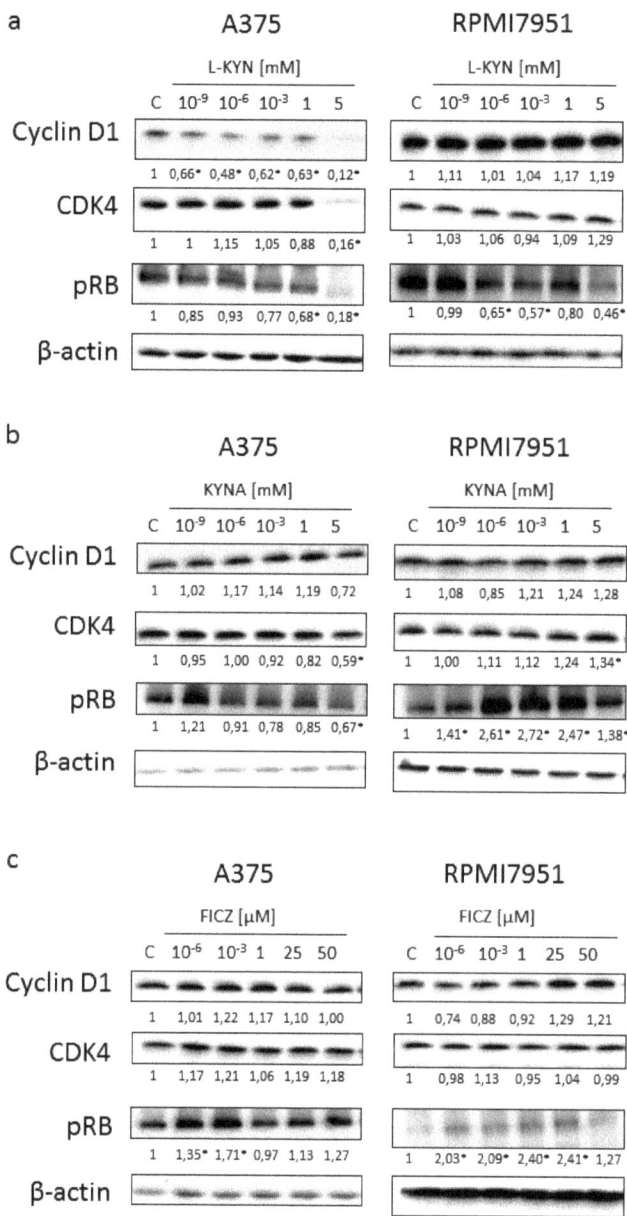

Figure 3. The effect of L-KYN (**a**), KYNA (**b**) and FICZ (**c**) on the protein level of selected cell cycle regulators in melanoma A375 and RPMI7951 cells. Western blot analysis of the protein level of cyclin D1, CDK4 and phosphorylation of Rb in A375 and RPMI7951 cells after treatment with L-KYN (**a**) and KYNA (**b**) in the range of concentrations 10^{-9}–5 mM and FICZ (**c**) in the range of concentrations 10^{-6}–50 μM for 24 h (C control; not treated). Western blots shown in the figure were selected as the most representative of the series of repetitions. The data were normalized relative to β-actin. The results of densitometric analysis are shown as % of control (the changes ≥30% were considered as significant (*)).

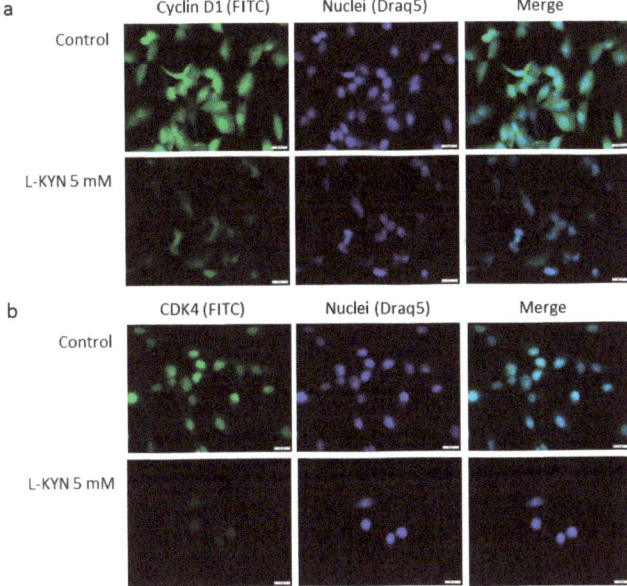

Figure 4. The effect of L-KYN on the protein level and cellular localization of cyclin D1 (**a**) and CDK4 (**b**) in melanoma A375 cells. Immunofluorescent staining of cyclin D1 (**a**) and CDK4 (**b**) in A375 cells treated with L-KYN 5 mM for 24 h (control; not treated). Cell nuclei were labeled with cell permeable fluorescent DNA dye DraQ5. Magnification 40×.

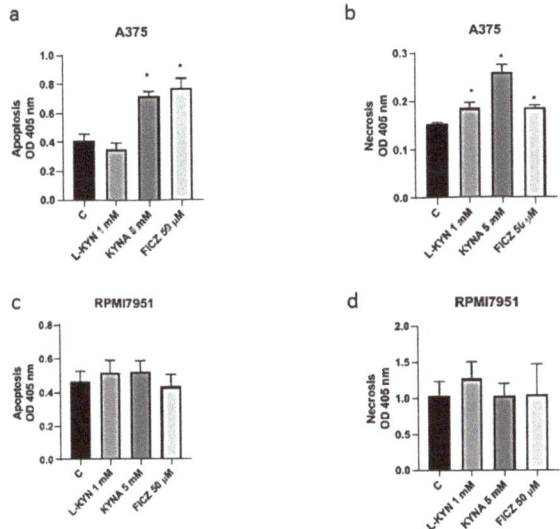

Figure 5. The effect of L-KYN, KYNA and FICZ on induction of apoptosis and necrosis in melanoma A375 (**a,b**) and RPMI7951 (**c,d**) cells. Human melanoma A375 and RPMI7951 cells were exposed to fresh medium (control, C) or selected tryptophan-derived AhR ligands: L-KYN (1 mM), KYNA (5 mM) and FICZ (50 µM) for 24 h. The effect of tested compounds on induction of apoptosis (**a,c**) and necrotic cell death (**b,d**) was determined by means of Cell Death Detection ELISA. Data represent a mean value ± SEM of three independent experiments. Values significant (*) in comparison to control with $p < 0.05$ (one-way ANOVA with Tukey post hoc test).

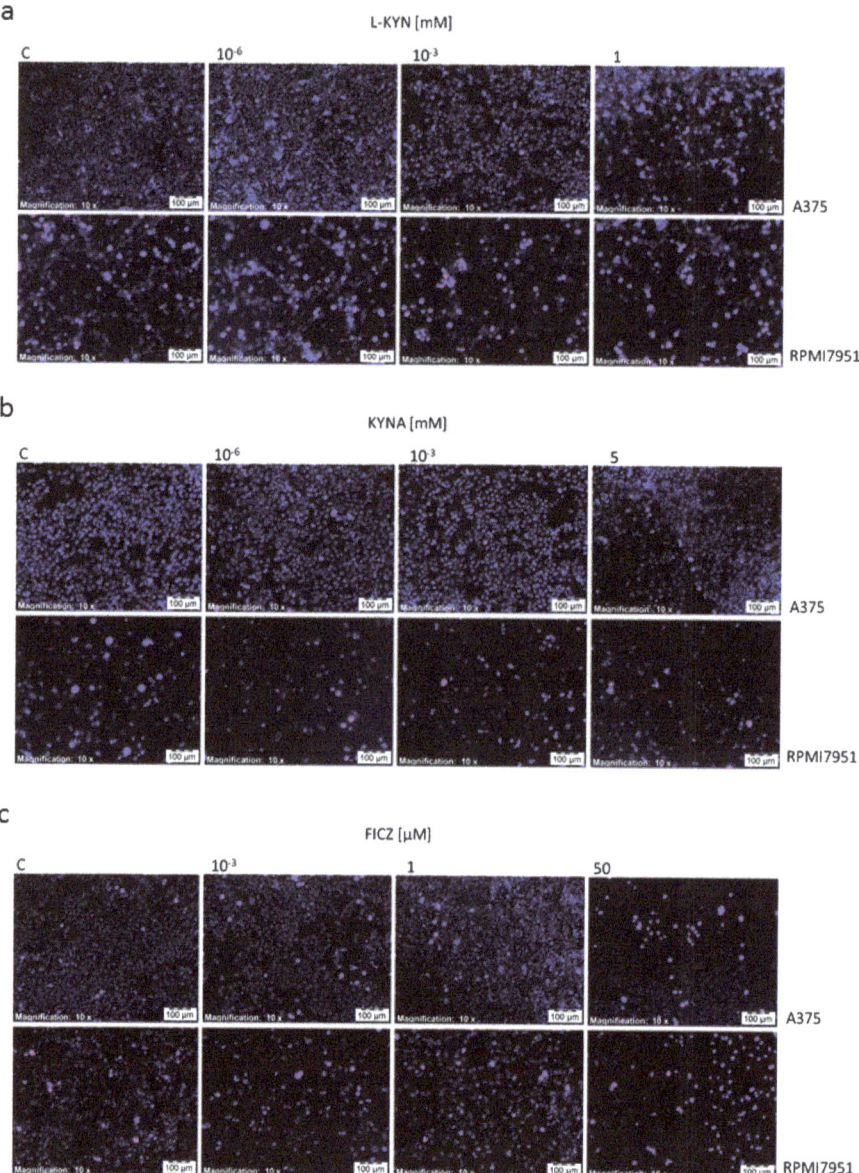

Figure 6. The effect of L-KYN, KYNA and FICZ on induction of cell death in melanoma A375 and RPMI7951 cells. Human melanoma A375 and RPMI7951 cells were exposed to fresh medium (control, C) or selected tryptophan-derived AhR ligands: L-KYN (1 mM) (**a**), KYNA (5 mM) (**b**) and FICZ (50 μM) (**c**) for 24 h. The effect of tested compounds on induction of apoptosis and necrotic cell death was determined by means of staining with Hoechst 33342 and propidium iodide. Cells with fragmented nuclei stained in an intense blue color were considered apoptotic cells while the pink staining of the nuclei was necrotic cells. Magnification 10×.

Because all tested compounds are tryptophan-derived AhR ligands, we decided to study the effect of L-KYN, KYNA and FICZ on the expression of AhR. Interestingly, all tested compounds in higher concentrations decreased the protein level of AhR in both melanoma cell lines (Figure 7). However, L-KYN 1 mM, KYNA 5 mM and FICZ 50 µM did not affect the gene expression of *AHR* (Figure 8) in melanoma A375 and RPMI7951 cells.

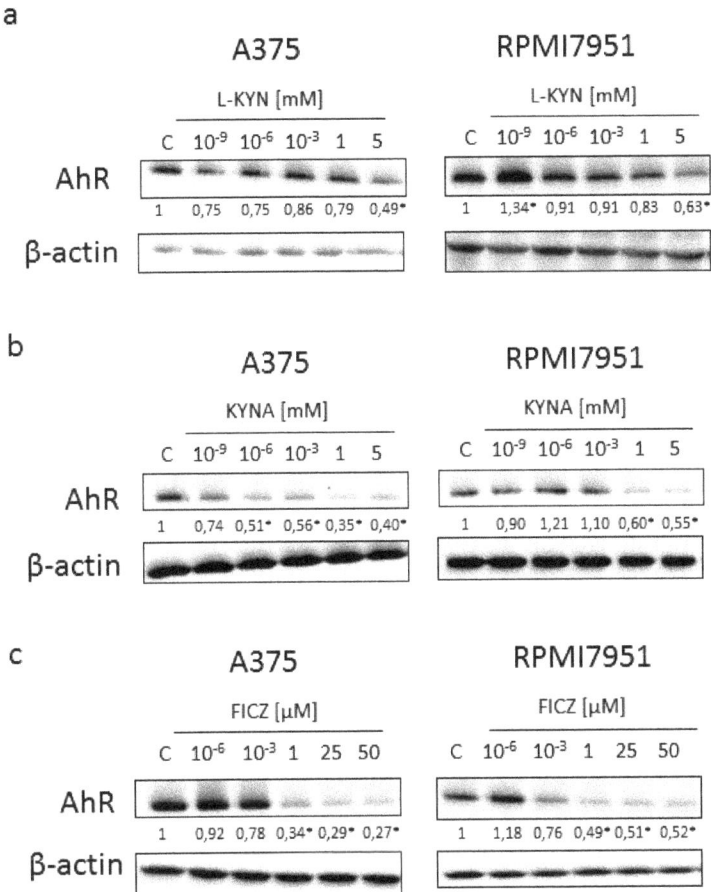

Figure 7. The effect of L-KYN, KYN A and FICZ on AhR protein level in melanoma A375 and RPMI7951 cells. Western blot analysis of protein level of AhR in A375 and RPMI7951 cells after treatment with L-KYN (**a**) and KYNA (**b**) in the range of concentrations 10^{-9}–5 mM and FICZ (**c**) in the range of concentrations 10^{-6}–50 µM for 24 h (C control; not treated). Western blots shown in the figure were selected as the most representative of the series of repetitions. The data were normalized relative to β-actin. The results of densitometric analysis are shown as % of control (changes ≥30% were considered as significant (*)).

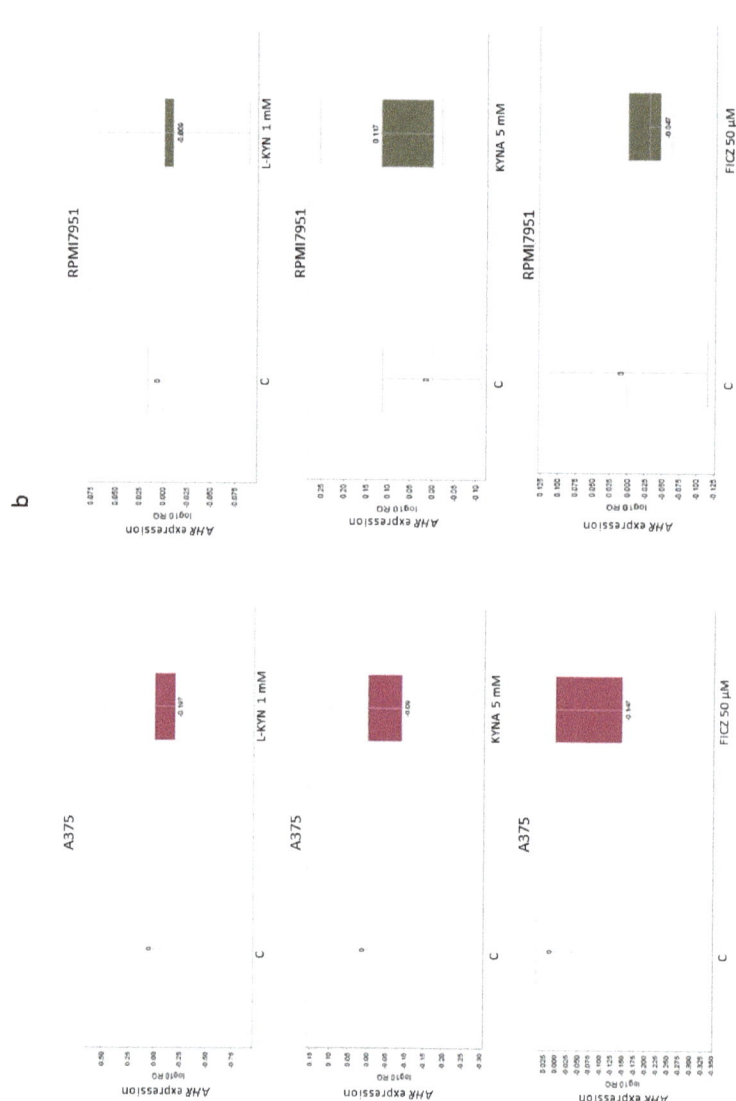

Figure 8. The effect of L-KYN, KYNA and FICZ on *AHR* gene expression in melanoma A375 (**a**) and RPMI7951 (**b**) cells. Real-Time PCR analyses were used to evaluate *AHR* gene expression in A375 and RPMI7951 cells after 24 h treatment with L-KYN, KYNA and FICZ at the concentration of 1mM, 5 mM and 50 µM, respectively (C, control–not treated). *ACTB* was used as the reference gene. For more reliable results, RQ values were analyzed after log transform to log10RQ. Data represent a mean value ± SEM of three independent experiments. The differences in AHR gene expression were not statistically significant in comparison to control (C) (values significant with $p < 0.05$, unpaired *t*-test).

3. Discussion

In this study, we showed that tryptophan-derived AhR ligands, including KYN, KYNA and FICZ, do not stimulate promotion and progression of melanoma in vitro, but simultaneously the tested compounds in higher concentrations inhibited proliferation and stimulated cell death of melanoma cells. This appears to be a priority as the skin is constantly exposed to KYN, KYNA and FICZ belonging to AhR ligands, and the activation of this receptor may be involved in several cellular processes including proliferation, migration or cell death [5,6]. KYN and KYNA are endogenously produced tryptophan metabolites present in various tissues and physiological fluids (reviewed in [15,27,28]). Additionally, previous studies revealed that several food products, herbs and beverages contain KYN [29–32] and KYNA [16–18,33,34], which may be considered as exogenous sources of these tryptophan metabolites. Taking into consideration the increasing use of plant extracts and herbs in cosmetics and skin care products, it can be assumed that the skin is also exposed frequently to certain amounts of KYN and KYNA. On the other hand, FICZ is considered as a photometabolite of tryptophan synthetized in skin after exposure to UV or visible irradiation [22], thus skin is constantly exposed to this compound.

The study was conducted on normal human adult melanocytes and two melanoma cell lines A375 and RPMI7951 differing from each other not only in origin but also in mutated genes. The A375 cell line represents primary malignant melanoma, whereas the RPMI7951 cell line was derived from lymph node metastasis and was reported as a strongly invasive melanoma. Both A375 and RPMI7951 melanoma cells bear the *BRAF* mutation (p.V600E) and *CDKN2A* mutation, however, only A375 cells harbor a homozygous mutation in the *BRAF* gene. Additionally, a mutant *TP53* and *PTEN* were identified only in RPMI7951 cells ([35]; product information ATCC). Thus, the obtained results make it possible to analyze the biological effect of KYN, KYNA and FICZ on melanoma cells representing the subsequent stages of the disease. However, at this stage of research, we are not able to clearly determine whether the differences in the effects of substances result from the stage of the disease or a specific gene mutation.

All selected tryptophan-derived compounds are considered as AhR ligands. AhR plays a role in various physiological and pathological processes in skin, including carcinogenesis [5,6]. Importantly, in the study we confirmed that exposure to KYN, KYNA and FICZ did not stimulate proliferation of melanoma A375 and RPMI7951 cells. On the contrary, KYN and FICZ inhibited DNA synthesis of melanoma cells (Figure 1a,c). Interestingly, more potent inhibitory activity was observed towards melanoma A375 cells (Figure 1). Antiproliferative activity of KYN towards A375 cells was already reported at a concentration of 10^{-9} mM, whereas KYN inhibited proliferation of RPMI7951 cells at a concentration of 5 mM (Figure 1a). Similarly, FICZ statistically significantly inhibited DNA synthesis in A375 cells at a concentration range of 10^{-6}–50 µM, but the inhibitory effect towards RPMI7951 cells was observed only in the highest tested micromolar concentrations. These results confirm that RPMI7951 cells, derived from melanoma node metastasis, are strongly invasive and more resistant to pharmacological treatment [36]. Importantly, KYN, KYNA and FICZ did not affect proliferation of normal human melanocytes (HEMa) (Figure 1). Only L-KYN in the highest concentration of 5 mM caused a significant inhibition of DNA synthesis in HEMa cells, therefore this dose was excluded from further studies. Surprisingly, KYNA with evidenced antiproliferative activity towards several cancer cell lines [37–41] did not affect DNA synthesis in melanoma A375 and RPMI7951 cells (Figure 1b). On the other hand, previous studies reported a more potent antiproliferative effect measured by means of MTT Assay, rather than BrdU Assay [38,40,41]. The difference may result from the mechanism of biological activity of KYNA, as the MTT Assay determines metabolic activity of cells, whereas BrdU Assay determines DNA synthesis. Similarly, despite no changes in DNA synthesis in KYNA-treated A375 cells (Figure 1b), the cytotoxicity assay confirmed statistically significant increase in LDH release in these cells (Figure 2b), which may suggest a different mechanism of KYNA biological activity towards melanoma cells. A reduction of LDH release in melanocytes exposed to KYN 5 mM probably results from the strong antiproliferative effect of the compound confirmed in the BrdU Assay (Figure 2b).

In conclusion, the presented results showed that KYNA, KYNA and FICZ, despite belonging to the same group of tryptophan-derived AhR ligands, showed different mechanisms of action.

This hypothesis was also confirmed in the molecular studies. The dysregulation of the p16-cyclinD-CDK4/6-Rb pathway is common in melanomas and occurs in 22–78% of cases [42]. Thus, we studied the effect of KYN, KYNA and FICZ on the protein level of cyclin D1, CDK4 and Rb phosphorylation in melanoma A375 and RPMI7951 cells (Figure 3). Cyclin D1, CDK4 and Rb control the G1/S transmission. Previous studies revealed the positive correlation between elevated CDK4 expression and increased therapeutic activity of CDK4/6 inhibitors undergoing clinical trials or currently used in anti-cancer therapy [42–45]. Our studies revealed that KYN inhibited the protein level of cyclin D1 and CDK4 in melanoma A375 cells, but not in more resistant RPMI7951 cells (Figure 3a). However, KYN affected the activation of Rb, one of the key regulators of cell cycle in both melanoma cell lines (Figure 3a). Interestingly, KYNA in the highest tested concentration (5 mM) decreased CDK4 protein level and phosphorylation of Rb in A375 cells (Figure 3b), despite no significant changes in proliferation (DNA synthesis) of KYNA-treated A375 melanoma cells. However, it cannot be excluded that the biological effect of the observed molecular changes in cell cycle regulators would be seen after a longer incubation time. The opposite effect of KYNA on the protein level and activation of cell cycle regulators was observed in RPMI7951 cells. Surprisingly, despite the increased phosphorylation of Rb in RPMI7951 cells exposed to KYNA (Figure 3b), no significant changes in proliferation was observed (Figure 1b). Similarly, enhanced phosphorylation of Rb was determined in A375 and RPMI7951 cells exposed to FICZ (Figure 3c). Although the exact biological effect of the Rb activity in melanoma A375 and RPMI7951 cells exposed to FICZ and RPMI8951 cells exposed to KYNA is unknown, it should be noted that Rb interacts not only with cyclins, CDKs, phosphatases and chromatin-associated proteins but is also involved in regulation of metabolic pathways [46–48]. Previous studies revealed that *RB1* mutant cells are characterized by reduced mitochondrial respiration, polarity and alternation with metabolic flux (reviewed in [46,48]). Bearing in mind the origin of these AhR ligands as tryptophan metabolites, it cannot be excluded that phosphorylation of Rb may be associated with metabolic changes in A375 and RPMI7951 cells exposed to KYNA and FICZ. Importantly, previous studies suggested a relationship between cellular metabolism and the anti-tumor activity of KYNA [15].

Additionally, we also revealed that tryptophan-derived AhR ligands affect not only proliferation of melanoma cells in vitro, but may also induce cell death in melanoma A375 cells (Figures 5 and 6). KYN, KYNA and FICZ increased necrosis in A375 cells, but a significant increase in apoptotic cells was determined only in A375 cells exposed to KYNA and FICZ. Interestingly, no positive or negative effect on cell death was observed in RPMI7951, representing a more resistant melanoma cell line (Figure 3). However, it cannot be excluded that the differences in the effects of selected tryptophan metabolites in A375 and RPMI7951 cells result from a specific gene mutation in tested cell lines. It should be noted that RPMI7951 cells harbor a *TP53* mutation, thus the crucial role of p53 in the induction of apoptosis in A375 cells may be suggested and needs further investigation.

KYN, KYNA and FICZ represent tryptophan-derived AhR ligands, so we decided to investigate their effect on the protein and gene expression of AhR. Surprisingly, although all tested compounds decreased the protein level of AhR in a dose-dependent manner, none of the tryptophan-derived AhR ligands affected gene expression of *AHR* (Figures 7 and 8). These results suggest that KYN, KYNA and FICZ may increase the proteolytic degradation of AhR but not interfere at the genome level. Similar results were reported previously by Mengoni et al. [7] in a panel of melanocytic and dedifferentiated human melanoma cell lines treated with FICZ. This mechanism may be a form of a negative regulation of AhR activity in melanoma cells protecting against excessive activity of AhR-dependent pathways. On the other hand, it cannot be excluded that it is one of the mechanisms of biological activity of tested tryptophan-derived AhR ligands leading to the inhibition of proliferation, increased toxicity and increased death of cancer cells.

Despite the common origin of KYN, KYNA and FICZ as tryptophan metabolites, the activity of these substances on melanoma cells and their molecular mechanism of biological activity are different.

Moreover, due to similarity of the activity of all tested tryptophan metabolites towards AhR, and the differences in the biological effects of their activity on proliferation, cell cycle regulation and cell death, it may be suggested that the molecular mechanism of KYN, KYNA and FICZ is not only dependent on AhR. Although further studies should be conducted to reveal the specific mechanism of activity of tested tryptophan metabolites towards melanoma cells, taking into consideration the obtained results, we may conclude that KYN, KYNA and FICZ do not promote melanoma A375 and RPMI7951 cell proliferation and growth in vitro. However, to exclude the pro-carcinogenic potential of tryptophan-derived AhR ligands, further in vivo and clinical trials are necessary. Despite the undisputed value of the results showing the biological activity of KYN, KYNA and FICZ towards melanoma cells, in vitro studies do not take into account the influence of the tested substances on the immune system and the additional influence of KYN and KYNA from food products. It should be noted that KYN is strongly associated with immunosuppression and cancer escape from immune surveillance [28]. Moreover, in order to exclude possible negative effects of frequent exposure to tryptophan-derived AhR ligands on the skin, further research into effects on other skin cells should also be performed.

In conclusion, the selected tryptophan-derived AhR ligands, KYN, KYNA and FICZ, produced endogenously in the skin and present in herbs and plant extracts used in skin care treatments, did not promote melanoma cell growth in vitro, and even in higher concentrations they may exhibit antiproliferative and cytotoxic activity and promote cell death in melanoma A375 and RPMI7951 cells. However, the biological activity and molecular mechanism was different for each compound and not strictly dependent on AhR.

4. Materials and Methods

4.1. Drugs

L-KYN and KYNA were obtained from Sigma Aldrich (St. Louis, MO, USA). L-KYN was dissolved in cell culture medium, whereas KYNA was dissolved in 1 N NaOH, and then phosphate buffered saline (PBS). FICZ, obtained from Tocris Bioscience (Bristol, UK), was dissolved in dimethyl sulfoxide (DMSO). The final concentration of DMSO in cell culture medium dilutions was less than 0.2%. No significant effects of solvents on melanoma and melanocyte cell proliferation and morphology were observed.

4.2. Cell Cultures

Normal human adult primary epidermal melanocytes (HEMa) and human melanoma A375 and RPMI7951 cells were obtained from American Type Culture Collection (ATCC; Manassas, VA, USA). HEMa cells were cultured in Dermal Cell Basal Medium supplemented with Adult Melanocyte Growth Kit (ATCC; Manassas, VA, USA). A375 cells were grown in Dulbecco's modified Eagle's medium (DMEM) supplemented with 10% heat inactivated fetal bovine serum (FBS), penicillin (100 U/mL) and streptomycin (100 µg/mL). RPMI7951 cells were grown in Minimum Essential Medium with Earle's salts supplemented with sodium pyruvate (final concentration 1 mM), 10% heat inactivated FBS, penicillin (100 U/mL) and streptomycin (100 µg/mL). All A375 and RPMI 7951 cell culture reagents were purchased from Sigma Aldrich (St. Louis, MO, USA). Cells were maintained in a humidified atmosphere of 95% air and 5% CO_2 at 37 °C.

4.3. Experiment Design

Previous studies revealed that KYN competes for a membrane transporter with tryptophan [49]. To exclude any interactions between tryptophan-derived AhR ligands and tryptophan included in the culture medium and to provide the maximum of bioavailability of the tested compounds, melanocytes or melanoma cells (~60–70% confluence) were treated with serial dilutions of L-KYN, KYNA and FICZ dissolved in Hanks' Balanced Salt solution (HBSS, Sigma Aldrich, St. Louis, MO, USA) for 1 h in standard conditions. After incubation time, HBSS was discarded, and cells were exposed to fresh medium (control) or serial dilutions of L-KYN, KYNA or FICZ in a fresh medium and incubated for 23 h in standard conditions. Then further analyses were performed (LDH assay, BrdU assay, Cell Death Detection ELISA, western blot, PI and Hoechst staining, RT-PCR).

4.4. BrdU Assay

BrdU assay, quantifying the incorporation of 5′-bromo-2′-deoxy-uridine (BrdU) into newly synthesized DNA of actively proliferating cells, was used to determine the effect of L-KYN, KYNA and FICZ on proliferation of melanocyte and melanoma cells according to the procedure described previously [38]. Briefly, HEMa, A375 and RPMI7951 cells were plated in 96-well plates (NUNC, Roskilde, Denmark) at the density of 4×10^4 cells/mL, 2×10^4 cells/mL and 4×10^4 cells/mL, respectively. Next day, the cells were exposed to serial dilutions of the tested compound (L-KYN: 10^{-9}, 10^{-6}, 10^{-3}, 1, 5 mM; KYNA: 10^{-9}, 10^{-6}, 10^{-3}, 1, 5 mM, FICZ: 10^{-6}, 10^{-3}, 1, 25, 50 µM) or fresh cell culture medium (control, C) according to the experiment design described in detail above. Cell proliferation was quantified after 24 h incubation according to the manufacturer's procedure (Cell Proliferation ELISA BrdU, Roche Diagnostics GmbH, Penzberg, Germany).

4.5. LDH Assay

The In Vitro Toxicology Assay Kit (TOX-7) (Sigma Aldrich, St. Louis, MO, USA), based on the reduction of NAD by lactic dehydrogenase (LDH) released from damaged cells, was applied to determine the cytotoxicity of L-KYN, KYNA and FICZ. HEMa, A375 and RPMI7951 cells were plated in 96-well plates (NUNC, Roskilde, Denmark) at the density of 4×10^4 cells/mL, 2×10^4 cells/mL and 4×10^4 cells/mL, respectively. The next day, the cells were exposed to serial dilutions of tested compound (L-KYN: 10^{-9}, 10^{-6}, 10^{-3}, 1, 5 mM; KYNA: 10^{-9}, 10^{-6}, 10^{-3}, 1, 5 mM, FICZ: 10^{-6}, 10^{-3}, 1, 25, 50 µM) or fresh cell culture medium (control, C) according to the experiment design described in detail above. The activity of released LDH in the supernatants was measured after 24 h incubation according to the manufacturer's procedure and was quantified at 490 nm (Epoch microplate reader (BioTek Instruments, Inc., Winooski, VT, USA) equipped with Gen5 software (v. 2.01, BioTek Instruments, Inc., Winooski, VT, USA)).

4.6. Western Blot

Melanoma A375 and RPMI7951 cells were exposed to serial dilutions of the tested compounds (L-KYN: 10^{-9}, 10^{-6}, 10^{-3}, 1, 5 mM; KYNA: 10^{-9}, 10^{-6}, 10^{-3}, 1, 5 mM, FICZ: 10^{-6}, 10^{-3}, 1, 25, 50 µM) or fresh cell culture medium (control, C) for 24 h according to the experiment design described in detail above. The protein level or its activation was determined by means of western blot procedure previously described in [50]. The following primary antibodies were used in the procedure: cyclin D1, CDK4, phospho-Rb (Ser 807/811), AhR and β-actin antibody (1:1000; Cell Signaling Technology, Danvers, MA, USA). In the procedure the secondary antibodies coupled to horseradish peroxidase were used (1:2000) (Cell Signaling Technology, Danvers, MA, USA). The blots were visualized by using enhanced chemiluminescence (Pierce Biotechnology, Waltham, MA, USA) and the Syngene G:BOX Chemi XT4 gel documentation system (Syngene, Cambridge, UK).

4.7. Immunofluorescence Staining

A375 cells cultured on LabTek Chamber Slides (Nunc, ThermoFisher Scientific, Roskilde, Denmark) were exposed to culture medium (control, C) or L-KYN 5 mM for 24 h in standard conditions according to the experiment design described in detail above. Then, cells were fixed with 3.7% paraformaldehyde, permeabilized with 0.2% Triton X-100, and treated with 5% bovine serum albumin (BSA, Sigma Aldrich, St. Louis, MO, USA). The cells were exposed to primary antibodies against cyclin D1 and CDK4 (1:100; Cell Signaling Technology, Danvers, MA, USA) overnight at 4 °C and then were incubated with secondary antibodies conjugated with fluorescein isothiocyanate (FITC) (1:100) (Sigma Aldrich, St. Louis, MO, USA) for 2 h at room temperature. Cell nuclei were labeled with cell permeable

4.8. Cell Death Detection ELISA

The effect of the tryptophan-derived AhR ligands, L-KYN, KYNA and FICZ, on apoptosis and necrosis of A375 and RPMI7951 melanoma cells was assessed by ELISA. The Cell Death Detection ELISA^{PLUS} photometric enzyme immunoassay was used for the quantitative in vitro determination of cytoplasmic histone-associated DNA fragments (mono- and oligonucleosomes) after induced cell death. A375 and RPMI7951 melanoma cells were plated in 96-well plates (NUNC, Roskilde, Denmark) at the density of 2×10^4 cells/mL and 4×10^4 cells/mL, respectively. Next day, the cells were exposed for 24 h to serial dilutions of tested compound (L-KYN 1 mM, KYNA 5 mM, FICZ 50 µM) or fresh cell culture medium (control, C) according to the experiment design described in detail above. ELISA test was performed according to the manufacturer's procedure. The colorful product was quantified spectrophotometrically at 405 nm using a microplate reader (Epoch, BioTek Instruments, Inc., Winooski, VT, USA) equipped with Gen5 software (v. 2.01, BioTek Instruments, Inc., Winooski, VT, USA).

4.9. Fluorescent Cell Death Analysis—Hoechst 33342 and Propidium Iodide Staining

Melanoma A375 and RPMI7951 cells were plated on Lab-Tek Chamber Slides (NUNC, Roskilde, Denmark) at the density of 4×10^4 cells/mL and 8×10^4 cells/mL, respectively. The next day, cells were exposed to serial dilutions of tested compound (L-KYN: 10^{-6}, 10^{-3}, 1 mM; KYNA: 10^{-6}, 10^{-3}, 5 mM, FICZ: 10^{-3}, 1, 50 µM) or fresh cell culture medium (control, C) according to the experiment design described in detail above. After 24 h treatment, the effect of L-KYN, KYNA and FICZ on induction of cell death was analyzed after fluorescence staining with Hoechst 33342 and propidium iodide at a concentration of 0.2 µg/mL and 0.4 µg/mL, respectively (5 min at 37 °C). Cell images were captured with fluorescence microscopy (Olympus IX83 System Microscope; Olympus Optical Co., Ltd. and CellSens RT software, Olympus Optical Co., Ltd., Tokyo, Japan) at 10× magnification.

4.10. Real-Time PCR

Melanoma A375 and RPMI7951 cells were exposed to the tested compounds (L-KYN 1 mM, KYNA 5 mM, FICZ 50 µM) or fresh cell culture medium (control, C) for 24 h according to the experiment design described in detail above. Extracted total RNA (High Pure RNA Isolation Kit (Roche Diagnostics GmbH, Penzberg, Germany)) was reverse-transcribed using a High Capacity cDNA Reverse Transcription Kit (Applied Biosystems, Foster City, CA, USA) according to the manufacturer's procedure. Real-time PCR analyses were performed using TaqMan Gene Expression Assays (IDs: Hs00169233_m1 for *AHR* and Hs99999903_m1 for reference gene *ACTB*; ThermoFisher Scientific, Waltham, MA, USA) and TaqMan Fast Universal PCR MasterMix (ThermoFisher Scientific, Waltham, MA, USA) as previously described [51] using a QuantStudio 12K Flex (Appllied Biosystems, Foster City, CA, USA). The expression of reference gene *ACTB* was used as an endogenous control. The relative expression was calculated by the formula $RQ = 2^{-\Delta\Delta Ct}$ [52] (QuantStudioTM 12K Flex Softwere v1.2.2, Applied Biosystems, Foster City, CA, USA).

4.11. Data Analysis

The data were plotted as the mean value ± standard error of the mean (SEM) and analyzed by means of GraphPad Prism 8 software (GraphPad Software, Inc., La Jolla, CA, USA). Statistical analysis was performed using one-way ANOVA with Tukey post hoc test or unpaired *t*-test (significance was accepted at $p < 0.05$). The western blots shown in the figures were selected as the most representative among repetitions $n \geq 3$. Western blots were quantified densitometrically using NIH ImageJ software (Wayne Rasband, Bethesda, MD, USA) and are shown as relative value of control (the fold changes in protein level ≥30% were considered as significant; qualitative analysis). The data were normalized relative to β-actin.

Author Contributions: Conceptualization, K.W.; methodology, K.W. and E.L.; investigation, K.W., E.L., S.M., K.S., M.S. and A.M.-K.; writing—original draft preparation, K.W. and A.M.-K.; writing—review and editing, K.W. and T.P.; visualization, K.W.; project administration, K.W.; funding acquisition, K.W. All authors have read and agreed to the published version of the manuscript.

Funding: This research was funded by National Science Centre, Poland 2015/17/D/NZ7/02170 (NCN, DEC-2015/17/D/NZ7/02170).

Conflicts of Interest: The authors declare no conflict of interest.

Abbreviations

AhR	aryl hydrocarbon receptor
BSA	bovine serum albumin
BrdU	5′-bromo-2′-deoxy-uridine
CDK	cyclin-dependent kinase
DMEM	Dulbecco's modified Eagle's medium
DMSO	dimethyl sulfoxide
FBS	fetal bovine serum
FICZ	6-formylindolo[3,2-b]carbazole
FITC	fluorescein isothiocyanate
HBSS	Hanks' Balanced Salt solution
HEMa	normal human adult primary epidermal melanocytes
IDO	indoleamine 2,3-dioxygenase
KYNA	kynurenic acid
KYN	kynurenine
LDH	lactic dehydrogenase
PBS	phosphate buffered saline
Rb	retinoblastoma tumor suppressor protein
SEM	standard error of the mean
TDO	tryptophan 2,3-dioxygenase

References

1. Miller, A.J.; Mihm, M.C., Jr. Melanoma. *N. Engl. J. Med.* **2006**, *355*, 51–65. [CrossRef] [PubMed]
2. Mabeta, P. Paradigms of vascularization in melanoma: Clinical significance and potential for therapeutic targeting. *Biomed. Pharmacother.* **2020**, *127*, 110135. [CrossRef] [PubMed]
3. Bertolotto, C. Melanoma: From melanocyte to genetic alterations and clinical options. *Scientifica* **2013**, *2013*, 635203. [CrossRef] [PubMed]
4. Guerrina, N.; Traboulsi, H.; Eidelman, D.H.; Baglole, C.J. The Aryl Hydrocarbon Receptor and the Maintenance of Lung Health. *Int. J. Mol. Sci.* **2018**, *19*, 3882. [CrossRef] [PubMed]
5. Vogeley, C.; Esser, C.; Tuting, T.; Krutmann, J.; Haarmann-Stemmann, T. Role of the Aryl Hydrocarbon Receptor in Environmentally Induced Skin Aging and Skin Carcinogenesis. *Int. J. Mol. Sci.* **2019**, *20*, 6005. [CrossRef] [PubMed]
6. Silbergeld, E.K.; Gasiewicz, T.A. Dioxins and the Ah receptor. *Am. J. Ind. Med.* **1989**, *16*, 455–474. [CrossRef] [PubMed]
7. Mengoni, M.; Braun, A.D.; Gaffal, E.; Tüting, T. The aryl hydrocarbon receptor promotes inflammation-induced dedifferentiation and systemic metastatic spread of melanoma cells. *Int. J. Cancer.* **2020**. [CrossRef]
8. Mezrich, J.D.; Fechner, J.H.; Zhang, X.; Johnson, B.P.; Burlingham, W.J.; Bradfield, C.A. An interaction between kynurenine and the aryl hydrocarbon receptor can generate regulatory T cells. *J. Immunol.* **2010**, *185*, 3190–3198. [CrossRef]
9. DiNatale, B.C.; Murray, I.A.; Schroeder, J.C.; Flaveny, C.A.; Lahoti, T.S.; Laurenzana, E.M.; Omiecinski, C.J.; Perdew, G.H. Kynurenic acid is a potent endogenous aryl hydrocarbon receptor ligand that synergistically induces interleukin-6 in the presence of inflammatory signaling. *J. Soc. Toxicol.* **2010**, *115*, 89–97. [CrossRef]

10. Terai, M.; Londin, E.; Rochani, A.; Link, E.; Lam, B.; Kaushal, G.; Bhushan, A.; Orloff, M.; Sato, T. Expression of Tryptophan 2,3-Dioxygenase in Metastatic Uveal Melanoma. *Cancers* **2020**, *12*, 405. [CrossRef]
11. Pilotte, L.; Larrieu, P.; Stroobant, V.; Colau, D.; Dolusic, E.; Frederick, R.; De Plaen, E.; Uyttenhove, C.; Wouters, J.; Masereel, B.; et al. Reversal of tumoral immune resistance by inhibition of tryptophan 2,3-dioxygenase. *Proc. Natl. Acad. Sci. USA* **2012**, *109*, 2497–2502. [CrossRef]
12. Zhai, L.; Ladomersky, E.; Lenzen, A.; Nguyen, B.; Patel, R.; Lauing, K.L.; Wu, M.; Wainwright, D.A. IDO1 in cancer: A Gemini of immune checkpoints. *Cell. Mol. Immunol.* **2018**, *15*, 447–457. [CrossRef]
13. Opitz, C.A.; Litzenburger, U.M.; Sahm, F.; Ott, M.; Tritschler, I.; Trump, S.; Schumacher, T.; Jestaedt, L.; Schrenk, D.; Weller, M.; et al. An endogenous tumour-promoting ligand of the human aryl hydrocarbon receptor. *Nature* **2011**, *478*, 197–203. [CrossRef]
14. Bekki, K.; Vogel, H.; Li, W.; Ito, T.; Sweeney, C.; Haarmann-Stemmann, T.; Matsumura, F.; Vogel, C.F. The aryl hydrocarbon receptor (AhR) mediates resistance to apoptosis induced in breast cancer cells. *Pestic. Biochem. Physiol.* **2015**, *120*, 5–13. [CrossRef]
15. Walczak, K.; Wnorowski, A.; Turski, W.A.; Plech, T. Kynurenic acid and cancer: Facts and controversies. *Cell Mol. Life Sci.* **2020**, *77*, 1531–1550. [CrossRef]
16. Turski, P.M.; Chwil, S.; Turska, M.; Chwil, M.; Kocki, T.; Rajtar, G.; Parada-Turska, J. An exceptionally high content of kynurenic acid in chestnut honey and flowers of chestnut tree. *J. Food Compos. Anal.* **2016**, *48*, 67–72.
17. Turski, M.P.; Turska, M.; Paluszkiewicz, P.; Parada-Turska, J.; Oxenkrug, G.F. Kynurenic Acid in the digestive system-new facts, new challenges. *Int. J. Tryptophan Res.* **2013**, *6*, 47–55. [CrossRef]
18. Turski, M.P.; Turska, M.; Zgrajka, W.; Kuc, D.; Turski, W.A. Presence of kynurenic acid in food and honeybee products. *Amino Acids* **2009**, *36*, 75–80. [CrossRef]
19. Swanbeck, G.; Wennersten, G.; Nilsson, R. Participation of singlet state excites oxygen in photohemolysis induced by kynurenic acid. *Acta Derm. Venereol.* **1974**, *54*, 433–436.
20. Wennersten, G.; Brunk, U. Cellular aspects of phototoxic reactions induced by kynurenic acid. I. Establishment of an experimental model utilizing in vitro cultivated cells. *Acta Derm. Venereol.* **1977**, *57*, 201–209.
21. Sheipouri, D.; Braidy, N.; Guillemin, G.J. Kynurenine Pathway in Skin Cells: Implications for UV-Induced Skin Damage. *Int. J. Tryptophan Res.* **2012**, *5*, 15–25. [CrossRef] [PubMed]
22. Jonsson, M.E.; Franks, D.G.; Woodin, B.R.; Jenny, M.J.; Garrick, R.A.; Behrendt, L.; Hahn, M.E.; Stegeman, J.J. The tryptophan photoproduct 6-formylindolo[3,2-b]carbazole (FICZ) binds multiple AHRs and induces multiple CYP1 genes via AHR2 in zebrafish. *Chem. Biol. Interact.* **2009**, *181*, 447–454. [CrossRef] [PubMed]
23. Syed, D.N.; Mukhtar, H. FICZ: A Messenger of Light in Human Skin. *J. Investig. Dermatol.* **2015**, *135*, 1478–1481. [CrossRef] [PubMed]
24. Bunaciu, R.P.; Jensen, H.A.; MacDonald, R.J.; LaTocha, D.H.; Varner, J.D.; Yen, A. 6-Formylindolo(3,2-b)Carbazole (FICZ) Modulates the Signalsome Responsible for RA-Induced Differentiation of HL-60 Myeloblastic Leukemia Cells. *PLoS ONE* **2015**, *10*, e0135668. [CrossRef] [PubMed]
25. Justiniano, R.; de Faria Lopes, L.; Perer, J.; Hua, A.; Park, S.L.; Jandova, J.; Baptista, M.S.; Wondrak, G.T. The Endogenous Tryptophan-derived Photoproduct 6-formylindolo[3,2-b]carbazole (FICZ) Is a Nanomolar Photosensitizer That Can Be Harnessed for the Photodynamic Elimination of Skin Cancer Cells in Vitro and in Vivo. *Photochem. Photobiol.* **2020**. [CrossRef]
26. Turski, M.P.; Turska, M.; Zgrajka, W.; Bartnik, M.; Kocki, T.; Turski, W.A. Distribution, synthesis, and absorption of kynurenic acid in plants. *Planta Med.* **2011**, *77*, 858–864. [CrossRef] [PubMed]
27. Moffett, J.R.; Namboodiri, M.A. Tryptophan and the immune response. *Immunol. Cell Biol.* **2003**, *81*, 247–265. [CrossRef]
28. Cervenka, I.; Agudelo, L.Z.; Ruas, J.L. Kynurenines: Tryptophan's metabolites in exercise, inflammation, and mental health. *Science* **2017**, *357*, 6349. [CrossRef]
29. Soto, M.E.; Ares, A.M.; Bernal, J.; Nozal, M.J.; Bernal, J.L. Simultaneous determination of tryptophan, kynurenine, kynurenic and xanthurenic acids in honey by liquid chromatography with diode array, fluorescence and tandem mass spectrometry detection. *J. Chromatogr. A* **2011**, *1218*, 7592–7600. [CrossRef]
30. Bochniarz, M.; Kocki, T.; Dąbrowski, R.; Szczubiał, M.; Wawron, W.; Turski, W.A. Tryptophan, kynurenine, kynurenic acid concentrations and indoleamine 2,3-dioxygenase activity in serum and milk of dairy cows with subclinical mastitis caused by coagulase-negative staphylococci. *Reprod. Domest. Anim.* **2018**, *53*, 1491–1497. [CrossRef]

31. Yılmaz, C.; Gökmen, V. Determination of tryptophan derivatives in kynurenine pathway in fermented foods using liquid chromatography tandem mass spectrometry. *Food Chem.* **2018**, *243*, 420–427. [CrossRef]
32. Muszyńska, B.; Sułkowska-Ziaja, K.; Ekiert, H. Analysis of indole compounds in methanolic extracts from the fruiting bodies of Cantharellus cibarius (the Chanterelle) and from the mycelium of this species cultured in vitro. *J. Food Sci. Technol.* **2013**, *50*, 1233–1237. [CrossRef] [PubMed]
33. Turska, M.; Pelak, J.; Turski, M.P.; Kocki, T.; Dukowski, P.; Plech, T.; Turski, W. Fate and distribution of kynurenic acid administered as beverage. *Pharmacol. Rep.* **2018**, *70*, 1089–1096. [CrossRef] [PubMed]
34. Turski, M.P.; Kaminski, P.; Zgrajka, W.; Turska, M.; Turski, W.A. Potato- an important source of nutritional kynurenic acid. *Plant Foods Hum. Nutr.* **2012**, *67*, 17–23. [CrossRef] [PubMed]
35. Pap, M.; Bátor, J.; Szeberényi, J. Sensitivity of Human Malignant Melanoma Cell Lines to Newcastle Disease Virus. *Anticancer Res.* **2015**, *35*, 5401–5406.
36. Mazzini, G.; Carpignano, F.; Surdo, S.; Aredia, F.; Panini, N.; Torchio, M.; Erba, E.; Danova, M.; Scovassi, A.I.; Barillaro, G.; et al. 3D Silicon Microstructures: A New Tool for Evaluating Biological Aggressiveness of Tumor Cells. *IEEE Trans. Nanobiosci.* **2015**, *14*, 797–805. [CrossRef]
37. Walczak, K.; Dabrowski, W.; Langner, E.; Zgrajka, W.; Pilat, J.; Kocki, T.; Rzeski, W.; Turski, W.A. Kynurenic acid synthesis and kynurenine aminotransferases expression in colon derived normal and cancer cells. *Scand. J. Gastroenterol.* **2011**, *46*, 903–912. [CrossRef] [PubMed]
38. Walczak, K.; Deneka-Hannemann, S.; Jarosz, B.; Zgrajka, W.; Stoma, F.; Trojanowski, T.; Turski, W.A.; Rzeski, W. Kynurenic acid inhibits proliferation and migration of human glioblastoma T98G cells. *Pharmacol. Rep.* **2014**, *66*, 130–136. [CrossRef]
39. Walczak, K.; Turski, W.A.; Rajtar, G. Kynurenic acid inhibits colon cancer proliferation in vitro: Effects on signaling pathways. *Amino Acids* **2014**, *46*, 2393–2401. [CrossRef]
40. Walczak, K.; Turski, W.A.; Rzeski, W. Kynurenic acid enhances expression of p21 Waf1/Cip1 in colon cancer HT-29 cells. *Pharmacol. Rep.* **2012**, *64*, 745–750. [CrossRef]
41. Walczak, K.; Zurawska, M.; Kis, J.; Starownik, R.; Zgrajka, W.; Bar, K.; Turski, W.A.; Rzeski, W. Kynurenic acid in human renal cell carcinoma: Its antiproliferative and antimigrative action on Caki-2 cells. *Amino Acids* **2012**, *43*, 1663–1670. [CrossRef]
42. Lee, B.; Sandhu, S.; McArthur, G. Cell cycle control as a promising target in melanoma. *Curr. Opin. Oncol.* **2015**, *27*, 141–150. [CrossRef]
43. Nathanson, K.L.; Martin, A.M.; Wubbenhorst, B.; Greshock, J.; Letrero, R.; D'Andrea, K.; O'Day, S.; Infante, J.R.; Falchook, G.S.; Arkenau, H.T.; et al. Tumor genetic analyses of patients with metastatic melanoma treated with the BRAF inhibitor dabrafenib (GSK2118436). *Clin. Cancer Res.* **2013**, *19*, 4868–4878. [CrossRef] [PubMed]
44. Pedace, L.; Cozzolino, A.M.; Barboni, L.; De Bernardo, C.; Grammatico, P.; De Simone, P.; Buccini, P.; Ferrari, A.; Catricalà, C.; Colombo, T.; et al. A novel variant in the 3' untranslated region of the CDK4 gene: Interference with microRNA target sites and role in increased risk of cutaneous melanoma. *Cancer Genet.* **2014**, *207*, 168–169. [CrossRef]
45. Kwong, L.N.; Costello, J.C.; Liu, H.; Jiang, S.; Helms, T.L.; Langsdorf, A.E.; Jakubosky, D.; Genovese, G.; Muller, F.L.; Jeong, J.H.; et al. Oncogenic NRAS signaling differentially regulates survival and proliferation in melanoma. *Nat. Med.* **2012**, *18*, 1503–1510. [CrossRef]
46. Dyson, N.J. RB1: A prototype tumor suppressor and an enigma. *Genes Dev.* **2016**, *30*, 1492–1502. [CrossRef]
47. Talluri, S.; Dick, F.A. Regulation of transcription and chromatin structure by pRB: Here, there and everywhere. *Cell Cycle* **2012**, *11*, 3189–3198. [CrossRef]
48. Lopez-Mejia, I.C.; Fajas, L. Cell cycle regulation of mitochondrial function. *Curr. Opin. Cell Biol.* **2015**, *33*, 19–25. [CrossRef]
49. Speciale, C.; Hares, K.; Schwarcz, R.; Brookes, N. High-affinity uptake of L-kynurenine by a Na+-independent transporter of neutral amino acids in astrocytes. *J. Neurosci.* **1989**, *9*, 2066–2072. [CrossRef]
50. Walczak, K.; Langner, E.; Szalast, K.; Makuch-Kocka, A.; Pożarowski, P.; Plech, T. A Tryptophan Metabolite, 8-Hydroxyquinaldic Acid, Exerts Antiproliferative and Anti-Migratory Effects on Colorectal Cancer Cells. *Molecules* **2020**, *25*, 1655. [CrossRef]

51. Langner, E.; Walczak, K.; Jeleniewicz, W.; Turski, W.A.; Rajtar, G. Quinaldic acid inhibits proliferation of colon cancer ht-29 cells in vitro: Effects on signaling pathways. *Eur. J. Pharmacol.* **2015**, *757*, 21–27. [CrossRef] [PubMed]
52. Livak, K.J.; Schmittgen, T.D. Analysis of relative gene expression data using real-time quantitative PCR and the $2^{-\Delta\Delta Ct}$ Method. *Methods* **2001**, *25*, 402–408. [CrossRef] [PubMed]

Publisher's Note: MDPI stays neutral with regard to jurisdictional claims in published maps and institutional affiliations.

 © 2020 by the authors. Licensee MDPI, Basel, Switzerland. This article is an open access article distributed under the terms and conditions of the Creative Commons Attribution (CC BY) license (http://creativecommons.org/licenses/by/4.0/).

MDPI
St. Alban-Anlage 66
4052 Basel
Switzerland
Tel. +41 61 683 77 34
Fax +41 61 302 89 18
www.mdpi.com

International Journal of Molecular Sciences Editorial Office
E-mail: ijms@mdpi.com
www.mdpi.com/journal/ijms

www.ingramcontent.com/pod-product-compliance
Lightning Source LLC
LaVergne TN
LVHW070211100526
838202LV00015B/2036